MATHEMATICAL DEVELOPMENTS
ARISING FROM
HILBERT PROBLEMS

PROCEEDINGS OF SYMPOSIA
IN PURE MATHEMATICS
Volume XXVIII

MATHEMATICAL DEVELOPMENTS ARISING FROM HILBERT PROBLEMS

AMERICAN MATHEMATICAL SOCIETY
PROVIDENCE, RHODE ISLAND
1976

PROCEEDINGS OF THE SYMPOSIUM IN PURE MATHEMATICS
OF THE AMERICAN MATHEMATICAL SOCIETY

HELD AT NORTHERN ILLINOIS UNIVERSITY
DEKALB, ILLINOIS
MAY 1974

EDITED BY
FELIX E. BROWDER

Prepared by the American Mathematical Society
with partial support from National Science Foundation grant GP-41994

Library of Congress Cataloging in Publication Data CIP

Symposium in Pure Mathematics, Northern Illinois University, 1974.
Mathematical developments arising from Hilbert problems.

(Proceedings of symposia in pure mathematics ; v. 28)
Bibliography: p.
1. Mathematics--Congresses. I. Browder, Felix E. II. American Mathematical Society. III. Title. IV. Title: Hilbert problems. V. Series.
QA1.S897 1974 510 76-20437
ISBN 0-8218-1428-1

AMS (MOS) subject classifications (1970). Primary 00-02.

CONTENTS

CONTENTS

Introduction

In May 1974, the American Mathematical Society sponsored a special Symposium on the mathematical consequences of the Hilbert problems which was held at Northern Illinois University in De Kalb, Illinois. The present volume contains the proceedings of that symposium and includes papers corresponding to all the invited addresses with one exception, that of Dr. J. R. Conway of Cambridge University, England. It contains as well the address of Professor G. Stampacchia that could not be delivered at the Symposium because of health problems.

The volume includes a number of other features to which we direct the attention of the reader. Thus, we have included photographs of the speakers (by the courtesy of Paul Halmos), and a translation of the text of the Hilbert Problems as published in the Bulletin of the American Mathematical Society of 1903. The papers are published in the order of the problems to which they are affiliated, and not in the alphabetical order of their authors.

We should remark explicitly that the central concern of this Symposium was not the explicit solutions (so far as they have been given) of Hilbert's problems nor commentary upon their details. It was rather an attempt to focus upon those areas of importance in contemporary mathematical research which can be seen as descended in some way from the ideas and tendencies put forward by Hilbert in his speech at the International Congress of Mathematicians in Paris in 1900. The connection was direct and unequivocal in a number of cases, and a good deal more ambiguous in others. The Organizing Committee of the Symposium consisted of P. R. Bateman (Secretary), F. E. Browder (Chairman), R. C. Buck, D. Lewis, and D. Zelinsky. Its basic objective was to obtain as broad a representation of significant mathematical research as possible within the general constraint of relevance to the Hilbert problems.

An additional unusual feature of the present volume is the article entitled *Problems of present day mathematics* which appears immediately after the text of Hilbert's article. The development of this material was initiated by Jean Dieudonné through correspondence with a number of mathematicians throughout the world. The resulting problems as well as others obtained by the Editor, appear in the form in which they were suggested by the mathematicians whose names are attached to them. The editing process as carried on by Professor Dieudonné or the present Editor consisted simply in selection of problems in cases where suggested problems overlapped seriously in their content.

The general reaction of the participants to the Symposium was one of great enthusiasm and approval, both by speakers and audience, except for one point on which comment is appropriate. This concerned the provisional text of the collection of new problems as distributed at the Symposium and discussed at a special hour session. Criticism was put forward of the method by which the problems had been collected and of the scarcity of problems in a number of important directions, particularly in analysis and geometry. The Editor also received correspondence criticizing the texts of several of the draft problems, particularly in so far as they impinged on mathematical physics, logic, and computer science. In the form in which the problems appear in the present volume, I believe that some of these criticisms have been met, particularly in terms of new sets of problems in various areas of geometry and analysis, as well as a new set of problems in mathematical physics. Despite efforts in the direction of logic and computer science, it turned out to be impracticable to obtain problems on the right level in these directions. Two sets of problems, those by Hugh Montgomery on the theory of prime numbers and those by Arthur Wightman on mathematical physics, were transposed from their submitted papers to the Problem Collection, and in the case of Montgomery's paper, this was the whole of the submitted manuscript. We believe that the final result meets most of the substantive criticisms which have been made. It does not meet the criticism that these problems were obtained by consulting a self-selecting elite. That criticism could have been made with much greater force against the problems presented by Hilbert (see the discussion of the circumstances of Hilbert's address by Constance Reid in her biography of Hilbert). In either case, the significance of the problems depends upon their intrinsic merit, not their origin.

The significance of this set of problems can only be judged in the future, in terms of their consequences and the role they may play in focusing the process of mathematical discovery. We should not believe that anyone, even Hilbert, could or can foresee the mysteries of the future in terms of new discoveries and new turns of interest. Despite the great role of the Hilbert problems as recorded in the present volume, we should not fail to note that there is no hint in them of such decisive developments in the following decade as the development of topology, both combinatorial and set-theoretic, or of functional analysis (as in the theory of linear integral equations to which Hilbert himself was to devote most of his own efforts). Nor, despite the scope of the Hilbert problems and the breadth of the present Symposium, should we ignore the fact that a rather different Symposium of equal importance could be organized around the interests and achievements of Hilbert's great contemporary, Henri Poincaré.

There is obviously an enormous difference between the collection of problems which is given below as originating from a number of leading mathematicians in different areas of mathematical research and the synthesis of problems made by Hilbert as the product of a single mind. There is no common statement of philosophy or attitude such as Hilbert's article begins with, nor a common conclusion. The whole thrust of the Symposium and of the new problems leads however to a re-endorsement of an essential part of Hilbert's own conclusion:

> "The problems mentioned are merely samples of problems, yet they will suffice to show how rich, how manifold and how extensive the mathematical science of to-day is, and the question is urged upon us whether mathematics is doomed to the fate of those other sciences that have split up onto separate branches, whose representatives scarcely understand one another and whose connection becomes ever more loose. I do not believe this nor wish it. Mathematical science is in my opinion an indivisible whole, an organism whose vitality is conditioned upon the connection of its parts. For with all the variety of mathematical knowledge, we are still clearly conscious of the similarity of the logical devices, the relationship of the ideas in mathematics as a whole and the numerous analogies in its different departments. We also notice that, the farther a mathematical theory is developed the more harmoniously and uniformly does its construction proceed, and unsuspected relations are disclosed between hitherto separate branches of the science. So it happens that, with the extension of mathematics, its organic character is not lost but only manifests itself the more clearly."

F. E. BROWDER

LIPMAN BERS

ENRICO BOMBIERI

HERBERT BUSEMANN

J. CONWAY

MARTIN DAVIS

NICHOLAS M. KATZ

STEVEN L. KLEIMAN

G. KREISEL

R. P. LANGLANDS

G. G. LORENTZ

DONALD A. MARTIN

J. MILNOR

HUGH L. MONTGOMERY

DAVID MUMFORD

O. T. O'MEARA

ALBRECHT PFISTER

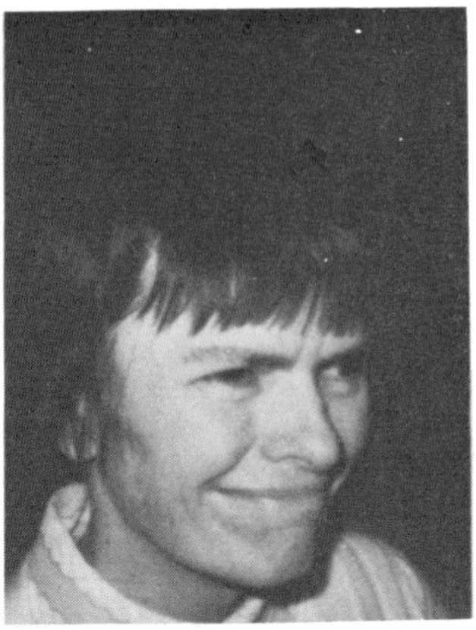

JULIA ROBINSON

JAMES SERRIN

J. TATE

R. TIJDEMAN

A. S. WIGHTMAN

C. T. YANG

Proceedings of Symposia in Pure Mathematics
Volume 28, 1976

MATHEMATICAL PROBLEMS.*†

LECTURE DELIVERED BEFORE THE INTERNATIONAL CONGRESS OF MATHEMATICIANS AT PARIS IN 1900.

BY PROFESSOR DAVID HILBERT.

WHO of us would not be glad to lift the veil behind which the future lies hidden; to cast a glance at the next advances of our science and at the secrets of its development during future centuries? What particular goals will there be toward which the leading mathematical spirits of coming generations will strive? What new methods and new facts in the wide and rich field of mathematical thought will the new centuries disclose?

History teaches the continuity of the development of science. We know that every age has its own problems, which the following age either solves or casts aside as profitless and replaces by new ones. If we would obtain an idea of the probable development of mathematical knowledge in the immediate future, we must let the unsettled questions pass before our minds and look over the problems which the science of to-day sets and whose solution we expect from the future. To such a review of problems the present day, lying at the meeting of the centuries, seems to me well adapted. For the close of a great epoch not only invites us to look back into the past but also directs our thoughts to the unknown future.

The deep significance of certain problems for the advance of mathematical science in general and the important rôle which they play in the work of the individual investigator are not to be denied. As long as a branch of science offers an abundance of problems, so long is it alive; a lack of problems foreshadows extinction or the cessation of independent development. Just as every human undertaking pursues certain objects, so also mathematical research requires its problems. It is by the solution of problems that the investigator tests the temper of his steel; he finds new methods and new outlooks, and gains a wider and freer horizon.

It is difficult and often impossible to judge the value of a problem correctly in advance; for the final award depends upon the gain which science obtains from the problem. Nevertheless we can ask whether there are general criteria which mark a good mathematical problem. An old French mathematician said: "A mathematical theory is not to be considered complete until you have made it so clear that you can explain it to the first man whom you meet on the street." This clearness and ease of comprehension, here

---

* Translated for the BULLETIN, with the author's permission, by Dr. MARY WINSTON NEWSON. The original appeared in the *Göttinger Nachrichten*, 1900, pp. 253-297, and in the *Archiv der Mathematik und Physik*, 3d ser., vol. 1 (1901), pp. 44-63 and 213-237.

†Reprinted from the Bulletin of the American Mathematical Society, vol. 8 (1902), pp. 437–479.

insisted on for a mathematical theory, I should still more demand for a mathematical problem if it is to be perfect; for what is clear and easily comprehended attracts, the complicated repels us.

Moreover a mathematical problem should be difficult in order to entice us, yet not completely inaccessible, lest it mock at our efforts. It should be to us a guide post on the mazy paths to hidden truths, and ultimately a reminder of our pleasure in the successful solution.

The mathematicians of past centuries were accustomed to devote themselves to the solution of difficult particular problems with passionate zeal. They knew the value of difficult problems. I remind you only of the "problem of the line of quickest descent," proposed by John Bernoulli. Experience teaches, explains Bernoulli in the public announcement of this problem, that lofty minds are led to strive for the advance of science by nothing more than by laying before them difficult and at the same time useful problems, and he therefore hopes to earn the thanks of the mathematical world by following the example of men like Mersenne, Pascal, Fermat, Viviani and others and laying before the distinguished analysts of his time a problem by which, as a touchstone, they may test the value of their methods and measure their strength. The calculus of variations owes its origin to this problem of Bernoulli and to similar problems.

Fermat had asserted, as is well known, that the diophantine equation

$$x^n + y^n = z^n$$

(x, y and z integers) is unsolvable—except in certain self-evident cases. The attempt to prove this impossibility offers a striking example of the inspiring effect which such a very special and apparently unimportant problem may have upon science. For Kummer, incited by Fermat's problem, was led to the introduction of ideal numbers and to the discovery of the law of the unique decomposition of the numbers of a circular field into ideal prime factors—a law which to-day, in its generalization to any algebraic field by Dedekind and Kronecker, stands at the center of the modern theory of numbers and whose significance extends far beyond the boundaries of number theory into the realm of algebra and the theory of functions.

To speak of a very different region of research, I remind you of the problem of three bodies. The fruitful methods and the far-reaching principles which Poincaré has brought into celestial mechanics and which are to-day recognized and applied in practical astronomy are due to the circumstance that he undertook to treat anew that difficult problem and to approach nearer a solution.

The two last mentioned problems—that of Fermat and the problem of the three bodies—seem to us almost like opposite poles—the former a free invention of pure reason, belonging to the region of abstract number theory, the latter forced upon us by astronomy and necessary to an understanding of the simplest fundamental phenomena of nature.

But it often happens also that the same special problem finds application in the most unlike branches of mathematical knowledge. So, for example, the problem of the shortest line plays a chief and historically important part in the foundations of geometry, in the theory of curved lines and surfaces, in mechanics and in the calculus of variations. And how convincingly has F. Klein, in his work on the icosahedron, pictured the significance which attaches to the problem of the regular polyhedra in elementary geometry, in group theory, in the theory of equations and in that of linear differential equations.

In order to throw light on the importance of certain problems, I may also refer to Weierstrass, who spoke of it as his happy fortune that he found at the outset of his scientific career a problem so important as Jacobi's problem of inversion on which to work.

Having now recalled to mind the general importance of problems in mathematics, let us turn to the question from what sources this science derives its problems. Surely the first and oldest problems in every branch of mathematics spring from experience and are suggested by the world of external phenomena. Even the rules of calculation with integers must have been discovered in this fashion in a lower stage of human civilization, just as the child of to-day learns the application of these laws by empirical methods. The same is true of the first problems of geometry, the problems bequeathed us by antiquity, such as the duplication of the cube, the squaring of the circle; also the oldest problems in the theory of the solution of numerical equations, in the theory of curves and the differential and integral calculus, in the calculus of variations, the theory of Fourier series and the theory of potential—to say nothing of the further abundance of problems properly belonging to mechanics, astronomy and physics.

But, in the further development of a branch of mathematics, the human mind, encouraged by the success of its solutions, becomes conscious of its independence. It evolves from itself alone, often without appreciable influence from without, by means of logical combination, generalization, specialization, by separating and collecting ideas in fortunate ways, new and fruitful problems, and appears then itself as the real questioner. Thus arose the problem of prime numbers and the other problems of number theory, Galois's theory of equations, the theory of algebraic invariants, the theory of abelian and automorphic functions; indeed almost all the nicer questions of modern arithmetic and function theory arise in this way.

In the meantime, while the creative power of pure reason is at work, the outer world again comes into play, forces upon us new questions from actual experience, opens up new branches of mathematics, and while we seek to conquer these new fields of knowledge for the realm of pure thought, we often find the answers to old unsolved problems and thus at the same time advance most successfully the old theories. And it seems to me that the numerous and surprising analogies and that apparently prearranged harmony which the mathematician so often perceives in the questions, methods and ideas of the various branches of his science, have their origin in this ever-recurring interplay between thought and experience.

It remains to discuss briefly what general requirements may be justly laid down for the solution of a mathematical problem. I should say first of all, this: that it shall be possible to establish the correctness of the solution by means of a finite number of steps based upon a finite number of hypotheses which are implied in the statement of the problem and which must always be exactly formulated. This requirement of logical deduction by means of a finite number of processes is simply the requirement of rigor in reasoning. Indeed the requirement of rigor, which has become proverbial in mathematics, corresponds to a universal philosophical necessity of our understanding; and, on the other hand, only by satisfying this requirement do the thought content and the suggestiveness of the problem attain their full effect. A new problem, especially when it comes from the world of outer experience, is like a young twig, which thrives and bears fruit only when it is grafted carefully and in accordance with strict horticultural rules upon the old stem, the established achievements of our mathematical science.

Besides it is an error to believe that rigor in the proof is the enemy of simplicity. On the contrary we find it confirmed by numerous examples that the rigorous method is at the same time the simpler and the more easily comprehended. The very effort for rigor forces us to find out simpler methods of proof. It also frequently leads the way to methods which are more capable of development than the old methods of less rigor. Thus the theory of algebraic curves experienced a considerable simplification and attained greater unity by means of the more rigorous function-theoretical methods and the consistent introduction of transcendental devices. Further, the proof that the power series permits the application of the four elementary arithmetical operations as well as the term by term differentiation and integration, and the recognition of the utility of the power series depending upon this proof contributed materially to the simplification of all analysis, particularly of the theory of elimination and the theory of differential equations, and also of the existence proofs demanded in those theories. But the most striking example for my statement is the calculus of variations. The treatment of the first and second variations of definite integrals required in part extremely complicated calculations, and the processes applied by the old mathematicians had not the needful rigor. Weierstrass showed us the way to a new and sure foundation of the calculus of variations. By the examples of the simple and double integral I will show briefly, at the close of my lecture, how this way leads at once to a surprising simplification of the calculus of variations. For in the demonstration of the necessary and sufficient criteria for the occurrence of a maximum and minimum, the calculation of the second variation and in part, indeed, the wearisome reasoning connected with the first variation may be completely dispensed with—to say nothing of the advance which is involved in the removal of the restriction to variations for which the differential coefficients of the function vary but slightly.

While insisting on rigor in the proof as a requirement for a perfect solution of a problem, I should like, on the other hand, to oppose the opinion that only the concepts of analysis, or even those of arithmetic alone, are susceptible of a fully rigorous treatment. This opinion, occasionally advocated by eminent men, I consider entirely erroneous. Such a one-sided interpretation of the requirement of rigor would soon lead to the ignoring of all concepts arising from geometry, mechanics and physics, to a stoppage of the flow of new material from the outside world, and finally, indeed, as a last consequence, to the rejection of the ideas of the continuum and of the irrational number. But what an important nerve, vital to mathematical science, would be cut by the extirpation of geometry and mathematical physics! On the contrary I think that wherever, from the side of the theory of knowledge or in geometry, or from the theories of natural or physical science, mathematical ideas come up, the problem arises for mathematical science to investigate the principles underlying these ideas and so to establish them upon a simple and complete system of axioms, that the exactness of the new ideas and their applicability to deduction shall be in no respect inferior to those of the old arithmetical concepts.

To new concepts correspond, necessarily, new signs. These we choose in such a way that they remind us of the phenomena which were the occasion for the formation of the new concepts. So the geometrical figures are signs or mnemonic symbols of space intuition and are used as such by all mathematicians. Who does not always use along with the double inequality $a > b > c$ the picture of three points following one another on a straight line as the geometrical picture of the idea "between"? Who does not make use of drawings of segments and rectangles enclosed in one another, when it is required to prove with perfect rigor a difficult theorem on the continuity of functions or the existence of points of condensation? Who could dispense with the figure of the triangle, the circle with its center, or with the cross of three perpendicular axes? Or who would give up the representation of the vector field, or the picture of a family of curves or surfaces with its envelope which plays so important a part in differential geometry, in the theory of differential equations, in the foundation of the calculus of variations and in other purely mathematical sciences?

The arithmetical symbols are written diagrams and the geometrical figures are graphic formulas; and no mathematician could spare these graphic formulas, any more than in calculation the insertion and removal of parentheses or the use of other analytical signs.

The use of geometrical signs as a means of strict proof presupposes the exact knowledge and complete mastery of the axioms which underlie those figures; and in order that these geometrical figures may be incorporated in the general treasure of mathematical signs, there is necessary a rigorous axiomatic investigation of their conceptual content. Just as in adding two numbers, one must place the digits under each other in the right order, so that only the rules of calculation, *i. e.*, the axioms of arithmetic, determine the correct use of the digits, so the use of geometrical signs is determined by the axioms of geometrical concepts and their combinations.

The agreement between geometrical and arithmetical thought is shown also in that we do not habitually follow the chain of reasoning back to the axioms in arithmetical, any more than in geometrical discussions. On the contrary we apply, especially in first attacking a problem, a rapid, unconscious, not absolutely sure combination, trusting to a certain arithmetical feeling for the behavior of the arithmetical symbols, which we could dispense with as little in arithmetic as with the geometrical imagination in geometry. As an example of an arithmetical theory operating rigorously with geometrical ideas and signs, I may mention Minkowski's work, Die Geometrie der Zahlen. *

Some remarks upon the difficulties which mathematical problems may offer, and the means of surmounting them, may be in place here.

If we do not succeed in solving a mathematical problem, the reason frequently consists in our failure to recognize the more general standpoint from which the problem before us appears only as a single link in a chain of related problems. After finding this standpoint, not only is this problem frequently more accessible to our investigation, but at the same time we come into possession of a method which is applicable also to related problems. The introduction of complex paths of integration by Cauchy and of the notion of the IDEALS in number theory by Kummer may serve as examples. This way for finding general methods is certainly the most practicable and the most certain ; for he who seeks for methods without having a definite problem in mind seeks for the most part in vain.

In dealing with mathematical problems, specialization plays, as I believe, a still more important part than generalization. Perhaps in most cases where we seek in vain the answer to a question, the cause of the failure lies in the fact that problems simpler and easier than the one in hand have been either not at all or incompletely solved. All depends, then, on finding out these easier problems, and on solving them by means of devices as perfect as possible and of concepts capable of generalization. This rule is one of the most important levers for overcoming mathematical difficulties and it seems to me that it is used almost always, though perhaps unconsciously.

Occasionally it happens that we seek the solution under insufficient hypotheses or in an incorrect sense, and for this reason do not succeed. The problem then arises : to show the impossibility of the solution under the given hypotheses, or in the sense contemplated. Such proofs of impossibility were effected by the ancients, for instance when they showed that the ratio of the hypotenuse to the side of an isosceles right triangle is irrational. In later mathematics, the question as to the impossibility of certain solutions plays a preëminent part, and we perceive in this way that old and

* Leipzig, 1896.

difficult problems, such as the proof of the axiom of parallels, the squaring of the circle, or the solution of equations of the fifth degree by radicals have finally found fully satisfactory and rigorous solutions, although in another sense than that originally intended. It is probably this important fact along with other philosophical reasons that gives rise to the conviction (which every mathematician shares, but which no one has as yet supported by a proof) that every definite mathematical problem must necessarily be susceptible of an exact settlement, either in the form of an actual answer to the question asked, or by the proof of the impossibility of its solution and therewith the necessary failure of all attempts. Take any definite unsolved problem, such as the question as to the irrationality of the Euler-Mascheroni constant C, or the existence of an infinite number of prime numbers of the form $2^n + 1$. However unapproachable these problems may seem to us and however helpless we stand before them, we have, nevertheless, the firm conviction that their solution must follow by a finite number of purely logical processes.

Is this axiom of the solvability of every problem a peculiarity characteristic of mathematical thought alone, or is it possibly a general law inherent in the nature of the mind, that all questions which it asks must be answerable? For in other sciences also one meets old problems which have been settled in a manner most satisfactory and most useful to science by the proof of their impossibility. I instance the problem of perpetual motion. After seeking in vain for the construction of a perpetual motion machine, the relations were investigated which must subsist between the forces of nature if such a machine is to be impossible;* and this inverted question led to the discovery of the law of the conservation of energy, which, again, explained the impossibility of perpetual motion in the sense originally intended.

This conviction of the solvability of every mathematical problem is a powerful incentive to the worker. We hear within us the perpetual call: There is the problem. Seek its solution. You can find it by pure reason, for in mathematics there is no *ignorabimus.*

The supply of problems in mathematics is inexhaustible, and as soon as one problem is solved numerous others come forth in its place. Permit me in the following, tentatively as it were, to mention particular definite problems, drawn from various branches of mathematics, from the discussion of which an advancement of science may be expected.

Let us look at the principles of analysis and geometry. The most suggestive and notable achievements of the last century in this field are, as it seems to me, the arithmetical formulation of the concept of the continuum in the works of Cauchy, Bolzano and Cantor, and the discovery of non-euclidean geometry by Gauss, Bolyai, and Loba-

---

*See Helmholtz, "Ueber die Wechselwirkung der Naturkräefte und die darauf bezüglichen neuesten Ermittelungen der Physik"; Vortrag, gehalten in Königsberg, 1854.

chevsky. I therefore first direct your attention to some problems belonging to these fields.

1. Cantor's Problem of the Cardinal Number of the Continuum.

Two systems, *i. e*, two assemblages of ordinary real numbers or points, are said to be (according to Cantor) equivalent or of equal *cardinal number*, if they can be brought into a relation to one another such that to every number of the one assemblage corresponds one and only one definite number of the other. The investigations of Cantor on such assemblages of points suggest a very plausible theorem, which nevertheless, in spite of the most strenuous efforts, no one has succeeded in proving. This is the theorem :

Every system of infinitely many real numbers, *i. e.*, every assemblage of numbers (or points), is either equivalent to the assemblage of natural integers, 1, 2, 3,··· or to the assemblage of all real numbers and therefore to the continuum, that is, to the points of a line ; *as regards equivalence there are, therefore, only two assemblages of numbers, the countable assemblage and the continuum.*

From this theorem it would follow at once that the continuum has the next cardinal number beyond that of the countable assemblage; the proof of this theorem would, therefore, form a new bridge between the countable assemblage and the continuum.

Let me mention another very remarkable statement of Cantor's which stands in the closest connection with the theorem mentioned and which, perhaps, offers the key to its proof. Any system of real numbers is said to be ordered, if for every two numbers of the system it is determined which one is the earlier and which the later, and if at the same time this determination is of such a kind that, if a is before b and b is before c, then a always comes before c. The natural arrangement of numbers of a system is defined to be that in which the smaller precedes the larger. But there are, as is easily seen, infinitely many other ways in which the numbers of a system may be arranged.

If we think of a definite arrangement of numbers and select from them a particular system of these numbers, a so-called partial system or assemblage, this partial system will also prove to be ordered. Now Cantor considers a particular kind of ordered assemblage which he designates as a well ordered assemblage and which is characterized in this way, that not only in the assemblage itself but also in every partial assemblage there exists a first number. The system of integers 1, 2, 3, ··· in their natural order is evidently a well ordered assemblage. On the other hand the system of all real numbers, *i. e.*, the continuum in its natural order, is evidently not well ordered. For, if we think of the points of a segment of a straight line, with its initial point excluded, as our partial assemblage, it will have no first element.

The question now arises whether the totality of all numbers may not be arranged in another manner so that every partial assemblage may have a first element, *i. e.*, whether the continuum cannot be considered as a well ordered assemblage—a question which Cantor thinks must be answered in the affirmative. It appears to me most desirable to obtain a direct proof of this remarkable statement of Cantor's, perhaps by actually giving an arrangement of numbers such that in every partial system a first number can be pointed out.

2. The Compatibility of the Arithmetical Axioms.

When we are engaged in investigating the foundations of a science, we must set up a system of axioms which contains an exact and complete description of the relations subsisting between the elementary ideas of that science. The axioms so set up are at the same time the definitions of those elementary ideas; and no statement within the realm of the science whose foundation we are testing is held to be correct unless it can be derived from those axioms by means of a finite number of logical steps. Upon closer consideration the question arises: *Whether, in any way, certain statements of single axioms depend upon one another, and whether the axioms may not therefore contain certain parts in common, which must be isolated if one wishes to arrive at a system of axioms that shall be altogether independent of one another.*

But above all I wish to designate the following as the most important among the numerous questions which can be asked with regard to the axioms: *To prove that they are not contradictory, that is, that a finite number of logical steps based upon them can never lead to contradictory results.*

In geometry, the proof of the compatibility of the axioms can be effected by constructing a suitable field of numbers, such that analogous relations between the numbers of this field correspond to the geometrical axioms. Any contradiction in the deductions from the geometrical axioms must thereupon be recognizable in the arithmetic of this field of numbers. In this way the desired proof for the compatibility of the geometrical axioms is made to depend upon the theorem of the compatibility of the arithmetical axioms.

On the other hand a direct method is needed for the proof of the compatibility of the arithmetical axioms. The axioms of arithmetic are essentially nothing else than the known rules of calculation, with the addition of the axiom of continuity. I recently collected them * and in so doing replaced the axiom of continuity by two simpler axioms, namely, the well-known axiom of Archimedes, and a new axiom essentially as follows: that numbers form a system of things which is capable of no further extension, as long as all the other axioms hold (axiom of completeness). I am convinced that it must be possible to find a direct proof for the compatibility of the arithmetical axioms, by means of a careful study and suitable modification of the known methods of reasoning in the theory of irrational numbers.

To show the significance of the problem from another point of view, I add the following observation: If contradictory attributes be assigned to a concept, I say, that *mathematically the concept oes not exist.* So, for example, a real number whose square is -1 does not exist mathe-

---

* *Jahresbericht der Deutschen Mathematiker-Vereinigung*, vol. 8 (1900), p. 180.

matically. But if it can be proved that the attributes assigned to the concept can never lead to a contradiction by the application of a finite number of logical processes, I say that the mathematical existence of the concept (for example, of a number or a function which satisfies certain conditions) is thereby proved. In the case before us, where we are concerned with the axioms of real numbers in arithmetic, the proof of the compatibility of the axioms is at the same time the proof of the mathematical existence of the complete system of real numbers or of the continuum. Indeed, when the proof for the compatibility of the axioms shall be fully accomplished, the doubts which have been expressed occasionally as to the existence of the complete system of real numbers will become totally groundless. The totality of real numbers, *i. e.*, the continuum according to the point of view just indicated, is not the totality of all possible series in decimal fractions, or of all possible laws according to which the elements of a fundamental sequence may proceed. It is rather a system of things whose mutual relations are governed by the axioms set up and for which all propositions, and only those, are true which can be derived from the axioms by a finite number of logical processes. In my opinion, the concept of the continuum is strictly logically tenable in this sense only. It seems to me, indeed, that this corresponds best also to what experience and intuition tell us. The concept of the continuum or even that of the system of all functions exists, then, in exactly the same sense as the system of integral, rational numbers, for example, or as Cantor's higher classes of numbers and cardinal numbers. For I am convinced that the existence of the latter, just as that of the continuum, can be proved in the sense I have described; unlike the system of *all* cardinal numbers or of *all* Cantor's alephs, for which, as may be shown, a system of axioms, compatible in my sense, cannot be set up. Either of these systems is, therefore, according to my terminology, mathematically non-existent.

From the field of the foundations of geometry I should like to mention the following problem:

3. The Equality of the Volumes of Two Tetrahedra of Equal Bases and Equal Altitudes.

In two letters to Gerling, Gauss * expresses his regret that certain theorems of solid geometry depend upon the method of exhaustion, *i. e.*, in modern phraseology, upon the axiom of continuity (or upon the axiom of Archimedes). Gauss mentions in particular the theorem of Euclid, that triangular pyramids of equal altitudes are to each other as their bases. Now the analogous problem in the plane has been solved.† Gerling also succeeded in proving the equality of volume of symmetrical polyhedra by dividing them into congruent parts. Nevertheless, it seems to me probable that a general proof of this kind for the theorem of Euclid just mentioned is impossible, and it should be our task to give a rigor-

* Werke, vol. 8, pp. 241 and 244.

† Cf., beside earlier literature, Hilbert, Grundlagen der Geometrie, Leipzig, 1899, ch. 4. [Translation by Townsend, Chicago, 1902.]

ous proof of its impossibility. This would be obtained, as soon as we succeeded in *specifying two tetrahedra of equal bases and equal altitudes which can in no way be split up into congruent tetrahedra, and which cannot be combined with congruent tetrahedra to form two polyhedra which themselves could be split up into congruent tetrahedra.*‡

4. Problem of the Straight Line as the Shortest Distance Between Two Points.

Another problem relating to the foundations of geometry is this: If from among the axioms necessary to establish ordinary euclidean geometry, we exclude the axiom of parallels, or assume it as not satisfied, but retain all other axioms, we obtain, as is well known, the geometry of Lobachevsky (hyperbolic geometry). We may therefore say that this is a geometry standing next to euclidean geometry. If we require further that that axiom be not satisfied whereby, of three points of a straight line, one and only one lies between the other two, we obtain Riemann's (elliptic) geometry, so that this geometry appears to be the next after Lobachevsky's. If we wish to carry out a similar investigation with respect to the axiom of Archimedes, we must look upon this as not satisfied, and we arrive thereby at the non-archimedean geometries which have been investigated by Veronese and myself. The more general question now arises: Whether from other suggestive standpoints geometries may not be devised which, with equal right, stand next to euclidean geometry. Here I should like to direct your attention to a theorem which has, indeed, been employed by many authors as a definition of a straight line, viz., that the straight line is the shortest distance between two points. The essential content of this statement reduces to the theorem of Euclid that in a triangle the sum of two sides is always greater than the third side—a theorem which, as is easily seen, deals solely with elementary concepts, *i. e.*, with such as are derived directly from the axioms, and is therefore more accessible to logical investigation. Euclid proved this theorem, with the help of the theorem of the exterior angle, on the basis of the congruence theorems. Now it is readily shown that this theorem of Euclid cannot be proved solely on the basis of those congruence theorems which relate to the application of segments and angles, but that one of the theorems on the congruence of triangles is necessary. We are asking, then, for a geometry in which all the axioms of ordinary euclidean geometry hold, and in particular all the congruence axioms except the one of the congruence of triangles (or all except the theorem of the equality of the base angles in the isosceles triangle), and in which, besides, the proposition that in every triangle the sum of two sides is greater than the third is assumed as a particular axiom.

---

‡ Since this was written Herr Dehn has succeeded in proving this impossibility. See his note: "Ueber raumgleiche Polyeder," in *Nachrichten d. K. Gesellsch. d. Wiss. zu Göttingen*, 1900, and a paper soon to appear in the *Math. Annalen* [vol. 55, pp. 465–478].

One finds that such a geometry really exists and is no other than that which Minkowski constructed in his book, Geometrie der Zahlen,* and made the basis of his arithmetical investigations. Minkowski's is therefore also a geometry standing next to the ordinary euclidean geometry; it is essentially characterized by the following stipulations:

1. The points which are at equal distances from a fixed point O lie on a convex closed surface of the ordinary euclidean space with O as a center.

2. Two segments are said to be equal when one can be carried into the other by a translation of the ordinary euclidean space.

In Minkowski's geometry the axiom of parallels also holds. By studying the theorem of the straight line as the shortest distance between two points, I arrived* at a geometry in which the parallel axiom does not hold, while all other axioms of Minkowski's geometry are satisfied. The theorem of the straight line as the shortest distance between two points and the essentially equivalent theorem of Euclid about the sides of a triangle, play an important part not only in number theory but also in the theory of surfaces and in the calculus of variations. For this reason, and because I believe that the thorough investigation of the conditions for the validity of this theorem will throw a new light upon the idea of distance, as well as upon other elementary ideas, *e. g.*, upon the idea of the plane, and the possibility of its definition by means of the idea of the straight line, *the construction and systematic treatment of the geometries here possible seem to me desirable.*

5. Lie's Concept of a Continuous Group of Transformations Without the Assumption of the Differentiability of the Functions Defining the Group.

It is well known that Lie, with the aid of the concept of continuous groups of transformations, has set up a system of geometrical axioms and, from the standpoint of his theory of groups, has proved that this system of axioms suffices for geometry. But since Lie assumes, in the very foundation of his theory, that the functions defining his group can be differentiated, it remains undecided in Lie's development, whether the assumption of the differentiability in connection with the question as to the axioms of geometry is actually unavoidable, or whether it may not appear rather as a consequence of the group concept and the other geometrical axioms. This consideration, as well as certain other problems in connection with the arithmetical axioms, brings before us the more general question: *How far Lie's concept of continuous groups of transformations is approachable in our investigations without the assumption of the differentiability of the functions.*

Lie defines a finite continuous group of transformations as a system of transformations

---

* Leipzig, 1896.

* *Math. Annalen*, vol. 46, p. 91.

$$x_i' = f_i(x_1, \cdots, x_n;\ a_1, \cdots, a_r) \qquad (i = 1, \cdots, n)$$

having the property that any two arbitrarily chosen transformations of the system, as

$$x_i' = f_i(x_1, \cdots, x_n;\ a_1, \cdots, a_r),$$

$$x_i'' = f_i(x_1', \cdots, x_n';\ b_1, \cdots, b_r),$$

applied sucessively result in a transformation which also belongs to the system, and which is therefore expressible in the form

$$x_i'' = f_i\{f_1(x, a), \cdots, f_n(x, a);\ b_1, \cdots, b_r\} = f_i(x_1, \cdots, x_n;\ c_1, \cdots, c_r),$$

where $c_1, \cdots, c_r$ are certain functions of $a_1, \cdots, a_r$ and $b_1, \cdots, b_r$. The group property thus finds its full expression in a system of functional equations and of itself imposes no additional restrictions upon the functions $f_1, \cdots, f_n;\ c_1, \cdots, c_r$. Yet Lie's further treatment of these functional equations, viz., the derivation of the well-known fundamental differential equations, assumes necessarily the continuity and differentiability of the functions defining the group.

As regards continuity: this postulate will certainly be retained for the present—if only with a view to the geometrical and arithmetical applications, in which the continuity of the functions in question appears as a consequence of the axiom of continuity. On the other hand the differentiability of the functions defining the group contains a postulate which, in the geometrical axioms, can be expressed only in a rather forced and complicated manner. Hence there arises the question whether, through the introduction of suitable new variables and parameters, the group can always be transformed into one whose defining functions are differentiable; or whether, at least with the help of certain simple assumptions, a transformation is possible into groups admitting Lie's methods. A reduction to analytic groups is, according to a theorem announced by Lie * but first proved by Schur,† always possible when the group is transitive and the existence of the first and certain second derivatives of the functions defining the group is assumed.

For infinite groups the investigation of the corresponding question is, I believe, also of interest. Moreover we are thus led to the wide and interesting field of functional equations which have been heretofore investigated usually only under the a umption of the differentiability of the functions involved. In particular the functional equations treated by Abel * with so much ingenuity, the difference equations, and other equations occurring in the literature of mathematics, do not directly involve anything which necessitates the requirement of the differentiability of the accompanying functions. In the search for certain existence proofs in the calculus of variations I came directly upon the problem: To prove the differen-

* Lie-Engel, Theorie der Transformationsgruppen, vol. 3, Leipzig, 1893, §§ 82, 144.

† "Ueber den analytischen Charakter der eine endliche Kontinuierliche Transformationsgruppen darstellenden Funktionen," *Math. Annalen*, vol. 41.

* Werke, vol. 1, pp. 1, 61, 389.

tiability of the function under consideration from the existence of a difference equation. In all these cases, then, the problem arises : *In how far are the assertions which we can make in the case of differentiable functions true under proper modifications without this assumption?*

It may be further remarked that H. Minkowski in his above-mentioned Geometrie der Zahlen starts with the functional equation

$$f(x_1 + y_1, \cdots, x_n + y_n) \leqq f(x_1, \cdots, x_n) + f(y_1, \cdots, y_n)$$

and from this actually succeeds in proving the existence of certain differential quotients for the function in question.

On the other hand I wish to emphasize the fact that there certainly exist analytical functional equations whose sole solutions are non-differentiable functions. For example a uniform continuous non-differentiable function $\varphi(x)$ can be constructed which represents the only solution of the two functional equations

$$\varphi(x + \alpha) - \varphi(x) = f(x), \qquad \varphi(x + \beta) - \varphi(x) = 0,$$

where α and β are two real numbers, and $f(x)$ denotes, for all the real values of x, a regular analytic uniform function. Such functions are obtained in the simplest manner by means of trigonometrical series by a process similar to that used by Borel (according to a recent announcement of Picard)† for the construction of a doubly periodic, non-analytic solution of a certain analytic partial differential equation.

6. Mathematical Treatment of the Axioms of Physics.

The investigations on the foundations of geometry suggest the problem : *To treat in the same manner, by means of axioms, those physical sciences in which mathematics plays an important part ; in the first rank are the theory of probabilities and mechanics.*

As to the axioms of the theory of probabilities,* it seems to me desirable that their logical investigation should be accompanied by a rigorous and satisfactory development of the method of mean values in mathematical physics, and in particular in the kinetic theory of gases.

Important investigations by physicists on the foundations of mechanics are at hand ; I refer to the writings of Mach,† Hertz,‡ Boltzmann § and Volkmann.‖ It is therefore very desirable that the discussion of the foundations of mechanics be taken up by mathematicians also. Thus Boltzmann's work on the principles of mechanics suggests the problem of developing mathematically the limiting processes, there merely indicated, which lead from the atomistic view to the

† "Quelques théories fondamentales dans l'analyse mathématique," Conférences faites à Clark University, *Revue générale des Sciences*, 1900, p. 22.

* Cf. Bohlmann, "Ueber Versicherungsmathematik", from the collection : Klein and Riecke, Ueber angewandte Mathematik und Physik, Leipzig, 1900.

† Die Mechanik in ihrer Entwickelung, Leipzig, 4th edition, 1901.

‡ Die Prinzipien der Mechanik, Leipzig, 1894.

§ Vorlesungen über die Principe der Mechanik, Leipzig, 1897.

‖ Einführung in das Studium der theoretischen Physik, Leipzig, 1900.

laws of motion of continua. Conversely one might try to derive the laws of the motion of rigid bodies by a limiting process from a system of axioms depending upon the idea of continuously varying conditions of a material filling all space continuously, these conditions being defined by parameters. For the question as to the equivalence of different systems of axioms is always of great theoretical interest.

If geometry is to serve as a model for the treatment of physical axioms, we shall try first by a small number of axioms to include as large a class as possible of physical phenomena, and then by adjoining new axioms to arrive gradually at the more special theories. At the same time Lie's a principle of subdivision can perhaps be derived from profound theory of infinite transformation groups. The mathematician will have also to take account not only of those theories coming near to reality, but also, as in geometry, of all logically possible theories. He must be always alert to obtain a complete survey of all conclusions derivable from the system of axioms assumed.

Further, the mathematician has the duty to test exactly in each instance whether the new axioms are compatible with the previous ones. The physicist, as his theories develop, often finds himself forced by the results of his experiments to make new hypotheses, while he depends, with respect to the compatibility of the new hypotheses with the old axioms, solely upon these experiments or upon a certain physical intuition, a practice which in the rigorously logical building up of a theory is not admissible. The desired proof of the compatibility of all assumptions seems to me also of importance, because the effort to obtain such proof always forces us most effectually to an exact formulation of the axioms.

So far we have considered only questions concerning the foundations of the mathematical sciences. Indeed, the study of the foundations of a science is always particularly attractive, and the testing of these foundations will always be among the foremost problems of the investigator. Weierstrass once said, "The final object always to be kept in mind is to arrive at a correct understanding of the foundations of the science. * * * But to make any progress in the sciences the study of particular problems is, of course, indispensable." In fact, a thorough understanding of its special theories is necessary to the successful treatment of the foundations of the science. Only that architect is in the position to lay a sure foundation for a structure who knows its purpose thoroughly and in detail. So we turn now to the special problems of the separate branches of mathematics and consider first arithmetic and algebra.

7. Irrationality and Transcendence of Certain Numbers.

Hermite's arithmetical theorems on the exponential function and their extension by Lindemann are certain of the admiration of all generations of mathematicians. Thus the task at once presents itself to penetrate further along the path here entered, as A. Hurwitz has already done in two interesting papers,* "Ueber arithmetische Eigenschaften gewisser transzendenter Funktionen." I should like, therefore, to sketch a class of problems which, in my opinion, should be attacked as here next in order. That certain special transcendental functions, important in analysis, take algebraic values for certain algebraic arguments, seems to us particularly remarkable and worthy of thorough investigation. Indeed, we expect transcendental functions to assume, in general, transcendental values for even algebraic arguments; and, although it is well known that there exist integral transcendental functions which even have rational values for all algebraic arguments, we shall still consider it highly probable that the exponential function $e^{i\pi z}$, for example, which evidently has algebraic values for all rational arguments z, will on the other hand always take transcendental values for irrational algebraic values of the argument z. We can also give this statement a geometrical form, as follows:

If, in an isosceles triangle, the ratio of the base angle to the angle at the vertex be algebraic but not rational, the ratio between base and side is always transcendental.

In spite of the simplicity of this statement and of its similarity to the problems solved by Hermite and Lindemann, I consider the proof of this theorem very difficult; as also the proof that

The expression α^β, for an algebraic base α and an irrational algebraic exponent β, e. g., the number $2^{\sqrt{2}}$ or $e^\pi = i^{-2i}$, always represents a transcendental or at least an irrational number.

It is certain that the solution of these and similar problems must lead us to entirely new methods and to a new insight into the nature of special irrational and transcendental numbers.

8. Problems of Prime Numbers.

Essential progress in the theory of the distribution of prime numbers has lately been made by Hadamard, de la Vallée-Poussin, Von Mangoldt and others. For the complete solution, however, of the problems set us by Riemann's paper "Ueber die Anzahl der Primzahlen unter einer gegebenen Grösse," it still remains to prove the correctness of an exceedingly important statement of Riemann, viz., *that the zero points of the function $\zeta(s)$ defined by the series*

$$\zeta(s) = 1 + \frac{1}{2^s} + \frac{1}{3^s} + \frac{1}{4^s} + \cdots$$

all have the real part $\frac{1}{2}$, except the well-known negative integral real zeros. As soon as this proof has been successfully established, the next problem would consist in testing more exactly Riemann's infinite series for the number of primes below a given number and, especially, *to decide whether the difference between the number of primes below a number x and the integral logarithm of x does in fact become infinite*

* *Math. Annalen*, vols. 22, 32 (1883, 1888).

*of an order not greater than $\frac{1}{2}$ in x.** Further, we should determine whether the occasional condensation of prime numbers which has been noticed in counting primes is really due to those terms of Riemann's formula which depend upon the first complex zeros of the function $\zeta(s)$.

After an exhaustive discussion of Riemann's prime number formula, perhaps we may sometime be in a position to attempt the rigorous solution of Goldbach's problem,† viz., whether every integer is expressible as the sum of two positive prime numbers; and further to attack the well-known question, whether there are an infinite number of pairs of prime numbers with the difference 2, or even the more general problem, whether the linear diophantine equation

$$ax + by + c = 0$$

(with given integral coefficients each prime to the others) is always solvable in prime numbers x and y.

But the following problem seems to me of no less interest and perhaps of still wider range: *To apply the results obtained for the distribution of rational prime numbers to the theory of the distribution of ideal primes in a given number-field k*—a problem which looks toward the study of the function $\zeta_k(s)$ belonging to the field and defined by the series

$$\zeta_k(s) = \sum \frac{1}{n(j)^s},$$

where the sum extends over all ideals j of the given realm k, and $n(j)$ denotes the norm of the ideal j.

I may mention three more special problems in number theory: one on the laws of reciprocity, one on diophantine equations, and a third from the realm of quadratic forms.

9. Proof of the Most General Law of Reciprocity in Any Number Field.

For any field of numbers the law of reciprocity is to be proved for the residues of the lth power, when l denotes an odd prime, and further when l is a power of 2 or a power of an odd prime.

The law, as well as the means essential to its proof, will, I believe, result by suitably generalizing the theory of the field of the lth roots of unity,* developed by me, and my theory of relative quadratic fields.†

10. Determination of the Solvability of a Diophantine Equation.

Given a diophantine equation with any number of unknown quantities and with rational integral numerical

---

* Cf. an article by H. von Koch, which is soon to appear in the *Math. Annalen* [Vol. 55, p. 441].

† Cf. P. Stäckel: "Über Goldbach's empirisches Theorem," *Nachrichten d. K. Ges. d. Wiss. zu Göttingen*, 1896, and Landau, *ibid.*, 1900.

* *Jahresber. d. Deutschen Math.-Vereinigung*, "Ueber die Theorie der algebraischen Zahlkörper," vol. 4 (1897), Part V.

† *Math. Annalen*, vol. 51 and *Nachrichten d. K. Ges. d. Wiss. zu Göttingen*, 1898.

coefficients : *To devise a process according to which it can be determined by a finite number of operations whether the equation is solvable in rational integers.*

11. Quadratic Forms With any Algebraic Numerical Coefficients.

Our present knowledge of the theory of quadratic number fields ‡ puts us in a position *to attack successfully the theory of quadratic forms with any number of variables and with any algebraic numerical coefficients.* This leads in particular to the interesting problem : to solve a given quadratic equation with algebraic numerical coefficients in any number of variables by integral or fractional numbers belonging to the algebraic realm of rationality determined by the coefficients.

The following important problem may form a transition to algebra and the theory of functions :

12. Extension of Kronecker's Theorem on Abelian Fields to any Algebraic Realm of Rationality.

The theorem that every abelian number field arises from the realm of rational numbers by the composition of fields of roots of unity is due to Kronecker. This fundamental theorem in the theory of integral equations contains two statements, namely :

First. It answers the question as to the number and existence of those equations which have a given degree, a given abelian group and a given discriminant with respect to the realm of rational numbers.

Second. It states that the roots of such equations form a realm of algebraic numbers which coincides with the realm obtained by assigning to the argument z in the exponential function $e^{i\pi z}$ all rational numerical values in succession.

The first statement is concerned with the question of the determination of certain algebraic numbers by their groups and their branching. This question corresponds, therefore, to the known problem of the determination of algebraic functions corresponding to given Riemann surfaces. The second statement furnishes the required numbers by transcendental means, namely, by the exponential function $e^{i\pi z}$.

Since the realm of the imaginary quadratic number fields is the simplest after the realm of rational numbers, the problem arises, to extend Kronecker's theorem to this case. Kronecker himself has made the assertion that the abelian equations in the realm of a quadratic field are given by the equations of transformation of elliptic functions with singular moduli, so that the elliptic function assumes here the same rôle as the exponential function in the former case. The proof of Kronecker's conjecture has not yet been

‡ Hilbert, "Ueber den Dirichlet'schen biquadratischen Zahlenkörper," *Math. Annalen*, vol. 45 ; "Ueber die Theorie der relativquadratischen Zahlkörper," *Jahresber. d. Deutschen Mathematiker-Vereinigung*, 1897, and *Math. Annalen*, vol. 51 ; "Ueber die Theorie der relativ-Abelschen Körper," *Nachrichten d. K. Ges. d. Wiss. zu Göttingen*, 1898 ; Grundlagen der Geometrie, Leipzig, 1899, Chap. VIII, § 83 [Translation by Townsend, Chicago, 1902]. Cf. also the dissertation of G. Rückle, Göttingen, 1901.

furnished; but I believe that it must be obtainable without very great difficulty on the basis of the theory of complex multiplication developed by H. Weber * with the help of the purely arithmetical theorems on class fields which I have established.

Finally, the extension of Kronecker's theorem to the case that, *in place of the realm of rational numbers or of the imaginary quadratic field, any algebraic field whatever is laid down as realm of rationality*, seems to me of the greatest importance. I regard this problem as one of the most profound and far-reaching in the theory of numbers and of functions.

The problem is found to be accessible from many standpoints. I regard as the most important key to the arithmetical part of this problem the general law of reciprocity for residues of lth powers within any given number field.

As to the function-theoretical part of the problem, the investigator in this attractive region will be guided by the remarkable analogies which are noticeable between the theory of algebraic functions of one variable and the theory of algebraic numbers. Hensel * has proposed and investigated the analogue in the theory of algebraic numbers to the development in power series of an algebraic function; and Landsberg † has treated the analogue of the Riemann-Roch theorem. The analogy between the deficiency of a Riemann surface and that of the class number of a field of numbers is also evident. Consider a Riemann surface of deficiency $p = 1$ (to touch on the simplest case only) and on the other hand a number field of class $h = 2$. To the proof of the existence of an integral everywhere finite on the Riemann surface, corresponds the proof of the existence of an integer a in the number field such that the number $\sqrt{a}$ represents a quadratic field, relatively unbranched with respect to the fundamental field. In the theory of algebraic functions, the method of boundary values (*Randwerthaufgabe*) serves, as is well known, for the proof of Riemann's existence theorem. In the theory of number fields also, the proof of the existence of just this number a offers the greatest difficulty. This proof succeeds with indispensable assistance from the theorem that in the number field there are always prime ideals corresponding to given residual properties. This latter fact is therefore the analogue in number theory to the problem of boundary values.

The equation of Abel's theorem in the theory of algebraic functions expresses, as is well known, the necessary and sufficient condition that the points in question on the Riemann surface are the zero points of an algebraic function belonging to the surface. The exact analogue of Abel's theorem, in the theory of the number field of class $h = 2$, is the equation of the law of quadratic reciprocity ‡

---

* Elliptische Functionen und algebraische Zahlen. Braunschweig, 1891.

* *Jahresber. d. Deutschen Math.-Vereinigung*, vol. 6, and an article soon to appear in the *Math. Annalen* [Vol. 55, p. 301]: "Ueber die Entwickelung der algebraischen Zahlen in Potenzreihen."

† *Math. Annalen*, vol. 50 (1898).

‡ Cf. Hilbert, "Ueber die Theorie der relativ-Abelschen Zahlkörper," *Gött. Nachrichten*, 1898.

$$\left(\frac{a}{j}\right) = +1,$$

which declares that the ideal j is then and only then a principal ideal of the number field when the quadratic residue of the number a with respect to the ideal j is positive.

It will be seen that in the problem just sketched the three fundamental branches of mathematics, number theory, algebra and function theory, come into closest touch with one another, and I am certain that the theory of analytical functions of several variables in particular would be notably enriched if one should succeed *in finding and discussing those functions which play the part for any algebraic number field corresponding to that of the exponential function in the field of rational numbers and of the elliptic modular functions in the imaginary quadratic number field.*

Passing to algebra, I shall mention a problem from the theory of equations and one to which the theory of algebraic invariants has led me.

13. Impossibility of the Solution of the General Equation of the 7th Degree by means of Functions of only Two Arguments.

Nomography* deals with the problem: to solve equations by means of drawings of families of curves depending on an arbitrary parameter. It is seen at once that every root of an equation whose coefficients depend upon only two parameters, that is, every function of two independent variables, can be represented in manifold ways according to the principle lying at the foundation of nomography. Further, a large class of functions of three or more variables can evidently be represented by this principle alone without the use of variable elements, namely all those which can be generated by forming first a function of two arguments, then equating each of these arguments to a function of two arguments, next replacing each of those arguments in their turn by a function of two arguments, and so on, regarding as admissible any finite number of insertions of functions of two arguments. So, for example, every rational function of any number of arguments belongs to this class of functions constructed by nomographic tables; for it can be generated by the processes of addition, subtraction, multiplication and division and each of these processes produces a function of only two arguments. One sees easily that the roots of all equations which are solvable by radicals in the natural realm of rationality belong to this class of functions; for here the extraction of roots is adjoined to the four arithmetical operations and this, indeed, presents a function of one argument only. Likewise the general equations of the 5th and 6th degrees are solvable by suitable nomographic tables; for, by means of Tschirnhausen transformations, which require only extraction of roots, they can be reduced to a form where the coefficients depend upon two parameters only.

* d'Ocagne, Traité de Nomographie, Paris, 1899.

Now it is probable that the root of the equation of the seventh degree is a function of its coefficients which does not belong to this class of functions capable of nomographic construction, *i. e.*, that it cannot be constructed by a finite number of insertions of functions of two arguments. In order to prove this, the proof would be necessary *that the equation of the seventh degree* $f^7 + xf^3 + yf^2 + zf + 1 = 0$ *is not solvable with the help of any continuous functions of only two arguments.* I may be allowed to add that I have satisfied myself by a rigorous process that there exist analytical functions of three arguments x, y, z which cannot be obtained by a finite chain of functions of only two arguments.

By employing auxiliary movable elements, nomography succeeds in constructing functions of more than two arguments, as d'Ocagne has recently proved in the case of the equation of the 7th degree.*

14. Proof of the Finiteness of Certain Complete Systems of Functions.

In the theory of algebraic invariants, questions as to the finiteness of complete systems of forms deserve, as it seems to me, particular interest. L. Maurer † has lately succeeded in extending the theorems on finiteness in invariant theory proved by P. Gordan and myself, to the case where, instead of the general projective group, any subgroup is chosen as the basis for the definition of invariants.

An important step in this direction had been taken already by A. Hurwitz,‡ who, by an ingenious process, succeeded in effecting the proof, in its entire generality, of the finiteness of the system of orthogonal invariants of an arbitrary ground form.

The study of the question as to the finiteness of invariants has led me to a simple problem which includes that question as a particular case and whose solution probably requires a decidedly more minutely detailed study of the theory of elimination and of Kronecker's algebraic modular systems than has yet been made.

Let a number m of integral rational functions $X_1, X_2, \cdots, X_m$ of the n variables $x_1, x_2, \cdots, x_n$ be given,

$$(S) \qquad \begin{aligned} X_1 &= f_1(x_1, \cdots, x_n), \\ X_2 &= f_2(x_1, \cdots, x_n), \\ &\cdots\cdots\cdots\cdots\cdots \\ X_m &= f_m(x_1, \cdots, x_n). \end{aligned}$$

Every rational integral combination of $X_1, \cdots, X_m$ must evidently always become, after substitution of the above expressions, a rational integral function of $x_1, \cdots, x_n$. Nevertheless, there may well be rational fractional functions of

* "Sur la résolution nomographique de l'équation du septième degré." *Comptes rendus*, Paris, 1900.

† Cf. *Sitzungsber. d. K. Acad. d. Wiss. zu München*, 1899, and an article about to appear in the *Math. Annalen*.

‡ "Ueber die Erzeugung der Invarianten durch Integration," *Nachrichten d. K. Gesellschaft d. Wiss. zu Göttingen*, 1897.

$X_1, \cdots, X_m$ which, by the operation of the substitution S, become integral functions in $x_1, \cdots, x_n$. Every such rational function of $X_1, \cdots, X_m$, which becomes integral in $x_1, \cdots, x_n$ after the application of the substitution S, I propose to call a *relatively integral* function of $X_1, \cdots, X_m$. Every integral function of $X_1, \cdots, X_m$ is evidently also relatively integral; further the sum, difference and product of relative integral functions are themselves relatively integral.

The resulting problem is now to decide whether it is always possible *to find a finite system of relatively integral function $X_1, \cdots, X_m$ by which every other relatively integral function of $X_1, \cdots, X_m$ may be expressed rationally and integrally.*

We can formulate the problem still more simply if we introduce the idea of a finite field of integrality. By a finite field of integrality I mean a system of functions from which a finite number of functions can be chosen, in terms of which all other functions of the system are rationally and integrally expressible. Our problem amounts, then, to this: to show that all relatively integral functions of any given domain of rationality always constitute a finite field of integrality.

It naturally occurs to us also to refine the problem by restrictions drawn from number theory, by assuming the coefficients of the given functions $f_1, \cdots, f_m$ to be integers and including among the relatively integral functions of $X_1, \cdots, X_m$ only such rational functions of these arguments as become, by the application of the substitutions S, rational integral functions of $x_1, \cdots, x_n$ with rational integral coefficients.

The following is a simple particular case of this refined problem: Let m integral rational functions $X_1, \cdots, X_m$ of one variable x with integral rational coefficients, and a prime number p be given. Consider the system of those integral rational functions of x which can be expressed in the form

$$\frac{G(X_1, \cdots, X_m)}{p^h},$$

where G is a rational integral function of the arguments $X_1, \cdots, X_m$ and p^h is any power of the prime number p. Earlier investigations of mine* show immediately that all such expressions for a fixed exponent h form a finite domain of integrality. But the question here is whether the same is true for all exponents h, *i. e.*, whether a finite number of such expressions can be chosen by means of which for every exponent h every other expression of that form is integrally and rationally expressible.

From the boundary region between algebra and geometry, I will mention two problems. The one concerns enumerative geometry and the other the topology of algebraic curves and surfaces.

15. Rigorous Foundation of Schubert's Enumerative Calculus.

The problem consists in this: *To establish rigorously and with*

---

* *Math. Annalen*, vol. 36 (1890), p. 485.

an exact determination of the limits of their validity those geometrical numbers which Schubert† *especially has determined on the basis of the so-called principle of special position, or conservation of number, by means of the enumerative calculus developed by him.*

Although the algebra of to-day guarantees, in principle, the possibility of carrying out the processes of elimination, yet for the proof of the theorems of enumerative geometry decidedly more is requisite, namely, the actual carrying out of the process of elimination in the case of equations of special form in such a way that the degree of the final equations and the multiplicity of their solutions may be foreseen.

16. Problem of the Topology of Algebraic Curves and Surfaces.

The maximum number of closed and separate branches which a plane algebraic curve of the nth order can have has been determined by Harnack.‡ There arises the further question as to the relative position of the branches in the plane. As to curves of the 6th order, I have satisfied myself—by a complicated process, it is true—that of the eleven branches which they can have according to Harnack, by no means all can lie external to one another, but that one branch must exist in whose interior one branch and in whose exterior nine branches lie, or inversely. *A thorough investigation of the relative position of the separate branches when their number is the maximum seems to me to be of very great interest, and not less so the corresponding investigation as to the number, form, and position of the sheets of an algebraic surface in space.* Till now, indeed, it is not even known what is the maximum number of sheets which a surface of the 4th order in three dimensional space can really have.*

In connection with this purely algebraic problem, I wish to bring forward a question which, it seems to me, may be attacked by the same method of continuous variation of coefficients, and whose answer is of corresponding value for the topology of families of curves defined by differential equations. This is the question as to the maximum number and position of Poincaré's boundary cycles (cycles limites) for a differential equation of the first order and degree of the form

$$\frac{dy}{dx} = \frac{Y}{X},$$

where X and Y are rational integral functions of the nth degree in x and y. Written homogeneously, this is

$$X\left(y\frac{dz}{dt} - z\frac{dy}{dt}\right) + Y\left(z\frac{dx}{dt} - x\frac{dz}{dt}\right) + Z\left(x\frac{dy}{dt} - y\frac{dx}{dt}\right) = 0,$$

where X, Y, and Z are rational integral homogeneous func-

† Kalkül der abzählenden Geometrie, Leipzig, 1879.

‡ *Math. Annalen*, vol. 10.

* Cf. Rohn, "Flachen vierter Ordnung," Preisschriften der Fürstlich Jablonowskischen Gesellschaft, Leipzig, 1886.

tions of the nth degree in x, y, z, and the latter are to be determined as functions of the parameter t.

17. Expression of Definite Forms by Squares.

A rational integral function or form in any number of variables with real coefficients such that it becomes negative for no real values of these variables, is said to be *definite.* The system of all definite forms is invariant with respect to the operations of addition and multiplication, but the quotient of two definite forms—in case it should be an integral function of the variables—is also a definite form. The square of any form is evidently always a definite form. But since, as I have shown,* not every definite form can be compounded by addition from squares of forms, the question arises—which I have answered affirmatively for ternary forms†—whether every definite form may not be expressed as a quotient of sums of squares of forms. At the same time it is desirable, for certain questions as to the possibility of certain geometrical constructions, to know whether the coefficients of the forms to be used in the expression may always be taken from the realm of rationality given by the coefficients of the form represented.‡

I mention one more geometrical problem :

18. Building up of Space from Congruent Polyhedra.

If we enquire for those groups of motions in the plane for which a fundamental region exists, we obtain various answers, according as the plane considered is Riemann's (elliptic), Euclid's, or Lobachevsky's (hyperbolic). In the case of the elliptic plane there is a finite number of essentially different kinds of fundamental regions, and a finite number of congruent regions suffices for a complete covering of the whole plane; the group consists indeed of a finite number of motions only. In the case of the hyperbolic plane there is an infinite number of essentially different kinds of fundamental regions, namely, the well-known Poincaré polygons. For the complete covering of the plane an infinite number of congruent regions is necessary. The case of Euclid's plane stands between these; for in this case there is only a finite number of essentially different kinds of groups of motions with fundamental regions, but for a complete covering of the whole plane an infinite number of congruent regions is necessary.

Exactly the corresponding facts are found in space of three dimensions. The fact of the finiteness of the groups of motions in elliptic space is an immediate consequence of a fundamental theorem of C. Jordan,§ whereby the number of essentially different kinds of finite groups of linear substitutions in n variables does not surpass a certain finite limit

* *Math. Annalen,* vol. 32.

† *Acta Mathematica,* vol. 17.

‡ Cf. Hilbert: Grundlagen der Geometrie, Leipzig, 1899, Chap. 7 and in particular § 38.

§ *Crelle's Journal,* vol. 84 (1878), and *Atti d. Reale Acad. di Napoli* 1880.

dependent upon n. The groups of motions with fundamental regions in hyperbolic space have been investigated by Fricke and Klein in the lectures on the theory of automorphic functions,* and finally Fedorov,† Schoenflies ‡ and lately Rohn § have given the proof that there are, in euclidean space, only a finite number of essentially different kinds of groups of motions with a fundamental region. Now, while the results and methods of proof applicable to elliptic and hyperbolic space hold directly for n-dimensional space also, the generalization of the theorem for euclidean space seems to offer decided difficulties. The investigation of the following question is therefore desirable: *Is there in n-dimensional euclidean space also only a finite number of essentially different kinds of groups of motions with a fundamental region?*

A fundamental region of each group of motions, together with the congruent regions arising from the group, evidently fills up space completely. The question arises: *Whether polyhedra also exist which do not appear as fundamental regions of groups of motions, by means of which nevertheless by a suitable juxtaposition of congruent copies a complete filling up of all space is possible.* I point out the following question, related to the preceding one, and important to number theory and perhaps sometimes useful to physics and chemistry: How can one arrange most densely in space an infinite number of equal solids of given form, *e. g.*, spheres with given radii or regular tetrahedra with given edges (or in prescribed position), that is, how can one so fit them together that the ratio of the filled to the unfilled space may be as great as possible?

If we look over the development of the theory of functions in the last century, we notice above all the fundamental importance of that class of functions which we now designate as analytic functions—a class of functions which will probably stand permanently in the center of mathematical interest.

There are many different standpoints from which we might choose, out of the totality of all conceivable functions, extensive classes worthy of a particularly thorough investigation. Consider, for example, *the class of functions characterized by ordinary or partial algebraic differential equations.* It should be observed that this class does not contain the functions that arise in number theory and whose investigation is of the greatest importance. For example, the before-mentioned function $\zeta(s)$ satisfies no algebraic differential equation, as is easily seen with the help of the well-known relation between $\zeta(s)$ and $\zeta(1-s)$, if one refers to the theorem proved by Hölder,* that the function $\Gamma(x)$ satisfies no algebraic differential equation. Again, the function of the two variables s and x defined by the infinite series

* Leipzig, 1897. Cf. especially Abschnitt I, Chapters 2 and 3.

† Symmetrie der regelmässigen Systeme von Figuren, 1890.

‡ Krystallsysteme und Krystallstruktur, Leipzig, 1891.

§ *Math. Annalen*, vol. 53.

* *Math. Annalen*, vol. 28.

$$\zeta(s, x) = x + \frac{x^2}{2^s} + \frac{x^3}{3^s} + \frac{x^4}{4^s} + \cdots,$$

which stands in close relation with the function $\zeta(s)$, probably satisfies no algebraic partial differential equation. In the investigation of this question the functional equation

$$x \frac{\partial \zeta(s, x)}{\partial x} = \zeta(s - 1, x)$$

will have to be used.

If, on the other hand, we are lead by arithmetical or geometrical reasons to consider the class of all those functions which are continuous and indefinitely differentiable, we should be obliged in its investigation to dispense with that pliant instrument, the power series, and with the circumstance that the function is fully determined by the assignment of values in any region, however small. While, therefore, the former limitation of the field of functions was too narrow, the latter seems to me too wide.

The idea of the analytic function on the other hand includes the whole wealth of functions most important to science, whether they have their origin in number theory, in the theory of differential equations or of algebraic functional equations, whether they arise in geometry or in mathematical physics; and, therefore, in the entire realm of functions, the analytic function justly holds undisputed supremacy.

19. Are the Solutions of Regular Problems in the Calculus of Variations Always Necessarily Analytic?

One of the most remarkable facts in the elements of the theory of analytic functions appears to me to be this: That there exist partial differential equations whose integrals are all of necessity analytic functions of the independent variables, that is, in short, equations susceptible of none but analytic solutions. The best known partial differential equations of this kind are the potential equation

$$\frac{\partial^2 f}{\partial x^2} + \frac{\partial^2 f}{\partial y^2} = 0$$

and certain linear differential equations investigated by Picard;* also the equation

$$\frac{\partial^2 f}{\partial x^2} + \frac{\partial^2 f}{\partial y^2} = e^f,$$

the partial differential equation of minimal surfaces, and others. Most of these partial differential equations have the common characteristic of being the lagrangian differential equations of certain problems of variation, viz., of such problems of variation

$$\iint F(p, q, z; x, y)\, dx\, dy = \text{minimum}$$

$$\left[p = \frac{\partial z}{\partial x}, q = \frac{\partial z}{\partial y}\right],$$

* *Jour. de l'Ecole Polytech.*, 1890.

as satisfy, for all values of the arguments which fall within the range of discussion, the inequality

$$\frac{\partial^2 F}{\partial p^2}\cdot\frac{\partial^2 F}{\partial q^2}-\left(\frac{\partial^2 F}{\partial p \partial q}\right)^2>0,$$

F itself being an analytic function. We shall call this sort of problem a *regular* variation problem. It is chiefly the regular variation problems that play a rôle in geometry, in mechanics, and in mathematical physics; and the question naturally arises, whether all solutions of regular variation problems must necessarily be analytic functions. In other words, *does every lagrangian partial differential equation of a regular variation problem have the property of admitting analytic integrals exclusively?* And is this the case even when the function is constrained to assume, as, *e. g.*, in Dirichlet's problem on the potential function, boundary values which are continuous, but not analytic?

I may add that there exist surfaces of constant *negative* gaussian curvature which are representable by functions that are continuous and possess indeed all the derivatives, and yet are not analytic; while on the other hand it is probable that every surface whose gaussian curvature is constant and *positive* is necessarily an analytic surface. And we know that the surfaces of positive constant curvature are most closely related to this regular variation problem: To pass through a closed curve in space a surface of minimal area which shall inclose, in connection with a fixed surface through the same closed curve, a volume of given magnitude.

20. The General Problem of Boundary Values.

An important problem closely connected with the foregoing is the question concerning the existence of solutions of partial differential equations when the values on the boundary of the region are prescribed. This problem is solved in the main by the keen methods of H. A. Schwarz, C. Neumann, and Poincaré for the differential equation of the potential. These methods, however, seem to be generally not capable of direct extension to the case where along the boundary there are prescribed either the differential coefficients or any relations between these and the values of the function. Nor can they be extended immediately to the case where the inquiry is not for potential surfaces but, say, for surfaces of least area, or surfaces of constant positive gaussian curvature, which are to pass through a prescribed twisted curve or to stretch over a given ring surface. It is my conviction that it will be possible to prove these existence theorems by means of a general principle whose nature is indicated by Dirichlet's principle. This general principle will then perhaps enable us to approach the question: *Has not every regular variation problem a solution, provided certain assumptions regarding the given boundary conditions are satisfied* (say that the functions concerned in these boundary conditions are continuous and have in sections one or more derivatives), *and provided also if need be that the notion of a solution shall be suitably extended?* *

* Cf. my lecture on Dirichlet's principle in the *Jahresber. d. Deutschen Math.-Vereinigung*, vol. 8 (1900), p. 184.

21. Proof of the Existence of Linear Differential Equations Having a Prescribed Monodromic Group.

In the theory of linear differential equations with one independent variable z, I wish to indicate an important problem, one which very likely Riemann himself may have had in mind. This problem is as follows: *To show that there always exists a linear differential equation of the Fuchsian class, with given singular points and monodromic group.* The problem requires the production of n functions of the variable z, regular throughout the complex z plane except at the given singular points; at these points the functions may become infinite of only finite order, and when z describes circuits about these points the functions shall undergo the prescribed linear substitutions. The existence of such differential equations has been shown to be probable by counting the constants, but the rigorous proof has been obtained up to this time only in the particular case where the fundamental equations of the given substitutions have roots all of absolute magnitude unity. L. Schlesinger has given this proof,* based upon Poincaré's theory of the Fuchsian ζ-functions. The theory of linear differential equations would evidently have a more finished appearance if the problem here sketched could be disposed of by some perfectly general method.

22. Uniformization of Analytic Relations by Means of Automorphic Functions.

As Poincaré was the first to prove, it is always possible to reduce any algebraic relation between two variables to uniformity by the use of automorphic functions of one variable. That is, if any algebraic equation in two variables be given, there can always be found for these variables two such single valued automorphic functions of a single variable that their substitution renders the given algebraic equation an identity. The generalization of this fundamental theorem to any analytic non-algebraic relations whatever between two variables has likewise been attempted with success by Poincaré,† though by a way entirely different from that which served him in the special problem first mentioned. From Poincaré's proof of the possibility of reducing to uniformity an arbitrary analytic relation between two variables, however, it does not become apparent whether the resolving functions can be determined to meet certain additional conditions. Namely, it is not shown whether the two single valued functions of the one new variable can be so chosen that, while this variable traverses the *regular* domain of those functions, the totality of all regular points of the given analytic field are actually reached and represented. On the contrary it seems to be the case, from Poincaré's investigations, that there are beside the branch points certain others, in general infinitely many other discrete exceptional points of the analytic field, that can be reached only by making the new variable approach certain limiting

* Handbuch der Theorie der linearen Differentialgleichungen, vol. 2, part 2, No. 366.

†*Bull. de la Soc. Math. de France*, vol. 11 (1883).

points of the functions. *In view of the fundamental importance of Poincaré's formulation of the question it seems to me that an elucidation and resolution of this difficulty is extremely desirable.*

In conjunction with this problem comes up the problem of reducing to uniformity an algebraic or any other analytic relation among three or more complex variables—a problem which is known to be solvable in many particular cases. Toward the solution of this the recent investigations of Picard on algebraic functions of two variables are to be regarded as welcome and important preliminary studies.

23. Further Development of the Methods of the Calculus of Variations.

So far, I have generally mentioned problems as definite and special as possible, in the opinion that it is just such definite and special problems that attract us the most and from which the most lasting influence is often exerted upon science. Nevertheless, I should like to close with a general problem, namely with the indication of a branch of mathematics repeatedly mentioned in this lecture—which, in spite of the considerable advancement lately given it by Weierstrass, does not receive the general appreciation which, in my opinion, is its due—I mean the calculus of variations.*

The lack of interest in this is perhaps due in part to the need of reliable modern text-books. So much the more praiseworthy is it that A. Kneser in a very recently published work has treated the calculus of variations from the modern points of view and with regard to the modern demand for rigor.†

The calculus of variations is, in the widest sense, the theory of the variation of functions, and as such appears as a necessary extension of the differential and integral calculus. In this sense, Poincaré's investigations on the problem of three bodies, for example, form a chapter in the calculus of variations, in so far as Poincaré derives from known orbits by the principle of variation new orbits of similar character.

I add here a short justification of the general remarks upon the calculus of variations made at the beginning of my lecture.

The simplest problem in the calculus of variations proper is known to consist in finding a function y of a variable x such that the definite integral

$$J=\int_a^b F(y_x,\, y\,;\ x)dx, \quad y_x=\frac{dy}{dx}$$

assumes a minimum value as compared with the values it

* Text-books: Moigno-Lindelöf, Leçons du calcul des variations, Paris, 1861, and A. Kneser, Lehrbuch der Variations-rechnung, Braunschweig, 1900.

† As an indication of the contents of this work, it may here be noted that for the simplest problems Kneser derives sufficient conditions of the extreme even for the case that one limit of integration is variable, and employs the envelope of a family of curves satisfying the differential equations of the problem to prove the necessity of Jacobi's conditions of the extreme. Moreover, it should be noticed that Kneser applies Weierstrass's theory also to the inquiry for the extreme of such quantities as are defined by differential equations.

takes when y is replaced by other functions of x with the same initial and final values.

The vanishing of the first variation in the usual sense

$$\delta J = 0$$

gives for the desired function y the well-known differential equation

$$\frac{dF_{y_x}}{dx} - F_y = 0, \tag{1}$$

$$\left[F_{y_x} = \frac{\partial F}{\partial y_x}, \quad F_y = \frac{\partial F}{\partial y} \right].$$

In order to investigate more closely the necessary and sufficient criteria for the occurrence of the required minimum, we consider the integral

$$J^* = \int_a^b \{F + (y_x - p)\, F_p\} dx,$$

$$\left[F = F(p, y\,;\, x),\ F_p = \frac{\partial F(p, y; x)}{\partial p} \right].$$

Now we inquire how p is to be chosen as function of x, y in order that the value of this integral J^* shall be independent of the path of integration, i. e., of the choice of the function y of the variable x. The integral J^* has the form

$$J^* = \int_a^b \{A y_x - B\} dx,$$

where A and B do not contain y_x and the vanishing of the first variation

$$\delta J^* = 0$$

in the sense which the new question requires gives the equation

$$\frac{\partial A}{\partial x} + \frac{\partial B}{\partial y} = 0,$$

i. e., we obtain for the function p of the two variables x, y the partial differential equation of the first order

$$\frac{\partial F_p}{\partial x} + \frac{\partial (pF_p - F)}{\partial y} = 0. \tag{1*}$$

The ordinary differential equation of the second order (1) and the partial differential equation (1*) stand in the closest relation to each other. This relation becomes immediately clear to us by the following simple transformation

$$\delta J^* = \int_a^b \{F_y \delta y + F_p \delta p + (\delta y_x - \delta p)\, F_p + (y_x - p) \delta F_p\}\, dx$$

$$= \int_a^b \{F_y \delta y + \delta y_x F_p + (y_x - p)\, \delta F_p\}\, dx$$

$$= \delta J + \int_a^b (y_x - p)\, \delta F_p dx.$$

We derive from this, namely, the following facts: If we construct any *simple* family of integral curves of the ordinary differential equation (1) of the second order and then form an ordinary differential equation of the first order

$$(2) \qquad y_x = p(x, y)$$

which also admits these integral curves as solutions, then the function $p(x, y)$ is always an integral of the partial differential equation (1*) of the first order; and conversely, if $p(x, y)$ denotes any solution of the partial differential equation (1*) of the first order, all the non-singular integrals of the ordinary differential equation (2) of the first order are at the same time integrals of the differential equation (1) of the second order, or in short if $y_x = p(x, y)$ is an integral equation of the first order of the differential equation (1) of the second order, $p(x, y)$ represents an integral of the partial differential equation (1*) and conversely; the integral curves of the ordinary differential equation of the second order are therefore, at the same time, the characteristics of the partial differential equation (1*) of the first order.

In the present case we may find the same result by means of a simple calculation; for this gives us the differential equations (1) and (1*) in question in the form

$$(1) \qquad y_{xx}F_{y_x y_x} + y_x F_{y_x y} + F_{y_x x} - F_y = 0,$$

$$(1^*) \qquad (p_x + pp_y)\, F_{pp} + pF_{py} + F_{px} - F_y = 0,$$

where the lower indices indicate the partial derivatives with respect to x, y, p, y_x. The correctness of the affirmed relation is clear from this.

The close relation derived before and just proved between the ordinary differential equation (1) of the second order and the partial differential equation (1*) of the first order, is, as it seems to me, of fundamental significance for the calculus of variations. For, from the fact that the integral J^* is independent of the path of integration it follows that

$$(3) \qquad \int_a^b \{F(p) + (y_x - p)\, F_p(p)\}\, dx = \int_a^b F(\bar{y}_x)\, dx,$$

if we think of the left hand integral as taken along any path y and the right hand integral along an integral curve $\bar{y}$ of the differential equation

$$\bar{y}_x = p(x, \bar{y}).$$

With the help of equation (3) we arrive at Weierstrass's formula

$$(4) \qquad \int_a^b F(y_x)\, dx - \int_a^b F(\bar{y}_x)\, dx = \int_a^b E(y_x, p)\, dx,$$

where E designates Weierstrass's expression, depending upon y_x, p, y, x,

$$E(y_x, p) = F(y_x) - F(p) - (y_x - p)\, F_p(p).$$

Since, therefore, the solution depends only on finding an integral $p(x, y)$ which is single valued and continuous in a certain neighborhood of the integral curve $\bar{y}$, which we are considering, the developments just indicated lead immediately —without the introduction of the second variation, but only by the application of the polar process to the differential equation (1)—to the expression of Jacobi's condition and to the answer to the question: How far this condition of Jacobi's in conjunction with Weierstrass's condition $E>0$ is necessary and sufficient for the occurrence of a minimum.

The developments indicated may be transferred without necessitating further calculation to the case of two or more required functions, and also to the case of a double or a multiple integral. So, for example, in the case of a double integral

$$J=\int F(z_x, z_y, z; x, y)d\omega, \quad \left[z_x=\frac{\partial z}{\partial x}, \quad z_y=\frac{\partial z}{\partial y}\right]$$

to be extended over a given region ω, the vanishing of the first variation (to be understood in the usual sense)

$$\delta J=0$$

gives the well-known differential equation of the second order

$$\text{(I)} \qquad \frac{dF_{z_x}}{dx}+\frac{dF_{z_y}}{dy}-F_z=0,$$

$$\left[F_{z_x}=\frac{\partial F}{\partial z_x}, \; F_{z_y}=\frac{\partial F}{\partial z_y}, \; F_z=\frac{\partial F}{\partial z}\right],$$

for the required function z of x and y.

On the other hand we consider the integral

$$J^*=\int\{F+(z_x-p)F_p+(z_y-q)F_q\}\,d\omega,$$

$$\left[F=F(p, q, z; x, y), \; F_p=\frac{\partial F(p, q, z; x, y)}{\partial p}, \quad F_q=\frac{\partial F(p, q, z; x, y)}{\partial q}\right],$$

and inquire, how p and q are to be taken as functions of x, y and z in order that the value of this integral may be independent of the choice of the surface passing through the given closed twisted curve, i. e., of the choice of the function z of the variables x and y.

The integral J^* has the form

$$J^*=\int\{Az_x+Bz_y-C\}\,d\omega$$

and the vanishing of the first variation

$$\delta J^*=0,$$

in the sense which the new formulation of the question demands, gives the equation

$$\frac{\partial A}{\partial x} + \frac{\partial B}{\partial y} + \frac{\partial C}{\partial z} = 0,$$

i. e., we find for the functions p and q of the three variables x, y and z the differential equation of the first order

$$\frac{\partial F_p}{\partial x} + \frac{\partial F_q}{\partial y} + \frac{\partial\,(pF_p + qF_q - F)}{\partial x} = 0.$$

If we add to this differential equation the partial differential equation

$$\text{(I*)} \qquad p_y + qp_z = q_x + pq_z,$$

resulting from the equations

$$z_x = p\,(x,\, y,\, z), \quad z_y = q\,(x,\, y,\, z),$$

the partial differential equation (I) for the function z of the two variables x and y and the simultaneous system of the two partial differential equations of the first order (I*) for the two functions p and q of the three variables x, y, and z stand toward one another in a relation exactly analogous to that in which the differential equations (1) and (1*) stood in the case of the simple integral.

It follows from the fact that the integral J^* is independent of the choice of the surface of integration z that

$$\int \{F\,(p,\, q) + (z_x - p)F_p(p,\, q) + (z_y - q)F_q(p,\, q)\}\,d\omega$$
$$= \int F(\bar{z}_x,\, \bar{z}_y)\,d\omega,$$

if we think of the right hand integral as taken over an integral surface $\bar{z}$ of the partial differential equations

$$\bar{z}_x = p(x,\, y,\, \bar{z}),\ \bar{z}_y = q(x,\, y,\, \bar{z})\ ;$$

and with the help of this formula we arrive at once at the formula

$$\text{(IV)} \quad \int F(z_x,\, z_y)\,d\omega - \int F(\bar{z}_x,\, \bar{z}_y)\,d\omega = \int E(z_x,\, z_y,\, p,\, q)\,d\omega,$$

$$[E(z_x,\, z_y,\, p,\, q) = F(z_x,\, z_y) - F(p,\, q) - (z_x - p)F_p(p,\, q)$$
$$- (z_y - q)\,F_q(p,\, q)],$$

which plays the same rôle for the variation of double integrals as the previously given formula (4) for simple integrals. With the help of this formula we can now answer the question how far Jacobi's condition in conjunction with Weierstrass's condition $E > 0$ is necessary and sufficient for the occurrence of a minimum.

Connected with these developments is the modified form in which A. Kneser,* beginning from other points of view, has presented Weierstrass'stheory. While Weierstrass employed to derive sufficient conditions for the extreme tehos

* Cf. his above-mentioned textbook, §§ 14, 15, 19 and 20.

integral curves of equation (1) which pass through a fixed point, Kneser on the other hand makes use of any simple family of such curves and constructs for every such family a solution, characteristic for that family, of that partial differential equation which is to be considered as a generalization of the Jacobi-Hamilton equation.

The problems mentioned are merely samples of problems, yet they will suffice to show how rich, how manifold and how extensive the mathematical science of to-day is, and the question is urged upon us whether mathematics is doomed to the fate of those other sciences that have split up into separate branches, whose representatives scarcely understand one another and whose connection becomes ever more loose. I do not believe this nor wish it. Mathematical science is in my opinion an indivisible whole, an organism whose vitality is conditioned upon the connection of its parts. For with all the variety of mathematical knowledge, we are still clearly conscious of the similarity of the logical devices, the *relationship* of the *ideas* in mathematics as a whole and the numerous analogies in its different departments. We also notice that, the farther a mathematical theory is developed, the more harmoniously and uniformly does its construction proceed, and unsuspected relations are disclosed between hitherto separate branches of the science. So it happens that, with the extension of mathematics, its organic character is not lost but only manifests itself the more clearly.

But, we ask, with the extension of mathematical knowledge will it not finally become impossible for the single investigator to embrace all departments of this knowledge? In answer let me point out how thoroughly it is ingrained in mathematical science that every real advance goes hand in hand with the invention of sharper tools and simpler methods which at the same time assist in understanding earlier theories and cast aside older more complicated developments. It is therefore possible for the individual investigator, when he makes these sharper tools and simpler methods his own, to find his way more easily in the various branches of mathematics than is possible in any other science.

The organic unity of mathematics is inherent in the nature of this science, for mathematics is the foundation of all exact knowledge of natural phenomena. That it may completely fulfil this high mission, may the new century bring it gifted masters and many zealous and enthusiastic disciples.

Proceedings of Symposia in Pure Mathematics
Volume 28, 1976

Problems of Present Day Mathematics

Contents

I. Foundations (Yu. I. Manin). In accordance with Hilbert's prophecy, we are living in Cantor's paradise. So we are bound to be tempted.

Most mathematicians nowadays do not see any point in banning infinity, nonconstructivity, etc. Gödel made clear that it takes an infinity of new ideas to understand all about integers only. Hence we need a creative approach to creative thinking, not just a critical one. Two lines of research are naturally suggested.

(a) To find out new axioms of (more or less naive) set theory, demonstrably efficient in number theory. Most advanced new methods (l-adic cohomology) should be explored thoroughly. Are they readily formalized in Zermelo-Fraenkel or Gödel-Bernays systems? Can we use in necessary categorical constructions only known axioms, or has something new already slipped in?

(b) We should consider possibilities of developing a totally new language to speak about infinity. Classical critics of Cantor (Brouwer *et al.*) argued that, say, the general choice axiom is an illicit extrapolation of the finite case.

I would like to point out that this is rather an extrapolation of commonplace physics, where we can distinguish things, count them, put them in some order, etc. New quantum physics has shown us models of entities with quite different behavior. Even "sets" of photons in a looking-glass box, or of electrons in a nickel piece are much less cantorian than the "set" of grains of sand. In general, a highly probabilistic "physical infinity" looks considerably more complicated and interesting than a plain infinity of "things".

Certainly there are no *a priori* reasons to choose fundamental concepts of mathematics so as to make them parallel to those of physics. Nevertheless it happened constantly and proved extremely fruitful.

The twentieth century return to Middle Age scholastics taught us a lot about formalisms. Probably it is time to look outside again. Meaning is what really matters.

II. Number theory.

(A) *Decidability of classical problems* (E. Bombieri). There are very many old problems in arithmetic whose interest is practically nil, e.g. the existence of odd perfect numbers, problems about the iteration of numerical functions, the existence of infinitely many Fermat primes $2^{2^n} + 1$, etc. Some of these questions may well be undecidable in arithmetic; the construction of arithmetical models in which questions of this type have different answers would be of great importance.

(B) *Diophantine geometry* (E. Bombieri).

(a) The Mordell conjecture on the finiteness of rational points on a curve of genus 2 defined over **Q**.

(b) The Hasse principle for nonsingular cubic threefolds. This may be considerably generalized; the search for obstructions to a "local-to-global" principle may be very profitable.

(C) *Cyclotomic fields* (K. Iwasawa). Let p be a prime number, and k a totally real number field. For each $n \geqslant 1$, let K_n denote the real subfield of the cyclotomic field of p^{n+1}th roots of unity (or 2^{n+2}th roots of unity if $p = 2$) with degree p^n over the rational field, and let p^{e_n} be the exact power of p dividing the class number of kK_n. Known examples show that the exponent e_n is bounded for all $n \geqslant 1$, and we propose the problem to investigate whether this is true for all p and k. The answer is not known even in such special cases where k is a real quadratic field or the maximal real subfield of the cyclotomic field of pth roots of unity with irregular p.

(D) *Transcendental numbers* (E. Bombieri). The *algebraic independence of logarithms of multiplicatively independent algebraic numbers* is a very interesting problem, which raises many questions. A related problem is the Schanuel conjecture: if $z_1, \ldots, z_n \in \mathbf{C}$ are linearly independent over **Q**, then

$$\text{tr.deg } \mathbf{Q}(z_1, \ldots, z_n, e^{z_1}, \ldots, e^{z_n}) \geqslant n.$$

III. Prime numbers (H. L. Montgomery). Writings in analytic number theory are generally restricted to presenting what the author is capable of proving, while more ambitious aims are rarely mentioned. On this occasion we take the opposite tack by listing the main conjectures and problems which motivate most of analytic number theory. We need mention only a few questions, because answers to these would provide solutions to a large number of familiar problems. One may feel that the list is not useful in that it does not provide readily accessible research problems, but I have often found it useful to have a larger picture of the suspected truth, against which I can test ideas.

1. *The Riemann zeta function.* We expect that the Riemann Hypothesis is true, namely that if ρ is a nontrivial zero of $\zeta(s)$ then Re $\rho = ½$. Hence we write $\rho = ½ + i\gamma$. We expect that all zeros are simple (so we let $0 < \gamma_1 < \gamma_2 < \cdots$ denote the ordinates of the zeros in the upper half-plane), and we ask for a precise lower bound for the differences $\gamma_{n+1} - \gamma_n$. In this direction we put

---

AMS (MOS) subject classifications (1970). Primary 10Hxx.

$$\alpha = \liminf_{n\to\infty} \frac{\log(\gamma_{n+1} - \gamma_n)}{\log n}$$

and suggest that $-1 < \alpha < 0$; perhaps $\alpha = -1/3$. More generally, we suppose that the numbers γ are linearly independent over Q, and we ask for lower bounds for linear forms in the γ's.

Let $N(T)$ denote the number of zeros ρ of $\zeta(s)$ with $0 < \gamma \leqslant T$. Then

$$N(T) = \frac{T}{2\pi} \log \frac{T}{2\pi} - \frac{T}{2\pi} + S(T) + \frac{7}{8} + O\left(\frac{1}{T}\right),$$

where $S(t) = \pi^{-1} \arg \zeta(\tfrac{1}{2} + it)$. The statistical distribution of $S(t)$ was determined by A. Selberg, but the maximum order of $S(t)$ remains unknown. Probably

$$S(t) = O\left(\left(\frac{\log t}{\log \log t}\right)^{1/2}\right).$$

Knowledge of the local distribution of the zeros would have implications concerning prime numbers. For example, it would be useful to know that for fixed $\alpha < \beta$,

$$\sum_{\substack{\gamma \leqslant T; \gamma' \leqslant T; \\ \frac{2\pi\alpha}{\log T} \leqslant \gamma - \gamma' \leqslant \frac{2\pi\beta}{\log T}}} 1 \sim \left(\int_\alpha^\beta 1 - \left(\frac{\sin \pi u}{\pi u}\right)^2 du + \delta(\alpha, \beta)\right) \frac{T}{2\pi} \log T,$$

where $\delta(\alpha, \beta) = 1$ if $\alpha \leqslant 0 \leqslant \beta$, $\delta(\alpha, \beta) = 0$ otherwise. We also expect that

$$\limsup_{n\to\infty} (\gamma_{n+1} - \gamma_n) \log \gamma_n = +\infty.$$

In applications it would also be very useful to understand the behavior of sums of the sort $\Sigma_{|\gamma| \leqslant T} x^{i\gamma}$, for x and T in various ranges.

For Dirichlet L-functions and Dedekind zeta functions the problems are much the same, but some new problems arise, such as that of demonstrating the integrality of, and finding useful formulae for Artin's L-functions.

2. *Arithmetic sequences.* We know (assuming the Riemann Hypothesis) that $\psi(x) - x$ is never larger than $Cx^{1/2}(\log x)^2$, and also that it is infinitely often as large as $cx^{1/2} \log \log \log x$. It would be interesting to know more precisely how large $\psi(x) - x$ becomes. Similarly we may inquire about the maximum order of $\Sigma_{n \leqslant x} u(n)$, of $Q(x) - 6/\pi^2$, where $Q(x)$ is the number of squarefree integers not exceeding x, and of $\Sigma_{n \leqslant x} d(n) - (x \log x + (2C - 1)x)$. Concerning prime numbers in arithmetic progressions we may conjecture that

$$\psi(x; q, a) = \frac{x}{\phi(q)} + O\left(\left(\frac{X}{q}\right)^{1/2+\epsilon}\right),$$

provided $(a, q) = 1$.

We are also interested in the distribution of primes in short intervals. Presumably

$$\psi(x + h) - \psi(x) = h + O(h^{1/2}x^{\epsilon})$$

for $1 \leqslant h \leqslant x$. This would imply that there is a prime in the interval $(x, x + h)$. This can be made more precise by conjecturing that

$$\lim_{x\to\infty} \sup_{p_n \leqslant x} \frac{p_{n+1} - p_n}{(\log p_n)^2} = 1;$$

here p_n denotes the nth prime. This is slightly stronger than Cramér's conjecture that $\lim \sup((p_{n+1} - p_n)/(\log p_n)^2) = 1$. Concerning the frequency with which $p_{n+1} - p_n$ is large, it is thought that for fixed $\alpha > 0$,

$$\sum_{\substack{p_n \leqslant x; \\ p_{n+1}-p_n > \alpha \log p_n}} 1 \sim e^{-\alpha}\pi(x).$$

In the opposite direction, we may ask how many primes can lie in an inteval $(x, x + h)$. Put $\rho(h) = \lim \sup_{x\to\infty}(\pi(x + h) - \pi(x))$. It seems that for large h, $\pi(h) < \rho(h) < 2h/\log h$; it would be instructive to know which of these inequalities is sharp, if either.

3. *Additive prime number theory.* Given a number of polynomials with integral coefficients, we expect that for infinitely many choices of the integral variables the polynomials will simultaneously take on prime values, unless some local condition holds which makes this obviously impossible. A very general hypothesis of this sort might very well be undecidable, but in the well-known special cases of Goldbach's conjecture, prime k-tuples, primes of the form $n^2 + 1$, etc., we not only expect positive results, but we suppose that the number of solutions is very accurately predicted by the major arc contribution in the Hardy-Littlewood circle method.

4. *Diophantine questions.* Are the numbers $\zeta(3)$, $\zeta(3)\pi^{-3}$, and C transcendental? Here C denotes Euler's constant. One may inquire similarly about the imaginary parts of the zeros of the zeta function. In a different direction, it would be useful to know that one can approximate to real numbers by rational numbers whose numerators and denominators are both prime. Specifically, is it true that for every irrational number θ and every $\epsilon > 0$ there are infinitely many prime numbers p, q such that

$$|\theta - p/q| < q^{-2+\epsilon}?$$

IV. What is a "motive"? (P. Deligne). Let X be a projective smooth algebraic variety over an algebraically closed field k of characteristic p. For each prime number $l \neq p$, one has l-adic cohomology groups $H^i(X, \mathbf{Q}_l)$ [SGA 4]; one also has crystalline cohomology groups ([2], [3]). The similarities between these theories lead to the conjectural existence of a "motivic" cohomology H_m from which both could be derived in some sense ([4], [6]).

As an example of similarity, for an endomorphism f of X (and even a class of correspondences of degree 0), $\mathrm{Tr}(f^*, H^i(X, \mathbf{Q}_l))$ is independent of the prime l. *Warning* (for $p \neq 0$). When X is not proper or has singularities, one does not know whether the Betti numbers are *independent of l*. For X smooth and projective, of dimension n, let

$$\begin{matrix} L \\ L' \end{matrix} : H^{n-i}(X, \mathbf{Q}_l) \underset{\eta'^i\Lambda}{\overset{\eta^i\Lambda}{\underset{\longrightarrow}{\overset{\longrightarrow}{\sim}}}} H^{n+i}(X, \mathbf{Q}_l)(i)$$

(where on the right-hand side, the . . . (i) means a Tate twist) be the isomorphisms defined by two polarizations; it is unknown if $\mathrm{Tr}(L^{-1}L', H^i(X, \mathbf{Q}_l))$ is a rational number.

Example. The groups $H^1(X)$ are all deduced from the Jacobian $J(X)$ of X. This suggests identification of abelian varieties with a class of "motives", and to look upon $J(X)$ as being $H^1_m(X)$.

For $k = \mathbf{C}$, one would hope that knowledge of $H^i_m(X)$ would be equivalent to knowledge of $H^i(X, \mathbf{Z})$, equipped with the Hodge decomposition

$$H^i(X, \mathbf{Z}) \otimes \mathbf{C} = \bigoplus_{p+q=i} H^{p,q}.$$

This leads to the following question, for $k = \overline{\mathbf{Q}}$. Let $\iota\colon k \longrightarrow \mathbf{C}$ be an embedding, and let $L(\iota) \subset H^{2p}(X, \mathbf{Q}_l)(p)$ (for $0 \leqslant p \leqslant \dim X$) be the group of classes whose image by ι in $H^{2p}(\iota X, \mathbf{Z}) \otimes \mathbf{Q}_l$ is integral of type (p, p).

Conjecture. $L(\iota)$ is independent of the embedding ι.

The *Hodge conjecture*: "over $\mathbf{C}$, every rational cohomology class of type (p, p) is the class of an algebraic cycle" implies the preceding one, which is sometimes more accessible. For instance, when X is an abelian variety, Piatetskiĭ-Shapiro has proved that $L(\iota)$ is independent of ι, when ι is arbitrary in the orbit of an *open* subgroup of $\mathrm{Gal}(k/\mathbf{Q})$. It would be exciting to deal completely with the case of abelian varieties. The essential case is that of abelian varieties of type *CM*.

Suppose X is defined over a field k_0 finitely generated over the prime field. The l-adic analog of the Hodge conjecture is the

Tate conjecture: $H^{2p}(X, \mathbf{Q}_l)(p)^{\mathrm{Aut}(k/k_0)}$ is generated by the cohomology classes of algebraic cycles defined over k_0. This conjecture has meaning only if one believes that the Frobenius endomorphisms in l-adic cohomology are *semisimple.* The cases of divisors on abelian varieties or $K3$ elliptic surfaces defined on a finite field are settled ([7], [1]).

REFERENCES

1. M. Artin and H. P. F. Swinnerton-Dyer, *The Shafarevich-Tate conjecture for pencils of elliptic curves on K3 surfaces.*

2. P. Berthelot (partly in collaboration with L. Illusie), C. R. Acad. Sci. Paris Sér. A–B **269** (1969), 297, 357, 397; ibid. **270** (1970), 1695, 1750; **272** (1971), 42, 141, 254, 1314, 1397, 1574. MR **40** #151; #2686; **41** #8432; **42** #4555; **43** #4831; **44** #220; #221; **44** #222; #223; #1671.

3. A. Grothendieck, *Crystals and the deRham cohomology of schemes,* Dix Exposés sur la Cohomologie des Schémas, North-Holland, Amsterdam; Masson, Paris, 1968, pp. 306–358. MR **42** #4558.

4. S. Kleiman, *Algebraic cycles and the Weil conjectures,* Dix Exposés sur la Cohomologie des Schémas, North-Holland, Amsterdam; Masson, Paris, 1968, pp. 359–386. MR **45** #1920.

5. I. I. Pjateckiĭ-Šapiro, *Interrelations between the Tate and Hodge conjectures for abelian varieties,* Mat. Sb. **85** (127) (1971), 610–620 = Math. USSR Sb. **14** (1971), 615–625. MR **45** #3422.

6. N. Saavedra, *Catégories Tanakiennes,* Lecture Notes in Math., vol. 265, Springer-Verlag, Berlin and New York, 1972. MR **49** #3769.

7. J. Tate, *Endomorphisms of abelian varieties over finite fields,* Invent. Math. **2** (1966), 134–144. MR **34** #5829.

[SGA 4], *Séminaire de géométrie algébrique du Bois Marie,* Lecture Notes in Math., vols. 269, 270, 305, Springer-Verlag, Berlin and New York, 1972/73. MR **50** #7130; #7131; #7132.

V. Non-abelian class field theory (E. Bombieri, P. Cartier, P. Deligne, Yu. I. Manin, J. P. Serre; text by Deligne).

(A) *Global theory*: "To understand the relation between "motives" defined over a number field, modular forms and their L-functions."

A "motive" M over $\mathbf{Q}$ defines a compatible system (in the sense of [14]) of l-adic representations $\mathrm{Gal}(\overline{\mathbf{Q}}/\mathbf{Q}) \longrightarrow \mathbf{GL}(M_l)$: the typical case is that in which M is the ith cohomology group of a nonsingular projective variety X defined over $\mathbf{Q}$, and for each l, $\mathrm{Gal}(\overline{\mathbf{Q}}/\mathbf{Q})$ acts on $M_l = H^i(X, \mathbf{Q}_l)$; i is the *weight* of M.

This gives rise to a function

$$L(M, s) = \prod_p (\det(1 - F_p p^{-s}, M_l))^{-1}$$

(product over all primes, F_p being the Frobenius endomorphism at p; the product is absolutely convergent for $\mathbf{R}(s)$ large enough; it is well defined except for a finite number of factors). *Conjectures* about the analytic properties of these L-functions have been put forward: existence of a *Meromorphic continuation* to the whole complex plane, local factors at the bad primes (and Γ factors at places at infinity) needed to obtain a nice *functional equation* [16], with precise "constants" (see [2]). For an L-function thus extended and modified, for M of weight i, a generalization of the *Riemann conjecture* is that the zeros are on the line $\mathbf{R}(s) = \frac{1}{2}(i + 1)$; a generalization of the *Artin conjecture* determines the poles and their multiplicities.

The only known results are proved by first linking $L(M, s)$ to a modular form. Rather than modular forms, one should speak of the irreducible components of the natural representation of $G(\mathbf{A})$ ($\mathbf{A}$ being the adele ring of $\mathbf{Q}$) in $L_\chi(G(\mathbf{A})/G(\mathbf{Q}))$ (G reductive algebraic group over $\mathbf{Q}$, χ a quasi-character of its center, L_χ a "sensible" function space) (see [3]). Such a component π defines functions $L(\pi, \rho, s)$ (ρ representation of the dual group $\hat{G}$; see [9]).

Conjecture. Every function $L(M, s)$ is of type $L(\pi, \rho, s)$. The ρ should be interpreted in the following way: one wants to have a functor $\rho \longrightarrow M(\pi, \rho)$ compatible with $\oplus$ and $\otimes$, with values in the $\otimes$-category deduced from the category of motives by extension of scalars from $\mathbf{Q}$ to $\mathbf{C}$, and such that $L(M(\pi, \rho), s) = L(\pi, \rho, s)$. The image of the Galois group in the l-adic representation in M_l should be "smaller" than $\rho(\hat{G})$.

The typical case of that general conjecture is the *Weil conjecture* [18]: "every elliptic curve over $\mathbf{Q}$, with conductor N, is (up to isogeny) a direct factor in the jacobian of the modular curve $X_0(N)$, and thus attached to a modular form of weight 2 for $\mathbf{GL}(2)$."

The case of motives of weight 0 (the Artin L-functions) is also very interesting, and seems inextricably linked to the Artin conjecture.

It is not known which modular forms should come from motives. When G is commutative, the modular forms are essentially Grössencharaktere of number fields. Only those of type A_0 seem linked to motives (for instance the zêta function of an abelian variety of type *CM* is a product of such L-functions).

The conjectured properties of L-functions of motives suggest similar properties for L-functions of arbitrary modular forms, and relations between modular forms relative to different groups.

Example. ([7, §16]) gives the relation between modular forms for $\mathbf{GL}(2)$ and for a quaternion algebra.

Conjectural example. Let $f(z) = \Sigma_n a_n e^{2\pi i n z}$ be a holomorphic parabolic modular form of weight k for $\mathbf{GL}(2)$, which is an eigenvector of the Hecke operators. Let $T^2 - a_p T + p^{k-1}$ be the characteristic polynomial of $F_p \in \mathbf{GL}(2, \mathbf{C})$. For $m \geqslant 1$, let

$$L(f, \mathrm{Sym}^m, s) = \prod_p \det(1 - \mathrm{Sym}^m(F_p)p^{-s}, \mathrm{Sym}^m(\mathbf{C}^2))^{-1}.$$

(a) One expects that this L-function has a holomorphic continuation to the whole plane, and a functional equation (where s goes to $m(k-1)+1-s$) the precise form of which is given in [2] and [16].

(b) One expects that this L-function comes from a modular form for $\mathbf{GL}(m+1)$.

For $m = 1, 2, 3$, see [6], [8] and [13]. For the relation between (a) and (b) see [4] and [5].

One would like to understand the values of L-functions at certain integers. For A an abelian variety and $L(H^1(A), s)$, one has the *Birch-Swinnerton-Dyer con-*

jecture relative to the behavior of L at $s = 1$; its precise form requires the conjectured *finiteness of the Tate-Shafarevich group* (A).

In an increasing number of cases, one also has p-adic L-functions (p-adic valued functions of p-adic variables), which are similar to the complex L-functions, and are sometimes defined by interpolation from the values of the latter. For L-functions of modular forms for **GL**(2), see [12]. For some Artin L-functions, or L-functions relative to Grössencharaktere of type A_0, see [1] and [15].

In the case of Artin L-functions, these p-adic functions should give rise to p-adic measures on Galois groups; see e.g. [15].

Problem. Show that these measures correspond to relations between ideal classes, generalizing Stickelberger's relations.

(B) *Local theory.* Let F be a local field and G a reductive group defined over F. Langlands conjectures that the admissible irreducible representations of $G(F)$ are naturally grouped in clusters, which are roughly indexed by the representations of the Galois group of F in $\hat{G}(\mathbf{C})$. For more precise statements and partial results, see [9], [10], [11] and [3, §3].

REFERENCES

1. R. M. Damerell, *L-functions of elliptic curves with complex multiplication.* I, Acta Arith. **17** (1970), 287–301. MR **44** #2758; erratum, **44**, p. 1633.

2. P. Deligne, *Les constantes des équations fonctionnelles des fonctions L,* Modular Functions of One Variable, II (Proc. Internat. Summer School, Univ. Antwerp, 1971), Lecture Notes in Math., vol. 349, Springer-Verlag, Berlin, 1973, pp. 501–597. MR **50** #2128.

3. ———, *Formes modulaires et représentations de GL*(2), Modular Functions of One Variable, II (Proc. Internat. Summer School, Univ. Antwerp, 1972), Lecture Notes in Math., vol. 349, Springer-Verlag, Berlin, 1973, pp. 55–105. MR **50** #240.

4. I. M. Gel'fand and D. A. Kazdan, *Representations of GL(n, K) where K is a local field,* Moscow, 1971 (preprint).

5. R. Godement and H. Jacquet, *Zêta functions of simple algebras,* Lecture Notes in Math., vol. 260, Springer-Verlag, Berlin and New York, 1972. MR **49** #7241.

6. E. Hecke, *Mathematische Werke,* Vandenhoeck & Ruprecht, Göttingen, 1959. MR **21** #3303.

7. H. Jacquet and R. Langlands, *Automorphic forms on GL*(2), Lecture Notes in Math., vol. 114, Springer-Verlag, Berlin and New York, 1970.

8. R. Langlands, *Euler products,* Yale University.

9. ———, *Problems in the theory of automorphic forms,* Lectures in Modern Analysis and Applications, III, Lecture Notes in Math., vol. 170, Springer-Verlag, Berlin, 1970, pp. 18–61. MR **46** #1758.

10. ———, *Representations of abelian algebraic groups,* Yale University.

11. ———, *On the classification of irreducible representations of real algebraic groups* (preprint).

12. Ju. I. Manin, *Parabolic points and zeta-functions of modular curves,* Izv. Akad. Nauk SSSR Ser. Mat. **36** (1972), 19–66 = Math. USSR Izv. **6** (1972), 19–64. MR **47** #3396 (see also: Séminaire Bourbaki, Exposé 414, 1971).

13. G. Shimura, *Modular forms of half integral weight,* Modular Functions of One Variable, I (Proc. Internat. Summer School, Univ. Antwerp, 1972), Lecture Notes in Math., vol. 320, Springer-Verlag, Berlin and New York, 1973.

14. J. P. Serre, *Abelian l-adic representations,* Benjamin, New York.

15. J. P. Serre, *Formes modulaires et fonctions zêta p-adiques,* Modular Functions of One Variable, III, Lecture Notes in Math., vol. 350, Springer-Verlag, Berlin and New York, 1973.

16. ———, *Facteurs locaux des fonctions zêta des variétés algébriques (définitions et conjectures)*, Séminaire Delange-Pisot-Poitou, 19, 1969/70.

17. J. Tate, *On the conjecture of Birch and Swinnerton-Dyer and a geometric analog,* Séminaire Bourbaki: vol. 1965/66, Exposé 306, Benjamin, New York, 1966. MR **34** #5605.

18. A. Weil, *Über die Bestimmung Dirichletscher Reihen durch Funktionalgleichungen,* Math. Ann. **168** (1967), 149–156. MR **34** #7473.

See also a forthcoming paper by Manin and Visik.

VI. Algebraic geometry (D. Mumford).

How rational is the moduli space of curves? For a considerable set of small g (e.g., $g \leqslant 10$), almost all curves C of genus g can be described by a suitable set of equations whose coefficients contain a certain number N of free parameters which can vary over $\mathbf{C}^N$-(some subvarieties): e.g., if $g = 3$, take $C = V(\Sigma_{i+j+k=4} c_{ijk} X_0^i X_1^j X_2^k)$, with 15 parameters c_{ijk}. This implies that the moduli space M_g is unirational. The problem is whether for large g, M_g remains "at all rational"? A weak rationality property M_g might have, would be that M_g is rationally connected, i.e., there are chains of rational curves on M_g connecting any 2 points; for that matter is the *generic* curve C of genus g in a pencil on any surface X not birational to $C \times \mathbf{P}^1$ (compare Hartshorne, Publ. I. H. E. S. No. 36)? A good step would be to check whether there are any holomorphic q-forms on M_g. The success of Hirzebruch (L'Enseignement Math., 19 (1973)) and Freitag (Int. Colloq., Tata Institute, 1973) in calculating the invariants of the Hilbert modular varieties associated to totally real number fields is certainly encouraging. The problem is that there are no very explicit ways of describing the generic curve of genus g. The nearest is Petri's equations (cf. Saint-Donat, Math. Annalen **206** (1973)) for the canonical curve. Three other ways are natural: (a) to write down explicitly the generic covering of $\mathbf{P}^1$ of order n with $2d$ branch points (i.e., to describe the Hurwitz schemes in Fultin, Ann. of Math., 1969). (b) To write down via theta functions the curve inside its jacobian. This might lead to a solution too of Schottky's problem: to characterize those principally polarized abelian varieties which are jacobians. (c) Analytically, it is natural to attack M_g via its universal covering Space T_g, the Teichmüller space. Royden's work (Annals of Math. Studies, No. 66) indicates how asymmetrical T_g is, however, and T_g is far from being described "explicitly" yet. Between T_g and M_g, there are the "higher level" moduli spaces: $T_g \longrightarrow M_g^* \longrightarrow M_g$. What type of a variety are the M_g^*'s, e.g., what is their Kodaira-dimension?

Integrating meromorphic vector fields. Let $G = \mathrm{Aut}_{\mathbf{C}}\mathbf{C}(X, Y)$ be the Cremona group. Very little is known about G except for Noether's theorem that "standard quadratic transformations" generate G, and classifications of (possibly reducible) algebraic subgroups $H \subset G$ up to conjugacy. It is certainly infinite-dimensional in that it contains arbitrarily large algebraic groups Aut(V), V a projective variety with $\mathbf{C}(X, Y)$ as function field. The problem is to topologize G and associate to it a Lie algebra consisting, roughly, of those meromorphic vector fields D on $\mathbf{P}^2$ which "integrate" into an analytic family of Cremona transformations.

There are many other problems about G: e.g., is G simple? For classical results on G, cf. Berzolari's article in the Encyclopädie (Chapter 11 of Part III_2, "Algebraische Geometrie"; especially §§46–72). For recent work on G, cf. Manin (1966 International Congress) on its Galois cohomology, and Demazure, Annales de L'Ecole Norm. Sup. 3 (1970), on algebraic subgroups.

Resolution of singularities. Well known and of fundamental importance is the famous problem: is every variety X birational to a complete nonsingular variety X'? When the ground field k has a characteristic 0, this has been solved by Hironaka. If $\mathrm{char}(k) = p$, $\dim X = 2$, this has been solved by Abhyankar. In general, it is open and very intriguing. Also open is the arithmetic case: is every integral scheme X of finite type over $\mathbf{Z}$ birational to a regular scheme X' proper over $\mathbf{Z}$?

Projective varieties of small codimensions. The extraordinary results of Barth (Amer. J. Math. 92 (1970)) showing that the Betti numbers of nonsingular projective varieties $X^{n-r} \subset \mathbf{P}^n$ are equal to those of $\mathbf{P}^n$ in a large range if $r \ll n$ have suggested a very elementary and striking conjecture: that there is a function $f(r)$ such that if $n > f(r)$ then X is a transversal intersection of r hypersurfaces. The lowest dimensional specific conjecture here would be that a 4-dimensional nonsingular variety in $\mathbf{P}^6$ is a transversal intersection of 2 hypersurfaces. This conjecture is superficially similar (but totally unrelated) to the older one as to whether a nonsingular curve in $\mathbf{P}^3$ is an intersection (*not* necessarily transversal) of 2 surfaces.

VII. Superposition of algebraic functions (V. Arnold, G. Shimura). Hilbert asked, as his 13th problem, whether or not an arbitrary analytic function of three variables can be obtained by finitely many substitutions of (continuous) functions of two variables. This was settled affirmatively, in fact against Hilbert's negative conviction, by V. I. Arnol'd and A. N. Kolmogorov.[1] Hilbert posed this question especially in connection with the solution of a general algebraic equation of degree 7. It is reasonable to presume that he formulated it in terms

[1] V. I. Arnol'd, Dokl. Akad. Nauk SSSR **114** (1957), 679–681; Mat. Sb. **48** (90) (1959), 3–74; English transl., Amer. Math. Soc. Transl. (2) **28** (1963), 51–54, 61–147. MR 22 #2668; #12191; 27 #3758; #3759.

A. N. Kolmogorov, Dokl. Akad. Nauk SSSR **114** (1957), 953–956; English transl., Amer. Math. Soc. Transl. (2) **28** (1963), 55–59. MR 22 #2669; 27 #3760.

of continuous functions partly because he had an interest in nomography and partly because he expected a negative answer. Now that it is settled affirmatively, one can ask an equally fundamental, and perhaps more interesting, question with algebraic functions instead of continuous functions.

To be specific, take an algebraically closed basic field k and n independent variables $x_1, \ldots, x_n$ over k. Let K be an algebraic closure of $k(x_1, \ldots, x_n)$. For each positive integer $r \leqslant n$, we can define the subfield M_r of K consisting of all the elements of K obtained by successive compositions of algebraic functions of r variables. More precisely, let $[t_1, \ldots, t_r]$ denote, for any $t_1, \ldots, t_r \in K$, the algebraic closure of $k(t_1, \ldots, t_r)$ in K. Let $M_r^1 = k(x_1, \ldots, x_n)$, and let M_r^{s+1} be the composite of $[t_1, \ldots, t_r]$ for all choices of $t_1, \ldots, t_r \in M_r^s$. Then M_r is defined to be the union $\bigcap_{s=1}^{\infty} M_r^s$. Obviously $M_1 \subset M_2 \subset \cdots \subset M_n = K$. Now we can ask:

Can it happen that $M_r = K$ for some $r < n$?

Or more generally:

How many distinct fields are there among $M_1, M_2, \ldots, M_n$, or among the M_r^s? It is also interesting to ask the smallest M_r containing a given element of K. For example:

Determine the smallest M_r or M_r^s that contains an element f satisfying

$$f^n + x_1 f^{n-1} + \cdots + x_{n-1} f + x_n = 0.$$

One can also formulate and ask the questions within the algebraic closure of $k(x_1, \ldots, x_n, \ldots)$ with infinitely many variables x_i. This may or may not make a difference in the nature of the problem.

VIII. The *A-D-E* classifications (V. Arnold). The Coxeter-Dynkin graphs A_k, D_k, E_k appear in many independent classification theorems. For instance:

(a) Classification of the platonic solids (or finite orthogonal groups in the euclidean 3-space).

(b) Classification of the categories of linear spaces and maps (Gabriel, Gelfand-Ponomarev, Roiter-Nasarova) (see Sém. Bourbaki, no. 444, 1974).

(c) Classification of the singularities of algebraic hypersurfaces with a definite intersection form of the neighboring smooth fiber (Tjurina).

(d) Classification of critical points of functions, having no modules (see Sém. Bourbaki, no. 443, 1974).

(e) Classification of the Coxeter groups generated by reflections, or of Weyl groups with roots of equal length.

The problem is *to find a common origin of all the A-D-E classification theorems, and to substitute a priori proofs to a posteriori verifications of the parallelism of the classifications.*

IX. Serre's conjecture (R. G. Swan, H. Bass). In [12], Serre proposed the following problem: *Let* $A = k[x_1, \ldots, x_n]$ *be a polynomial ring in* n *variables over a field. Is every finitely generated projective* A*-module free*? This problem was suggested by an analogy with topology. Serre showed that projective A-modules correspond to algebraic vector bundles over n-dimensional affine space. For $k = \mathbf{R}$ or $\mathbf{C}$ such bundles are topologically trivial. This problem has proved to be a very fruitful one as it has inspired a great deal of fundamental work in algebraic K-theory and related areas. However, the problem itself, in spite of its simple appearance, has turned out to be extremely intractible. Until very recently, the only known cases in which it was solved were $n \leqslant 1$ (classical) and $n = 2$ (Seshadri [15]). In the general case, Grothendieck and Serre [13] showed that any finitely generated projective A-module P is stably free, i.e., $P \oplus A^s$ is free for some finite s. By combining this with his cancellation theorem [1], Bass showed that P is free if rk $P > n$ and also that nonfinitely generated projectives are free [2]. Since A is factorial, it is also easy to see that P is free if rk $P = 1$. Things stood at this point for about ten years. However, in the past year or so there has been quite a bit more progress made. Murthy and Towber [8] showed that the answer is affirmative for $n = 3$ and k algebraically closed. M. Roitman [11] showed that P is free if rk $P \geqslant n$ while A. A. Suslin [19] showed that P is free for rk $P \geqslant n/2 + 1$ (and, according to the latest reports for rk $P \geqslant (n + 4)/3$ if k is infinite). Suslin and Vaserštein [19], [6] then showed that all projectives are free for $n = 3$ and also for $n = 4$ and 5 in case char $k \neq 2$ (the restriction to finite k for $n = 5$ in [19] now seems to have been removed). The restriction to char $k \neq 2$ comes from a theorem of Karoubi [7], [6]: If $W(A)$ is the symplectic Witt group, then $W(A[x]) \xrightarrow{\approx} W(A)$ if $\frac{1}{2} \in A$. A proof of this without the condition $\frac{1}{2} \in A$ would certainly be a welcome contribution. The methods used to obtain the results for $n = 3$, 4, and 5 are rather special and give no indication of a way to handle the general case. In particular, for $n \leqslant 5$, using Suslin's bound on the rank and results of Bass [4] one can reduce to the case rk $P = 2$. In this case P has a symplectic structure and strong use is made of this fact.

There have naturally been some attempts to generalize Serre's problem to a wider class of rings. For example, in [5] Bass asks, for a commutative regular ring A, whether all finitely generated projective $A[x]$-modules are extended, i.e. of the form $A[x] \otimes_A P$. If A is not regular, this will not be so since $K_0A[x] \neq K_0A$ in general, but one can still ask whether all stably free $A[x]$-modules are extended (or, more generally, if all $A[x]$-modules stably isomorphic to extended modules are extended). As a special case of this we can consider finitely generated projective $A[x]$-modules P such that $A[x, x^{-1}] \otimes_{A[x]} P$ is free. These are easily seen to be stably free and we can ask whether P is free. If this is so, we could deduce an affirmative answer to Serre's problem in $n + 1$ variables for k

algebraically closed from the general case for n variables. A few special cases are treated in [17] and [18]. In view of the difficulty of Serre's problem, it is generally felt that these more general problems will have a negative solution but as yet no counterexample has been produced. Murthy has pointed out one reason why this may be so: all known examples of stably free modules which are not free have been shown not to be free by topological methods, i.e., either by reducing to the case of vector bundles [16] or by using methods of etale cohomology over more general fields [10]. Since the topological analogue of the above problem has an affirmative answer (all bundles on $X \times I$ are obtained from bundles over X), it is only to be expected that topological methods will be of no help. We must look for purely algebraic ways of showing that stably free modules are not free. One could also ask whether there exists an effective algorithm for deciding whether a projective module is free. If P has rank r, then P is free if and only if the augmentation ideal I of the symmetric algebra $S(P)$ has r generators. (If $P \oplus Q \approx A^s$ we can consider instead $S(P \oplus Q)I$ in $S(P \oplus Q) = A[y_1, \ldots, y_s]$.) Therefore we can ask for nontrivial lower bounds for the number of generators of an ideal and whether there is an effective method to find the minimal number of generators. This is closely related to the problem of complete intersections in algebraic geometry [14]. Finally mention should be made of some generalizations of Serre's problem which do not work. In [9], Ojanguren and Sridharan show that for any noncommutative division ring D there are nonfree projectives over $D[x, y]$. (In the generalization given in [5], D should be assumed to be a domain.) Also, Eisenbud has remarked that there can be nonfree projectives over localizations of $k[x_1, \ldots, x_n]$, in particular, over $\mathbf{R}[x, y, z, s^{-1}]$ where $s = x^2 + y^2 + z^2$, in spite of the fact that $K_0 = \mathbf{Z}$ for this ring.

ADDED IN PROOF (APRIL 1976). Recently Quillen and Suslin have independently obtained complete (affirmative) solutions to Serre's problem. Their methods are completely different. Quillen makes use of earlier work of Horrocks [5], [8], [4], [2] while Suslin uses ideas related to [18]. The extension to $A[x]$ modules mentioned above remains open for dim $A > 1$.

REFERENCES

1. H. Bass, *K-theory and stable algebra*, Inst. Hautes Études Sci. Publ. Math. No. 22 (1964), 5–60. MR **30** #4805.

2. ———, *Big projective modules are free*, Illinois J. Math. **7** (1963), 24–31. MR **26** #1341.

3. ———, *Algebraic K-theory*, Benjamin, New York, 1968. MR **40** #2736.

4. ———, *Modules which support nonsingular forms*, J. Algebra **13** (1969), 246–252. MR **39** #6875.

5. ———, *Some problems in "classical" algebraic K-theory*, Algebraic K-Theory. II: "Classical" Algebraic K-Theory and Connections with Arithmetic, Lecture Notes in Math., vol. 342, Springer-Verlag, Berlin and New York, 1973. MR **48** #3656b.

6. H. Bass, *Libération des modules projectifs sur certain anneaux de polynomes*, Séminaire Bourbaki, Juin 1974.

7. M. Karoubi, *Périodicité de la K-théorie hermitienne*, Algebraic K-Theory. III: K-Theory and Geometric Applications, Lecture Notes in Math., vol. 343, Springer-Verlag, Berlin and New York, 1973. MR **48** #3656c.

8. M. P. Murthy and J. Towber, *Algebraic vector bundles over* $\mathbf{A}^3$ *are trivial*, Invent Math. (to appear).

9. M. Ojanguren and R. Sridharan, *Cancellation of Azumaya algebras*, J. Algebra **18** (1971), 501–505. MR **43** #2018.

10. M. Raynaud, *Modules projectifs universels*, Invent. Math. **6** (1968), 1–26. MR **38** #4462.

11. M. Roitman, *On Serre's problem on projective modules* (to appear).

12. J.-P. Serre, *Faisceaux algébriques cohérent*, Ann. of Math. (2) **61** (1955), 197–278. MR **16**, 953.

13. ———, *Modules projectif et sespaces fibres à fibre vectorielle*, Séminaire Dubreil, M.-L. Dubreil-Jacotin et C. Pisot lle année: 1957/58, Secrétariat mathématique, Paris, MR **21** #7222.

14. ———, *Sur les modules projectifs*, Séminaire Dubreil, M.-L. Dubreil-Jacotin et C. Pisot 14ième année: 1960/61, Secrétariat mathématique, Paris, 1963. MR **28** #3911.

15. C. S. Seshadri, *Triviality of vector bundles over the affine space* K^2, Proc. Nat. Acad. Sci. U.S.A. **44** (1958), 456–458. MR **21** #1318.

16. R. G. Swan, *Vector bundles and projective modules*, Trans. Amer. Math. Soc. **105** (1962), 264–277. MR **26** #785.

17. ———, *A cancellation theroem for projective modules in the metastable range*, Invent. Math. (to appear).

18. R. G. Swan and J. Towber, *A class of projective modules which are almost free* (to appear).

19. L. N. Vaserstein and A. A. Suslin, *Serre's problem on projective modules over polynomial rings and algebraic K-theory* (announcement) (to appear) (Russian).

X. Finite groups (M. Asbacher). The major problem in the theory of finite groups is the classification of finite simple groups. For this purpose, the class of finite groups may be subdivided into the subclasses of groups of odd and even characteristic, resembling the Chevalley groups of odd and even characteristic, respectively.

In the classification of groups of odd characteristic, there are two major outstanding problems. The first is to establish the B-conjecture, or more generally, to determine the unbalanced groups. The second is to solve the standard form problems for the known simple groups, most particularly the Chevalley groups of even characteristic.

The situation with groups of even characteristic is less clear. One possible approach has been suggested by D. Gorenstein and is derived from Thompson's work on N-groups. Basically it involves the interplay between 2-local and 3-local structure. Major problems in this program include determining the groups in which the maximum 3-rank of 2-locals is one or two, developing a theory of nonsolvable signalizer functors, and classifying groups with a strongly 3-imbedded subgroup.

XI. Representations of finite Chevalley groups (P. Cartier). What is required is the general construction of the characters of these groups, at least for the classical ones (orthogonal, unitary and symplectic groups), a general method (using perhaps Tits' buildings) to construct the supercuspidal representations, and new ideas to tackle the decomposition of the principal series.

This program should be feasible within a few years. It can be extended to cover the p-adic reductive groups, but here really new ideas seem necessary.

The application of the aforementioned method to automorphic functions has barely been begun by Hecke and Kloosterman.

XII. Constructive algebra (P. Cartier). Most of the standard constructions in modern algebra lead to quite unfeasible algorithms. It is very painful to substitute polynomials into polynomials (or power series into power series); it is often difficult to determine a Galois group explicitly, the decomposition law of prime numbers in number fields, or to classify algebraic singularities. The new generation of pocket-size computers makes the search of algorithms of this kind very compelling, since many new computational problems shall be well in the range of nonexperts in programming. More generally, there is the problem of implementing efficient high level programming languages to deal with algebraic structures, like matrices, polynomials, etc. as global objects without resorting at every step of the calculation to their definition. The recent development in implementing APL might provide a clue.

XIII. Real algebraic geometry (the 16th Hilbert problem) (V. Arnold). It is still unknown *how many connected components can have the complement of an algebraic hypersurface of degree n in the real projective k-dimensional space, even for* $k = 3$.

It follows from a theorem of Herrmann on the nodes of vibrating membranes (mentioned in Courant-Hilbert, *Methoden der math.*, Physik, Vol. I, Chapter VI, §6) that this number of components does exceed $1 + \binom{n+k-2}{k}$. This maximum is attained for plane curves (for n straight lines) and for surfaces of degree 4 (some smooth ones). However, the general theorem on membranes is not true, so the above estimation is only proved for curves.

Another still open problem is *to find all possible configurations of ovals of a plane curve of degree n with the maximum number of ovals* (i.e. $1 + (n-1)(n-2)/2$).

There is only one configuration of the 4 ovals of a curve of degree 4 (all exterior to one another), and 3 configurations of the 11 ovals of a curve of degree 6 (only one oval contains others, and it can contain 1, 5, or 9 of them). For the 22 ovals of a curve of degree 8, there are 145 configurations which verify all the known restrictions, and algebraic realizations are known for 9 configurations only.

In contrast to the recent progress with the algebraic part of the 16th Hilbert problem (due to Goudkov, Rohlin, Harlamov and others), there is not much progress with the second part, dealing with limit cycles: *it is still unknown whether a plane vector field, given by two polynomials of degree* 2, *can have more than* 3 *limit cycles.*

Profound ideas of I. G. Petrovskiĭ, concerning the relation of real limit cycles to the topology of the corresponding 2-dimensional foliation of the complex projective plane have not yet found their realizations.

XIV. Complex geometry[2] (S. S. Chern).

1. *Biholomorphic equivalence.* Perhaps the most natural problem in complex geometry is to decide whether two complex spaces or manifolds are biholomorphically equivalent. The following are simple samples:

Problem 1. Let M be a compact Kählerian manifold of positive sectional curvature. Is M biholomorphically equivalent to the complex projective space?

Andreotti and Frankel proved that the answer is affirmative if M is of dimension 2 [8]; the proof depends on the classification of rational surfaces. The condition "positive sectional curvature" can be replaced by "positive holomorphic bisectional curvature", in which formulation the problem becomes one belonging entirely to complex geometry; cf. [10]; the problem can also be formulated as one on algebraic geometry:

Problem 1a. Let M be a compact complex manifold with positive tangent bundle. Is it biholomorphically equivalent to the complex projective space?

Problem 2. Let $z_j = x_j + iy_j$, $1 \leqslant j \leqslant n$, be the coordinates in the complex number space C_n. The domain defined by

$$\sum_j (a_j x_j^2 + b_j y_j^2) < 1,$$

where a_j, b_j are positive real numbers, is the interior of an ellipsoid and is strictly pseudo-convex. Classify these domains under biholomorphic equivalence.

I. Naruki and S. Webster [11] showed that there are such domains which are biholomorphically inequivalent to the ball ($a_j = b_j = 1$). Recently Charles Fefferman proved that a biholomorphic equivalence of two strictly pseudo-convex domains with smooth boundaries in C_n is smooth up to the boundary [5]. The latter is a Cauchy-Riemann manifold whose local invariants have been studied by E. Cartan, N. Tanaka, and more recently, by S. S. Chern and J. K. Moser [3]. These results should provide some tools for Problem 2. In general the study of invariants of pseudo-convex, but not necessarily strictly pseudo-convex, domains is a problem of great importance.

2 It is my pleasure to thank P. Griffiths, S. Kobayashi, R. Narasimhan, and H. Wu for suggestions and help in the preparation of these problems.

2. *Holomorphic maps and imbeddings.* In C_n let

$$B_n = \{z \mid |z_1|^2 + \cdots + |z_n|^2 < 1\},$$

$$\Delta_n = \{z \mid |z_1| < 1, \ldots, |z_n| < 1\},$$

be the ball and the polydisc respectively.

Problem 3. Does there exist a proper holomorphic map $B_2 \longrightarrow \Delta_2$?

In the other direction it is well known that there is no proper holomorphic map $\Delta_2 \longrightarrow B_2$. More general and related to Problem 3 are the problems:

Problems 3a. Does every bounded domain of holomorphy admit a proper holomorphic map into some Δ_n? Is any proper holomorphic map of B_n into itself an automorphism? Is it at least given by rational functions?

Following earlier works of Remmert, E. Bishop, and R. Narasimhan on the imbedding of Stein manifolds in C_k, O. Forster proved that [6] any Stein manifold of dimension $n \geqslant 6$ can be imbedded as a closed submanifold of C_k for $k = 2n + 1 - [(n-2)/3]$. It will be interesting to know whether this result can be improved:

Problems 4. Can a Stein manifold of dimension $n \geqslant 6$ be imbedded as a closed submanifold of C_k, with $k = 2n + 1 - [n/2]$? Are there topological restrictions on the manifold (such as being diffeomorphic to R^{2n}) which ensure imbeddings in still lower dimension? Which open Riemann surfaces can be properly imbedded in C_2?

3. *Hermitian geometry.* Many results of geometric complex function theory can be traced to a theorem of Picard that the complement of three distinct points of the complex projective line $P_1(C)$ has as universal covering surface the disc Δ_1, and hence has a complete Kählerian metric of constant negative curvature. The generalizations of this theorem to high dimensions take different forms [9]. The following seems to be a problem not yet susceptible to methods developed on the subject:

Problem 5. Let $P_n(C)$ be the complex projective space of dimension n and $H_0, H_1, \ldots, H_{2n}$, $2n+1$ hyperplanes in $P_n(C)$ in general position. Does the manifold $P_n(C) - (H_0 \cup \cdots \cup H_{2n})$ have a Kählerian metric with holomorphic sectional curvature $\leqslant -1$?

Because of its crucial role in the understanding of the Ricci curvature we will state the Calabi conjecture [2]. Let M be a complex manifold of dimension n with the Kählerian metric

$$ds^2 = \sum g_{\alpha\bar{\beta}}\, dz^\alpha\, \overline{dz^\beta}, \qquad 1 \leqslant \alpha, \beta \leqslant n,$$

so that the Kähler form

$$\omega = \frac{i}{2}\sum g_{\alpha\bar{\beta}}\, dz^\alpha \wedge \overline{dz^\beta}$$

is closed. Let $g = \det(g_{\alpha\bar{\beta}})$ and

$$R_{\alpha\bar{\beta}} = \partial^2 \log g/\partial Z^\alpha \partial \bar{Z}^\beta.$$

Then the Ricci form

$$\sum = \frac{i}{2\pi}\sum R_{\alpha\bar{\beta}}\, dz^\alpha \wedge \overline{dz^\beta}$$

is a real closed form of bidegree (1, 1) and belongs to the first Chern class of M.

Problem 6 (Calabi conjecture). Let M be a compact Kählerian manifold with Kähler form ω and Ricci form Σ. If Σ' is a closed real form of bidegree (1, 1) which is cohomologous to Σ, there is a unique Kählerian metric whose Kähler form ω' is cohomologous to ω and whose Ricci form is Σ'.

Aubin proved that the conjecture is true if M has nonnegative curvature [1].

4. *Analytic cycles on noncompact algebraic varieties.* Cornalba and Griffiths have recently studied analytic cycles on noncompact algebraic varieties [4]. Among their results is the theorem that on a Stein manifold M the even-dimensional homology $H_{\text{ev}}(M, Q)$ is generated by analytic subvarieties. For an affine variety these subvarieties are not necessarily algebraic. This use of transcendental objects to study algebraic varieties is of great attractiveness. The degree of transcendency is measured by the order of growth, defined in [4]. We mention two simple problems:

Problems 7. (a) Is every even-dimensional homology on a smooth affine variety represented by an analytic cycle of finite order? (b) Let C be the twisted cubic curve in C_3. By a theorem of Forster-Ramspott [7], C is the complete intersection of two analytic surfaces S_1 and S_2. What growth is necessary for the latter?

REFERENCES

1. T. Aubin, *Métriques riemanniennes et courbure,* J. Differential Geometry **4** (1970), 383–424. MR **43** #5452.

2. E. Calabi, *On Kähler manifolds with vanishing canonical class,* Algebraic Geometry and Topology (A Sympos. in Honor of S. Lefschetz), Princeton Univ. Press, Princeton, N. J., 1957, pp. 78–89. MR **19**, 62.

3. S. Chern and J. K. Moser, *Real hypersurfaces in complex manifolds,* Acta Math. **133** (1974), 219–271.

4. M. Cornalba and P. A. Griffiths, *Analytic cycles and vector bundles on noncompact algebraic varieties,* Invent. Math. **28** (1975), 1–106.

5. C. Fefferman, *The Bergman kernel and biholomorphic mappings of pseudo-convex domains,* Invent. Math. **26** (1974), 1–65. MR **50** #2562.

6. O. Forster, *Plongements des variétés de Stein,* Comment Math. Helv. **45** (1970), 170–184. MR **42** #4773.

7. O. Forster and K. J. Ramspott, *Über die Darstellung analytischer Mengen,* Bayer Akad. Wiss. Math.-Natur. Kl. S.-B. **1963**, Abt. II, 89–99 (1964). MR **29** #4912.

8. T. Frankel, *Manifolds with positive curvature,* Pacific J. Math. **11** (1961), 165–174. MR 23 #A600.

9. P. Kiernan and S. Kobayashi, *Holomorphic mappings into projective space with lacunary hyperplanes,* Nagoya Math. J. **50** (1973), 199–216. MR **48** #4353.

10. T. Ochiai, *On compact Kähler manifolds with positive holomorphic bisectional curvature,* Proc. Sympos. Pure Math., vol. 27, part II, Amer. Math. Soc., Providence, R. I., 1975, pp. 113–123.

11. S. Webster, *Real hypersurfaces in complex spaces,* Ph.D. Thesis, Berkeley, Calif. 1975.

XV. Differential Geometry.[3] (J. Milnor).

1. *The (self-intersecting) soap bubble problem.* According to Plateau[4] in 1873: "Very probably the sphere is the only closed surface of constant mean curvature." For the boundary of a star-shaped region this had been proved by Jellett in 1853. For immersed surfaces of genus zero it has been proved by Hopf in 1950, and for embedded surfaces of arbitrary genus by Aleksandrov in 1958. However, the case of an immersed surface of higher genus remains open. Chern remarks that the situation is the same for elliptic W-surfaces, that is, surfaces with a relation $W(k_1, k_2) = 0$ between principal curvatures, where the equation defines k_2 as a monotone decreasing function of k_1.

2. *Understanding the scalar curvature* $R = \Sigma g^{ik}g^{jl}R_{ijkl}$. In 1960, Yamabe claimed the following theorem: *For every smooth compact manifold of dimension at least* 3 *with Riemannian metric* g_{ij} *there exists a conformally equivalent metric* $e^{2\rho}g_{ij}$ *which has constant scalar curvature.* Yamabe's proof was defective, and his theorem may well be false in some cases. However, his paper has led to important work by Aubin, Trudinger, Eliasson, Fischer and Marsden, and Kazdan and Warner. In particular Eliasson showed that every manifold of dimension $\geqslant 3$ admits a metric of constant negative scalar curvature. It is not known which manifolds admit metrics of positive scalar curvature (constant or not), or zero scalar curvature. One important contribution is Lichnerowicz's 1963 proof that a $4k$-dimensional spin manifold M with $\hat{A}[M] \neq 0$ cannot admit a metric with $R \geqslant 0, R \not\equiv 0$. Furthermore, if the first Betti number is nonzero, then the case $R \equiv 0$ cannot occur either. (See Kazdan and Warner, "*Prescribing curvatures*", §5.4. For more on spin manifolds with $\hat{A}[M] \neq 0$, see Atiyah and Hirzebruch.) Hitchin has shown that an exotic sphere which does not bound a spin manifold cannot admit a metric of positive scalar curvature.

Closely related is a problem which has been posed by R. Geroch in connection with general relativity theory. Consider a Riemannian metric on the real coordinate space $\mathbf{R}^n$ which coincides with the Euclidean metric δ_{ij} outside of a bounded open set. If $R \geqslant 0$ everywhere, does it follow that this metric is flat?

---

[3] I am indebted to S. S. Chern, J. Nitsche, J. O'Sullivan, N. Wallach, and S. T. Yau for useful suggestions and references.

[4] This problem was perhaps first posed by Delaunay in 1843.

Can it happen that $R > 0$ throughout the open set? (The case $R < 0$ can occur. Compare Aubin.)

3. *Understanding the Ricci tensor* $R_{ik} = \Sigma g^{jl} R_{ijkl}$. If a compact Riemannian manifold has R_{ik} identically zero, does it follow that it is flat? In this connection we note Calabi's conjecture that any Kähler manifold with $c_1 = 0$ admits a Kähler metric with $R_{ik} \equiv 0$.

Which compact manifolds admit metrics of positive definite (or negative definite) Ricci curvature? A key result is Meyer's classical theorem that positive Ricci curvature implies finite fundamental group. Here are some examples of manifolds of negative Ricci curvature. Nagano and Smyth consider a nonsingular hyperplane section of an abelian variety of complex dimension $n \geqslant 3$. The resulting manifold has negative definite Ricci tensor almost everywhere, hence by a theorem of Ehrlich the metric can be deformed so that the Ricci tensor will be negative definite everywhere. Yet the fundamental group is free abelian of rank $2n$, and the universal covering space is not contractible, so this manifold cannot admit a metric of nonpositive sectional curvature. Yau has shown that the connected sum of two manifolds of constant nonpositive curvature admits a metric of negative Ricci curvature. Again the universal covering space is not contractible provided that the dimension is 3 or more. It is not known whether or not the sphere S^3 or the torus $S^1 \times S^1 \times S^1$ admits a metric of negative Ricci curvature. For further information see Bochner and Yano, Fischer and Wolf, Kobayashi, Milnor, and Yau.

4. *Manifolds of positive sectional curvature.* In spite of important work by Synge, Rauch, Berger, Gromoll, Cheeger, Wallach and many others, there remain a number of unanswered questions concerning compact Riemannian manifolds for which the sectional curvatures $K = \Sigma R_{ijkl} u^i v^j u^k v^l$ are all positive. Can such a manifold split topologically as a product? For example, does the product $S^m \times S^n$, with $m, n \geqslant 2$, admit a metric of strictly positive sectional curvature? (This manifold does of course admit a metric of positive Ricci curvature.) If the manifolds M and N have odd dimension, note that the product $M \times N$ has Euler number zero, and carries two linearly independent vector fields u and v which commute: $[u, v] = 0$. Thus the following two questions are related to the question just raised.

Can any compact manifold of positive sectional curvature admit two linearly independent commuting vector fields? According to Lima, the sphere S^3 does not admit two such vector fields, but the corresponding question for S^7, $S^{11}, \ldots$ remains open.

Does every even-dimensional compact manifold of positive sectional curvature have positive Euler number? In the 4-dimensional case this can be proved using the generalized Gauss-Bonnet formula. (See Chern.) However the corresponding argument in the 6-dimensional case fails, since the Gauss-Bonnet integrand can be negative at a point of positive sectional curvature. (Geroch, unpublished.)

Similarly one can ask which manifolds admit metrics of sectional curvature < 0 (or $\leqslant 0$). Compare Eberlein (or Wolf).

REFERENCES

A. D. Aleksandrov, *Uniqueness theorems for surfaces in the large*. V, Vestnik Leningrad. Univ. **13** (1958), no. 19, 5–8; English transl., Amer. Math. Soc. Transl. (2) **21** (1962), 412–416. MR **21** #909; **27** #698e.

S. Aloff and N. Wallach, *An infinite family of compact simply-connected positively curved 7-manifolds*, Bull. Amer. Math. Soc. **81** (1975), 93–97.

M. Atiyah and F. Hirzebruch, *Spin-manifolds and group actions*, Essays on Topology and Related Topics (Mém. Dédiés à Georges deRham), Springer, New York, 1970, pp. 18–28. MR **43** #4064.

T. Aubin, *Métriques riemanniennes et courbure*, J. Differential Geometry **4** (1970), 383–424. MR **43** #5452.

R. L. Bishop and R. J. Crittenden, *Geometry of manifolds*, Pure and Appl. Math., vol. 15, Academic Press, New York, 1964. MR **29** #6401.

K. Yano and S. Bochner, *Curvature and Betti numbers*, Ann. of Math. Studies, no.32, Princeton Univ. Press, Princeton, N. J., 1953. MR **15**, 989.

E. Calabi, *On Kähler manifolds with vanishing canonical class*, Algebraic Geometry and Topology (A Sympos. in Honor of S. Lefschetz), Princeton Univ. Press, Princeton, N. J., 1957, pp. 78–89. MR **19**, 62.

J. Cheeger, *Some examples of manifolds of nonnegative curvature*, J. Differential Geometry **8** (1973), 623–628. MR **49** #6085.

J. Cheeger and D. Gromoll, *On the structure of complete manifolds of nonnegative curvature*, Ann. of Math. (2) **96** (1972), 413–443. MR **46** #8121.

S. Chern, *On curvature and characteristic classes of a Riemann manifold*, Abh. Math. Sem. Univ. Hamburg **20** (1955), 117–126. MR **17**, 783.

C. Delaunay, *Mémoire sur le calcul des variations*, J. École Polytech. **17** (1843), 37–120 (See p. 111).

P. Eberlein, *Some properties of the fundamental group of a Fuchsian manifold*, Invent. Math. **19** (1973), 5–13.

P. E. Ehrlich, *Local convex deformations of Ricci and sectional curvature on compact Riemannian manifolds*, Proc. Sympos. Pure Math., vol. 27, part I, Amer. Math. Soc., Providence, R. I., 1975, pp. 69–71.

H. I. Eliasson, *On variations of metrics*, Math. Scand. **29** (1971), 317–327. MR **47** #985.

A. E. Fischer and J. E. Marsden, *Manifolds of Riemannian metrics with prescribed scalar curvature*, Bull. Amer. Math. Soc. **80** (1974), 479–484. MR **49** #11561.

A. E. Fischer and J. A. Wolf, *The Calabi construction for compact Ricci flat Riemannian manifolds*, Bull. Amer. Math. Soc. **80** (1974), 92–97.

R. Geroch, *General relativity*, Proc. Sympos. Pure Math., vol. 27, part II, Amer. Math. Soc., Providence, R. I., 1975, pp. 401–441. (see Conjecture 3).

D. Gromoll, W. Klingenberg and W. Meyer, *Riemannsche Geometrie im Grossen*, Lecture Notes in Math., no. 55, Springer-Verlag, Berlin and New York, 1968. MR **37** #4751.

N. Hitchin, *The space of harmonic spinors* (to appear) (see Chap. 4, Section 3).

H. Hopf, *Über Flächen mit einer Relation zwischen den Hauptkrümmungen*, Math. Nachr. **4** (1950), 232–249. MR **12**, 634.

M. Jellett, *Sur la surface dont la courbure moyenne est constante*, J. Liouville **18** (1853), 163–167.

J. L. Kazdan and F. W. Warner, *Scalar curvature and conformal deformation of Riemannian structure*, J. Differential Geometry (to appear).

J. L. Kazdan and F. W. Warner, *Prescribing curvatures,* Proc. Sympos. Pure Math., vol. 27, part II, Amer, Math. Soc., Providence, R. I., 1975, pp. 309–320.

———, *Existence and conformal deformation of metrics with prescribed Gaussian and scalar curvatures,* Ann. of Math. (to appear).

S. Kobayashi, *On compact Kaehler manifolds with positive definite Ricci tensor,* Ann. of Math. (2) **74** (1961), 570–574. MR **24** #A2922.

A. Lichnerowicz, *Spineurs harmoniques,* C. R. Acad. Sci. Paris **257** (1963), 7–9. MR **27** #6218.

E. Lima, *Commuting vector fields on* S^3, Ann. of Math. (2) **81** (1965), 70–81. MR **30** #1517.

J. Milnor, *A note on curvature and fundamental group,* J. Differential Geometry **2** (1968), 1–7. MR **38** #636.

T. Nagano and B. Smyth, *Minimal varieties in tori,* Proc. Sympos. Pure Math., vol. 27, part I, Amer. Math. Soc., Providence, R. I., 1975, pp, 189–190.

J. Plateau, *Statique experimentale et théorique des liquides,* Gauthier-Villars, Paris, 1873. (see Vol. 1, p. 438).

N. S. Trudinger, *Remarks concerning the conformal deformation of Riemannian structures on compact manifolds,* Ann. Scuola Norm. Sup. Pisa (3) **22** (1968), 265–274. MR **39** #2093.

N. Wallach, *Compact homogeneous Riemannian manifolds with strictly positive curvature,* Ann. of Math. (2) **96** (1972), 277–295. MR **46** #6243.

J. A. Wolf, *Growth of finitely generated solvable groups and curvature of Riemannian manifolds,* J. Differential Geometry **2** (1968), 421–446. MR **40** #1939.

H. Yamabe, *On a deformation of Riemannian structures on compact manifolds,* Osaka Math. J. **12** (1960), 21–37. MR **23** #A2847.

S. T. Yau, *Compact flat Riemannian manifolds,* J. Differential Geometry **6** (1971/72), 395–402. MR **46** #4439.

XVI. Study of singularities.

(A) *Stratifications of analytic sets* (R. Thom). Given an analytic set A defined by an ideal **J**, define the minimal family of differential operators Δ (generating and generalizing the Boardman symbols) such that if one, starting from **J**, considers the extensions obtained by adding to **J** all functions of type $\Delta(f_1, \ldots, f_r)$ with the $f_j \in \mathbf{J}$, one gets a family of subsets $A_\Delta \subset A$ which defines the minimal stratification of A.

(B) *Singularities of differentiable maps* (R. Thom).

(1) If $f(x)$, for $x \in \mathbf{R}^n$, is a real analytic function in a neighborhood U of the origin $0 \in \mathbf{R}^n$, every trajectory of the vector field $X = \operatorname{grad} f$ which tends to 0 has a tangent at that point.

(2) If 0 is an isolated singular point of the complexification of f, whose Milnor number is μ, there exists a deformation of f (by addition of a linear form in the coordinates?) which has exactly μ real nondegenerate quadratic critical points.

(3) More generally: when the complex type of an isolated singularity of a real analytic function f is given, determine all signatures of the critical points which appear in all deformations of f which are Morse functions.

(4) *Conjecture.* Every stratum of the universal unfolding of an isolated singularity of a function is contractible (in the real case), and a $K(\pi, 1)$ in the complex case.

(5) *Conjecture.* Any (complex) analytical set is locally homeomorphic to an algebraic set.

(6) *Problem.* To realize an Eilenberg-Mac Lane space $K(\pi, n)$ by an infinite-dimensional manifold, in such a way that the fundamental cohomology class $\iota \in H^n(K, \pi)$ be the Poincaré dual of the fundamental cycle of a stratified set of finite codimension n.

(C) *Pfaff systems* (R. Thom).

(1) Let $\mathbf{G}_p^n$ be the fibre bundle with base $\mathbf{R}^n$, whose fibre at a point consists of the p-planes through that point; a section σ: $\mathbf{R}^n \longrightarrow \mathbf{G}_p^n$ is a field of p-planes; let $j^r(\sigma)(x_0)$ be its jet of order r at x_0. The group of jets of order r of the local automorphisms of $\mathbf{R}^n$ fixing x_0 denoted $L^r(n)$, operates algebraically in the space $J^r(\sigma)$ of the jets of order r of the section σ. Define (and construct) the stratification Σ defined in $J^r(\sigma)$ by that action.

(2) *Conjecture.* Every Pfaff system of dimension p is defined by a section σ. The local properties of that system (from the view point of the existence of germs of solutions of given dimension) are entirely defined by the jet of order r of σ, if the derivative $j^r(\sigma)$ is locally transversal to Σ.

(D) *Analytical properties of the thermodynamical functions at critical states* (V. Arnold). The thermodynamical functions are defined, with the help of some limit procedure, by the interaction potential of particles. *To find the singularities of these functions, knowing the interaction potential* is a very difficult analytical problem. The solution is not known for generic potential, nor for such simple models as the Ising lattice model in 3 dimensions.

(E) *Calculation of the order of the wildest generic singularities in l-parametric families of functions* (V. Arnold). For every integer $l > 0$ there is a rational number β_l, describing the asymptotics of the l-dimensional generalization of the Airy function, and the monodromy of the wildest singularity, generic in the l-parametric families of functions.

One can find an exact definition of the order of the wildest singularity β_l, in *Remarks on the stationary phase method and Coxeter numbers,* Russian Math. Surveys, 1973, no. 5, 17–44. The first 10 values of β_l are 1/6, 1/4, 1/3, 3/8, 5/12, 1/2, 1/2, 13/24, 9/16, 2/3. The optimistic formualtion of the problem: *to find all the* β_l; the pessimistic one: *to find* β_{1000}.

This problem is one of the ways to formulate precisely the fundamental but vague "problem of classification of singularities".

(F) *Topological properties of the bifurcation subsets* (V. Arnold). The bifurcation subsets are subsets in spaces of smooth or holomorphic functions on jets. There are two bifurcation subsets. A function belongs to the first (resp. second) subset, iff it is not a Morse function (resp. iff 0 is a critical value).

It seems very interesting *to study the topological properties of the complement of the bifurcation subsets* (for functions and also for differentiable mappings in general).

Near holomorphic functions with simple critical points of types A_k, D_k, E_k the complements to both bifurcation subsets have locally the homotopy type of a $K(\pi, 1)$ space (Brieskorn, Deligne, Ljaschko, etc.). The cohomology of the complement in the simplest case A_k is that of the second loop space of the 3-sphere (G. Segal, D. Fuks). On the other hand, in the real case there exist interesting relations of the topology of the bifurcation subsets to pseudo-isotopy and algebraic K-theory (Cerf, Wagoner, Volodin, etc.). See also the talk of Thom in *Manifolds,* Amsterdam 1970, Lecture Notes no. 197, Springer 1971.

XVII. Dynamical systems and differential equations.

(A) *Qualitative dynamics* (R. Thom).

Conjecture. Let M^n be a compact differentiable manifold $\mathbf{X}^r(M)$ the Fréchet space of vector fields on M^n, with the C^r-topology. There exists a dense open subset $U \subset \mathbf{X}^r(M)$ such that, if ξ is a vector field in U, almost every trajectory of ξ has as ω-limit set an *attractor A,* and each of these attractors (but not the set they form) is topologically structurally stable.

(B) *Is the stability problem for stationary points algorithmically decidable?* (V. Arnold). The well-known Lyapounov theorem solves the problem in the absence of eigenvalues with zero real parts. In more complicated cases, where the stability depends on higher order terms in the Taylor series, there exists no *algebraic* criterion.

Let a vector field be given by polynomials of a fixed degree, with rational coefficients. Does an algorithm exist, allowing to decide, whether the stationary point is stable?

A similar problem: Does there exist an algorithm to decide, whether a plane polynomial vector field has a limit cycle?

A model problem: Does there exist an algorithm to decide, whether a period of a given abelian integral is positive (the equation of the curve and the integrand have rational coefficients and are of fixed degree). The simplest model problem is that of positivity on the real line of sums of products of exponentials and of polynomials. For this problem there exists an algorithm. This follows from the solution of Hilbert's problem on transcendental numbers (Matijasevich).

(C) *To prove the instability of equilibrium, which is not a minimum point for the potential energy* (V. Arnold). The equation of motion is that of Newton,

$x'' = -\partial U/\partial x$ with an analytical potential energy U defined in the n-dimensional real space. The problem goes back to Lagrange and Dirichlet, who proved that points of minimum are stable. The instability seems obvious, but is not always the case for infinitely differentiable potentials.

(D) *Problems on dynamical systems and celestial mechanics* (S. Smale).

(1) Is every expanding map of a compact manifold induced by an automorphism of a nilpotent Lie group?

An expanding map $f\colon M \longrightarrow M$ is characterized by the property that there are constants $c > 0$, $\lambda > 1$, such that for each x in M, $\|Df^n(x)(v)\| \geqslant c\lambda^n \|v\|$, for all positive integers n, and tangent vectors v at x. Here $Df^n(x)$ is the derivative of the nth iterate of f at x.

An example is the map $z \longrightarrow z^2$ for complex numbers z of absolute value 1. More generally let N be a simply connected nilpotent Lie group and Γ a discrete subgroup with N/Γ compact. Let $g\colon N \longrightarrow N$ be an automorphism such that $g\Gamma \subset \Gamma$ and $\|Dg(e)^{-1}\| < 1$, where e is the identity. Then $\tilde{g}\colon N/\Gamma \longrightarrow N/\Gamma$ is an expanding map (in the first example take $N = \mathbf{R}$, $\Gamma = \mathbf{Z}$, and g multiplication by 2). The problem becomes: Is every expanding map of this type, up to a finite covering; or Given an expanding map $f\colon M \longrightarrow M$ is there a lift $\tilde{f}$: $\tilde{M} \longrightarrow \tilde{M}$ to a finite covering of M with a homeomorphism h,

$$\begin{array}{ccc} \tilde{M} & \xrightarrow{\ h\ } & N/\Gamma \\ \downarrow{\scriptstyle \tilde{f}} & & \downarrow{\scriptstyle \tilde{g}} \\ \tilde{M} & \xrightarrow{\ h\ } & N/\Gamma \end{array}$$

for g induced by some commuting automorphism $\tilde{g}\colon N \longrightarrow N$ as above?

The basic reference for the above is Shub's thesis [7]. See Franks, Hirsch, Shub in [1] for a further background and partial results.

This problem is the first of many toward understanding the structure of the nonwandering set of an Axiom A dynamical system. Another problem in this direction is to discover whether every Anosov flow of a compact manifold is Lie group induced in the sense of Tomter [1].

(2) The "closing lemma problem". Is it a generic property of discrete dynamical systems on compact manifolds that the periodic points are dense in the nonwandering set?

Represent the discrete dynamical systems on M by diffeomorphisms $M \longrightarrow M$ and let Diff(M) be the space of C^∞ diffeomorphisms with the C^∞ topology. A *generic property* of $f \in$ Diff(M) is one that is true for a Baire set in Diff(M).

A point $x \in M$ is said to be in the *nonwandering* set $\Omega = \Omega(f)$ of $f \in$ Diff(M) provided that for each neighborhood U of x there is some positive integer n with $f^n(U) \cap U \neq \varnothing$.

If C^∞ is replaced by C^1, Pugh has given an affirmative answer [6].

(3) Does Ω-stability imply Axiom A? An affirmative answer will settle the problem of finding necessary and sufficient conditions for a dynamical system to be structurally stable or Ω-stable.

We say that $f \in \mathrm{Diff}(M)$ is *Ω-stable* provided there is some neighborhood $N(f)$ of f in $\mathrm{Diff}(M)$ so that if $g \in N(f)$ there is a homeomorphism h and

$$\begin{array}{ccc} \Omega(f) & \xrightarrow{\ h\ } & \Omega(g) \\ \Big\downarrow f & & \Big\downarrow g \\ \Omega(f) & \xrightarrow{\ h\ } & \Omega(g) \end{array}$$

In other words f is Ω-stable provided that perturbations have the same dynamical structure on the nonwandering set. It is a weaker condition than structural stability.

Axiom A postulates that the periodic points are dense in Ω and that Ω has a hyperbolic structure. A *hyperbolic* structure on Ω (for $f \in \mathrm{Diff}(M)$) is a splitting of the tangent bundle of M over Ω into the sum of subbundles E^s, E^u with the following property. Df on E^u is expanding as in problem (1) (i.e., with $x \in \Omega$ and $v \in E^u$) and Df^{-1} is expanding on E^s. While technical sounding at first, hyperbolicity is an important unifying idea in dynamical systems. For background on this problem, one can see [1], [2] and [5].

(4) Does the three-sphere $\mathbf{S}^3$ carry the structure of a minimal set of a smooth flow (defined by a C^∞ vector field)?

Here $\mathbf{S}^3$ is a *minimal set* provided every orbit of the flow is dense. The question generalizes to: What compact manifolds can be minimal sets? Of course the Euler Characteristic must be 0. These questions are valid also for discrete flows and for C^0 flows. A background reference is Gottschalk [3].

(5) Is having an attracting periodic orbit a generic property for diffeomorphisms of the two-sphere $\mathbf{S}^2$? [8].

(6) Problem of relative equilibria in the n-body problem of celestial mechanics.

Suppose that one is given positive masses $m_1, \ldots, m_n$ in the planar n-body problem. A configuration $x_1, \ldots, x_n$, each x_i a point in the plane, no two the same, is called a relative equilibrium if a rotation of these bodies at some constant angular velocity satisfies Newton's equations. Identify two relative equilibria which differ by a rotation or a positive scalar multiplication. An old problem, mentioned by Wintner [10] is: Whatever the masses, must there be at most a finite number of relative equilibria? To state a further problem, we state a proposition from [9].

PROPOSITION. *Let* $m = (m_1, \ldots, m_n)$ *and* $V_m \cdot \mathbf{S}_k - \Delta \longrightarrow \mathbf{R}$ *be the potential energy* $V_m(x_1, \ldots, x_n) = \Sigma_{i>j} i/\|x_i - x_j\|$ *restricted to the excised sphere*

$$\mathbf{S}_k - \Delta = \left\{(x_1, \ldots, x_n) \in \frac{(E^2)^n}{\sum m_i \|x_i\|^2} = 1,\ x_i \neq x_j,\ \sum m_i x_i = 0\right\}.$$

Let $\widetilde{V}_m: (\mathbf{S}_k - \Delta)/\mathbf{SO}(2) \longrightarrow \mathbf{R}$ *be the induced function on the quotient space by the rotation group. Then the relative equilibria correspond naturally in a* 1-1 *fashion to the critical points of* $\widetilde{V}_m$.

Say that $(m_1, \ldots, m_n)$ is critical if $\widetilde{V}_m$ has a degenerate critical point (i.e., $\widetilde{V}_m$ fails to be a Morse function).

Problem. What is the nature of the (closed) set Σ_n of critical masses in $(\mathbf{R}^+)^n$? Has it measure 0, finite Betti numbers? Palmore [4] has shown that $\Sigma_3 = \emptyset$, $\Sigma_4 \neq \emptyset$.

REFERENCES

1. S. S. Chern and S. Smale (Editors), *Global analysis,* Proc. Sympos. Pure Math., vol. 14, Amer. Math. Soc., Providence, R. I., 1970. MR **41** #7686.

2. J. Franks, *Differentiability Ω-stable diffeomorphisms,* mimeographed, Northwestern University.

3. W. Gottschalk, *Minimal Sets: An introduction to topological dynamics,* Bull. Amer. Math. Soc. **64** (1958), 336–351. MR **20** #6484.

4. J. Palmore, Thesis, Univ. of California, Berkeley, 1973.

5. M. Peixoto (Editor), *Dynamical systems* (Proc. Sympos., Bahia, Salvador, Brazil, 1971), Academic Press, New York and London, 1973. MR **48** #1255.

6. C. Pugh, *An improved closing lemma and a general density theorem,* Amer. J. Math. **89** (1967), 1010–1021. MR **37** #2257.

7. M. Shub, *Endomorphisms of compact differentiable manifolds,* Amer. J. Math. **91** (1969), 175–199. MR **39** #2169.

8. S. Smale, *Dynamical systems and the topological conjugacy problem for diffeomorphisms,* Proc. Internat. Congress Math. (Stockholm, 1962), Inst. Mittag-Leffler, Djursholm, 1963, pp. 490–496. MR **31** #759.

9. ———, *Topology and mechanics.* I, II, Invent. Math. **10** (1970), 305–331; ibid. **11** (1970), 45–64. MR **46** #8263; **47** #9671. See also: *Problems on the nature of relative equilibria in celestial mechanics,* Manifolds-Amsterdam 1970 (Proc. Nuffic Summer School), Lecture Notes in Math., vol. 197, Springer-Verlag, Berlin, 1971, pp. 194–198. MR **43** #4429.

10. A. Wintner, *The analytical foundations of celestial mechanics,* Princeton Math. Ser., vol. 5, Princeton Univ. Press, Princeton, N. J., 1941. MR **3**, 215.

XVIII. Geometric topology (L. C. Siebenmann).

(I) THE POINCARÉ CONJECTURE. *Every compact n-manifold homotopy equivalent to the n-sphere* S^n *is homeomorphic to* S^n.

Comments. This is known to be true for $n \neq 3, 4$ by the work of Smale, Stallings and Newman, 1959–1965. The classical case $n = 3$ has resisted a continual onslaught since the 1930's. The 4-dimensional case $n = 4$ has remained out of reach even for differentiable class C^∞ manifolds.

See references [6], [8].

(II) HILBERT CUBE CONJECTURES.

(1) *The Hilbert cube* $Q = [0, 1]^\infty$ *is the only homogeneous contractible infinite dimensional metric compactum.* (Homogeneous means that, given points x and y in Q, there exists a homeomorphism of Q onto itself carrying x to y.)

(2) *Every retract of the Hilbert cube is a cartesian factor of the Hilbert cube.*

Comments.

(i) The first conjecture is given to situate the second. It has been inaccessible thus far, like the related conjecture of Borsuk that every homogeneous ANR(= absolute neighborhood retract) is a manifold with one of the models R^n (euclidean n-space), $Q = [0, 1]^\infty$, or l_2 (the Hilbert space).

(ii) One consequence of (2) depending on results of J. West and T. Chapman is that every compact ANR X would have finite homotopy type (as Borsuk conjectured in 1954). In fact there would exist a finite simplicial complex K with $X \times Q$ homeomorphic to $K \times Q$. Recently H. Toruńczyk has proved that every retract of Hilbert space l_2 is a cartesian factor of l_2.

See references [1], [3], [9].

(III) TO CONSTRUCT NEW 4-MANIFOLDS. It is only in the dimension of space-time, dimension 4, that the work of this century has badly failed to give an overview of the classification of manifolds.

(a) *Does there exist a smooth (= differentiable class* C^∞*) 4-manifold that is homotopy equivalent to* $S^3 \times S^1$ *but not diffeomorphic to* $S^3 \times S^1$?

(b) *Does a nonsmoothable 4-manifold exist? More specifically does there exist a closed*[6] *topological 4-manifold* M^4 *verifying the following conditions on its characteristic classes*: $w_1 = w_2 = 0$, $1/3p_1 \equiv 8$ *modulo* 16?

Comments on (b). V. A. Rohlin proved that for any smooth closed 4-manifold with $w_1 = w_2 = 0$ one has $1/3p_1 \equiv 0$ modulo 16. Concerning the topological M^4 envisaged: either M^4 is not triangulable as a simplicial complex or the classical Poincaré conjecture is violated by a link of some vertex of triangulated M^4.

(c) *What cohomology rings occur for closed smooth 4-manifolds?* (One might impose simple connectivity, i.e. $\pi_1 = 0$.)

See references [4], [5], [7].

(IV) THE TRIANGULATION CONJECTURE: *Every (metrizable finite dimensional) topological manifold is homeomorphic to a simplicial complex.*

Comments. A homeomorphism to a simplicial complex is called a *triangulation.* T. Rado gave a proof of the triangulation conjecture for dimension $\leqslant 2$ in 1924; E. Moise gave one for dimension 3 in 1951. For a topological n-manifold M^n, $n \geqslant 5$ (without boundary if $n = 5$), Kirby and Siebenmann found

[6] Closed means compact with empty boundary.

in 1968–1969 a single obstruction in $H^4(M; Z_2)$ to triangulating M as a combinatorial manifold (= simplicial complex that is piecewise-linearly homogeneous). By examining examples for which this obstruction is nonzero one observes readily that the truth of the triangulation conjecture would imply the existence of a smooth homology 3-sphere M^3 (= smooth compact 3-manifold with the integral homology of S^3) such that:

(i) The double suspension $\Sigma^2 M^3$ (= the join of M^3 with the circle S^1) is homeomorphic to S^5. (For no smooth homology n-sphere M^n distinct from S^n is it yet known whether or not $\Sigma^2 M^n$ is homeomorphic to S^{n+2}.)

(ii) The Milnor connected sum $M \# M$ (preserving an orientation) is the boundary of a smooth homology n-disc.

(iii) M itself is the boundary of a compact smooth 4-manifold verifying these conditions on its characteristic classes: $w_1 = w_2 = 0$, $1/3p_1 \equiv 8$ modulo 16. (No smooth homology 3-sphere is known that verifies both (ii) and (iii); $SO(3)/G$ satisfies (iii) if G is the group of 60 rotations respecting the dodecahedron.)

Given a smooth homology 3-sphere M^3 verifying (i)–(iii) it may be a reasonable task to prove the triangulation conjecture.

See reference [7].

REFERENCES

1. R. D. Anderson and Nelly Kroonenberg, *Open problems in infinite dimensional topology,* mimeographed, Lousiana State U., Baton Rouge, 1973.

2. S. Cappell and J. L. Shaneson, *On four-dimensional surgery and applications,* Comment. Math. Helv. **46** (1971), 500–528. MR **46** #905.

3. A. Fathi and Y. Visetti, *A reduction of the fundamental conjecture about* ANR*'s* Proc. Amer. Math. Soc. (preprint Orsay 1973).

4. J. Milnor and M. Kervaire, *Bernoulli numbers, homotopy groups and a theorem of Rohlin,* Proc. Internat. Congress Math. (Edinborough 1958), Cambridge Univ. Press, New York, 1960, pp. 454–458. MR **22** #12531.

5. J. Milnor, *On simply connected 4-manifolds,* Sympos. Internacional Topologiá Algebraica, Universidad Nacional Autónoma de México and UNESCO, Mexico City, 1958, pp. 122–128. MR **21** #2240.

6. ———, *Lectures on the h-cobordism theorem,* Math. Notes, Princeton Univ. Press, Princeton, N. J., 1965. MR **32** #8352.

7. L. Siebenmann, *Topological manifolds,* Proc. Internat. Congress Math. (Nice, 1970), vol. 2, Gauthier-Villars, Paris, 1971, pp. 133–163.

8. J. Stallings, *Group theory and 3-dimensional manifolds,* Yale Math. Monographs, no. 4, Yale Univ. Press, New Haven, Conn., 1971.

9. H. Toruńczyk, *Compact absolute retracts as factors of the Hilbert space,* Fund. Math.

XIX. Geometric and differential topology (C. T. C. Wall). In geometric topology the major problems concern 4-dimensional manifolds, where the techniques of geometric topology almost all break down. A good account of this corpus of problems was given by Kirby in his lecture to the Nice congress. Among them are: the Poincaré conjecture in dimensions 3 and 4, the annulus problem in dimension 4, the question whether transversality holds for topological manifolds of dimension 4, and whether there exists a closed topological manifold with vanishing Stiefel-Whitney classes and signature 8.

Several questions in topology reduce to word problems in finite representations of groups. Call two presentations *strictly equivalent* if one can be obtained from the other by operations of the following types, and their inverses:

(1) Adjoin a generator x and relator xw (w any word in the other generators);

(2) Multiply one relator by another relator;

(3) Replace a relator by any conjugate of itself.

The problem consists in studying strict equivalence of presentations: for example, are any two presentations of the trivial group strictly equivalent?

One very interesting result is Mather's theorem C°-stable maps are dense. The proof, however, yields only a very partial understanding of C°-stable maps. One would like more effective criteria for C°-stability, and some method of classification. For example, are C°-stable map-germs with the same local algebra C°-equivalent? If so, when can different algebras yield C°-equivalent germs?

Recent work of many authors, particularly Haefliger and Thurston, has yielded extensive information concerning foliations, but reduced several problems to calculations in the classifying spaces usually denoted $B\Gamma_n$. The problem consists in obtaining an effective calculation of homotopy properties of these spaces.

One interesting development of the last 15 years has been the growth from topological motivations of algebraic K-theory: first K_0 and K_1 then (more slowly) K_2 and now Quillen's spectacular theory of the higher K groups. Using hermitian forms, other theories have been developed which I call KU-theory and L-theory: they are of course closely related. To give a complete concordance of different approaches seems only to be possible if a certain conjecture due to Karoubi holds (see papers in *Lecture Notes* (Springer) Vol. 343: his formulation and my account of it). This would clean up the general theory: there remains the need for calculation, particularly of the surgery obstruction groups $L_*(\mathbf{Z}\pi)$ for π a group. If π is a Poincaré duality group, and so has a classifying space $B\pi$ dominated by a finite complex, one can apply Quinn's homology theory $\mathbb{L}$ (represented by G/Top). I conjecture that there is a natural isomorphism $L_*(\mathbf{Z}\pi) \cong \mathbb{L}_*(B\pi)$. This is already known for poly-infinite-cyclic groups, and is a major source of motivation for further work.

Related to this is the problem of classifying free uniform actions of discrete groups π on $\mathbf{R}^n$. Further questions in this area are: are such actions always smoothable? If ρ contains π as subgroup of finite index, does the action of π necessarily extend to one of ρ? If M is compact and the universal cover $\widetilde{M}$

contractible, need $\widetilde{M}$ be homeomorphic to euclidean space $\mathbf{R}^m$ (if $m > 4$, it is enough to ask that $\widetilde{M}$ be 1-connected at infinity)?

If G is finitely generated, and has infinitely many ends, then according to Stallings, G decomposes as an amalgamated free product (or HNN structure) where the subgroup in question is finite. Does this process of decomposition necessarily terminate? A closely related problem is: is $H^1(G; \mathbf{Z}_2 G)$ finitely generated as module over the mod 2 group ring $\mathbf{Z}_2 G$?

Many problems on 3-manifolds depend on the Poincaré conjecture, but there are many more general questions, e.g. concerning the genus that can be formulated. Perhaps the most interesting is: Let M^3 be compact, with infinite fundamental group. Need M have a finite covering space which contains an incompressible closed surface T: i.e. T has infinite fundamental group, and $\pi_1(T) \longrightarrow \pi_1(M)$ is injective.

XX. Fixed points of symplectic diffeomorphisms (V. Arnold). The problem goes back to the "last geometric theorem" of Poincaré. The simplest case is the following problem: *Does every symplectic diffeomorphism of a 2-dimensional torus, which is homologous to the identity, have a fixed point*?

A symplectic diffeomorphism is a diffeomorphism which preserves some nondegenerate closed 2-form (the area in the 2-dimensional case). It is homologous to the identity iff it belongs to the commutator subgroup of the group of symplectic diffeomorphisms homotopical to the identity. With coordinates, such a diffeomorphism is given by $x \longrightarrow x + f(x)$, where x is a point of the plane and f is periodic. It is symplectic iff the Jacobian $\det(D(x + f(x))/Dx)$ is identically 1, and it is homologous to the identity iff the mean value of f is 0.

The "last geometric theorem" of Poincaré (proved by G. D. Birkhoff) deals with a circular ring. The existence of 2 geometrically different fixed points for symplectic diffeomorphisms of the 2-sphere is also proved (A. Shnirelman, N. Nikishin). In the general case, one may conjecture that the number of fixed points is bounded from below by the number of critical points of a function (both algebraically and geometrically).

XXI. Partial differential equations.

(A) *The regularity problem for elliptic systems* (E. Bombieri). A recent discovery of the last years shows that solutions of a regular analytic problem in the calculus of variations need not be analytic. Very little is known about the structure of the singular set, except for estimates of the dimension. Is the singular set semi-analytic? If not, what would be a good admissible class of singularities for the solutions? Cases of particular interest are uniformly elliptic nonlinear systems of the first order and the problem of singularities of minimal varieties of any dimension and codimension.

(B) *To what extent can one recover a compact manifold from the spectrum of the Laplacian*? (E. Bombieri). This is a problem about which much progress has been made in the last few years. It is likely that further investigation will lead to interesting developments.

(C) *Complex Monge-Ampere equations* (F. Browder, L. Nirenberg). Decisive advances have been made in recent years in the study of the real Monge-Ampere equation and its generalizations of the form $\mathrm{Det}(\partial^2 u/\partial x_j \partial x_k) = f(x, u, Du)$ by Pogorelov and more recently by Calabi-Nirenberg (see Nirenberg's report at the Vancouver Congress). The extension of these results to the complex case in which one seeks to solve the Dirichlet problem for a plurisubharmonic function u on a strongly pseudo-convex domain G of C^n satisfying the partial differential equation

$$\mathrm{Det}(\partial^2 u/\partial z_j \partial \bar{z}_k) = f \geqslant 0$$

would be of great significance to the study of the holomorphic and biholomorphic mappings of such domains. Such a study has been initiated by Kerzman, Kohn, and Nirenberg but with very sharp limitations upon the results already obtained.

(D) *Linear partial differential operators* (F. Browder). Important advances have been made in recent years in the characterization of the subclasses of linear partial differential operators of principal normal type having such properties as local solvability and hypoellipticity by the extension of the theory of pseudo-differential operators and the creation of the machinery of Fourier integral operators. The extension of these results to more general linear partial differential operators (which are no longer of principal normal type) presents many difficult problems of formulation and technique which are just beginning to be attacked. The interest of such problems is seriously increased by their connection with such "degenerate" problems as the restriction of the Cauchy-Riemann operators in n-variables to the boundary of pseudo-convex domains.

(E) *What is a mathematical equivalent to physical "turbulence"*? (V. Arnold). The well-known (at least since 1962) conjecture is that one can define turbulence of viscous incompressible fluids with the help of such finite dimensional attractors, in the infinite dimensional phase space of initial conditions, that the dynamical systems on these attractors are themselves exponentially unstable ("hyperbolic"); the dimension of an attractor is finite for every finite Reynolds number and grows with it. However there exist other points of view, e.g. that mathematical expression of turbulence is the nonexistence or non-uniqueness of solutions of the Navier-Stokes equations, and that these equations are not an adequate model of physical motion. I hope the last is not true; *to find good existence and uniqueness theorems for the 3-dimensional Navier-Stokes equations* is only one of the aspects of the problem (which can as well be considered for the 2-dimensional case, where existence and uniqueness are known).

(F) *The Korteweg-de Vries equation* (Yu. I. Manin). Recently an unexpected breakthrough was made in the theory of some very concrete nonlinear wave equations. The Korteweg-de Vries equation $u_t - 6uu_x + u_{xxx} = 0$ was first to attract attention, after the pioneering work of Gardner-Green-Kruskal-Muira. An infinity of conservation laws was discovered, a "quasilinear" composition law of solitary waves was demonstrated. Then V. Zakharov and L. Faddeev proved an exciting result: The Korteweg-de Vries equation describes an infinite dimensional hamiltonian system which is completely integrable. At the moment a number of physically interesting equations with similar properties are exhibited. V. Zakharov was able to explain in this way astonishing numerical results on the Fermi-Pasta-Ulam phenomenon: absence of stochastization in a system of nonlinear oscillators. I believe this to be a very important new field of research. After Boltzmann, Gibbs, Bohr, we understand better the basically statistical nature of our Universe. Still there is a lot of organization in it. "Solutions" are seen everywhere, and condensation is no less important than dissipation. If sufficiently many differential equations of mathematical physics, properly delinearized, show the same pattern as the Korteweg-de Vries equation, this might help us to balance our philosophy.

XXII. Nonlinear functional analysis (F. Browder). The very intensive development of the study of nonlinear operators of various types in Banach spaces over the past decade (especially in connection with applications to problems in partial differential equations, integral equations, optimization theory, and control theory, among other areas) has led to the crystallization of a number of significant problems of great technical difficulty in connection with specific classes of nonlinear operators.

(A) *Generalizations of the Schauder Theorem.* Let X be a Banach space, f a continuous mapping of X into itself. Suppose that there exists a compact subset A of X such that for each compact set K of X and each neighborhood U of A, there exists an integer $n(K, U)$ such that for $n \geqslant n(K, U)$, $f^n(K) \subset U$. Does f have a fixed point in A?

No counterexamples are known, and various special cases have been established. It was announced in [20] that if there exists $k < 1$ such that for all x in X, $\operatorname{dist}(f(x), A) \leqslant k \operatorname{dist}(x, A)$, then f has a fixed point. The result was established in Hilbert spaces in [10] and for general Banach spaces in [28], in both cases in the following slightly sharper form: Suppose that there exist sequences $\{\epsilon_j\}$ and $\{\delta_j\}$ tending to zero with $\epsilon_j < \delta_j$ for each j such that f maps the δ_j-neighborhood of A into the ϵ_j-neighborhood of A. Then f has a fixed point in A. If f is locally compact, the existence of a fixed point was established in [10], [11] and if f is locally condensing in [27]. For f of class C^1, the existence of a fixed point was shown in [28] if there exists an iterate f^m which is condensing on a neighborhood of A.

(B) *Nonexpansive mappings.* Let X be a Banach space, C a closed convex subset of X, f a mapping of C into X which is nonexpansive, i.e. for all x and u in C,

$$\|f(x) - f(u)\| \leqslant \|x - u\|.$$

The following two questions can be posed:

(1) Suppose that C is weakly compact and f maps C into C. Does f have a fixed point?

(2) Suppose that X is reflexive and C is weakly compact and satisfies the condition of normal structure (i.e. for each nontrivial closed convex subset C_1 of C there exists a point y of C_1 such that for some $d < \operatorname{diam}(C_1)$, C_1 is contained in the d-ball about y). Then: Is $(I - f)(C)$ closed in X?

Question 1 was answered in the affirmative in [12], [21], and [24] if X is uniformly convex or more generally if X is reflexive and C has normal structure. Similarly, the answer to Question 2 is affirmative if X is uniformly convex. For affine mappings an affirmative answer to Question 1 without assumptions upon X is provided by a well-known theorem of Ryll-Nardzewski.

(C) *Maximal monotone mappings in reflexive Banach spaces.* Let X be a Banach space, T a mapping of X into 2^{X^*}, with X^* the conjugate space of X. T is said to be monotone if for each w in $T(u)$, y in $T(x)$, $\langle w - y, u - x\rangle \geqslant 0$ where $\langle w, u\rangle$ denotes the pairing between X^* and X. A monotone mapping T is said to be maximal monotone if it is maximal in the sense of inclusion of graphs among monotone maps from X to 2^{X^*}. If X is reflexive, it has to be shown that a maximal monotone mapping T is surjective if and only if T^{-1} is locally bounded ([15], [30]). In order to apply this conclusion to the proof of existence theorems for variational inequalities for nonlinear parabolic and hyperbolic differential operators, it is of importance to determine general criteria for the sum $T + T_1$ of two maximal monotone mappings to be itself maximal monotone ([5]). Sufficient conditions have been established involving the quasi-boundedness of T_1, or local inequalities of the form

$$\|T_1(u)\| \leqslant k\|T(u)\| + k_1$$

with $k < 1$. Stronger results are necessary in order that the abstract theory be able to treat the whole range of cases now treated by ad hoc arguments.

A related problem concerns the extension of the notion of pseudomonotonicity. Let T be a maximal monotone mapping from X to 2^{X^*}, and let T_0 be a bounded finitely continuous mapping from X to X^*. Suppose that T_0 is

T-pseudomonotone in the following sense: For any sequence $\{u_j\}$ in X converging weakly to u in X and a bounded sequence $\{y_j\}$ with y_j in $T(u_j)$, suppose that $\overline{\lim}(T_0(u_j), u_j - u) \leqslant 0$. Then $T_0(u_j)$ converges weakly to $T_0(u)$ and $(T_0(u_j), u_j - u) \longrightarrow 0$.

Suppose that $(T + T_0)$ is coercive. Is it then true that $(T + T_0)$ is surjective?

An affirmative result was established by Brezis [4] if T is linear. The general case remains open.

(D) *Maximal monotone mappings in nonreflexive Banach spaces.* Let X be a general Banach space (not necessarily reflexive) and let T be a maximal monotone mapping from X^* to 2^X. Suppose that T^{-1} is locally bounded. Is T surjective?

Under the somewhat stronger assumption that T is coercive (i.e. that the set

$$\{u \mid \text{There exists } w \text{ in } T(u) \text{ such that } \langle w, u\rangle \leqslant c\|u\|\}$$

is bounded for each constant c), it was proved in [8] that T is surjective. In [6] it was shown that the conjectured result is valid if T is also maximal monotone from X^* to $2^{X^{**}}$. The conjectured result (like the results we have just mentioned and a number of generalizations to mappings between paired Banach spaces X and Y) would have interesting applications to the theory of strongly nonlinear Hammerstein integral equations. (See also [22], [23].)

In [7], the following theorem was proved in reflexive Banach spaces. Let L_0 and L_1 be monotone linear mappings from X to X^* with $L_0 \subseteq (L_1)^*$. Then there exists a maximal monotone linear mapping L with $L_0 \subseteq L \subseteq (L_1)^*$. The question can now be posed: Are there results of this type valid for every Banach space X? An affirmative answer would have important applications to the study of singular Hammerstein integral equations.

(E) *Ergodic theory for nonlinear mappings.* Let X be a uniformly convex Banach space, U a nonexpansive mapping of the unit ball B of X into itself. Form

$$S_n(x) = \frac{1}{n+1} \sum_{j=0}^{n} U^j(x).$$

Does $S_n(x)$ converge weakly to a fixed point of U?

For X a Hilbert space, this result was obtained by Baillon [1] and a corresponding result [2] obtained for one-parameter semigroups $U(t)$ of nonexpansive mappings in a Hilbert space. The proofs use the Hilbert space structure intensively. (See also the results of Bruck [17] on the method of steepest descent.)

(F) *φ-accretive mappings.* The study of monotone mappings from a Banach space X to its conjugate space X^* is paralleled by the study of accretive mappings from a Banach space X to itself. A continuous mapping f of X into X is said to be accretive if for all $\lambda > 0$ and all x and u,

$$\|(u - x) + \lambda(f(x) - f(u))\| \geqslant \|x - u\|.$$

The theory of nonlinear semigroups in Banach spaces (which generalizes the corresponding Hille-Yosida theory in the linear case) identifies the accretive mappings and their multivalued generalizations as negatives of infinitesimal generators of continuous one-parameter semigroups of nonexpansive mappings. The concept of accretivity may be defined in a fashion more akin to the conventional definition for monotone mappings by the inequality

$$\langle f(x) - f(u), J(x - u)\rangle \geqslant 0$$

where J is the duality mapping of X into X^* given by the conditions

$$\|J(x)\| = \|x\|,$$
$$\langle J(x), x\rangle = \|x\|^2, \quad \text{(at least if } X^* \text{ is strictly convex, and with } J \text{ multi-valued in more general cases).}$$

Let X and Y be Banach spaces, and suppose given a mapping φ of X into Y^* such that φ is surjective and uniformly continuous on bounded sets, $\|\varphi(x)\| = \|x\|$, and $\varphi(\lambda x) = \lambda\varphi(x)$ for $\lambda > 0$. A continuous mapping f of X into Y is said to be (strongly) φ-accretive if for all x and u in X,

$$\langle f(x) - f(u), \varphi(x - u)\rangle \geqslant c\|x - u\|^2$$

for a fixed positive constant c. Is each strongly φ-accretive mapping f surjective?

Affirmative answers have been given in [10] and [14] in the special cases where f is locally Lipschitzian or if f satisfies a Lipschitz condition of order ½ and Y has a C^2 norm on the complement of the origin. Some of these results depend upon the theory of normally solvable mappings. The general case remains open and is of special interest because it furnishes a potential link between the two special cases of monotone and accretive mappings whose study is based upon essentially different modes of argument.

(G) *Constructive solvability.* Let X be a reflexive Banach space, f a continuous monotone mapping of X into X^* which is bounded and coercive. It is known that there exists a solution u_0 of the equation $f(u_0) = 0$. Suppose that f^{-1} is single-valued and has a known modulus of continuity. Does there exist a way of proving the existence theorem constructively, i.e. giving a sequence $\{x_n\}$ such that $\|x_m - x_n\| \leqslant \epsilon_{m,n}$ with $\epsilon_{m,n}$ given explicitly such that $x_{m,n}$

converges to u_0? In particular, is this true for the Galerkin approximations if X is separable and f is strongly monotone in some suitable sense?

If X is a Hilbert space H, this has been done in [10], [16] as well as the corresponding generalization to accretive mappings by using a semigroup approach. This method does not extend to the monotone case if X is not a Hilbert space.

(H) *Nonconvex functions and sets.* Find a single systematic theory which includes and illuminates a variety of diverse results concerning the passage from convex functions and sets in Banach spaces to nonconvex functions and closed sets. This includes such cases as: the normal solvability theorems [13], [14], the conditions for finding trajectories of a differential equation in a closed nonconvex set [26], the existence of generic extrema for nonconvex functions [3], [19], and others ([18], [9], [25], [29]). The hope is to find a principle of more general value than the particular mutual implications of the individual results, and the fact that all are derived by applying arguments on ordered sets in Banach spaces.

BIBLIOGRAPHY

1. J. B. Baillon, *Un théorème de type ergodique pour les contractions non-linéaires dans un espace de Hilbert,* C. R. Acad. Sci. Paris (1975).

2. J. B. Baillon and H. Brezis, *Une remarque sur les comportement asymptotique des semigroupes non-linéaires,* C. R. Acad. Sci. Paris (1975).

3. J. Baranger, *Existence des solutions pour les problèmes d'optimization nonconvexe,* J. Math. Pures Appl. **52** (1973), 377–405.

4. H. Brezis, *Perturbations non linéaires d'opérateurs maximaux monotones,* C. R. Acad. Sci. Paris Sér. A–B **269** (1969), A566–A569. MR **40** #3351.

5. ———, *Opérateurs maximaux monotones et semigroupes de contractions dans les espaces de Hilbert,* North-Holland Math. Studies, no. 5, Notas de Matemática (50), North-Holland, Amsterdam; American Elsevier, New York, 1973. MR **50** #1060.

6. H. Brezis and F. E. Browder, *Nonlinear integral equations and systems of Hammerstein type,* Advances in Math. **18** (1975), 115–147.

7. ———, *Singular Hammerstein equations and maximal monotone operators,* Bull. Amer. Math. Soc. (to appear).

8. ———, *Maximal monotone operators in nonreflexive Banach spaces and nonlinear integral equations of Hammerstein type,* Bull. Amer. Math. Soc. **80** (1974), 82–88.

9. A. Brøndsted, *On a lemma of Bishop and Phelps,* Pacific J. Math. **55** (1974), 335–341.

10. F. E. Browder, *Nonlinear operators and nonlinear equations of evolution in Banach spaces,* Proc. Sympos. Pure Math., vol. 18, part 2, Amer. Math. Soc., Providence, R. I., 1976.

11. ———, *Asymptotic fixed point theorems,* Math. Ann. **185** (1970), 38–60. MR **43** #1165.

12. ———, *Nonexpansive nonlinear operators in a Banach space,* Proc. Nat. Acad. Sci. U.S.A. **54** (1965), 1041–1044. MR **32** #4574.

13. ———, *Normal solvability for nonlinear mappings into Banach spaces,* Bull. Amer. Math. Soc. **77** (1971), 73–77. MR **42** #5114.

14. ———, *Normal solvability and φ-accretive mappings of Banach spaces,* Bull. Amer. Math. Soc. **78** (1972), 186–192. MR **46** #6113.

15. ———, *Nonlinear monotone and accretive operators in Banach spaces,* Proc. Nat. Acad. Sci. U.S.A. **61** (1968), 388–393. MR **44** #7389.

16. F. E. Browder, *On the constructive solvability of nonlinear equations*, J. Functional Analysis (to appear).

17. R. H. Bruck, *Asymptotic convergence of nonlinear contraction semigroups in Hilbert space*, J. Functional Analysis **18** (1975), 15–26.

18. I. Ekeland, *Sur les problèmes variationnels*, C. R. Acad. Sci. Paris Sér. A–B **275** (1972), A1057–A1059. MR **46** #9768.

19. I. Ekeland and G. Lebourg, *Generic Fréchet-differentiability and perturbed optimization problems in Banach spaces* (to appear).

20. R. L. Frum-Ketkov, *Mappings into a Banach sphere*, Dokl. Akad. Nauk SSSR **175** (1967), 1229–1231 = Soviet Math. Dokl. **8** (1967), 1004–1006. MR **36** #3181.

21. D. Göhde, *Zum prinzip der kontraktiven Abbildung*, Math. Nachr. **30** (1965), 251–258. MR **32** #8129.

22. J. P. Gossez, *Opérateurs monotones non linéaires dans les espaces de Banach non reflexifs*, J. Math. Anal. Appl. **34** (1971), 371–395. MR **47** #2442.

23. ———, *On the range of a coercive maximal monotone operator in a non-reflexive Banach space*, Proc. Amer. Math. Soc. **35** (1972), 88–92. MR **45** #7544.

24. W. A. Kirk, *A fixed point theorem for mappings which do not increase distances*, Amer. Math. Monthly **72** (1965), 1004–1006. MR **32** #6436.

25. W. A. Kirk and J. Caristi, *Mapping theorems in metric and Banach spaces* (to appear).

26. R. H. Martin, *Differential equations on closed subsets of a Banach space*, Trans. Amer. Math. Soc. **179** (1973), 399–414. MR **47** #7537.

27. R. D. Nussbaum, *Asymptotic fixed point theorems for local condensing mappings*, Math. Ann. **191** (1971), 181–195. MR **45** #7554.

28. ———, *Some asymptotic fixed point theorems*, Trans. Amer. Math. Soc. **171** (1972), 349–375. MR **46** #9817.

29. R. R. Phelps, *Support cones in Banach spaces and their applications*, Advances in Math. **13** (1974), 1–19. MR **49** #3505.

30. R. T. Rockafellar, *Local boundedness for nonlinear, monotone operators*, Michigan Math. J. **16** (1969), 397–407. MR **40** #6229.

XXIII. Singular integral operators (A. P. Calderón, Y. Meyer). Let $a(y)$ be a Lipschitz function with compact support on the line, and for a nonnegative integer n, consider the singular integral operator $T_n(f)$ given by

$$T_n(f)(x) = \int \left\{ \frac{a(x) - a(y)}{x - y} \right\}^n \frac{f(y)}{x - y}\, dy$$

for functions f in L_2. If $n = 0$, this is simply the Hilbert transform which is a bounded linear mapping on L_2. The corresponding result for $n = 1$ was established by Calderón in 1965 using complex methods, and more recently, Coiffman and Meyer have established this result for $n = 2$. The methods applied in these cases do not extend to higher n, and new ideas and techniques will have to be applied.

XXIV. Fourier analysis (C. Fefferman).

Problem. Develop substitutes for Lebesgue measure for use in Fourier analysis on R^n.

In one variable, Lebesgue measure is a key concept on which essential results of classical Fourier analysis are based: Lebesgue's theorem on differentiation of integrals, L^p boundedness of the Hilbert transform, the Littlewood-Paley inequalities, etc. In particular, the Hilbert transform is so closely tied to Lebesgue measure that an arbitrary function and its Hilbert transform may be said to behave more or less the same outside a set of small Lebesgue measure.

However, Lebesgue measure seems far less suited to n-dimensional Fourier analysis. Already differentiation of the integral in its sharp form fails in R^2; in fact, $\lim_{R\to\{x\}} (1/|R|) \int_R f(y)\,dy$ may exist almost nowhere, even for $f \in L^\infty$, where R denotes a small rectangle of arbitrary eccentricity and direction. This sad fact comes from the Kakeya phenomenon, i.e., the existence of sets $E \subseteq R^2$ of arbitrarily small area containing unit line segments in all directions; and it leads directly to the future of L^p boundedness of natural analogues of the Hilbert transform in R^n. Thus, much of our deep knowledge of Fourier series must so far be confined to one dimension.

On the other hand, the Fourier transform has good properties in R^n that do not show up in R^1. For instance, the Fourier transform of a function $f \in L^{4/3-\epsilon}(R^2)$ may be restricted to the unit circle $S^1 \subseteq R^2$, and even belongs to $L^{4/3}(S^1)$. The analogous "restriction theorem" is false if we replace the circle by a straight line-segment. This suggests that we are nowhere near a sharp answer to the elementary question: How big is the Fourier transform of an L^p function on R^n?

One possible way to understand R^n is to develop notions of "size" of sets and functions in R^n, defined in terms of efficient coverings by long thin rectangles. Measured by "size", the Kakeya set E might be quite large; this might remove some flavor of paradox from the Kakeya phenomenon, and perhaps lead to boundedness of n-dimensional Hilbert transforms on suitable function spaces. In terms of coverings with thin rectangles, circles are clearly larger than line-segments, so that "restriction" phenomena might also be explained.

Of course, to be useful, any notion of "size" has to be computable in nonpathological cases.

XXV. Banach spaces (P. Enflo).

(A) In the theory of general spaces one of the most interesting problems is to decide whether every infinite-dimensional Banach space has infinite-dimensional subspaces with certain regularity properties. Dvoretsky's theorem which says that an infinite-dimensional Banach space has subspaces almost isometric to n-dimensional Euclidean space for every n, is one of the deepest and most useful results in Banach space theory. So, to better understand general Banach spaces, some generalization of this result to infinite-dimensional subspaces is important. The most obvious generalization that every infinite-dimensional Banach space contains a subspace isomorphic to l_p or c_0 is not true as was shown by Tzirelson. It is likely that something weaker is true and the following problems seem approachable.

(1) Does every infinite-dimensional Banach space contain a subspace with an unconditional basis–a seq. $\{e_n\}$ s.t. for every x, $x = \Sigma a_n e_n$ no matter in which order the summation is done?

(2) Does every infinite-dimensional Banach space contain an isomorph of l_1 or c_0 or a reflexive Banach space?

A recent characterization by Rosenthal of spaces containing l_l has reduced both these problems and shown that they have strong connections with problems in hard classical analysis.

(B) In the theory of classical Banach spaces, it is hard to single out particular problems as being important for the development of the theory. Recent development of the theory has revealed strong connections with harmonic analysis, probability theory, topological dynamics, integral geometry, and so on, and an even stronger interplay between questions in Banach spaces and questions in more classical analysis seems to be a very fruitful direction to continue. As examples, two problems could be mentioned; they are arbitrary choices and many other choices are possible.

(1) Does H_1 have an unconditional basis?

(2) If $X \subset L_1(0, 1)$ and there is a continuous linear projection onto X, is X isomorphic to l_1 or $L_1(0, 1)$?

(C) The well-known problem whether every operation on a Banach space has a nontrivial invariant subspace is obviously an important problem in operator theory. There are also Banach space aspects of this problem. To what extent does the regularity of the Banach space affect the answer? Several positive partial results are true in all Banach spaces. Some classes of operators for which the answer is positive are not well defined in all Banach spaces, and there are also indications that conditions like reflexivity of the Banach space play a role in this problem. It should also be mentioned that there are many other important directions for Banach space theory.

XXVI. Probability (P. Cartier). So far only two classes of stochastic processes are well understood: Markov processes and Gaussian processes. At the intersection of these two classes lies Brownian motion, one of the most studied processes. Here are a few possible directions for future inquiries.

(1) *Morkov processes in more than one "time" parameter.* The recent discovery of Nelson's process (free Markov field), the earlier investigations by P. Levy and MacKean of Brownian process with many parameters point out the necessity of studying Markov processes parametrized by points in Euclidean spaces. One should possibly begin with Gaussian processes.

The main challenge is to define and exploit good Markov properties and to find substitutes for the powerful methods of martingales and stochastic integrals. Various classes of random distributions ought to be studied in some detail.

(2) *Nonlinear differential and integral equations.* There are alrady some indications (work done by Ueno, Nagasawa and others) that some nonlinear functional equations might be efficiently solved by probabilistic methods. To that class of equations certainly belongs Boltzmann's equation, whose connection with the "master equation" in statistical mechanics is well known. There is also the solution by H. Rost of equations of the type $\partial\psi/\partial t = \Delta\psi$ in the region where ψ is positive (free boundaries) by means of his "methode de remplissage".

(3) *Regularity properties of non-Gaussian processes.* The local properties of Gaussian processes are well understood after recent work by Fernique, Dudley *et al.* Comparatively little is known for non-Gaussian processes. This field should provide a good test for the new methods of L. Schwartz (radonification problem). Among these processes I include random distributions of various kinds.

(4) *Random measure, point processes.* We know barely more than a few basic facts and scattered results. Recent work by Lenard for instance indicates that even the foundations need reexamination.

What is required is the isolation of good classes of such processes, distinguished by their potential usefulness in the numerous applications (biology, statistics, statistical mechanics) and their tractability. Connections with problems headed under (2) above may be expected.

XXVII. Mathematical physics (A. Wightman). The following problems have been chosen somewhat arbitrarily but have in common that they are likely to be of some significance for both mathematics and physics.

Problems of classical statistical mechanics. In statistical mechanics the success of a macroscopic description of the behavior of systems composed of microscopic particles is attributed to three causes: (a) the predictions of the statistical theory become valid in the *thermodynamic limit* when the number of particles becomes arbitrarily large; (b) one restricts one's attention to *macroscopic observables* which are not sensitive to the chaotic microscoptic complexity of the system; and (c) the laws governing the motion of the particles have only a few integrals of motion. Otherwise their behavior is characterized by *ergodicity*. The relative importance of these three factors for interesting Hamiltonian systems is still not understood. We know that in many Hamiltonian systems there are extra integrals of motion whose existence has been established by Kolmogorov, Arnold, and Moser [1], [2]. Thus we have the problem:

(1) *What is the role of the Kologorov-Arnold-Moser extra integrals of*

motion in the thermodynamic limit? In particular, when are the predictions of statistical mechanics invalidated by the existence of such extra integrals of motion?

The typical model of classical statistical mechanics is a (reversible!) Hamiltonian system. There exist numerous macroscopic transport theories which describe irreversible behavior in terms of a small number of parameters. However, with few exceptions the latter have been derived from the former only by heuristic arguments. Hence, the problem:

(2) *What is the relation between a microscopic reversible theory of particles and a macroscopic irreversible deterministic transport theory with a state expressible in terms of a small number of parameters?*

The existence of the thermodynamic limit has been established for the equilibrium states of certain classical Hamiltonian systems, and it is believed to hold for many more [3]. The uniqueness of the equilibrium state at high temperatures has been fairly generally established [3, Chapter 4], and in a few cases the existence of a phase transition to a two-phase region has been proved. However, in general little is known. Thus, the problem:

(3) *What is the qualitative character of the set of equilibrium phases for a classical Hamiltonian system? For example, is it usually a piecewise smooth manifold?*

Problems of n-body quantum mechanics. The problem of asymptotic completeness has already been cited in connection with relativistic quantum field theory. In the Schrödinger n-body problem, asymptotic completeness has been established for repulsive forces and for forces differing from repulsive by sufficiently small perturbations. However, for general forces the spectrum of bound states of subsystems becomes quite involved and asymptotic completeness is open.

(1) *Prove asymptotic completeness in the n-body problem for reasonably general forces.*

Problems of quantum statistical mechanics. Quantum statistical mechanics includes the many-body theory applicable to ordinary matter which shows a rich collection of properties, some simple, some intricate. It is characteristic of the current situation that very few of these properties have been derived in a mathematically satisfactory way directly from first principles. I single out two characteristic problems of this kind:

(1) *Derive the existence of crystals from first principles assuming some reasonable set of forces, for example, the Coulomb forces occurring in a finite set of species of atoms.*

(2) *Derive the salient properties of the solid and liquid states from first principles, for example, metallic conductivity, superconductivity and superfluidity.*

Problems of relativistic quantum field theory. In the present state of constructive quantum field theory, there seems to be a general feeling that the necessary ideas are either in hand or near at hand for a mopping up of all super-renormalizable models. However, this program has so far out turned out to be as easy to carry out as it appeared several years ago. Even if this were done (and it would be very educational indeed to carry out the program to the end to see the qualitative nature of the models in the strong coupling limit), it would leave untouched the great problems of renormalizable theories in four-dimensional space-time, for which it is far from obvious what ideas will be necessary. I list as my first problem a typical theory in four dimensions, the best quantum field theory we know from the point of view of agreement with experiment.

(1) *Prove the existence of a solution of the quantum electrodynamics of massive spin one-half particles in four-dimensional space-time.*

In the general discussion of the relation between local quantum theory and relativistic quantum field theory above, it was pointed out that there is no general theory relating them.

(2) *Establish a relation between local quantum theory and relativistic quantum field theory.*

There is a problem of local quantum theory put forward by Dyson [4], which is so intriguing and of such potential significance to a possible quantum theory of gravitation that we repeat it here.

(3) *Find a form of local quantum theory compatible with general relativity.*

Renormalization is the key technical complication of relativistic quantum field theory. Dynamical instability, and its accompanying symmetry breaking are among the features of the theory that made it most promising as a possible description of Nature. The Euclidean functional integral for the Schwinger functions of Lagrangian field theories is the most direct expression we have of the content of the theories. This suggests:

(4) *Find a form of renormalization theory and the theory of dynamical instability expressible directly in terms of the Euclidean functional integral for the Schwinger functions of Lagrangian field theories.*

One of the crowning achievements of the axiomatic quantum field theory that starts with Green's functions and retarded functions is so-called structure analysis [5], [6]. This theory comes in two stages. First, it analyzes the contributions to the perturbation series for the Green's functions of Lagrangian field theories. Such contributions can be labelled by graphs (the so-called Feynman diagrams) and the graphs in turn classified by their connectivity (for a connected graph, this is characterized by how many lines have to be cut to disconnect it). Crudely, the partial sum of the contributions with nth-order connectivity is called the n-particle irreducible Green's functions. (Refinements which would distinguish different kinds of lines are being ignored.) Second, structure analysis gives sets of nonlinear functional equations for the n-particle irreducible Green's functions. These nonlinear equations have the property that their iterative solutions yield precisely the partial sums of contributions to the perturbation series appearing in the first part of the analysis. The nonlinear functional equations are therefore candidates for a nonperturbative analysis of the structure of Green's functions. Specific Lagrangian field theories would be characterized by the vanishing of all irreducible Green's functions beyond a certain order. That sets up the last problem:

(5) *Derive Lagrangian field theory by an application of structure analysis.*

BIBLIOGRAPHY

1. V. I. Arnol'd and A. Avez, *Problèmes ergodiques de la méchanique classique*, Monographies Internat. Math. Modernes, no. 9, Gauthier-Villars, Paris, 1967; English transl., Benjamin, New York, 1968. MR **35** #334; **38** #1233.

2. J. Moser, *Stable and random motions in dynamical systems*, Ann. of Math. Studies, no. 77, Princeton Univ. Press, Princeton, N. J., 1973.

3. D. Ruelle, *Statistical mechanics: rigorous results*, Chapter 7, Benjamin, New York, 1969. MR **44** #6279.

4. F. J. Dyson, *Missed opportunities*, Bull. Amer. Math. Soc. **78** (1972), 635–652.

5. K. Symanzik, *On the many-particle structure of Green's functions in quantum field theory*, J, Mathematical Phys. **1** (1960), 249–273. MR **26** #3406.

6. ———, *Grundlagen und gegenwärtiger Stand der feldgleichungsfreien Feldtheorie*, in Werner Heisenberg und die Physik unserer Zeit, Vieweg, Braunschweig, 1961, pp. 275–298.

Proceedings of Symposia in Pure Mathematics
Volume 28, 1976

HILBERT'S FIRST PROBLEM: THE CONTINUUM HYPOTHESIS

Donald A. Martin

ABSTRACT

The present status of the continuum hypothesis and the generalized continuum hypothesis is discussed. Both independence results and recent positive theorems are listed. An analysis is given of the bearing on the continuum problem of work on large cardinals and projective sets.

1. INTRODUCTION

Hilbert's First Problem is in the curious position that there is serious disagreement as to whether it has been solved and there is related disagreement as to whether the problem, in the natural way of understanding it, is a mathematical problem at all.

The First Problem is to settle the famous continuum hypothesis (CH) of Cantor. CH asserts that the cardinal number of the continuum (the set of all real numbers) is $\aleph_1$, the smallest uncountable cardinal number. Equivalently, CH states that there are the same number of real numbers as countable ordinal numbers.

The generalized continuum hypothesis (GCH) asserts that, for every infinite cardinal number $\aleph_\alpha$, $2^{\aleph_\alpha} = \aleph_{\alpha+1}$. In other words, the cardinal number of the collection of all subsets of a set of cardinality $\aleph_\alpha$ is the smallest cardinal number greater than $\aleph_\alpha$. CH is the special case $2^{\aleph_0} = \aleph_1$.

AMS(MOS) subject classifications (1970). Primary 02K25, 04A30; Secondary 02K05, 02K30, 02K35, 04A15.

Hilbert [13] himself attempted to prove CH, but he was unable to carry the proof through to completion. In 1938 Gödel [9,10] attacked the problem in a very surprising manner. He showed that, if the standard Zermelo-Fraenkel (ZFC) axioms for set theory are consistent, then there can be no refutation of GCH from these axioms. (There is a relation -- however tenuous -- between Gödel's proof and Hilbert's unsuccessful attempt to prove CH.) Gödel [11] conjectured that the formal ZFC axioms do not suffice to prove CH either and thus that CH is formally undecidable in the theory ZFC. In 1963 Cohen [4] proved that this is the case.

Where do these results leave Hilbert's First Problem? From Hilbert's formalist standpoint CH is an assertion of ideal mathematics rather than of real mathematics. Hilbert, however, presumably thought that the formal axioms of set theory were strong enough to settle such propositions as CH. It is unclear whether he would regard the Gödel and Cohen results as a solution to his problem. Gödel [11] holds that the meaning of CH is independent of formal axioms and that independence proofs only show the weakness of our current axioms. In his view, CH is either true or false, and the problem of discovering which is still with us. Cohen [5] espouses a formalist position but still holds that we may yet be led to a decision about CH, specifically that we may be led to accept the negation of CH as an axiom.

2. CH AND GCH IN ZFC

I shall return to these essentially philosophical questions later, but first I wish to discuss the mathematical work related to CH and GCH which grew out of the independence proofs.

Gödel's work, for reasons which are unclear, was followed by twenty years of stagnation in the field of set theory. Though his proof had introduced new concepts and techniques in set theory and an interesting new proposition, the axiom of constructibility, little use was put to these new ideas until after the work of Cohen. Instead his theorem had a negative effect on some branches of set theory. For example, Gödel showed that certain basic propositions considered in classical descriptive set theory could not be proved in ZFC. Hence there was some danger that many of the important questions in this area were formally undecidable. This tended as a practical matter to discourage work in the field, and few basic advances were made (excepting Addison [1,2] and work of Choquet) until recently.

Cohen's proof had quite the opposite effect. It ushered in a period of intense activity in set theory. Cohen's methods were applied to every imaginable set theoretic question, and a great number of questions were shown to be undecidable in ZFC. There was even a revival of interest in Gödel's axiom of constructibility, and many important consequences of this axiom were deduced (mostly by Ronald Jensen).

The effect of independence proofs on mathematics is not entirely negative. For example, there are several cases of theorems having been proved assuming CH where independence proofs allow one to eliminate the hypothesis CH. For (a class of) example(s), given any proposition ϕ of second order number theory (loosely speaking: any proposition about integers and reals only -- e.g., not about arbitrary sets of reals) if ϕ is provable from ZFC + CH then ϕ is provable from ZFC (Platek [24], S. Kripke, J. Silver). A similar theorem is true about the negation of CH, though one must restrict ϕ to be an assertion about integers alone. Such theorems are special cases of a general absoluteness technique. Suppose one can prove in ZFC that a proposition ψ implies another proposition ϕ. It is sometimes the case that, given any model of ZFC, one can extend it by Cohen's methods to a model of ZFC + ψ. In the larger model ϕ must be true. But if ϕ can be proved to be sufficiently absolute, one can conclude that ϕ must be true in the original model and hence that ZFC implies ϕ. Absoluteness of ϕ means that the truth value of ϕ is preserved under all or a wide class of such Cohen extensions of models.

Before the work of Cohen and Gödel, one important restriction on the operation $\aleph_\alpha \to 2^{\aleph_\alpha}$ was known. This result of König [15] states that $2^{\aleph_\alpha}$ cannot have cofinality $\leq \aleph_\alpha$. The cofinality of a cardinal κ is the least cardinal λ such that a set of cardinality κ is always the union of λ sets of cardinality smaller than κ. In particular $2^{\aleph_0}$ cannot equal $\aleph_\omega$, where ω is the first infinite ordinal number.

Cohen and R. Solovay showed (using models constructed by Cohen) that König's restriction is the only restriction on $2^{\aleph_0}$. Easton [7] attacked the GCH using Cohen's methods. He almost showed that the operation $\aleph_\alpha \to 2^{\aleph_\alpha}$ can be anything consistent with König's theorem. The "almost" basically involves the problem of singular cardinals. κ is singular if the cofinality of κ is smaller than κ. Easton could produce models of ZFC with $\aleph_\alpha \to 2^{\aleph_\alpha}$ whatever he wished on the class of regular (non-singular) $\aleph_\alpha$ but could not at the same time control the values at singular $\aleph_\alpha$. For example, it is not known whether it is consistent with ZFC that $2^{\aleph_n} = \aleph_{n+1}$ for $n < \omega$ and $2^{\aleph_\omega} > \aleph_{\omega+1}$. (One restriction in addition to König's has been known for some time: If κ is singular and $2^\gamma = \lambda$ for all sufficiently large $\gamma < \kappa$, then $2^\kappa = \lambda$.)

One of the major problems in post-Cohen set theory has been this <u>singular cardinals problem</u>. Most workers have felt that the theorems of Easton could be extended to singular cardinals. Work of Prikry, Silver, and Magidor has led to some consistency results, but they are partial and the "consistency" is relative to theories much stronger than ZFC. Very recently Jack Silver astonished the set-theoretic world by essentially settling the singular cardinals problem for cardinals of cofinality greater than $\aleph_0$ -- and settling it in the "wrong" direction. A consequence of Silver's theorem (a theorem of ZFC alone!) is that if $\aleph_\alpha$ is singular of cofinality $> \aleph_0$ and if $2^{\aleph_\beta} = \aleph_{\beta+1}$, for all $\beta < \alpha$ then $2^{\aleph_\alpha} = \aleph_{\alpha+1}$. Galvin and Hajnal [8] have extended Silver's work to compute, in a sense, absolute bounds on $2^{\aleph_\alpha}$ for certain singular $\aleph_\alpha$. The problem of singular cardinals of cofinality $\aleph_0$ remains, despite this breakthrough, as puzzling as ever.

3. LARGE CARDINAL AXIOMS

Although the ZFC axioms are insufficient to settle CH, there is nothing sacred about these axioms, and one might hope to find further axioms which seem clearly true of our notion of set (in the same way the ZFC axioms appear clearly true) and which do settle CH.

In the time since Cohen, there has been a great deal of research on one class of candidates for new axioms: the so-called <u>large cardinal</u>

axioms. (This area by no means began after Cohen, however. The subject is much older, and its revival occurred before Cohen's work. For example, [14] and [25] are pre-Cohen.) A large cardinal axiom is, roughly speaking, an assertion that cardinal numbers exist having some property P, such that one can prove that only very large cardinals can have P. Examples of such P are inaccessibility and measurability. κ is inaccessible if κ is regular and $\lambda < \kappa$ implies $2^{\lambda} < \kappa$. κ is measurable if there is a set A of cardinality κ and a function μ defined on all subsets of A such that μ takes only 0 and 1 as values, $\mu(A) = 1$, μ is 0 on singletons, $\mu(A-X) = 1 - \mu(X)$, and if $\mu(A_i) = 0$ for each $i \in I$ and cardinal $(I) < \kappa$ then $\mu(\cup A_i) = 0$.

The usual large cardinal axioms cannot be proved in ZFC if it is consistent. There are basically three sorts of arguments for accepting them: analogy with $\aleph_0$, reflection principles, and plausibility of consequences.

The argument from analogy with $\aleph_0$ goes as follows: $\aleph_0$ is inaccessible, measurable, etc. For each of these properties P it would be only by accident, as it were, that $\aleph_0$ should be definable as the unique infinite cardinal such that P (as it is an accident that man = featherless biped). Hence one would expect that larger cardinals having property P exist.

The argument from reflection principles starts with the usual notion of sets as generated by iteration of the power set operation. We start with R_0 = the empty set. Given R_α, for an ordinal number α, $R_{\alpha+1}$ is the collection of all subsets of R_α. For limit ordinals λ, $R_\lambda = \bigcup_{\beta<\lambda} R_\beta$. A set is anything which is a member of some R_α. The formal axioms of set theory are an attempt to describe this iterative construction. One thing we want the axioms to assert is that the construction does not stop too soon because the ordinal numbers are exhausted prematurely. The axiom of replacement is intended to assert that the ordinals do not run out at an unnatural point. Reflection principles are based on the idea that the class On of ordinal numbers is so large that, for any reasonable property P of the universe of all sets R_{On}, On is not the first stage α such that R_α has P. Examples of "reasonable" properties are first, second, and higher order properties. If this is right, then there should be stages R_α which look very much like R_{On}. It follows that there should be stages R_α and R_β which look very much alike. All important large cardinal axioms which have been studied are derivable from assertions that R_α and R_β exist which are difficult to distinguish. Of course, as the axioms become stronger their link with the basic principle becomes more and more tenuous.

A third argument for large cardinal axioms is that the theory their adoption gives is plausible and appealing. This is almost an empirical argument. I shall say more about this view later when I discuss another kind of axiom to which it better applies.

The reasons advanced for adopting large cardinal axioms are -- as the reader has surely noticed -- less compelling than the reasons for adopting ZFC. On the other hand, they are not completely negligible, and one should bear in mind that the axioms of ZFC (and the notion of set which they supposedly describe) are less compelling than the axioms of number theory.

What do large cardinal axioms tell us about the continuum hypothesis? Unfortunately they tell us very little. Large cardinal axioms tend to be absolute in the sense discussed earlier. If they are true in a model of ZFC they tend to be true in Cohen extensions of that model. This assertion can be made more precise as follows. A Cohen extension of a model M arises from an element P of M which is a partial ordering in M. A Cohen extension is <u>mild</u> with respect to a cardinal κ of M if cardinal $(P) < \kappa$ is true in M. All the standard large cardinal properties of κ are preserved under Cohen extensions mild with respect to κ. The truth value of CH, on the other hand can be changed by Cohen extensions where P has very small cardinal in M. This means that, for a large cardinal axiom A, there are models of ZFC + A + CH and models of ZFC + A + not CH, if there are models of ZFC + A at all.

Large cardinal axioms have, nonetheless, been successful in giving partial results about the GCH. Solovay [27] has shown that the existence of a so-called compact cardinal implies that $2^{\aleph_\alpha} = \aleph_{\alpha+1}$ for all sufficiently large singular strong limit cardinals (i.e., singular cardinals κ such that $\lambda < \kappa$ implies $2^\lambda < \kappa$) $\aleph_\alpha$.

4. PROJECTIVE SETS

We have at present no likely candidate for a new axiom which would settle CH. Let us despite this adopt for the moment the point of view that CH is a meaningful proposition and ask whether there is any information available which counts as evidence for or against the truth of CH.

Gödel [11] cites some facts which he believes are evidence against CH. He lists a number of known consequences of CH which he thinks are intuitively implausible. These consequences assert that very thin subsets of the real line exist of cardinality the continuum. Gödel says that such assertions are counterintuitive in a sense different from that in which the existence of Peano curves is counterintuitive. While Gödel's intuitions should never be taken lightly, it is very hard to see that the situation is different from that of Peano curves, and it is even hard for some of us to see why the examples Gödel cites are implausible at all.

Another way to look for evidence concerning CH is to examine simple cases. CH says that every set of reals is countable or has cardinality $2^{\aleph_0}$. We might test that assertion by looking at sets of reals which are, in some sense simple. If such simple sets do not provide counterexamples, we can try to see whether there are reasons to suspect that the simple sets considered are in a relevant way different from arbitrary sets of reals.

The simple sets I wish to consider are the projective sets. A set of reals is projective if it can be gotten from a Borel set via the operations of continuous image and complementation. The projective sets can be divided into a hierarchy as follows: Σ^1_1 (or analytic) sets are continuous images of Borel sets. Π^1_n sets are complements of Σ^1_n sets; Σ^1_{n+1} sets are continuous images of Π^1_n sets.

Now every Borel set is countable or has cardinal $2^{\aleph_0}$. The same is true of every Σ^1_1 set. Σ^1_2 sets are always unions of $\aleph_1$ Borel sets, so Σ^1_2 sets are countable or have cardinal $\aleph_1$ or $2^{\aleph_0}$. So far CH is confirmed. It is known to be consistent with ZFC that $2^{\aleph_0} > \aleph_1$ and that there are Σ^1_2 (even Π^1_1) sets of power $\aleph_1$. It is also consistent [12] that $2^{\aleph_0}$ be as large as you wish and that there are Π^1_2 sets of every cardinality $\leq 2^{\aleph_0}$. Thus ZFC furnishes us with no information about higher levels of the projective hierarchy.

Let us see if large cardinal axioms help. Let MC be the assertion that measurable cardinals $\aleph_0$ exist. Solovay has shown that MC implies that every infinite Σ^1_2 set has cardinal $\aleph_0$ or $2^{\aleph_0}$ [26]. Concerning Σ^1_3 sets, MC implies [18] that every Σ^1_3 set has cardinal $\aleph_0$, $\aleph_1$, $\aleph_2$ or $2^{\aleph_0}$.

Can we regard these facts about Σ^1_1 and Σ^1_2 sets as evidence for CH? I do not think we can. For the results about the cardinalities of Σ^1_i sets, i = 1, 2, are corollaries to stronger results. Every Σ^1_i set, i = 1, 2, is countable or has a perfect subset (assuming MC for i = 2). Now, by a simple application of the axiom of choice, there exists an uncountable set with no perfect subset. Thus, while our simple sets have the cardinalities required by CH, this is so because they have an <u>atypical</u> property, the perfect subset property.

We might try different formulations of CH. For example, CH says that every well-ordering of a subset of the reals has order type $< \omega_2$, the second uncountable initial ordinal. The notions of projective and Σ^1_n relations can be defined in the obvious way, and we can ask about the order type of Σ^1_n well-orderings. The relevant theorems are that Σ^1_i well-orderings are countable for i = 1 or 2, assuming MC for i = 2. In other words, the evidence suggests that simple well-orderings are countable and cannot even have order type ω_1. Once again our simple sets have proved atypical.

There is a third formulation of CH which is more promising. The negation of CH says that there is a <u>surjection</u>

$$f : R \rightarrow \omega_2 \ ;$$

where R is the reals and ω_2 is thought of as the set of its predecessors. Now a function

$$f : R \rightarrow \text{Ordinals}$$

is essentially the same as a prewellordering of R. To prewellorder a set, divide it into equivalence classes and well-order the equivalence classes. By the length of a prewellordering, I mean the order-type of the well-ordering of the equivalence classes. Now every Σ_1^1 prewell-ordering has countable length but there is a Π_1^1 prewellordering of R of length ω_1 (essentially the Lebesgue decomposition of R). This already shows that our simple sets are more typical with respect to pre-wellorderings than with respect to well-orderings.

Let δ_n^1 be the least ordinal > 0 not the length of a Σ_n^1 prewell-ordering of R. Then we have

$$\delta_1^1 = \omega_1 \quad ;$$

$$\delta_2^1 \leq \omega_2 \quad ;$$

$$MC \rightarrow \delta_3^1 \leq \omega_3 \quad .$$

(See [18] for the last two results.) Neither $\leq$ can be improved (in ZFC + MC) to $=$ since CH implies that $\delta_n^1 < \omega_2$ for all n. On the other hand $\delta_2^1 = \omega_2$ is consistent with ZFC. This follows from a result in [19]. Thus, while our simple sets have not provably given us a counter-example to CH, the possibility that they are counterexamples definitely arises.

Related theorems give a similar picture:

Every Σ_2^1 set is a union of $\aleph_1$ Borel sets.
MC $\rightarrow$ Every Σ_3^1 set is a union of $\aleph_2$ Borel sets.

Once again it is consistent that these results are not best possible, but there is no reason to believe they are not best possible.

Measurable cardinals do not give information about higher levels of the projective hierarchy, but there is another sort of "axiom" which does. This is the assertion projective determinacy. Let 2^ω be the collection of all infinite sequences of 0's and 1's. Regard 2^ω as a product of ω copies of the discrete space $\{0,1\}$ and give it the product topology. Given $A \subseteq 2^\omega$, the game G_A is defined as follows. Players I and II take turns picking 0 or 1, thus producing an element of 2^ω. I wins if this element belongs to A. The notion of a winning strategy for I or II is defined in the obvious way. G_A is determined if one of the players has a winning strategy. Using the axiom of choice, one can construct an A such that G_A is not determined (See [21]). On the other hand we have recently proved that G_A is determined for every Borel set A (the best previous result [23] concerned $F_{\sigma\delta\sigma}$ sets), and MC implies [17] that G_A is determined for every Π^1_1 set A. (The projective hierarchy is defined in the same way as for the reals.) Projective determinacy (PD) is the assertion that G_A is determined for every projective A.

There is no a priori evidence for PD, but there is a good deal of a posteriori evidence for it. PD has pleasing consequences about the behavior of projective sets, such as: Every projective set is Lebesgue measurable [22]; Every uncountable projective set has a perfect subset [6]. More impressive is the fact that PD allows one to extend the classical structural theory of projective sets, which dealt only with the first two levels of the projective hierarchy, to a very elegant and essentially complete theory of the projective sets (See [3], [16], [20]). PD cannot be proved in ZFC (or ZFC + MC, though it may be provable from large cardinal axioms), but it is not unreasonable to suspect that it may be true.

All the consequences of MC concerning the projective sets also follow from PD. Furthermore [18]

$$\text{PD} \rightarrow \delta^1_4 \leq \omega_4 \quad ;$$

$$\text{PD} \rightarrow \text{Every } \Sigma^1_4 \text{ set is a union of } \aleph_3 \text{ Borel sets.}$$

Concerning higher levels one has

$$\text{PD} \rightarrow \delta^1_{2n+2} \leq (\delta^1_{2n+1})^+ \quad ,$$

where α^+ is the least initial ordinal greater than α. Thus PD extends the pattern derived from MC (which extends that derived in ZFC alone) and reinforces the suggestion that the answer to CH may be negative.

Throughout the latter part of my discussion, I have been assuming a naive and uncritical attitude toward CH. While this is in fact my attitude, I by no means wish to dismiss the opposite viewpoint. Those who argue that the concept of set is not sufficiently clear to fix the truth-value of CH have a position which is at present difficult to assail. As long as no new axiom is found which decides CH, their case will continue to grow stronger, and our assertion that the meaning of CH is clear will sound more and more empty.

BIBLIOGRAPHY

1. J. W. Addison, Separation principles in the hierarchies of classical and effective descriptive set theory, Fund. Math. 46 (1959), 123-135.

2. J. W. Addison, Some consequences of the axiom of constructibility, Fund. Math. 46 (1959), 337-357.

3. J. W. Addison and Y. N. Moschovakis, Some consequences of the axiom of definable determinateness, Proc. Nat. Acad. Sci. U.S.A. 59 (1968), 703-712.

4. P. J. Cohen, The independence of the continuum hypothesis, Parts I, II, Proc. Nat. Acad. Sci. U.S.A. 50 (1963), 1143-1148; 51 (1964), 105-110.

5. J. Cohen, Comments on the foundations of set theory, Proc. A.M.S. Sympos. Pure Math. XIII, Part 1, Axiomatic Set Theory, 9 -15.

6. Morton Davis, Infinite games of perfect information, Ann. Math. Studies No. 52 (1964), 85 - 101.

7. W. Easton, Powers of regular cardinals, Ann. Math. Logic 1 (1970), 137-178.

8. F. Galvin and A. Hajnal, Inequalities for cardinal powers (to appear).

9. K. Gödel, Consistency proof for the generalized continuum hypothesis, Proc. Nat. Acad. Sci. U.S.A. 25 (1939), 220-224.

10. K. Gödel, The consistency of the axiom of choice and of the generalized continuum hypothesis, Ann. Math. Studies No. 3, Princeton, N. J., 1940.

11. K. Gödel, What is Cantor's continuum problem? Amer. Math. Monthly 54 (1947), 515-25.

12. L. Harrington, Long projective wellorderings (to appear).

13. D. Hilbert, Über das Unendliche, Math. Annalen 95 (1926), 161-192.

14. H. J. Keisler and A. Tarski, From accessible to inaccessible cardinals, Fund Math. 53 (1964), 225-308.

15. J. König, Zum Kontinuumproblem, Math. Annalen 60 (1905), 177-180.

16. D. A. Martin, The axiom of determinateness and reduction principles in the analytical hierarchy, Bull. Amer. Math. Soc. 74 (1968) 687-689.

17. D. A. Martin, Measurable cardinals and analytic games, Fund. Math. 66 (1970), 287-291.

18. D. A. Martin, Projective sets and cardinal numbers, Jour. Symb. Logic (to appear).

19. D. A. Martin and R. M. Solovay, Internal Cohen extensions, Ann. Math. Logic 2 (1970), 143-178.

20. Y. N. Moschovakis, Uniformization in a playful universe. Bull. Amer. Math. Soc. 77 (1970), 731-736.

21. J. Mycielski, On the axiom of determinateness, Fund. Math. 53 (1964), 205 - 224.

22. J. Mycielski and S. Swierczkowski, On the Lebesgue measurability and the axiom of determinateness, Fund. Math. 54 (1964), 67-71.

23. J. Paris, ZF + Σ_4^0 determinateness, Jour. Symb. Logic 37 (1972), 661-667.

24. R. Platek, Eliminating the continuum hypothesis, Jour. Symb. Logic 34 (1969), 219-225.

25. D. Scott, Measurable cardinals and constructive sets, Bull. Acad. Polon. Sci. Ser. Sci. Math. Astronom. Phys., 9 (1961), 521-524.

26. R. M. Solovay, On the cardinality of Σ_2^1 sets of reals, Foundations of Mathematics, Symposium papers commemorating the 60th birthday of Kurt Gödel, Springer-Verlag, 1966, 58-73.

27. R. M. Solovay, Strongly compact cardinals and the GCH. (To appear in the Proceedings of the Tarski Symposium.)

THE ROCKEFELLER UNIVERSITY

Proceedings of Symposia in Pure Mathematics
Volume 28, 1976

WHAT HAVE WE LEARNT FROM HILBERT'S SECOND PROBLEM?

G. Kreisel[1]

ABSTRACT

Section I of this article traces the development of Hilbert's second problem, about the consistency of the axioms of arithmetic, to his program. Its narrow form requires so-called finitary proofs of the (formal) consistency of formal arithmetic, an easy consequence of the broader claim that, quite generally, infinitistic notions are used in mathematics only as a manner of speaking, and must therefore be eliminable. Section II describes familiar branches of mathematics and elementary logic for which the program or at least a modification can be carried out, and others (arithmetic and metamathematics) where the broad claim behind the program is refuted, and the program and its relevance are dubious. Consequently by-products of work on Hilbert's original program are generally given principal emphasis. Thus traditional formulations are replaced in terms of recursion theory which treats general formal systems, and of definability theory concerning models of theories which are not necessarily formalized. Section II concludes with a preview of a genuine theory of proofs which -- in contrast to most work on Hilbert's program (proof theory) -- concerns the structure of proofs and not only the set of provable theorems.

I. FROM THE SECOND PROBLEM TO HILBERT'S PROGRAM

Hilbert's main point in his second problem was this: It should be possible to exploit the finiteness of (all) proofs and so establish the consistency of the axioms of arithmetic without the use of (familiar) <u>infinite</u> <u>models</u>; even though the theorems proved are ordinarily intended

---

AMS (MOS) subject classifications (1970). Primary 02.18, 02.65; Secondary 02.32, 02.74.

[1]Research supported by the National Science Foundation under grant GP 43901.

to be about infinite sets. This idea guided his further analysis of the problem, during the first decades of the century, and led to his program. The analysis removed several, apparently serious ambiguities in his presentation of the second problem (which, in contrast to most of his problems, dealt with notions that had not been previously analyzed). In particular, he had left open (i) exactly which axioms are to be considered, except that arithmetic was to include the theory of real numbers (as in analytic number theory), (ii) exactly which chains of inferences (from the axioms) are to be proved free from contradiction except that, besides being finite, they were to be 'strictly logical'. As to the significance of (mere) consistency, as a sufficient condition on axioms, he realized no doubt that this would be absurd in daily life or in, say, the natural sciences, but said that in mathematics consistency would ensure the 'existence of a concept' (satisfying the axioms in question; for limitations, see pp. 4, 5).

Rules and propositions of 'finitary character'. The first essential step towards removing the gaps (i) and (ii) was to consider formal rules for arithmetic, where, by intention, 'formal' or equivalently 'mechanical' rules are to have, hereditarily, finitary character (f.c.). Hilbert gave specific examples; much later, the notion was made precise in recursion theory and remarkably simple normal forms for formal rules were found [26]. As typical examples of f.c. propositions Hilbert used such things as Fermat's conjecture or, generally, propositions $(\forall n \in \omega)\ P(n)$ where, for each n, $n = 0, 1, \ldots$ it can be decided mechanically whether or not $P(n)$ holds. By use of recursion theoretic analysis, work on Hilbert's program is not sensitive to the exact meaning of f.c. proposition; one uses normal forms

$$\forall n_1 \cdots \forall n_{13}[p(n_1, \ldots, n_{13}) \neq 0]$$

for p ranging over polynomials with integral coefficients, $\forall\vec{n}[p(\vec{n}) \neq 0]$ for short [11]. Though he did not develop any kind of 'arithmetization' of sequences of symbols on which formal rules operate, Hilbert recognized and stressed that, for any formal system, formal consistency is (expressed by a proposition) of finitary character. Pedantically, given formal rules and hence, in particular, the (finite) list of symbols to be used, one can decide of any sequence of 'words' (= formulas) if it is built up according to the rules, and does not contain words of the form a and $\neg a$ where the symbol $\neg$ is used for negation. (We shall not be equally pedantic below.)

The second important f.c. property of formal rules to be used below is completeness for numerical arithmetic. A formalization of numerical computation (of polynomials p) by formal rules $\mathcal{F}$ is given by the following 'data':

Two f.c. maps σ and π: mapping polynomials p (determined by their coefficients say c_p) and arguments $\vec{n}$ into formulas (representing $p(\vec{n}) = 0$) and derivations of $\mathcal{F}$, and a proof of the f.c. proposition

$$(*) \quad \text{if} \quad \begin{matrix} p(\vec{n}) = 0 \\ p(\vec{n}) \neq 0 \end{matrix} \quad \text{then } \pi(c_p, \vec{n}) \text{ is a derivation in } \mathcal{F} \text{ of } \begin{matrix} \sigma(c_p, \vec{n}) \\ \neg\sigma(c_p, \vec{n}) \end{matrix}$$

(The idea is that π mimics the computation of $p(\vec{n})$ according to the recursion equations for $+$ and $\times$. Hence, for the usual systems which contain these computation rules, the proof of (*) is patently 'elementary' or, in technical jargon, primitive recursive. Clearly for given f.c. operations σ and π and for any c_p, $\vec{n}$, one can decide mechanically if (*) holds, and so (*) has f.c.)

At this stage a 'trivial' version of Hilbert's program can be stated for numerical arithmetic (not for all f.c. propositions). If $\mathcal{F}$ is consistent and complete for numerical arithmetic -- without any restriction at all on the (metamathematical) methods of proof (of consistency and completeness)

if $\sigma(c_p, \vec{n})$ has a formal derivation in $\mathcal{F}$ then $p(\vec{n}) = 0$

or equivalently

if $\sigma(c_p, \vec{n})$ has a formal derivation in $\mathcal{F}$ then $p(\vec{n}) = 0$ can be computed from the recursion equations.

In other words, 'mere' consistency of $\mathcal{F}$ is enough to ensure the truth of formally derivable numerical equations. No other requirement is needed, say, on the truth of the axioms of $\mathcal{F}$ for some intended interpretation of $\mathcal{F}$; if $\mathcal{F}$ is axiomatic set theory nothing is needed of the semantic analysis of set which may have originally led to the axioms of $\mathcal{F}$. Though trivial, this gives a good idea of the kind of irrelevance of set theoretic notions which Hilbert had in mind. More significant examples of complete formal systems (for geometry, the field of real numbers) in common use will be given in Section II.

Before stating the full program, we need a final preparatory step, concerning the formalization of universal numerical quantification or, equivalently, of free number variables, say $\underline{a}$. It is given by the following data:

Two f.c. maps φ and τ (where $\varphi(c_p)$ is to represent $\forall\vec{n}[p(\vec{n}) \neq 0]$ or $p(\vec{a}) \neq 0$ in $\mathcal{F}$[2])

and a proof of the <u>substitution property</u>:

for each sequence <u>d</u> and each $\vec{n}$, if d is a derivation in $\mathcal{F}$ of $\varphi(c_p)$ then $\tau(d, \vec{n})$ is one of $\neg\sigma(c, \vec{n})$.

Note that this is an f.c. property which -- to anticipate the main aim of Hilbert's scheme -- adequately expresses the meaning of the infinitistic idea of universal quantification.

<u>Significance of consistency for formal systems of arithmetic</u> (which satisfy completeness for numerical arithmetic and the substitution property). Hilbert deduced from consistency of $\mathcal{F}$ (and the additional f.c. properties):

$$\text{if } d \text{ derives } \varphi(c_p) \text{ then } p(\vec{n}) \neq 0$$

as follows. By completeness,

$$\text{if } p(\vec{n}) = 0 \text{ then } \pi(c_p, \vec{n}) \text{ derives } \sigma(c_p, \vec{n})$$

and by the substitution property

$$\text{if } d \text{ derives } \varphi(c_p) \text{ then } \tau(d, \vec{n}) \text{ derives } \neg\sigma(c_p, \vec{n}) .$$

So $\pi(c_p, \vec{n})$ followed by $\tau(d, \vec{n})$ derives an inconsistency. Hence, whatever class $\mathcal{P}$ of elementary or 'finitary' proofs one may have in mind, if $\mathcal{P}$ includes the steps above, and if there are proofs in $\mathcal{P}$ of consistency (completeness, and of the substitution property), we have

$$\text{if } d \text{ derives } \varphi(c_p) \text{ then there is a proof in } \mathcal{P} \text{ of } \forall\vec{n}[\vec{p}(n) \neq 0] .$$

(The reason for stressing consistency and mentioning the other properties parenthetically is that the usual systems contain the required rules. So, as mentioned earlier, the latter properties have proofs of the same character as the deduction above).

Note at once the following <u>limitation on the significance of consistency</u> (for systems containing negations of universally quantified formulas).

---

[2] An <u>inequality</u> $p \neq 0$ is used to have a <u>typical</u> example of an f.c. proposition.

Suppose c_p is such that, for each $\vec{n}$,

neither $\varphi(c_p)$ nor $\sigma(c_p, \vec{n})$ can be derived in $\mathcal{F}$,

in other words, the diophantine problem $p(\vec{n}) = 0$ has no solution in integers, but this fact cannot be (formally) derived in $\mathcal{F}$. Then $\mathcal{F} + \{\neg\varphi(c_p)\}$ is consistent, but $\varphi(c_p)$ is true over the integers. To use the language of Hilbert's second problem, even if the consistency of $\mathcal{F} + \{\neg\varphi(c_p)\}$ ensures the existence of some concept satisfying all the theorems of this system, it does not ensure that the particular concept (of the natural numbers), for which axioms of arithmetic are intended, satisfies those theorems.

Hilbert's program is nothing else but this. For the formal systems of arithmetic $\mathcal{F}$ (which he had obtained according to a systematic scheme described in II.1 and) which he had proved to be complete for numerical arithmetic and to have the substitution property:

To give an f.c. proof of the consistency of $\mathcal{F}$.

Hilbert gave only examples of this notion (of f.c. proof; as he had done in the case of f.c. rules and f.c. propositions). But the idea was familiar, for example, in number theory from the business of elementary as opposed to analytic proofs, or, more generally, from Kronecker's opposition to proofs which are not of f.c. These f.c. proofs are required to be not only (of course) finite, but hereditarily about finite configurations and f.c. operations. Hilbert gave a very simple example of an f.c. consistency proof for a formal system which has only infinite models, namely predicate logic with the successor axioms

$$0 \neq sa \qquad\qquad sa = sb \longleftrightarrow a = b$$

with free variables a and b, the constant 0 and the (successor) function symbol s. Clearly, here was a genuine idea for proving consistency without the use of infinite models (as required by the second problem), and it made obviously essential use of the then relatively new discovery of formalization (since without the latter the notion of 'strictly logical inference' would not be sufficiently analyzed to have f.c.). This kind of combination of an attractive idea and a new tool, of formalization, is often enough to make imaginative use of traditional 'philosophical' ideas (which come to mind when we know little, and seem hopelessly simple minded or vague when we know a bit more.) -- NB. F.c. proofs are also called 'finitist'; finitist rules and propositions are not only f.c., but established by means of f.c. proofs.

Finitism and Hilbert's program. Quite simply, the idea is that in our use of infinitistic, in particular, set theoretic notions in mathematics, these notions serve as a façon de parler, and, as such, they can be more or less directly eliminated. A minimal condition would be that for any f.c. theorem A the elimination provides a proof of A not containing any reference to infinitistic notions, when applied to an arbitrary formal derivation of A. If our use really involves only a façon de parler there is every reason to suppose that the elimination procedure would be easy to establish if we look for it at all. Furthermore, the requirement of eliminability would not constitute any restriction on the methods of proof we use freely: the eliminability is thought of as a fact (to be discovered), not a doctrinaire restriction. Perhaps we should also expect that at least occasionally, the elimination of set theoretic notions from a proof of an f.c. theorem A would increase our conviction in A. In short, if the view of infinitistic notions as a façon de parler is valid, the elimination should be, at least occasionally, a useful scientific tool.

Clearly, Hilbert's program, provides at least an approximation to an elimination scheme. The program does not mention explicitly an elimination procedure; but given the (hypothetical) f.c. proof of consistency etc. and a formal derivation of (the formal translation of) $\forall\vec{n}[p(\vec{n}) \neq 0]$ then, according to the program, we get effectively some f.c. proof of $\forall\vec{n}[p(\vec{n}) \neq 0]$. (And the details of the consistency proof may provide a more 'direct' elimination procedure.)

It would be futile at this state to go into ambiguities and variants (of the finitist 'philosophy' or of the program) because the basic results, in Section II, are general enough to be insensitive to the exact choice of formulation. However, the reader is strongly recommended to look at the following summary: it contains lessons which we have learnt from work on Hilbert's program (and which can be stated without further analysis).

Facts and issues. 1. Finiteness of proofs or, more precisely, f.c. operations for constructing (formal representations of) proofs. (a) There is no doubt that the structure of any proof is adequately represented by derivations built up according to suitably chosen formal (f.c.) rules, and that these formalizations are a basis for a structural theory of proofs; cf. Section II.5(c). (b) There is doubt whether formalization, by itself, is significant for Hilbert's particular purpose, of establishing consistency of axioms and rules of inference, that is, for a study of principles of proof (not primarily of the proofs themselves). The alternative to Hilbert's idea is of course, not to avoid infinite models for the proofs of consistency but rather to analyze the meaning of (infinitistic) concepts; to establish not only consistency but more generally, soundness of axioms for the intended meaning. (When examining finitism it would be a petitio principii to assume that the idea of infinite models must be avoided).

NB. It is by no means perverse to consider this alternative which is just what the program aimed to avoid! the alternative is natural and used; that's what makes Hilbert's program so 'grand'.

Historical remark (amplified in II.4(a) below). Despite a wide spread misunderstanding, it was not formalization (of axiomatic set theory) as such which helped one avoid[3] the paradoxes; Zermelo's analysis of what sets we are talking about [37] (nowadays called: the cumulative hierarchy of sets) really helped (him) find a consistent formalization.

2. Finitary character. (a) There is no doubt of the useful role of f.c. operations (and f.c. propositions about f.c. operations), both in analyzing the idea of computation and in mathematics generally, for example, group theory [19] or number theory [11]. (b) There is doubt about the role of f.c. proofs, where, roughly speaking, not only the objects we handle (in operations), but the objects we think about are required to be finite. This doubt is largely independent of any closer analysis of f.c. proofs. Once again, it would be a petitio principii to require a restriction to f.c. proofs.

Historical example. At various times number theorists were interested in elementary proofs, broadly similar to f.c. proofs (though the sharper notion of direct elementary proof has been more fruitful [30]). In the debate about the matter it is almost always forgotten that there should be elementary proofs of, say, analytically proved f.c. theorems if -- anything like -- the assumptions behind Hilbert's program are valid. So the logical interest of elementary proofs is so to speak negative: any doubts about those elementary proofs cast doubt on finitism .

3. F.c. proofs and others. (a) There is no doubt that, in particular parts of mathematics or at least of mathematical practice at the present stage of development, there are candidates for the following variant of Hilbert's program: There is an accepted class $\mathcal{P}$ of proofs and there

[3]Formalization, by Frege, perhaps helped one derive paradoxes (not to correct them). Even this was not needed since, back in 1885, in a review [9], Cantor explained very clearly why Frege's formulation could not be expected to hold for the concept of set described and used by Cantor.

are extensions $\mathcal{P}'$ which are either genuinely problematic or appear logically unnecessary. One looks for a class $\mathcal{F}_{\mathcal{P}}$ of propositions ('corresponding' to Hilbert's f.c. propositions) such that, for $F \in \mathcal{F}_{\mathcal{P}}$,

> if there is a proof in $\mathcal{P}'$ of F then there is a proof of F in $\mathcal{P}$ too;

for examples, see II.4(b) and II.5a(i). (b) There is doubt whether we have a candidate for any 'foundational' or universal class (for the whole of mathematics) which the class of f.c. proofs was intended to be; cf. particularly II.4(a). The obvious alternative to be considered here is this: we recognize, after having become familiar with the principles currently accepted in mathematics, that all of them are unproblematic, and that our knowledge is increasing significantly. Then the subject is not ripe for foundational analysis at the present stage (though, by what has just been said, it is a proper subject for reflection).

Loaded terminology: a warning. Many of the current formulations of results in proof theory (as Hilbert called the subject he developed for his program) implicitly assume that the restriction to elementary or f.c. proofs is significant; for example, (familiar) relative consistency results of the form:

> there is an f.c. proof of the consistency of S' from the consistency of S,

when, in fact, there is a convincing proof of the consistency of S' itself. In view of the doubts 2(b) and 3(b), it is a principal problem to analyze relative consistency proofs and reformulate their content. This explains why some of the formulations below are unfamiliar.

The next section describes the basic work on Hilbert's program and uses this work in II.5 to review the issues I.1 - I.3 above.

II. BASIC ADEQUACY AND INADEQUACY RESULTS

Summary. II.1 explains the precise mathematical sense of Hilbert's conviction that 'nothing is lost by formalization', and the evidence (which we can now present) for it. This work must be distinguished from the purely empirical case studies of, say, PM which provided evidence for the adequacy of particular formal rules for (then current) mathematical practice. II.2 formulates the semi-adequacy of formal rules for elementary logic, also called 'first order predicate logic' or 'lower functional calculus'. (Readers are assumed to know this easy subject; if not they can look it up in any modern text on logic.) II.3 - II.4 set out the principal inadequacy results; II.3 w.r.t the mathematical theory of the ring of integers (and, by [11], even w.r.t. the class of diophantine problems); II.4 w.r.t. to metamathematical practice in the sense that no formal system $\mathcal{F}$ actually used is adequate to formalize all results (about $\mathcal{F}$) established in current metamathematical practice. The critical inadequacy results of II.3 are compatible with empirical case studies, such as those of PM mentioned above, because not all diophantine problems have been decided in practice. II.5 analyzes the bearing of II.3 and II.4 on Hilbert's program.

1. From traditional axiomatizations to formalizations. The best known axiomatizations -- in 1900 and perhaps even now -- are the axioms (a) for the ordered field of real numbers (of Dedekind), (b) for geometry with or without the parallel axiom (for example, Hilbert), (c) for the natural numbers (for example, Peano). Each of these axioms involves arbitrary sets or properties (of the objects considered): (a) arbitrary (Dedekind) cuts and correspondingly, (b) arbitrary sets of points in (Hilbert's) continuity axioms, (c) arbitrary non-empty sets of natural numbers in Peano's induction axiom (well-foundedness of the successor relation). Far from rejecting the notion of such arbitrary sets, more precisely, of arbitrary subsets of the structures considered, one asks whether the notion can be eliminated. Precisely; are there formal rules for generating precisely those formulae, not involving set variables (technically: in the first order languages of the structures considered), which are true in the classes of structures satisfying (a)-(c) resp.?

(a) Ordered (real) closed fields, also called 'maximal' ordered fields. The familiar axioms, together with the (familiar) rules of predicate logic provide a positive solution. Instead of demanding the existence of reals, say x_S, for arbitrary cuts S

(*) $$\exists x_S \, \forall x (x < x_S \Longleftrightarrow x \in S),$$

one limits (*) to the formal list of cuts defined by (first order) formulas in the language of (ordered) fields (technically: with parameters).

THEOREM: A (first order) formula F is derivable by means of the formal rules mentioned if and only if F is true for the field of real numbers. Very little detailed knowledge of $\mathbb{R}$ is needed here because of the following formal (f.c.) property of the axioms considered: Tarski established a decision method for all (closed) formulas F of the language by means of an elimination procedure ϵ such that either

$\epsilon(F)$ is a formal derivation (from the axioms) of F, or

$\epsilon(F)$ is a formal derivation of $\neg F$.

In short, each F is either true in all ordered real closed fields or else false in all such fields. No axiom system can do more (as far as the language here considered goes)! So all we need to know is that $\mathbb{R}$ is a real closed (ordered) field.

Historical remark. Tarski's method is based on Sturm's algebraic theory of polynomials in a real variable [32]. Sturm's original aim was foundational as he explicitly stressed. He wanted to avoid continuity considerations which, he believed, required (dubious) infinitesimals. Thus, within this limited area of algebra, his aim was parallel to Hilbert's program, inasmuch as algebra has 'finitary character' and continuity arguments and infinitesimals do not, at least, as they are intended. For reference below, in connection with the permanent value of Hilbert's program: Sturm's foundational aim is no longer topical because of an appropriate analysis by Cauchy, of the meaning of continuity; Sturm's mathematical development suggested by his aim has become an important scientific tool, particularly since Artin recognized its relevance to Hilbert's 17th problem (which was originally stated only for the specific fields $\mathbb{Q}$ and $\mathbb{R}$). NB. By [27] this problem showed also the possibility of using work on Hilbert's program as a scientific tool, independently of the original foundational aims.

Warning: Contrary to a wide spread misunderstanding, a complete formalization or decision method for a branch of mathematics by no means 'finishes' it; cf. the end of II.2(a), and, again, the end of II.3.

(b) Geometry, of congruence and betweenness, with and without the axiom of parallels. As far as formulas in the first order language are concerned, the sets involved in the continuity axiom can be replaced, as in (a) above, by a formal list of those sets which are definable in the first order language of congruence and betweenness considered here. For details on this adequacy of the formalization, see Hilbert's book on the foundations of geometry or, for a modern careful exposition, Tarski's writings on the subject, for example [34].

The novel point, compared to (a), is this: without the 'axiom' of parallels, not every formula F (in particular, the parallel axiom) is either formally derivable or refutable (where 'refutable' means that

$\neg F$ is derivable). It should be stressed that by completeness of the formalization, not every such F is decided by the set theoretic, informal axiomatization either: familiar models of euclidean and non-euclidean geometry satisfy the continuity axiom for arbitrary sets (of points). This consequence of the proof of adequacy is absolutely central for correcting the popular but loose comparison between (i) the independence of the parallel axiom, and (ii) (what Gödel explicitly called in [17]) formal independence results of II.3 and II.4. The same applies to the independence of the continuum hypothesis from formal versions of, say, Zermelo's set theory.[4]

(c) Peano's axioms for zero 0 and the successor function s (or, if preferred, the successor relation): $\forall x(sx \neq 0)$, $\forall x\, \forall y(x = y \longleftrightarrow sx = sy))$ and induction. They characterize, up to isomorphism, the structure of the positive integers with a first element and successor function. When convenient we shall also speak of the analogous axioms for all integers.

The novel point, compared to (a) and (b) is this: the first order language (of the structure considered) consisting of formulas built up from equations between terms 0, (variables) x, $s(x)$, $ss(x)$ etc., by means of elementary logic, is obviously inadequate for any kind of arithmetic. Exercise: Verify that induction restricted to sets defined in the baby language above is sufficient to decide formally every formula of the language.

There is a natural scheme for extending the language by recursive definitions, of relations (for example, strict order: $\forall x(\neg\, x < 0)$, $\forall x\, \forall y[x < sy \longleftrightarrow (x = y \vee x < y)])$ or functions (for example, addition). Together with Peano's axioms, these 'implicit definitions' characterize the enriched structures uniquely. Furthermore for the particular relations and functions of order and addition, we still have the analogue to the Exercise above.

Every first order formula in the language, say L^+, of those (enriched) structures is formally decided by the recursion equations together with Peano's axioms when induction is restricted to sets defined in L^+.

For a proof, see (logical) texts on discretely ordered additive groups or on decision methods for the theory of addition of integers (later extended to all abelian groups in [33] and [12]).

---

[4] In Zermelo's own axiomatizations [36] or [37], the comprehension axiom was not formulated only for subsets (of any a) defined by formulas say $A(x)$ in the particular language of axiomatic set theory: $\forall a\, \exists y\, \forall x(x \in y \Longleftrightarrow [x \in a \wedge A(x)])$ but for arbitrary properties (of the elements of a).

Principal questions: Will the completeness result above for the formal systems (for successor, order, and addition) obtained by applying the extension scheme, extend to all recursive definitions? or at least, to suitably chosen, natural ones?

Even though, as far as I can determine, Hilbert did not explicitly formulate the scheme (nor the question), it is patently evident that the formal systems he considered fall under the scheme. The question is clearly relevant to the general program: if these particular sets, defined in the language of the structure itself, cannot replace the arbitrary sets of the informal axiomatizations adequately, what simple principle of replacement is there? As to the evidence (for a positive answer) provided by (a)-(c), it is certainly purely external -- though it must be admitted that the usual reasons for the choice of 'serious' systems, of set theory, are no more sophisticated.[5] For the 'hard nosed' mathematician, the evidence is weak, not so much because it consists of few instances, but because they are too simple to appear typical of actual mathematical experience.

Evidently the weaker version of the question leaves open which recursive definitions are 'suitable' or 'natural'. A negative answer is given in II.3. for the theory of the ring of integers, on the 'assumption' that addition and multiplication are natural.

2. Semi-adequate formalization of general (first order) logical theory. Here the intended principal notions, of

a structure satisfies a formula A

(or, equivalently, is a model of A), and A is valid in all structures, $\mathrm{Val}(A)$ for short, are defined in terms of the notion of arbitrary sets; there is no restriction on the structures. For a formula A with the relation symbols R_i of n_i arguments, and function symbols f_j of m_j arguments, the structures considered are given by an arbitrary (non-empty) domain or universe D, arbitrary relations $\bar{R}_i \subset D^{n_i}$ and functions $\bar{f}_j \subset D^{m_j} \to D$. We examine Hilbert's general idea, that 'nothing is lost' by formalization, in II.2(a) for validity with 'positive' results, and in II.2(b) for satisfiability (with 'negative' results).

The first 'crude' result is the theorem of Skolem-Loewenheim, which (for the present purpose) is best stated as follows:

---

[5] See, however, the brief discussion by Gödel [18] of extensions of the language $L_\in$ of set theory and their relation to socalled axioms of infinity formulated in $L_\in$.

Let C be any class of structures including the class C_ω of all countable structures, and let $Val(A;C)$ mean that A is valid in (all structures $\in$) C. Then $Val(A;C) \Longleftrightarrow Val(A;C_\omega)$ and similarly of course for satisfiability.

Thus provided $C \supset C_\omega$, first order logical validity (in C) is insensitive to the exact extent of C.

(a) The crude result above was first refined by Gödel's completeness theorem for derivability, $\vdash_1$, by means of the particular rules for predicate calculus which Frege had written down some 50 years earlier (the subscript 1 in $\vdash_1$ standing for 'first order'). Then, if $\neg A$ is consistent, that is, $\vdash_1 A$ is not true, there is a countable structure -- or 'concept' in the language of the second problem -- which satisfies $\neg A$. Put differently,

$$(*) \qquad Val(A;C_\omega) \Rightarrow \vdash_1 A \quad .$$

Actually, more is proved than is stated here -- and this is essential for Hilbert's aim. (*) can be sharpened by restricting the class C_ω of arbitrary countable structures to structures on ω defined in the (first order) language of the ring of integers; in technical jargon, Δ^0_2 definitions are enough. The refinement can now be stated in this language and formally derived from the axioms for addition and multiplication given in II.1(c), modulo a familiar representation (arithmetization) of the property $\{A : \vdash_1 A\}$ (of 'Gödel numbers' of formulas in predicate logic).

Digression on (massive) developments arising from the completeness theorem or, more precisely, from a remark by Gödel, at the end of [15], which is strictly irrelevant to formalization: for arbitrary (countable, later extended to uncountable) sets $\mathscr{A}$ of first order conditions, call A a consequence of $\mathscr{A}$, for short

$C(\mathscr{A} \vdash A)$ iff every structure which satisfies every formula of $\mathscr{A}$ also satisfies A.

For finite subsets $\mathscr{A}_1$ (of $\mathscr{A}$), let $\hat{\mathscr{A}}_1$ be the conjunction of formulas in $\mathscr{A}_1$. Then

$$C(\mathscr{A} \vdash A) \Longleftrightarrow (\exists \mathscr{A}_1) \vdash_1 (\hat{\mathscr{A}}_1 \to A) \quad .$$

This has a corollary which does not refer to formal rules at all, the finiteness (also called compactness) theorem of logic

$$C(\mathscr{A} \vdash A) \Longleftrightarrow (\exists \mathscr{A}_1)\ Val(\hat{\mathscr{A}}_1 \to A) \quad ,$$

(and even $Val(\hat{\mathscr{A}}_1 \to A, C)$ for C containing all structures of some infinite cardinal $\geq$ card $\mathscr{A}$). The remarkable discovery which led to model theory (begun by Tarski) was this: Results which had first been proved and stated for formal theories, in particular, sets of axioms $\mathscr{A}$ generated by formal rules, extended naturally to arbitrary $\mathscr{A}$; and the extensions were demonstrably useful. This discovery is of central importance for examining the value of formalization for mathematical reasoning -- in contrast to a doctrinaire requirement of formalization. Recently research has gone a step further: the 'language' of predicate logic, used to determine certain classes of structures, has finite formulas in accordance with the literal meaning of 'formula', and particularly with the intentions of formalization. But analysis of the uses made in model theory, showed that finiteness is not essential, whereas something like 'first order character' is (at the present stage of the subject, cf. [5] and [25]).

A second point, concerning the limitation of (*) is this. Though Frege's rules have now been available for about 100 years, as time went on mathematicians have used them less, not more! Specifically, many results in axiomatic mathematics, for example about ordered fields or division algebras of a given dimension over a real closed field can be stated as theorems, A, of logic. So there is what one might call a logical proof of A (by use of Frege's rules). But instead one looks for mathematical proofs, of Val(A) using set theoretic notions (Val(A) being defined set theoretically); such proofs can be demonstrably simpler, in the precise sense of II.5(c).

In short, to conclude the digression, formalization as such has not been used. Occasionally, quantitative refinements can be obtained by special, for example, (Herbrand's or Gentzen's) 'direct' rules which generate also exactly the valid theorems but have a more manageable structure; and can sometimes be used to get 'explicit' bounds; cf. pp. 361-362 of [27].

(b) Hilbert [21] wanted a complete formalization, also for satisfiability, namely definitions say $\bar{R}_i$, $\bar{f}_i$ of relations and predicates over ω such that $\neg \bar{A}$ can be formally derived in arithmetic if A is not derivable by Frege's rules. (By the refinement of (*), there are such $\bar{R}_i$, $\bar{f}_i$ -- which can even be effectively found from A and -- for which $\bar{A}$ is false.) A complete formalization, of validity and satisfiability, as required by Hilbert, would decide logical validity by the following 'decision' method formulated in formal arithmetic: for any A, run simultaneously through (formal) derivations in predicate logic and in arithmetic until you find either a derivation of A or of some $\neg \bar{A}$; one but not both of these cases will occur. ($\{A : \vdash_1 A\}$ is entscheidungsdefinit in the language of [17]). From a remark of Gödel, also in [17], the method would also decide all f.c. propositions, by the following connection between solubility of diophantine equations and logical validity (or -- as we now know: equivalently -- between f.c. propositions and logical satisfiability).

Let Z be the conjunction of elementary successor axioms and the recursion equations for addition and multiplication (which, in the terminology of I, are complete for numerical arithmetic). Then $\exists\vec{x}[p(\vec{x}) = 0]$ is true over the integers iff $Z \to \exists\vec{x}[p(\vec{x}) = 0]$ is logically valid. --The reader may wish to recover here Hilbert's observation in Section I on the significance of consistency for formal derivations of f.c. propositions. Suppose Z above is formally derivable from otherwise arbitrary axioms A. Then

$$(\exists\vec{n} \in \omega)\ [p(\vec{n}) = 0] \Rightarrow \vdash_1 (A \to \exists\vec{x}[p(\vec{x}) = 0]) \quad .$$

Hence, taking the contrapositive, if $A \to \exists\vec{x}[p(\vec{x}) = 0]$ is not derivable, that is, if $\forall\vec{x}[p(\vec{x}) \neq 0]$ is formally consistent (with A), then $\forall\vec{n}[p(\vec{n}) \neq 0]$.

Historical note: After the notion of recursiveness was introduced, Church used the connection above to establish that there is no recursive decision method for logical validity at all. Here it should perhaps be added that, at present and, probably, *sub specie aeternitatis*, questions of recursive undecidability of purely mathematical problems (in group theory, topology or number theory) are more rewarding. But the solutions above, for logical problems which arise from Hilbert's program, were easier to find, and helped one establish the subject of undecidability.

Remark: It should not be assumed that ordinary (first order) logic is the richest useful logical language possessing a semi-adequate formalization, that is, adequate for validity, but not for satisfiability; nor that semi-adequate formalization is dependent on the 'reduction' to countable C_ω

$$(\forall C \supset C_\omega)\ [\mathrm{Val}(A,C) \Longleftrightarrow \mathrm{Val}(A, C_\omega)] \ .$$

For a detailed exposition of such a richer language containing the additional operation 'for uncountably many' and its uses see the monograph by Keisler [24]. (The quantifiers 'for finitely many' and hence 'for infinitely many' do not admit a semi-adequate formalization.)

(c) Many sorted predicate calculus and maximal models. In accordance with 'grand' foundational schemes of universal languages with one 'universal' kind of mathematical object, predicate logic is formulated with a single sort of variable. This is no restriction 'in principle': given (the sets of) cabbages and kings, one forms the union of the two and recovers the two sorts or types of objects by introducing monadic predicates, say $T_C(x)$ and $T_K(x)$. But for manageable formulations it is better to introduce several sorts of variables, x^σ, especially if one has to do with a domain, D_0, of 'type' 0, and functions, of type $\sigma \to \tau$, mapping objects of type σ into objects of type τ. For objects x,y of type σ and $\sigma \to \tau$ resp. one has the socalled application (or 'evaluation') operation $E : E(y,x) = z$ where z, of type τ, is interpreted as the value of y at the argument x.

One immediate consequence of Hilbert's idea that 'nothing is gained' by considering arbitrary sets, for example, of functions: $D_\sigma \mapsto D_\tau$, is to consider all collections (subsets of $D_\sigma \mapsto D_\tau$ itself) which satisfy explicitly stated (closure) conditions. Now, the notion of model of a formula A in the many sorted language sketched above is just this: given a domain D_0, one considers all collections (of functions) $\subset D_\sigma \mapsto D_\tau$ which satisfy the conditions explicitly stated in A.

This notion is called 'general' model to distinguish it from maximal models. Here D_0 is arbitrary, but, for $\sigma \neq 0$, D_σ^{max} is uniquely determined: hereditarily,

$$D_{\sigma \to \tau}^{max} \text{ is the set of all mappings: } D_\sigma^{max} \mapsto D_\tau^{max} .$$

The traditional axiomatizations of II.1 were stated in 2-sorted languages (where the functions of type $0 \mapsto 0$ are characteristic functions, of sets). They were intended to apply to maximal models and the formalizations consisted, in effect, in postulating that $D_0 \to D_0$ contains (at least) the functions defined in the relevant first order language. This postulate is expressed by first order conditions (while, by II.2(a) there are no such conditions which characterize any infinite maximal models).

NB. Literally speaking, maximal models are rarely considered in mathematics; for example, if D_σ and D_τ are two topological spaces, one does not consider all mappings, but only continuous ones. But this restriction is not defined by first order conditions either -- and the same applies to most classes of functions occurring in traditional analysis (except in those parts which have been subjected to thorough 'algebraization').

Remark on an 'empirical' version of Hilbert's program, involving a refinement of PM. In contrast to Hilbert's mathematical aim, in II.1, of a complete formalization w.r.t. the whole first order language of the structures considered, some formalization of any particular proof is indeed obtainable from the mere fact that proofs are finite. Whatever (set theoretically defined) class of sets or functions is intended, any one proof can only use finitely many properties, usually some kind of closure property. A little care, that is, a refinement of PM, shows in effect that in the bulk of current mathematical practice only very 'weak' closure properties are used (in technical jargon, arithmetic or hyper-arithmetic comprehension axioms, but variables of 'higher type', as stressed above). Consequently, the bulk of current theorems generalize: wherever the notion of set is used, explicitly or implicitly, it may be interpreted to mean: set of a (so to speak 'elementary') collection of sets satisfying the particular 'weak' closure conditions. This line of research has great appeal for logicians who have become familiar with the many hierarchies of sets and funtions (recursive, arithmetic, hyperarithmetic, admissible, etc.) introduced in the last 25 years. Those hierarchies (of sets and functions) provide models satisfying the 'closure conditions' used -- also by logicians -- to state the 'refinements' of PM mentioned above. The connection is, perhaps, not altogether surprising, because both the defining conditions on the models and the refinements (or 'subsystems' in the terminology of [27]) are formulated in similar terms of logical complexity. The few genuine attempts to establish any general significance of these hierarchies and subsystems are not convincing.[6] However, as might be expected, occasionally this type of analysis has a pedagogic value; by suggesting possible defects of current practice, and, above all, by providing simple solutions overlooked in most traditional versions of socalled constructive foundations; for a significant foundational use, see II.5(b).

3. Inadequacy of formal rules for the theory of the ring of integers and hence of any mathematical theory containing that theory . Recall the 'principal question' (and the ambiguities) mentioned at the end of II.1(c). This question is answered negatively, on minimal assumptions, by a series of researches described in more detail in [11]; starting with Gödel's negative answer for the formal rules of PM and 'related systems', the possibility of extending the result to arbitrary formal rules (by Turing's analysis and normal forms for the notion of formal or equivalently, mechanical rule), Kleene's general 'symmetric' formulation [26], and the present record, by J. Robinson and Matyasevic, for diophantine equations in 13 variables.

A (weak) corollary of this work is of course the inadequacy or incom-

---

[6] Except of course for the constructible hierarchy of sets which has turned out to be a mathematically strikingly manageable object [23].

pleteness <u>of formal rules w.r.t. validity in maximal models</u> (in II.2c), inasmuch as there is no (consistent) <u>semi</u> adequate formalization of $\mathrm{Val}^{\max}(P \to F)$ where P are Peano's axioms with the recursion equations for addition and multiplication and F are formulas of the form $\forall x_1 \cdots \forall x_{13}\ (p \neq 0)$.

Evidently, this <u>theoretical</u> inadequacy is perfectly compatible with the <u>empirical</u> adequacy for current mathematics, in the sense of PM, of such formal rules as those of axiomatic set theory: current mathematics leaves many diophantine problems unsolved too.

There are some easy corollaries which are by no means surprising in the light of mathematical experience, but which could previously not have been stated precisely, let alone be established.

(a) There are formal theorems, of elementary logic, directly obtained from the law of the excluded middle

$$(*) \qquad \forall x\, \exists \vec{y}\, \forall \vec{z}\, [p(x,\vec{y}) = 0 \vee p(x,\vec{z}) \neq 0]$$

(obtained from $\forall x(\exists \vec{y}[p(x,\vec{y}) = 0] \vee \neg \exists \vec{y}[p(x,\vec{y}) = 0]))$ such that there is no mechanical formal procedure ρ: $x \mapsto \vec{y}$, for which

$$\forall x\, \forall \vec{z}\, [p(x, \rho x) = 0 \vee p(x, \vec{z}) \neq 0] \quad ,$$

is true. As a practical consequence: inasmuch as logical form (here: the combination $\forall\, \exists\, \forall$) is significant, a theorem of number theory of the form (*), for example (cf. [3] for contrast)

$$\forall n\, \exists q_0\, \forall p\, \forall q\, (q > q_0 \to |\sqrt[3]{2} - \frac{p}{q}| > \frac{1}{q^{2+n^{-1}}})$$

may not possess an 'effective' bound $q_0(n)$. This bound concerns the <u>size</u> of 'exceptional' q for which $|\sqrt[3]{2} - p/q| < q^{-2-n^{-1}}$ (for some p); some bound for the <u>number</u> of exceptions can be found by use of logic. However the latter does not seem to suffice for applications [4], which use a better bound found by mathematics [10].

(b) A <u>positive counterpart</u> to the inadequacy result: if not all true propositions about the ring of integers can be formally derived by means of given formal rules, we expect to formulate <u>what more we know about a formally derived theorem F than if we merely know that F is true.</u>

One answer, in recursion theoretic terms, is given by <u>(re)formulating</u> F by, say, F' such that

F' is recursively satisfiable if (and only if) F is true,

but F' is satisfiable in a proper subclass $\mathcal{R}^-_{\mathcal{F}}$ (of the type of recursive operations considered) if F is formally derivable by means of the rules $\mathcal{F}$.

There is one obvious formulation if F is of the form $\forall\vec{x}\ \exists\vec{y}\ R(\vec{x},\vec{y})$ where R is recursively decidable for each tuple $\vec{n}$, $\vec{m}$ of integers, namely

$$F' \text{ is } \exists Y\ \forall x\ R(\vec{x}, Yx)$$

and Y is a function variable: $\vec{x} \mapsto \vec{y}$. (If F is true, there is always a recursive Y: take the first $\vec{m}$ s.t. $R(\vec{n}, \vec{m})$.) If F is logically more complex, several socalled 'functional interpretations' have been used, which restrict the operations implicit in the logical symbolism; cf. pp. 377-383 of [27].

If F is derivable by means of $\mathcal{F}$, then

$$(\exists Y \in \mathcal{R}^-_{\mathcal{F}})\ \forall\vec{x}\ R(\vec{x}, Yx)\ .$$

To illustrate possible pedagogic uses consider some decidable property P such as being (i) an odd perfect number, or (ii) an even perfect number, or a prime Mersenne number; and contrast two questions. Is $\mathcal{F}$ consistent with (i) there are no, (ii) there are only finitely many numbers $\in P$? Question (i) is essentially mathematical inasmuch as (i) is consistent iff (i) is true. Question (ii) or, equivalently, the underivability in $\mathcal{F}$ of there are infinitely many numbers $\in P$, may need a mixture of logical and mathematical analysis; logical analysis provides a lower bound on the density of P if (ii) is not consistent since the function $Y \in \mathcal{R}^-_{\mathcal{F}}$ where $Y : n \mapsto m_n$ and m_n is the nth number $\in P$; mathematical analysis could provide a sufficiently small upper bound to show $Y \notin \mathcal{R}^-_{\mathcal{F}}$ without necessarily deciding whether or not there are infinitely many numbers $\in P$.

Historical note (on Hilbert's foundational aims): At this stage it is possible to refute an idea which he stressed repeatedly, though in less precise terms, in his later writings [22]; to give a final solution of all foundational questions by purely mathematical means, specifically, by a general method for deciding whether or not any given arbitrary formal system is consistent. He was convinced that the notion of formal system was sharp enough for mathematical analysis (as was later verified by Turing). Such a method would provide a final solution, at least 'in principle': we should know, here and now, how to decide the consistency of any formal rules we encounter, whatever their source (semantic analysis or formal experimentation). Nor did he expect to need any -- presumably philosphical-- analysis of f.c. proofs: only one such proof would have to be inspected, the one used to establish that the method in question fulfills its purpose. NB. As we know now [11], such a method could be derived from a positive solution of the 10th problem (for which he demanded a solution, not a decision whether it is effectively solvable!), since by II.2(b), $\forall \vec{n}[p(\vec{n}) \neq 0]$ is true iff $A \wedge \forall \vec{x}[p(\vec{x}) \neq 0]$ is consistent where A are the axioms of arithmetic mentioned loc cit. -- Conversely nothing short of such a decision method could really be 'final' or even 'essentially mathematical'. One alternative, considered by Gödel [17], is to suppose that, as our ordinary mathematical experience develops and is codified in systems S, we shall find new methods of f.c. proofs to establish the consistency of such S. This solution, involving continued analysis of the notion of f.c. proof, would certainly not be 'final' in the sense intended by Hilbert -- and so Hilbert's idea is rigorously refuted. Perhaps it should be added here that, contrary to inept attempts at creating dramatic issues, such a decision method would not necessarily have 'finished' mathematics! it could leave open the problem of actually finding convincing decisions. Moreover, statistically speaking, such genuine decisions may need just as much ingenuity as solving a problem which does not fall into a known decidable class of problems; cf. the end of II.2(a) concerning the search for proofs of Val(A).

We now turn to consistency proofs for particular systems (actually used in practice), continuing to formulate results which are independent of any close analysis of the notion of f.c. proof.

4. Inadequacy of formal rules w.r.t. metamathematical practice: some consequences. In contrast to the inadequacies of II.3 w.r.t. theory, Gödel's well-known second incompleteness theorem establishes the inadequacy of formal rules $\mathcal{F}$ (of mathematical practice) w.r.t. metamathematical practice, namely the metamathematics of $\mathcal{F}$ itself.

On the one hand, using terms introduced in Section I (to be made a little more precise in a moment) we have Gödel's general result:

> No consistent system can demonstrate both its own consistency and its adequacy for numerical arithmetic and for substitution (for universal quantifiers).

On the other hand, for formal systems $\mathcal{F}$ (of arithmetic, set theory, etc.) accepted in mathematical practice, we have not only (actually: f.c.) proofs of their adequacy for numerical arithmetic and for substitution, but also obviously convincing consistency proofs, usually by semantic analysis. So $\mathcal{F}$ is inadequate w.r.t. the part of metamathematics which provides this semantic analysis. Evidently -- once again: for examining finitist claims -- it would be petitio principii to reject the semantic analysis merely because its results cannot be formally derived in $\mathcal{F}$. To avoid not only 'possible', but actually wide spread misunderstanding on this score, II.4(a) will go into the actual state of affairs in detail. In particular, II.4a will compare the reliability of consistency proofs by means of (i) semantic analysis and (ii) of f.c. methods envisaged by Hilbert for systems $\mathcal{F}$ which admit both types of consistency proof (i) and (ii). This comparison is relevant to the interest (or good sense) of Hilbert's program; as mentioned already in the Historical note of II.3, we cannot expect to decide whether the program can be carried out without some analysis of what f.c. proofs are; we do have something (negative) to say on what they are intended for, namely for increasing reliability. In view of these negative conclusions, II.4(b) and II.4(c) describe pragmatic uses of 'consistency' proofs independent of (dubious) epistemological claims.

Remarks on a precise formulation of Gödel's second incompleteness theorem; cf., for example, [28] for an up-to-date exposition. There are two points which need attention in all formalization of metamathematics: (i) Which data determine a formal system, say $\mathcal{F}$? (ii) Which formal expressions of $\mathcal{F}$ express metamathematical propositions (about $\mathcal{F}$ or other systems)? that is, propositions about sequences of symbols which are well-formed formulas or derivations of $\mathcal{F}$. As to (ii) the adequacy conditions on computation and substitution (in Section I, by means of the 'maps' σ, π, φ, and τ) for the formalization of number theoretic (f.c.) propositions give a good idea of what is done for metamathematical propositions; the main difference is that $\mathcal{F}$ is usually[7] taken to contain the language (of the ring) of integers, while metamathematical propositions have to be 'arithmetized': numbers are assigned to the sequences of symbols in $\mathcal{F}$. (The reader may wish to consider conditions on such numberings and adequacy conditions which ensure uniqueness -- demonstrable in $\mathcal{F}$ -- that is, uniqueness up to isomorphism for functions and relations, and up to equivalence for propositions). As to (i), it is evident that, for consistency questions, the proper data are determined by the rules of $\mathcal{F}$ and not, for example, by the set of theorems generated by $\mathcal{F}$. Normal forms for such rules, such as Post systems [26], make it possible to develop a general

[7]Modern texts -- in contrast to Gödel's original exposition [17] -- sometimes develop the work for $\mathcal{F}$ in the language of set theory, not arithmetic, and exploit familiar representations of finite sequences. In view of the place of set theory in current mathematics this is more natural. Note, however, that some devices introduced in [17] are of value in [11].

theory. As would be expected from the role of the (other) two adequacy conditions used in Section I to establish the significance of consistency, there are obvious examples of $\mathcal{F}$, at the end of [28], which do demonstrate their own consistency, but of course not the other adequacy conditions. (For every familiar system there is such an $\mathcal{F}$ with exactly the same derivations, only different rules for building up derivations). -- Here it should be remarked that there is a less well-known, but, as will appear in II.4(c), more useful property than consistency which expresses directly the significance of the latter, namely soundness of $\mathcal{F}$ for (f.c. propositions) $\forall n[p(n) \neq 0]$; in terms of Section I

$$\forall d\, \forall \vec{n}\{\mathrm{Der}_{\mathcal{F}}[d, \neg\, \sigma(c_p, \vec{n})] \Rightarrow p(\vec{n}) \neq 0\} ,$$

where $\mathrm{Der}_{\mathcal{F}}$ formalizes adequately the relation between derivations d of $\mathcal{F}$ and the end formula of d. Modulo the other adequacy conditions, soundness is equivalent to consistency. So, for consistent $\mathcal{F}$, for some $p_{\mathcal{F}}$ (obtained effectively from $\mathcal{F}$), soundness for $p_{\mathcal{F}}$ is not formally derivable in $\mathcal{F}$.

(a) Consistency proofs for formal axiomatic theories accepted in current mathematics: a comparison. In view of the role of set theory for the exposition of contemporary mathematics, we consider formal rules $\mathcal{F}$ in the language of set theory (containing some or all the familiar axioms and all the usual rules of inference of predicate logic). As was stated earlier on, near the beginning of II.4, if $\mathcal{F}$ is accepted in practice, then, as a matter of empirical fact, there is a semantic analysis of notions (here: of particular collections of sets) which satisfy the axioms of $\mathcal{F}$, and so ensure the consistency of $\mathcal{F}$. -- NB. This fact is of course perfectly consistent with another (empirical) fact: many individual mathematicians may not have found those notions for themselves nor have learnt about them -- and, quite properly, they will, at most, be fairly convinced about the consistency of $\mathcal{F}$ without being at all certain. All that is asserted is that somebody has formulated a good semantic analysis. I believe, it pays to be patient here, and recall briefly the order of events. Some 100 years ago mathematicians found Cantor's explanations insufficient (and he considered himself a victim of misunderstanding). Until about 50 years ago, some version of Zermelo's axioms was accepted. They are satisfied by the structure consisting of the domain $C_{\omega+\omega}$ ('C' for 'cumulative hierarchy') with the membership relation, where $C_0 = \emptyset$,

$C_{n+1} = \mathfrak{P}(C_n)$ for all $n < \omega$ and $C_\omega = \bigcup_n C_n$;

$C_{\omega+n+1} = \mathfrak{P}(C_{\omega+n})$ and $C_{\omega+\omega} = \bigcup_n C_{\omega+n}$.

Then people began to think about the replacement property which asserts that, for any functional relation (defined in the language of set theory), if the domain of the relation is (a set) in the structure considered, so is the range. (This is clearly true for C_ω and equally clearly false for $C_{\omega+\omega}$). Von Neumann [33] considered structures in which the replacement axiom is satisfied, but not the power set axiom, for example, $C^-_{\omega_1}$ where the hierarchy C^-_α of sets is built up like the C_α, except that the power set operation is replaced by $\mathfrak{P}^-: x \mapsto$ collection of countable subsets of x (and ω_1 is the first uncountable ordinal). In the early thirties, Zermelo [37] explained the properties of the 'full' hierarchy C_κ for inaccessible cardinals κ and, slowly, the formal axioms satisfied by these C_κ became familiar to mathematicians; even more slowly, the explanations became familiar to logicians. (The question, whether the explanation would be particularly useful to mathematical practice will be quite central for II.5 below.)

In short, for the systems $\mathcal{F}$ accepted in practice both completeness (for numerical arithmetic and for substitution) and consistency are established convincingly, but $\mathcal{F}$ is inadequate w.r.t. the second of these metamathematical results. Not only can the reasoning not be simulated by a formal derivation in $\mathcal{F}$, but there is no derivation of the consistency of $\mathcal{F}$ in $\mathcal{F}$ itself.

NB. Of course, there are formal systems $\mathcal{F}$ which are studied, particularly in metamathematics, but which are simply not known to be consistent (and certainly not accepted, realistically speaking).

Historical remark: In footnote 48a on p. 191 of his work on incompleteness results [17], Gödel stressed particularly the passage to 'higher types', that is, to larger C_α as the 'reason for incompleteness' or, equivalently, as the intended means of establishing consistency. This passage uses the expressive power of the set theoretic language as is seen by looking at Zermelo's axioms. The property, $C_{\omega+\omega}(x)$, of x being the set $C_{\omega+\omega}$ can be properly expressed in the language and then the axiom $\exists x\, C_{\omega+\omega}(x)$ can be added without extending the language. The new set of axioms is of course no longer satisfied by $C_{\omega+\omega}$, but by all C_α for limit ordinals $\alpha \geq \omega+\omega+\omega$. It is perhaps significant that later ([18]) Gödel expected the use of such 'axioms of infinity', properly generalized, to solve all set theoretic problems (the term 'axiom of infinity' is intended to suggest that they are satisfied by C_α only for large α).

At this stage people have been tempted to 'defend' set theory by rhetoric about the 'needs' of mathematics or (Hilbert) about depriving a boxer of the use of his fists and the rest of us of Cantor's paradise. (In view of the 'refinement of PM' referred to at the end of II.2, the rhetoric is objectively dubious). But one rarely takes seriously the more radical question, in the case of those $\mathcal{F}$ for which we have both kinds of consistency proof, by (i) semantic analysis of notions as above, and (ii) by the kind of f.c. analysis of derivations envisaged by Hilbert:

(*) Are proofs (ii) really more reliable than proofs (i)?

True, according to the finitist doctrine they <u>ought</u> to be; but then it would be <u>most</u> <u>uncritical</u> to accept the doctrine without question. The proper empirical procedure is to look at all the evidence available (not to let doctrine decide what ought to be accepted). A good example is axiomatic set theory without the axiom of infinity, where we have (i) the consistency proof by use of C_ω, and (ii) an f.c. consistency proof <u>via</u> an f.c. mapping of the formal theory of C_ω into arithmetic [2] together with an f.c. consistency proof for arithmetic. (The first published attempt at such a proof (ii) was simply wrong [1]. Instead of C_ω, we could have used $(\omega, +, \times)$ itself.)

If the answer to (*) is negative, the proper foundational conclusion is skepticism concerning the finitist doctrine; and the proper mathematical conclusion is, I think, to ask a pragmatic question:

> Now that we have some undoubtedly clever -- if not particularly 'reliable' -- combinatorial consistency proofs, what can we do with them?

(b) Consistency proofs as a metamathematical tool. We consider a general 'scheme' suggested by the following corollary to Gödel's second incompleteness theorem (for the usual axiomatic theories); the corollary establishes the consistency of an <u>extension</u> $\mathcal{F}'$ of $\mathcal{F}$ which is <u>not</u> obviously consistent: If $\mathrm{Con}_{\mathcal{F}}$ expresses the consistency of $\mathcal{F}$ then $\mathcal{F}'$, that is,

$$\mathcal{F} \text{ together with the additional axiom } \neg\,\mathrm{Con}_{\mathcal{F}},$$

is also consistent. (The reader should compare this result with the weaker result at the end of (a) when $\mathcal{F}$ is set theory without the axiom of infinity; proof (ii) shows only that $\mathcal{F}$ together with $\neg\,\exists x\, C_\omega(x)$ is consistent.) In fact, an easy exercise shows that any f.c. proposition which can be derived from $\neg\, \mathrm{Con}_{\mathcal{F}}$ by use of the system $\mathcal{F}$, can already be derived in $\mathcal{F}$ itself.[8]

The scheme in question is to prove the consistency of other extensions of $\mathcal{F}$ which are not obviously consistent; specifically, for any P such that

$$P \to \mathrm{Con}_{\mathcal{F}} \text{ can be derived in } \mathcal{F},$$

if $\mathcal{F}$ is consistent so is $\mathcal{F} \cup \{\neg P\}$. There is a wide choice of -- obviously true -- propositions P since consistency is a kind of minimal condition. (NB. Concerning the relevance of f.c. proofs for this kind of application it is not at all obvious that anything is gained by requiring f.c. proofs of $P \to \mathrm{Con}_{\mathcal{F}}$: the obviously relevant property is that they be formalized in $\mathcal{F}$. Attempts so far made to establish that, for all formal systems $\mathcal{F}$, such proofs can be brought to a normal form independent of $\mathcal{F}$ -- as an analogue to the 'fundamental' class of f.c. proofs -- have been refuted.)

Gentzen's famous consistency proof for formal number theory $\mathcal{Z}$, obtained by the scheme of II.1(c) (applied to + and x), provides such a P, say P_G, usually but misleadingly described as (an instance of) ϵ_0-induction. What is meant is a certain f.c. version of that principle, comparable to the méthode de déscente at the time of Fermat when one reasoned only about polynomial equations and other f.c. propositions; not about predicates of arbitrary logical complexity (to which ordinary, that is, ω-induction is applied in $\mathcal{Z}$). Thus while the modern version of (transfinite) induction on an ordering $<$ is

$$TI(A; <) : \text{derive } \forall x\, A(x) \text{ from } \forall x[\forall y\{y < x \to A(y)\} \to A(x)],$$

the f.c. version is (formally) weaker; for an f.c. term ρ

$$TI_{FC}(A,\rho; <) : \text{derive } \forall x\, A(x) \text{ from } \forall x[\{\rho(x) < x \to A[\rho(x)]\} \to A(x)]$$

where $\forall x A$ is again an f.c. proposition.

More importantly, in Gentzen's proof ϵ_0 is not the bare ordering of the order type of that name, namely the least ordinal $\alpha\ (> \omega)$: $\alpha = \omega^{\alpha}$ (or equivalently, $\alpha = 2^{\alpha}$ where we shall write $\exp_2 \alpha$ for 2^{α}), but an enrichment containing, besides 0 and ω, the build-up functions (i) and retracing

[8] This is perhaps the simplest example of the kind of pragmatic analogue to Hilbert's program mentioned in I.3(a), where $\mathcal{P}$ loc. cit. corresponds to the proofs formalized in $\mathcal{F}$ and the 'genuinely problematic' extension is $\mathcal{F} \cup \neg\mathrm{Con}_{\mathcal{F}}$. For more useful examples see [29] where $\mathcal{P}'$ is the extension of axiomatic set theory $\mathcal{P}$ by the continuum hypothesis or its negation.

functions (ii) below:

(i) successor s, (ordinal or Hessenberg's natural) +, $\exp_2$

(ii) predecessor p, (cut-off) subtraction $\dot{-}$, $\log_2$

with the following conventions (e.g. for px when x is a limit element):

$$\forall x(spx = x \vee px = x),\ \forall x\, \forall y(x = sy \rightarrow spx = x)\ ;$$

and so

$$\forall x\, \forall y[px = x \longleftrightarrow \forall y(x \neq sy)]\ ;$$

$$\forall x\, \forall y(y + (x\text{-}y) = x \vee x\text{-}y = 0),\quad \forall x\, \forall y\, \forall z(y+z = x \rightarrow y + (x\text{-}y) = x);$$

and so

$$\forall x\, \forall y(x\text{-}y = 0 \longleftrightarrow x \leq y)\ ;$$

$$\log_2 x = y \longleftrightarrow \exp_2 y \leq x < \exp_2 sy\ .$$

The equations above together with the familiar defining properties of 0, ω and the build-up functions, which consist of the recursion equations and continuity conditions on + and exp, determine all the functions (i) and (ii). NB. The retracing functions allow one to express the continuity properties by means of f.c. propositions.

An arithmetization of ϵ_0 is determined, by definition, by (formulas and terms expressing) an ordering $\dot{<}$ of the non-negative integers together with

elements (numbers) $\dot{0}$ and $\dot{\omega}$ and functions $\dot{s}$, $\dot{+}$, $\dot{\exp}_2$, $\dot{p}$, $\dot{-}$, $\dot{\log}_2$

which demonstrably satisfy the defining properties (for orderings etc.) mentioned above. These arithmetizations are adequate in the sense that

any two arithmetizations $\dot{<}$, $\dot{<}'$ of ϵ_0 are demonstrably isomorphic.

(For the isomorphism the retracing functions, applied to any element of $\dot{<}$, give its Cantor normal form to the base 2, and the build-up functions, applied to this notation, give the corresponding element of $\dot{<}'$. Thus, adequacy requires only that this isomorphism be demonstrable in the system considered.)

Gentzen's result can now be stated as follows:

THEOREM: The structure ϵ_0 has an f.c. (in technical jargon, primitive recursive) arithmetization, say $\dot{<}_0$;

$$\mathrm{Con}_{\mathcal{Z}} \text{ is derivable by means of } TI_{FC}(A, \rho; \dot{<}_0)$$

for suitable f.c. A and ρ (and P_G is that instance of TI_{FC});

For each proper segment of ϵ_0, $\{x : x \dot{<}_0 \bar{n}\}$, where $n = 0, 1, \ldots$ $TI(A, <_n)$ is derivable in $\mathcal{Z}$ for arbitrary predicates A of $\mathcal{Z}$.

Corollaries: The additional structure of ϵ_0, indicating how ϵ_0 is built up, is not only plausible, but demonstrably relevant: $\mathrm{Con}_{\mathcal{Z}}$ and, in fact, any true f.c. proposition can be derived by induction on some (f.c.) ordering of order type ω (which, by above, cannot satisfy the adequacy conditions mentioned; cf., [27], pp. 333-334, where the result is extended to all true propositions in the language of $\mathcal{Z}$ for orderings of order type $< \omega^{\omega}$)-- Gentzen's theorem gives another answer to the question of II.3(b), on what 'more' we know if F can be formally derived in $\mathcal{Z}$ than if we merely know that F if true: F can be derived by induction on some proper segment of ϵ_0 (not merely on some 'unstructured' well ordering), applied to predicates of the same logical complexity as F itself.

It is easy to find -- far too many -- points of mild interest in $TI_{FC}(\epsilon_0)$ as compared to $TI(\omega)$, some of which are explained in the Remark at the end of II.4(c). None has led to any sufficiently striking development to be compelling. At present it is more rewarding to apply the scheme explained at the beginning of II.4(b): provided the formal independence of P_G in $\mathcal{Z}$ is of interest. But the brutal fact remains that, though P_G has a more mathematical 'look' than the metamathematical proposition $\mathrm{Con}_{\mathcal{Z}}$, no convincing number theoretic interest of P_G has yet been found. II.4(c) below introduces some modifications which have led to results of (modest) interest in other parts of mathematics.

Historical note. The ingenuity needed at the present time to discover genuine applications of Gentzen's ideas may well be the price to pay for the haphazard genesis of his consistency proof. He introduced ϵ_0-induction after quite unfounded objections had been made against an earlier version, which he withdrew (and which is now available in [14]). That version used higher type language, not unlike one of the functional interpretations alluded to in II.3(a); and the objections involved the most elementary misconceptions concerning higher type, socalled abstract language (of II.2c) generally and particularly the effect of the logical complexity of the predicates to which abstract principles are applied (as illustrated by induction principles in II.4b); for an account of this comedy of errors, see the review of [14].

(c) Second thoughts on Gentzen's proof: extensions to arbitrary formulas (beyond those expressing f.c. propositions) and to many-sorted languages (of higher type). As emphasized throughout this article, consistency is significant because it ensures the soundness of formal theorems expressing f.c. propositions. So if a meaning or 'interpretation' is given to all formulas F of a system $\mathscr{F}$ then the proper metamathematical requirement on $\mathscr{F}$ is this:

For all F, if F is derivable in $\mathscr{F}$ then F is true

(for the interpretation in question). Now, as it stands this 'global' condition for all F cannot be expressed in the language of $\mathscr{F}$ at all, in the following precise sense first analyzed by Tarski. A predicate, say $T(x)$, is called a truth predicate for a class $\mathscr{C}$ of (closed) formulas if for each $F \in \mathscr{C}$, $T(\sigma_F) \longleftrightarrow F$, where σ_F is the canonical definition of F in the sense explained on p. 21. Though this condition is patently a minimal requirement (independent of the particular interpretation considered), Tarski observed that there is no truth predicate for $\mathscr{C}$ such that (each instance $T(\sigma_F)$ of) T is itself in $\mathscr{C}$ provided only $\mathscr{C}$

> is closed under negation and each F is completely defined, not, for example, 'three-valued', and so not: $F \leftrightarrow \neg F$.

To extend the scheme of II.4(b) for getting metamathematical results about $\mathscr{F}$ itself, the global soundness requirement must be reformulated in the language of $\mathscr{F}$; for example, by a schema for individual formulas F. The best one can do by way of generality is to allow free variables in F. So suppose σ_F (with the variable n) is a term of $\mathscr{F}$ which defines the nth formula F(0), F(1), ... Then, soundness is expressed by

$$(*) \qquad \forall n\, \forall d\,[\mathrm{Der}_{\mathscr{F}}(d, \sigma_F) \rightarrow F] \quad .$$

This extends Hilbert's scheme since F is not required to be f.c. If F does not contain any variable then (*) is equivalent to

$$\forall d\,[\neg F \rightarrow \neg\, \mathrm{Der}_{\mathscr{F}}(d, \sigma_F)]$$

that is: The formula F cannot be derived in $\mathcal{F}$ from $\neg F$; in other words, $\mathcal{F} \cup \{\neg F\}$ is consistent. So proofs of soundness provide formal consistency proofs (from $\neg F$) for (such) finite extensions of $\mathcal{F}$ -- besides having a clear intrinsic meaning. In the particular case of Gentzen's consistency proof mentioned in II.4(b) it is not hard to see that it establishes soundness too. In fact we have the following neat equivalence:

> The same theorems are derivable in $\mathcal{Z}$ from (i) the soundness schema (*) above and (ii) ϵ_0-induction,

where both schemata are applied to arbitrary formulas of $\mathcal{Z}$ (of about the same logical complexity). More usefully:

> The result applies, mutatis mutandis, to the extension of the language of arithmetic to all finite types in the sense of II.2(c).

Here the principle of (ω -) induction is applied to all formulas of the extended language -- but no other axioms (about higher types) are added. By above, soundness automatically 'takes care' of any finite addition of axioms.

The observation above is useful because the extended language is quite well adapted to formulate branches of mathematics where ordinals do occur (in contrast to the case of pure number theory where nothing 'like' Gentzen's ordinal structures, of ordinal ϵ_0, has found a natural application); for example, in the descriptive theory of sets of points when measuring the lengths of sequences of derived sets in the theorem of Cantor-Bendixson. An immediate application of the extension of (*) to higher types for suitable additional axioms and definitions of closed subsets X of $\mathbb{R}$ is this: For X (and additional axioms) of suitably restricted logical complexity,[9] if X can be proved to have a perfect kernel then the length of the derived sequence of X is $< \epsilon_0$. Earlier applications of this type include bounds on the ordinals of orderings for which induction can be proved for all predicates of given complexity (by suitably restricted means) [27], and bounds on the lengths of hierarchies in recursion theory, mentioned at the end of II.2; cf. [13].

---

[9] For specialists: The axioms involved are elementary except for the so-called Σ_1^1-axiom of choice and the sets X are socalled Σ_1^1-closed, that is, the sets G_X of complementary intervals i with rational end points are Π_1^1 : $X = \{\xi : \forall i(\xi \in i \to i \notin G_X)\}$. (The axioms considered imply the principle of the least upper bound formulated in the language of arithmetic with types 0, $0 \to 0$ and $(0 \to 0) \to 0$).

Remark on a different application of the f.c. version TI_{FC} (of transfinite induction TI) as used in Gentzen's work, II.4(b). One considers (partial) orderings $<$ which are not well founded (and so TI($<$) is not generally valid), but pseudo-founded; in the sense that every descending recursive sequence in $<$ is finite, and every non-empty recursive subset of the field of $<$ has a first element (and so $TI_{FC}(<)$ is valid). Examples of such $<$, pertinent to the present article, arise in the natural extension of the completeness theorem in II.2(a) to validity in recursive models, say Val(A, Rec). Here the exact choice of data determining a model (of A) is relevant; a model is given by the satisfaction relation for all formulas in the language (of A). Let $\mathfrak{C}$ be any class of sets closed under recursive operations, for example, $\mathfrak{C}$ = Rec. Then $\mathrm{Val}(A, \mathfrak{C}) \Longleftrightarrow \models_1^{\mathfrak{C}} A$, where $\models_1^{\mathfrak{C}} A$ means that there is a (recursive) $\mathfrak{C}$-founded tree, built up locally by use of Frege's rules and having A as its end formula. Incidentally, this completeness theorem requires additional structure (as in Gentzen's ϵ_0, but here related to the rules of inference) connecting nodes of the tree, and not only the 'bare' tree-ordering of the nodes (where premises of inferences precede conclusions). This generalized completeness theorem is also a good example of the kind of 'definability theory' mentioned in the abstract of this article.

II.5. Hilbert's program in perspective. It is now possible to sharpen the general conclusions of I.1 - I.3 by referring to the details in II.1 - II.4. Though the main stress is on the mathematical side, it would be misleading to ignore the matter of finitist foundations (in I.3b) since this was - and remains - the most familiar aim of Hilbert's program--and it is the aim as much as the failure which demands critical analysis. I.1(a), concerning a structural theory of proofs, will be considered last, since this is farthest removed from Hilbert's proof theory, which concerns the validity of principles of proof.

(a) The mathematical significance of proof theory was touched on in (i)I.1(b) and I.3(a) which concern consistency and related properties and in (ii)I.2 which concerns f.c. operations and proofs.

(i) By II.4(a) the consistency of 'basic' theories such as current axiomatic set theory is established by semantic analysis. This does not depend on formalization at all, but applies to (second order) axiomatizations as explained in II.1 and II.2. In contrast incomplete formalizations and restrictions on the methods of the consistency proof are used essentially for the pragmatic version of Hilbert's program explained in II.4(b); for example, in the case of set theory $\mathcal{S}$ and the continuum hypothesis CH, the applications require $\mathrm{Con}(\mathcal{S} \cup \{\pm \mathrm{CH}\})$ to be proved in $\mathcal{S} \cup \{\mathrm{Con}\,\mathcal{S}\}$.

Warning. Granted that semantic analysis and hence in particular the question whether CH is true (in $C_{\omega+3}$ or-equivalently, as a moment's thought should convince the reader - in each $C_{\omega+\alpha}$ for $\alpha \geq 3$), are meaningful it is open whether semantic analysis is often useful for mathematical practice. After all, consistency questions are a kind of singularity in the body of mathematics and it cannot be assumed that the semantic methods particularly appropriate for establishing consistency are generally useful. For example, by [29], CH and $\neg$ CH are both in principle irrelevant to number theory and to logically simple statements in analysis (in technical jargon, $\Pi_2^1 \cup \Sigma_2^1$ statements). But experience shows that both CH and $\neg$ CH can be useful; in model theory several decidability results were first found by 'assuming' CH, and in degree theory several incomparability results were first proved simply by 'assuming' $\neg$ CH. Put differently, one embedded the structures involved (natural numbers, ordinals) in models of set theory which do, resp. do not satisfy CH. It would be simple minded to assume that, for the branches of mathematics considered, the full cumulative hierarchy C_α is as useful, let alone more useful than those other models (even though the latter are defined by means of the C_α). The fact remains that those models are more manageable than the C_α, roughly speaking, because our present knowledge of the C_α allows us to decide 'more' questions about those models than about the C_α.

(ii) The general notion of formal rule or of recursive function obviously has a permanent place in pure mathematics; both in connection with definability (of sets or functions of hereditarily finite objects such as elements of finitely generated groups) and in connection with validity or provability (of first order theorems). For applications, both outside mathematics and within it, the general or, as one sometimes says, 'idealized' notion is of quite limited interest or even misleading; since practical applications in the literal sense, for example, computations are concerned with a finite number of problems, and with actual decisions, not the possibility 'in principle' of deciding an infinite set of problems.

Historical note. The popular literature on socalled constructive mathematics uses a highly idealized notion of 'applicability' depending on a quite problematic view of the nature of mathematical reasoning; cf. II.5(b). As a result it rarely goes into the possibility of a conflict between actual applicability and the demand for constructive proofs. For actual applications (which requires efficiency and intelligibility) the difference between linear and exponential speed-up is more significant than the so to speak non-archimedean difference between exponential and non-recursive speed-up.

Apart from the mathematically precise general notion of formal rule the vague ideas which led originally to f.c. mathematics are used a lot in modern work since they attracted such mathematicians as Poincaré and Brouwer (whose dissertation consisted largely of a diatribe against set theoretic foundations). These ideas led them and others to the algebraization of analysis mentioned in II.2(c), and more generally to the use of topological methods involving operations on finitely determined neighborhoods. Their work allows the conclusive solution of problems about f.c. operations by use of quite superficial properties of these operations. Specifically, for negative solutions concerning f.c. operations, for example, on $\mathbb{R}$ one needs often only the fact that such operations are continuous (for the topology appropriate to the data used to present real numbers). If there is no continuous solution of the problem at all, there can be no f.c. solution; and so one need not even ask whether there is an f.c. proof. Again, for positive solutions one has a device familiar from the literature on category theory, the choice of 'richer' (realistically available) data. Here one uses 'enrichments,' for example, of the graph of a function by its range to solve a functional equation by restricted, for example, f.c. operations when this is demonstrably not possible without the enrichments.[10]

Historical note. Thus, to some extent, ordinary mathematics has 'preempted' the subject of (potential) applications of proof theory, having by now solved just those problems which appeared to be prime candidates for such applications in the twenties, the heyday of Hilbert's program. (As a practical consequence, the reader will find that much of Bishop's work [8] has a very familiar ring, and, perhaps therefore, immediate appeal; in any case it is quite intelligible without reference to specifically logical matters such as those appearing in the introduction to [8]). As mentioned already in II.3(b) and developed in detail in [27] pp. 361-362 and the literature cited there, proof theory has been used for getting some explicit bounds. The uses were pedagogic, proving easy mathematical results

---

[10] For examples of negative results, consider (i) the construction of a zero for cubic equations (with leading coefficient 1) from the coefficients, say $c \mapsto x$ for $x^3 - 3x = c$, (ii) the construction of a fixed point ζ for continuous mappings f of the unit circle S^2 into itself. There is clearly no mapping $f : c \mapsto x$ which is continuous for the usual topology on $\mathbb{R}$ in case (i), and no $F : f \mapsto \zeta$ in case (ii) where F is continuous on the uniform convergence topology on the class of mappings $f : S^2 \to S^2$. On the other hand, we get positive results in case (i) if the data c are 'enriched' by specific representations, say binary expansions with the corresponding Baire space topology on: $\omega \to 2$. (In case (ii), it appears, useful enrichments are more recondite giving, for example, domains in which there is at most one fixed point of f.)

easily by use of general logical devices; this is comparable to the uses of model theory until 10 or 15 years ago.

As to the place of f.c. proofs in mathematical practice, what requires justification is not its extension by abstract methods, but the restriction to f.c. proofs, for example, of TI to TI_{FC} in the Remark at the end of II.4(b); to be compared to our present view of Sturm's restricted algebraic methods, described at the end of I.1(a), or for that matter of many other temporary uses of Ockham's razor.

(b) Finitist foundations and the general issues of I.3(b). Tacitly, but quite deliberately the work on Hilbert's program was presented largely independently of Hilbert's own aims, and hence with a different emphasis; for example, what was treated in Section I as *the* significance of consistency for systems of arithmetic occurs as a mere aside in one of Hilbert's presentations. The original - and still most familiar - aim of the program is in the socalled *critical tradition* in the foundations of mathematics which, being preoccupied with the possibility of systematic *error in our ordinary view* of things, tries to discover or avoid such errors by taking *extraordinary precautions*; specifically, by questioning precisely those notions which are used most freely in practice. Naturally, conflicts with practice, for example those in II.5(a) above, by no means count against a foundational scheme in the critical tradition; they almost constitute a *raison d'être*.

Historical note. The familiar dramatic or pious presentations of the set theoretic paradoxes conceal - in effect if not in intention - the fact that they are a Godsend to the critical tradition (which, without them, decidedly lacks lustre). For the same reason, skepticism concerning this tradition leads to emphasis on the *reliability* of our ordinary practical tests, for example, to the aside in footnote 3 on Cantor's instantaneous criticism of Frege's axioms and to the account in II.4(a) of the actual development of set theory.

Further - and contrary to a perhaps almost universal view - the present exposition does not dwell on pragmatic weaknesses of (any restriction to) finitist foundations, since this is a delicate *statistical* matter. (Does the good done by the trend towards algebraization, associated in the minds of some of its founders with finitist foundations, outweigh the harm done by the neglect of set theoretic aspects of, say, functional analysis?) On the contrary, as stressed particularly at the end of II.4(a), the generally 'accepted' element of finitist foundations is most problematic, the claim to greater *reliability* or, in technical jargon, the claim that the finitist notion of mathematical rigor constitutes an *appropriate idealization of reliability*. It is a *bad* idealization if, for example, our ordinary non-finitist principles are equally reliable, say 100%, but the probability of error is greater for f.c. proofs than for others (of the same theorem). Put differently, what is at issue is this: Are the tacit assumptions behind finitist foundations, in particular, concerning the general character of our intellectual apparatus (involved in mathematical reasoning) even remotely as plausible as our ordinary views? On p.9 of [21] Hilbert expressed some of those tacit assumptions in the clearest possible terms, citing Aristotle (who-it must be admitted in view of II.1 - made quite a fair guess on the basis of _the mathematics of his time_).

Historical note. As an example of what might be called an 'uncritical view of the critical tradition,' one quite popular reaction to Gödel's incompleteness theorem is perhaps worth mentioning, namely neo-formalist foundations, as propounded, for example, in Bourbaki. Here the consistency of formal rules is to be established by a statistical analysis of their uses in actual practice. This ignores not only the fact in II.4(a) concerning the discovery of (new) axioms for set theory but--more significantly--the fact that practice does not even attempt to _apply the standard tests which the new proposal would require_; in particular, a first step would be to see if the statistical support applies to the usual systems or only to the 'subsystems,' restricted by logical complexity, considered at the end of II.2(c). In short, the proposal pays lip service to the finitist _assumption_ that, say, set theoretical principles are significantly less reliable than finitist ones, but _does nothing about it even if it could do so_ (specifically by stating explicitly the subsystem in which the results of practice can be derived).

Naturally, the skepticism concerning critical foundations expressed here by no means rejects foundational aims or questions as such. It criticized some common-place views not for being philosophical but for being bad or banal philosophy; and above all for _assuming_ that our present day mathematics is an adequate basis for the foundational questions considered. The stress on f.c. proofs is connected with the present topic: socalled predicativity is no better.

(c) Structural complexity of proofs is the subject of what is sometimes called _theory of proofs_ to distinguish it from Hilbert's proof theory (which, as memtioned before, is primarily concerned with the validity of principles of proofs, not with the way they are combined). What do we want to know about proofs? Obviously, the number of _steps_ is a factor, particularly when we have to do with proofs in which the individual formulas are all intelligible, for example, because of a simple abstract meaning; if not (or for automatic proof checking by machines), the number of _symbols_ used or the number of steps needed to verify that the proof is correctly built up is more realistic. A more imaginative factor, not relevant to deterministic computations, but to logical proofs which contain interlocking arguments, was introduced by Statman [31], the topological genus of proof figures built up according to rules of natural deduction (introduced by Gentzen for the express purpose of a faithful representation of natural reasoning). The need for something like this last measure is convincingly illustrated by Statman's analysis in [31] of the effect of explicit definitions (together with suitable 'axioms', as in successful uses of 'abstract' mathematics) which patently reduce complexity of proofs in the ordinary sense of the word, but affect the number of steps relatively little. Work in the theory of proofs establishes in a precise sense the conflict discussed in II.5(b) between actual reliability, that is, low probability of error, and the finitist idealization of relability; not only in 'principle' but for the methods used in ordinary practice. The theory of proofs has benefitted from work in proof theory both in an obvious and in a subtler way.

First of all, the theory of proofs applies some of the methods of proof theory, for example those used in Gentzen's consistency proof in II.4(b) but not mentioned in the statements of his results (which refer only to the sets of theorems provable by different kinds of rules). Indeed, even casual inspection of Gentzen's work shows that the bulk of the proof concerns certain 'reduction' procedures for transforming derivations; the principle $TI_{FC}(\epsilon_0)$ which is central to his own formulation of the (metamathematical) significance of the proof, is used only to prove the termination or 'convergence' of those procedures. (Consistency follows because derivations of purely numerical propositions are reduced to numerical computations of the same end formula). Without the foundational preoccupations about the unreliability of abstract principles, one would not think of mentioning only the method of the convergence proofs,[11] but would analyze the internal features of the derivations (other than the end formula) which are preserved by the transformations. The theory of proofs does just this by providing concepts and problems about the derivations themselves to which such an analysis is relevant.

Historical note. The shift of interest, from the 'results' stated in proof theory to what were - considered to be - 'auxiliary' constructions, is to be expected from general experience with 'grand' theories

---

The situation may be compared to that stage of the theory of differential equations when people were preoccupied with convergence proofs, say by the method of majorizing power series, and did not have the appropriate (topological) notions to state structural properties of the solution.

(here: the foundational theory behind Hilbert's program). When such grand theories are refuted, incidentally quite often by easy general arguments as in II.3, some of the methods developed in work on the grand theory may retain an interest; but the most essential step is to make the interest independent of the discarded assumptions, that is, to _reinterpret old proofs_ and _formulate new theorems_; cf. the warning at the end of Section I.

It may well be that one day the _negative_ results on Hilbert's program will be seen to have been more useful to a theory of proofs than the mathematical tools inherited from proof theory. Quite naively, principles come first and applications afterwards; so as long as the 'critical' doubts about abstract principles mentioned in II.5(b) persist, the first - and of course most glamorous - job is to remove them; all the more if there is a _conflict_ between the 'critical' ideal and realistic measures of reliability of proofs. Besides, the discovery of such measures is obviously a much more delicate business, even if less spectacular than the 'grand' aim (which, as mentioned before, is the kind of aim that occurs to us when we know very little).

Biographical note. Though the aims of Hilbert's program and of the theory of proofs are usually unrelated and often in conflict, it should not be assumed that the aims of that theory are in conflict with Hilbert's own interests! Both from his writings, [20], and from conversations quoted in [7], it appears that at least at one time, his principal interest was to attract the attention of mathematicians to the topic of _mathematical proofs as an object of_ (mathematical) _study_: and the 'grand' program was to have been the bait.

REFERENCES (to SECTIONS I and II)

[1] W. Ackermann, Begrundung des "Tertium non datur", Math. Ann. 93 (1924), 1-36.

[2] __________, Die Widerspruchsfreiheit der allgemeinen Mengenlehre, Math. Ann. 114 (1937), 305-315.

[3] A. Baker, Rational approximations to $\sqrt[3]{2}$ and other algebraic numbers, Quarterly J. Math. Oxford 15 (1964), 375-383.

[4] __________, On Mahler's classification of transcendental numbers, Acta. Math. 111 (1964), 97-120.

[5] J. Barwise and P. Eklof, Lefshetz's principle, J. Algebra 13 (1969), 554-570.

[6] P. Bernays, Sur les questions méthodologiques actuelles de la théorie hilbertienne de la démonstration, les entretiens de Zurich sur les fondements et la méthode des sciences mathématiques, Zurich 1941.

[7] ________, Mathematics as a domain of theoretical science and also of mental experience, Logic Colloquium Proc. Bristol 1973 (to appear).

[8] E. Bishop, Foundations of constructive analysis, N.Y. 1967.

[9] C. Cantor, Review of Frege's Grundlagen der Arithmetik, Deutsche Literaturzeitung 6 (1885), 728-729; reprinted in Cantor's Gesammelte Abhandlungen, ed. Zermelo, Berlin 1932.

[10] H. Davenport and K. F. Roth, Rational approximations to algebraic numbers, Mathematika 2 (1955), 160-167.

[11] M. Davis, Y. Matyasevic and J. Robinson, this volume.

[12] P. C. Eklof and E. R. Fisher, The elementary theory of abelian groups, Ann. Math. Logic 4 (1972), 115-171.

[13] H. Friedman, Iterated inductive definitions and Σ_2^1 - AC, pp. 435-442 in: Intuitionism and proof theory, ed. Myhill, et al., Amsterdam, 1970.

[14] G. Gentzen, The Collected Papers of Gerhard Gentzen, ed. Szabo, Amsterdam 1969; reviewed in J. of Philosophy 68 (1971), 238-265.

[15] K. Gödel, Die Vollständigkeit der Axiome des logischen Funktionenkalküls, Mh. Math. Phys. 37 (1930), 349-360.

[16] ________, Einige metamathematische Resultate über Entscheidungsdefinitheit und Widerspruchsfreiheit, Akad. Wiss. Wien, Math. Naturw. Klasse, Anzeiger, 67 (1930), 214-215.

[17] ________, Über formal unentscheidbare Sätze der Principia Mathematica und verwandter Systeme, Mh. Math. Phys. 38 (1931), 173-198.

[18] ________, Remarks before the Princeton Bicentennial Conference on problems in mathematics, pp. 84-88 in: The Undecidable, ed. Davis, N.Y. 1964.

[19] G. Higman, Subgroups of finitely presented groups, Proc. Roy. Soc. A 262 (1961), 455-474.

[20] D. Hilbert, Axiomatisches Denken, Math. Ann. 78 (1918), 405-515.

[21] ________, Probleme der Grundlegung der Mathematik, Math. Ann. 102 (1930), 1-9.

[22] ________, Grundlegung der elementaren Zahlentheorie, Math. Ann. 104 (1931), 485-494.

[23] R. B. Jensen, The fine structure of L, Ann. Math. Logic 4 (1972), 229-308.

[24] H. J. Keisler, Logic with the quantifier "there exist uncountably many", Ann. Math. Logic 1 (1970), 1-93.

[25] ________, Model theory for infinitary logic; logic with countable conjunctions and finite quantifiers, Amsterdam, 1971.

[26] S. C. Kleene, Introduction to metamathematics, Princeton, 1950.

[27] G. Kreisel, A survey of proof theory, JSL 33 (1968), 321-388.

[28] G. Kreisel and G. Takeuti, Formally self-referential propositions for cut-free classical analysis and related systems, Diss. Math. 118 (1974), 4-50.

[29] D. A. Martin. This volume.

[30] J. C. Shepherdson, Non-standard models for fragments of number theory, pp. 342-358 in: The theory of models, ed. Addison et al., Amsterdam 1965.

[31] R. Statman, Structural complexity of proofs, Dissertation, Stanford, 1974.

[32] C. Sturm, Mémoire sur la résolution des équations numériques, Mémoires présentés par divers savants étrangers à l' Acad. Roy. Sc. Inst. France, Sc. Math. Phys. 6 (1835).

[33] W. Szmielew, Elementary properties of Abelian groups, Fund. Math. 41 (1955), 203-271.

[34] A. Tarski, What is elementary geometry?, pp. 16-29 in: The axiomatic method, ed. Henkin et al., Amsterdam 1959.

[35] J. von Neumann, Eine Axiomatisierung der Mengenlehre, J. f. Math. 154 (1925), 219-240.

[36] E. Zermelo, Untersuchungen über die Grundlagen der Mengenlehre, Math. Ann. 65 (1908), 261-281.

[37] ________, Über Grenzzahlen und Mengenbereiche, Fund. Math. 16 (1930), 28-47.

P.S. Since this paper was completed (in Autumn 1974), the details of the extended completeness theorem, mentioned in the Remark at the end of II.4, have been published in [38], pp. 52-59, with an example on pp. 61-68 establishing the significance of the choice of data (determining a model).

[38] G. Kreisel, G.E. Mints and S.G. Simpson, The use of abstract language in elementary metamathematics: some pedagogic examples, Logic Colloquium, ed. Parikh, Springer Lecture Notes 453 (1975), 38-131.

Proceedings of Symposia in Pure Mathematics
Volume 28, 1976

PROBLEM IV: DESARGUESIAN SPACES

Herbert Busemann

The Fourth Problem concerns the geometries in which the ordinary lines, i.e. lines of an n-dimensional (real) projective space P^n or pieces of them are the shortest curves or geodesics. Specifically, Hilbert asks for the construction of all these metrics and the study of the individual geometries. It is clear from Hilbert's comments that he was not aware of the immense number of these metrics, so that the second part of the problem is not a well posed question and has inevitably been replaced by the investigation of special, or special classes of, interesting geometries. Already the historically first (non-Riemannian) of these metrics, the Minkowskian, exhibited the phenomenon of non-symmetric distances.

In giving a precise modern version of Problem IV we therefore begin with the definition of a <u>not necessarily symmetric metric space R</u>:

A distance xy is defined on $R \times R$ with $xx = 0$, $xy > 0$ if $x \neq y$, $xy + yz \geqq xz$, and $xx_\nu \to 0$ if and only if $x_\nu x \to 0$.

There are two types of balls:

$$B^+(p,\rho) = \{x : px \leqq \rho\},\ B^-(p,\rho) = \{x : xp \leqq \rho\}\ (\rho > 0).$$

For a symmetric distance we use $B(p,\rho)$. A <u>segment</u> $T(x,y)$ from x to y is a curve from x to y isometric to an interval of length xy on the real axis, i.e. representable as

$$z(t),\ \alpha \leqq t \leqq \beta = \alpha + xy,\ z(\alpha) = x,\ z(\beta) = y \text{ and}$$
$$z(t_1)\, z(t_2) = t_2 - t_1 \text{ if } t_1 < t_2.$$

R is an <u>n-dimensional Desarguesian space if R is a nonempty open subset of P^n, metrized by a not necessarily symmetric distance</u> xy (with the correct topology, of course) such that

(a) <u>The $B^+(p,\rho)$, but not necessarily the $B^-(p,\rho)$, are compact.</u>

(b) $T(x,y)$ <u>exists for any</u> x,y.

(c) $xy + yz > xz$ <u>unless</u> x, y, z <u>are collinear in</u> P^n.

Since $T(x,y)$ exists it must by (c) fall on the intersection of a projective line L with R and $L \cap R$ must be connected. Because R is an open subset of P^n either $L \cap R = L$ or $L \cap R$ is an open connected subset of L. In the sequel a line, plane, hyperplane etc. in R will always mean a nonempty intersection of a projective line, plane, hyperplane, etc. with R.

The subject is clearly related to three fields: the <u>foundations of geometry</u> on which

Hilbert put the principal emphasis, the calculus of variations and differential geometry.

Hilbert does not mention the latter for an obvious historical reason. The only metrics considered in differential geometry before 1900 were Riemannian, and for these the problem had been solved long before (1866) by Beltrami who showed that a Riemannian metric whose geodesics fall on ordinary lines has constant curvature. This in conjunction with the compactness of the $B(p,\rho)$ yields the three well known solutions: the entire P^n with an elliptic metric, the P^n with a hyperplane omitted, or the n-dimensional affine space A^n, with a euclidean metric and (with a proper choice of the ideal locus) the interior of an ellipsoid in A^n with a hyperbolic metric.

As to the calculus of variations, we have a so-called inverse problem. Usually an integrand F depending on point and direction is given and the curves giving the integral over F extremal values are sought. In an inverse problem the extremals are given (in our case the lines) and the corresponding F are to be found. The first contribution to Problem IV, Hamel's Thesis [1], was written under Hilbert's guidance and concentrates on the calculus of variations aspect with nonsymmetric distances. The later version [2] emphasizes principally symmetric distances.

From the point of view of the foundations of geometry it is mandatory to have an intrinsic geometric property of the geodesics which guarantees that they may be regarded as parts of projective lines, in fact, Hilbert's formulation implies this approach.

We keep all the axioms except the strong congruence axiom and replace this by the requirement that ordinary lines be the shortest connections. For $n > 2$ the situation is very simple. One axiom is that three points lie in a plane and this suffices. The space can then be mapped into P^n such that the given lines fall on projective lines. It is not necessary to postulate that any $r + 1$ points lie in a r-flat when $n - 1 \geqq r > 2$ because this follows.

For $n = 2$ everything lies in a plane and an additional requirement becomes necessary. The validity of the Theorem of Desargues and its converse in those cases where the necessary intersections exist (which we denote as the Desargues Property) is clearly necessary and proves also sufficient. Detailed proofs of these facts can be found, for instance, in [3, Sections 13 and 14]. It is a curious fact that Hilbert forgot to mention Desargues in his formulation of Problem IV, although he himself discovered and emphasized this role of the Desargues Theorem. The omission is repeated in [1], but corrected in [2].

The two-dimensional result is strongly used in the proof for $n > 2$, because a plane in a space of dimension exceeding 2 has automatically the Desargues Property. This justifies the brief term "Desarguesian Spaces" used here as title.

There are very few general theorems on Desarguesian spaces. The most important goes back to Hamel [2] and is formulated there only for symmetric distances, but it also holds in the general case, see [4, p. 37]:

If R is Desarguesian ($n \geqq 2$) then either $R = P^n$ and all lines have the same length (traversed either way) or R leaves out an entire hyperplane and may therefore be regarded as an open convex set in A^n. The theorem implies, for example, that the exterior of a conic in P^2 (i.e. the set of points lying on two tangents) cannot be metrized as a Desarguesian space.

This theorem exhibits one of two phenomena which apply to many statements in the theory, namely that the Desarguesian character is either not necessary or deducible from other hypothses. We give an example for the first.

Define a geodesic in a symmetric metric space as a locally isometric map of the entire real axis. If the $B(p, \rho)$ are compact and the geodesic through two distinct points is unique, then all geodesics are either isometric to the real axis or to one circle C (with its intrinsic distance). Moreover, in the latter case, if the dimension exceeds 1, the space has a two-sheeted universal covering space, whose geodesics are isometric to a circle with twice the length of C and share with the ordinary sphere the property that all geodesics through one point meet again at a second, see [3, p. 201]. The omission of a hyperplane requires, of course, the Desarguesian character. We will not return to this point later.

A second general theorem on Desarguesian spaces which is obvious in the Riemannian, but not the general case and has so far only been proved for symmetric distances is this (see [4, p. 32]):

An n-dimensional Desarguesian space can be embedded (with preservation of the metric) as a hyperplane in an $(n+1)$-dimensional Desarguesian space.

Our next task is to construct some interesting special Desarguesian metrics and to exhibit general methods of generating an abundance of Desarguesian metrics of which very few have interesting geometric properties.

For this purpose it is useful, although unsatisfactory from the purist's point of view, to introduce an auxiliary euclidean metric $\epsilon(x, y)$ in an affine space A^n with affine coordinates $x^1, \ldots, x^n$ by

$$\epsilon(x, y) = [\sum (x^i - y^i)^2]^{1/2}.$$

Besides the topology given by $\epsilon(x, y)$ we use only the following simple, but fundamental fact: If the pairs of points a, b and c, d lie on parallel lines then

$$\epsilon(a, b) : \epsilon(c, d) \text{ is an affine invariant,}$$

i.e. does not change if we replace the x^i by other affine coordinates.

The first non-Riemannian Desarguesian space and by far the most important one was constructed around 1890 by Minkowski for applications in number theory. Let C be a bounded nonempty open convex set in A^n with the (compact) boundary B and $p \in C$. For any two points x, y define the Minkowski distance $m(x, y)$ by $m(x, x) = 0$ and for $x \neq y$ as follows: let the ray with origin p parallel to the ray from x through y intersect B at q. Then

$$m(x, y) = \epsilon(x, y) : \epsilon(p, q).$$

If C has an affine center and this is p, then always $m(x, y) = m(y, x)$ and we obtain what is now known as a finite-dimensional Banach space. The condition c) is satisfied if and only if B is strictly convex. Both the $B^+(p, \rho)$ and $B^-(p, \rho)$ are compact.

The importance of Minkowski spaces is for non-Riemannian smooth spaces (Finsler spaces) the same as that of euclidean spaces for Riemannian geometry: every Finsler space behaves locally like a Minkowski space.

Consequently there are very many characterizations of the Minkowskian spaces. Besides the obvious one by translation invariance we mention here only two. A third and fourth will appear later in connection with the Funk geometry and the concept of curvature. The Minkowskian geometries are the only Desarguesian metrics defined in all of A^n and compact convex $B^{\pm}(p,\rho)$, see Zaustinsky [5, p. 46]. They are the only Desarguesian spaces in which the loci equidistant to a hyperplane H (i.e. $xH = \rho$ and $Hx = \rho$) are pairs of hyperplanes, see Phadke [6]. The principal point in the far from simple proof is showing that the metric must be defined in all of A^n. It is rather easily seen that the boundary cannot contain a strictly convex piece and is therefore polyhedral. In the differential geometric approach to Problem IV it is always assumed that the boundary has nonvanishing curvature, so that the problem becomes rather trivial. This applies to many situations and explains why the contributions of the differential geometers, which will be discussed later, have little relation to the foundations of geometry.

A non-Riemannian analogue $h(x,y)$ to hyperbolic geometry was discovered by Hilbert in 1894, see [7, Anhang 1]. With the same notation as above and R( ) denoting cross-ratio let x, y be points of C. Put $h(x,x) = 0$ and for $x \neq y$ let the line through x and y intersect B at u and v in the order v, x, y, u. Then with a positive constant k

$$h(x,y) = k \log R(x,y,u,v) = k \log\left[\frac{\epsilon(x,u)}{\epsilon(y,u)} + \log\frac{\epsilon(y,v)}{\epsilon(x,v)}\right] = h(y,x)$$

The triangle inequality follows from the convexity of C, but $xy + yz > xz$ for noncollinear x, y, z requires only that a plane which intersects C, intersects B in a curve which does not have two noncollinear segments. The set $\{x : xT \leqq \rho\}$ is convex for every segment T (see Pedersen [8] and [3, Section 18]). In Riemannian geometry this is equivalent to nonpositive curvature, see [8] and [3, Section 41].

The requirement that in a triangle with vertices a, b, c and midpoints b' of a, b and c' of a, c always $2h(b',c') \leqq h(b,c)$ is in Riemannian geometry also equivalent to nonpositive curvature [3, l. c.], but implies for a Hilbert geometry that it is hyperbolic (see Kelly and Straus [9]).

Finally we come to a most interesting metric $f(x,y)$ discovered by Funk [10] in 1929. It is so-to-say half a Hilbert geometry. Its main features are that not all $B^-(p,\rho)$ are compact and that it resembles euclidean geometry in some and hyperbolic geometry in other respects. Keeping the previous notations put $f(x,x) = 0$ and for $x \neq y$

$$f(x,y) = k \log \frac{\epsilon(x,u)}{\epsilon(y,u)}.$$

For $\epsilon(y,u) \to 0$ we have $f(x,y) \to \infty$ so that the $B^+(p,\rho)$ are compact, but for $\epsilon(x,v) \to 0$

$$f(x,y) \to k \log \frac{\epsilon(v,u)}{\epsilon(y,u)} < \infty ,$$

so that $B^-(p,\rho)$ is not compact for fixed p and large ρ.

The condition c) is equivalent to the strict convexity of B. The proof uses the

Theorem of Menelaus (compare [5, Appendix I, where the triangle fa'z must be replaced by xyz]).

Obviously the $B^+(p, \rho)$ are all homothetic to $B \cup C$ and hence to each other. Requiring that the $B^+(p, \rho)$ be homothetic distinguishes the geometries of Minkowski and Funk among all Desarguesian spaces; see [11].

With the above notation the spheres through x about y (i.e. $f(y, x) = \rho$) tend for $\epsilon(y, u) \to 0$ to a curve homothetic to B and touching B at u. Thus we encounter the hyperbolic phenomenon of limitspheres. There are several other facts showing this ambivalent behaviour of the Funk geometry.

There does not seem to be any special interesting non-Riemannian geometry defined in all of P^n, although we will see that there are very many.

The abundance of Desarguesian metrics is perhaps best understood by beginning with a very simple example. We modify $\epsilon(x, y)$:

$$xy = \epsilon(x, y) + |e^{x^n} - e^{y^n}|.$$

This defines a Desarguesian metric in all of A^n because the absolute value satisfies the triangle equality and e^t is monotone. Any other monotone function would have done, many such functions can be added, and watching convergence even infinitely many. Also, the addition can be replaced by an integration over a continuous family.

Of course, it is very easy to generate nonsymmetric distances by this method, for example by modifying a nonsymmetric Minkowski distance.

It seems clear that Hilbert was not aware of these simple possibilities. For, speaking of Desarguesian geometries defined in all of A^n he says: "One finds that such a geometry really exists and is no other than that which Minkowski constructed in his book, Geometrie der Zahlen...".

The method of addition of quasimetrics was not invented merely for destructive purposes, but to answer a variety of questions. For example, it allows us to show in a few lines, that infinitely many essentially different Desarguesian metrics exist for any given open convex set in A^n, see [3, p. 111]. Here one example will suffice to exhibit the method : the strip $-\pi/2 < x^n < \pi/2$ is metrized by

$$xy = \epsilon(x, y) + |\tan x^n - \tan y^n| .$$

The only property used is that tan t increases monotonically from $-\infty$ to ∞ in $(-\pi/2, \pi/2)$.

It is very easily seen that a metric in a set $C \subset A^2$ which possesses all rotations about two distinct points is euclidean or hyperbolic, but there is much liberty in constructing Desarguesian metrics with all rotations about one point. The above method applied with integration yields among others Desarguesian metrics of this type in all of A^2 such that for any two parallel lines L_1, L_2 always $xL_2 \to 0$ (or always $xL_2 \to \infty$) when x traverses L_1 in either direction, see [4, p. 29].

Integral geometry leads to a completely different way of generating Desarguesian spaces. In integral geometry one considers measures on sets of geometric objects other than points, for example sets of hyperplanes. For significant results, in E^n say, congruent sets of hyperplanes must have the same measure, but this aspect is irrevelant in the present context. Important is that the number or measure of hyperplanes intersecting a curve, each counted as often as it meets the curve, equals the length of the curve, usually with a constant factor depending on the unit of the measure. The author observed in [12] that this method can be reversed as it were and then automatically yields Desarguesian metrics.

Since we concentrated on A^n so far, let us consider P^n. Take a countably additive nonnegative set function $w(S)$ defined on all Borel sets S of hyperplanes H in P^n. For any pointset M put

$$H_M = \{H : H \cap M \neq \emptyset\}$$

and require

(1) $0 < w(G) < \infty$ for a nonempty open set G of H.

(2) $w(H_p) = 0$ for any point p.

Then, for any line L,

$$0 < w(H_L) = w(H_{P^n}) = 2k < \infty,$$

because $w(\{H \supset L\}) = 0$ by (2). If A is an arc of a projective line divided by the point q into the subarcs A_1, A_2 then, by (2),

$$w(H_{A_1}) + w(H_{A_2}) = w(H_A).$$

Consider a triangle with sides A, B, C such that $A \cup B \cup C$ bounds a simply connected domain in the plane of the triangle. Then

$$w(H_A) + w(H_B) > w(H_C).$$

For, any hyperplane intersecting C intersects A or B, and there is an open set of hyperplanes cutting $A \cup B$ without intersecting C.

This means that the projective lines are locally shortest connections and hence geodesics. We now define a distance xy as follows: $xx = 0$ and if $x \neq y$ we take that arc A (or one of the two) on the line L through x and y for which $w(H_A) \leqq k$ and put

$$(*) \qquad xy = w(H_A).$$

Then xy is a Desarguesian metrization of P^n. Thus each w yields a Desarguesian geometry, but notice that the distance is always symmetric.

The same can be done in A^n provided we admit ∞ as possible value for $w(S)$. A domain for which the distances are finite is automatically convex. For, choose p arbitrarily and consider the set of all points x for which $px < \infty$. This set is open if it is n-dimensional and is then convex. For, if $px < \infty$, $py < \infty$ then $xy \leqq xp + py < \infty$. Hence, if z lies on the segment from x to y then $zx < \infty$ and $zp \leqq px + xp < \infty$.

This leads naturally to the question of how much more general a symmetric Desarguesian space can be. In [13], entitled "A complete solution of Hilbert's Fourth Problem", Pogorelov announced the very beautiful result, that each symmetric

Desarguesian space can be obtained in this way. The proof is very sketchy and a detailed version does not seem to be available as yet.

The argument consists of two parts: in the first a smooth function $g(H)$ is determined such that for a smooth Desarguesian distance $w(S) = \int_S g(H)$ and (*) hold. Then for a given not necessarily smooth distance xy a sequence w_ν, derived from smooth distances, converging to a set function w is constructed by an averaging process such that (*) holds.

The second part, although many details are missing, seems clear enough. In the first part the case $n = 2$ is treated in some detail, but that there must be a solution for $n = 2$ is evident from an old paper of Blaschke [14] who solved the analogous problem for general two-dimensional variational problems, i.e. also those in which the extremals are not necessarily straight lines, giving a measure for the extremals such that a given curve has a length (in terms of the variational problem) which equals the number of extremals intersecting it.

Unfortunately Pogorelov passes the cases $n > 2$ over by a remark to the effect that they are treated in the same way as the plane case. However, it seems to the author that $n > 2$ offers essentially new difficulties and he was not able to reconstruct a complete proof from the brief observation of Pogorelov.

The title of [13] is somewhat misleading in as much as the integral geometric method yields only symmetric distances, whereas in Problem IV nonsymmetric distances have been considered from the very beginning (already by Minkowski). For these the first paper [1] by Hamel (which is restricted to $n = 2, 3$ and smooth metrics) provides until now the only solution. We do not reproduce it here, since it is not based on a general idea like integral geometry and the expression given by Hamel is not illuminating. Also, inevitably in view of the time it was written, the discussion of completeness does not satisfy our modern requirements, the necessary concepts did not yet exist. However, the immense number of possibilities becomes quite clear from [1].

Finally we discuss the contributions of differential geometry to Problem IV. These could not begin before a general theory of non-Riemannian spaces was created by Finsler [15] in 1918. Taking all necessary smoothness conditions for granted one considers on a differentiable manifold a function $F(x, \xi) = F(x^1, \dots, x^n; \xi^1, \dots, \xi^n)$ where ξ is a contravariant or tangent vector. The length of a curve $x(t)$ is given by $\int F(x, \dot{x})dt$. To yield meaningful results F must satisfy certain conditions. $F(x, \xi) > 0$ for $\xi \neq 0$ to make length positive, $F(x, k\xi) = kF(x, \xi)$ for $k \geqq 0$ to assure independence of the parametrization. This homogeneity implies with the summation convention

$$F^2(x, \xi) = \tfrac{1}{2} F^2_{\xi^i \xi^k}(x, \xi)\, \xi^i \xi^k = g_{ik}(x, \xi)\, \xi^i \xi^k \quad \text{for } \xi \neq 0.$$

In the Riemannian case the $g_{ik}(x, \xi)$ are independent of ξ. Our previous considerations show the enormous increase in generality caused by this at first sight innocuous looking difference.

The Legendre Condition that the matrix $(g_{ik}(x, \xi))$ be positive definite for $\xi \neq 0$

expresses that the locus $F(x,\xi) = 1$ in ξ-space has for fixed x nowhere vanishing curvature. It is always assumed in the theory of Finsler spaces. The Jacobi Condition, which plays a considerable role in the calculus of variations and hence in some aspects of Finsler geometry is irrelevant in our case, because it deals with conjugate points, which do not occur when the straight lines are the extremals.

Unfortunately there are many different theories of Finsler spaces; for an excellent detailed discussion see Rund [16]. It turns out that many Riemannian concepts can be generalized to Finsler spaces in different ways so that they coincide with the Riemannian when the g_{ik} do not depend on ξ. Fortunately for our subject, only one of these theories has been seriously applied to Problem IV, namely the historically first with Berwald and Funk as principal representatives, which flourished roughly in the decade from 1928 to 1938 in the then existing German University at Prague.

Following Berwald [17] (which generalizes results obtained by Funk [10] for $n = 2$ and gives a brief introduction to Berwald's general theory), we choose the parameter such that $F(x,\dot{x}) = 1$. Then the extremals are the same for F and F^2. The Euler equations for F^2 have the form

$$\ddot{x}^k + 2\Gamma^k(x,\dot{x}) = 0,\ k = 1,\ldots,n,$$

where the Γ^k are positive homogenous of order 2 in $\dot{x}$. With the notations

$$\frac{\partial \Gamma^k}{\partial \dot{x}^l} = \Gamma^k_l, \quad \frac{\partial^2 \Gamma^k}{\partial \dot{x}^l \partial \dot{x}^m} = \Gamma^k_{lm}$$

the curvature tensor is given by

$$K_{k\cdot lm}^{\ i} = \left(\frac{\partial \Gamma^i_{kl}}{\partial x^m} - \frac{\partial \Gamma^i_{kl}}{\partial \dot{x}^r}\Gamma^r_m\right) - \left(\frac{\partial \Gamma^i_{km}}{\partial x^l} - \frac{\partial \Gamma^i_{km}}{\partial \dot{x}^r}\Gamma^r_l\right) + \Gamma^r_{kl}\Gamma^i_{rm} - \Gamma^r_{km}\Gamma^i_{rl}.$$

As in Riemannian geometry we put

$$K_{iklm} = g_{kp}K^{\ p}_{i.lm},\quad K_{il} = K^{\ p}_{i.lp}$$

and define

$$s(x,\dot{x},\pi) = \frac{K_{iklm}\dot{x}^i\dot{x}^l\pi^k\pi^m}{(g_{il}g_{km} - g_{im}g_{kl})\,\dot{x}^i\dot{x}^l\pi^k\pi^m}$$

as the sectional curvature (Krümmungsmass) of the space at the line element $x,\dot{x}$ and the plane element π. The curvature scalar (or Ricci curvature) at $x,\dot{x}$ is

$$q(x,\dot{x}) = (n-1)^{-1}K_{ik}\dot{x}^i\dot{x}^k.$$

Among the many results of Berwald we mention in particular: If the extremals are straight lines then $s(x,\dot{x},\pi)$ has the same value for each plane π containing $\dot{x}$ and hence coincides with $q(x,\dot{x})$:

$$s(x,\dot{x},\pi) = q(x,\dot{x}) \text{ if } \dot{x} \text{ is parallel to } \pi.$$

If the extremals are straight lines and $q(x,\dot{x})$ is independent of $\dot{x}$ (or $s(x,\dot{x},\pi)$ is independent of $\dot{x}$) then $q(x)$ is constant.

A particularly attractive result of Berwald is, that the Desarguesian spaces of constant curvature are characterized by the property

P: An isometry of one line on another or itself is (the restriction of) a projectivity.

The three special geometries of Minkowski, Hilbert and Funk discussed above have this property and constant curvatures $0, -k^{-2}/4, -k^{-2}/4$.

The Minkowskian and Hilbert geometries are jointly characterized by the properties: they are defined in subsets of A^n, have compact $B^{\pm}(p, \rho)$ and constant curvature.

If we replace the last property by P we obtain a theorem whose assertion does not involve differentiability and can, in fact, be proved very easily without smoothness assumptions, compare [4, p. 38].

It should be emphasized that both the hypotheses that the space be noncompact and the $B^{\pm}(p, \rho)$ compact are important. If either of them is omitted there are many solutions, most of them uninteresting. Berwald determined in [17] all n-dimensional Desarguesian spaces with curvature 0 and Funk [18] all two-dimensional Desarguesian metrics with constant positive curvature. Even the latter depend on an essentially arbitrary function, where "essential" means that the function is restricted by an inequality only.

We conclude with some remarks on the future of Problem IV. The problem of generating all not necessarily symmetric Desarguesian metrics, preferably in some elegant way like Pogorelov's, remains still open. Also, it is quite possible that some new interesting special metrics will be discovered. A desideratum is, in particular, a nonelliptic metric in P^2 or P^n with intuitively appealing properties.

The present development seems, however, to point in a different direction. In spite of the generality of Desarguesian spaces some very interesting spaces, which ought to be considered under this heading, do not fall under our difinition, namely those where the lines are geodesics, but not necessarily the only ones. A general theory comprising this case is outlined in [19].

The simplest problem, where this case appears naturally, arises when we ask, in which symmetric Minkowski planes the unit-circle has maximal or minimal length. The maximal length is 8 and obtained when the circles are parallelograms; the minimal length is 6 and obtained when the circles are affinely equivalent to regular euclidean hexagons, see Petty [20].

There are many Minkowski planes in which the perpendicularity of lines is symmetric, compare [2, p. 104]. The only Hilbert planes with symmetric perpendicularity are those in which the boundary is either an ellipse (i.e. the metric is hyperbolic) or a triangle; see Kelly and Page [21].

A most interesting new two-dimensional (generalized Desarguesian) metric where the domain is a triangle and the circles are hexagons was recently discovered by Phadke [22]. It is besides the Minkowskian the only metric in which the loci equidistant to lines are again pairs of lines.

Added in proof.

At the time of my lecture (referred to below as R) only the Doklady note [13] by Pogorelov, claiming to solve Problem IV for symmetric distances, existed, with no proofs for dimension $n > 2$. In the meantime a detailed treatment by Pogorelov has appeared as the booklet

(1) The Fourth Problem of Hilbert, Moscow 1974

which, unfortunately, I only succeeded to see too late to amend R.

This is to inform the reader about a serious discrepancy between [13] and (1).

The most important tool is a "density" function $g(H)$ defined on the hyperplanes. In [13] it is asserted to be positive for all n. However, I had proved in

(2) Areas in affine spaces III — The integral geometry of affine area, Rend. Cont. Circ. Mat. Palermo Ser. II, 9(1960), 1-17

that a positive $g(H)$ solving Problem IV, which is translation invariant does not exist for $n > 2$ in every Minkowskian geometry, but that a unique $g(h)$ which may also take negative values does (it is positive if and only if the spheres are projection bodies). For the purposes of (2) which aims at general integral geometry, densities which take negative values had to be rejected.

Now Pogorelov's $g(H)$ is translation invariant in the Minkowski case, so that his claim in [13] could not be valid, but (2) suggested that $g(H)$ with some negative values might work for the general Problem IV. Very misleadingly I stated in R only that I could not complete Pogorelov's "proof" for $n > 2$, trying to give Pogorelov the privilege of revealing his error.

However, (1) does not mention [13] at all and replaces the correct reference [12] by an incorrect one to [3]. In (1) Pogorelov allows $g(H)$ indeed to take negative values giving necessary and sufficient conditions for the extent to which $g(H)$ may be negative. Thus he constructs, in fact, all symmetric Desarguesian spaces.

Nonsymmetric distances which entered the problem from its inception are completely disregarded in (1) so that the earliest contribution to Problem IV, Hamel's thesis [1] (written under Hilbert) remains to this day the only one in the general case.

Perhaps the two omissions of Pogorelov are due to the fact that the series in which (1) appeared is aiming at an elementary level.

Herbert Busemann

REFERENCES

[1] G. Hamel, Ueber die Geometrien, in denen die Graden die Kürzesten sind, Dissertation, Göttingen, 1901.

[2] G. Hamel, same title, Math. Ann. 57 (1903), 231-264.

[3] H. Busemann, The Geometry of Geodesics, New York, 1955.

[4] H. Busemann, Recent Synthetic Differential Geometry, Ergebnisse der Mathematik, Vol. 54, 1970.

[5] E.M. Zaustinsky, Spaces with nonsymmetric distance, Mem. Amer. Math. Soc., No 34, 1959.

[6] B.B. Phadke, Equidistant loci and the Minkowskian geometries, Canad. J. Math. 24 (1972), 312-327.

[7] D. Hilbert, Grundlagen der Geometrie, 8th ed. Stuttgart, 1956.

[8] F.P. Pedersen, On spaces with negative curvature, Mat. Tidsskrift B (1952), 66-89.

[9] P. Kelly and E.G. Straus, Curvature in Hilbert geometry, Pacific J. Math. 8 (1958), 119-125.

[10] P. Funk, Über Geometrien, bei denen die Geraden die Kürzesten sind, Math. Ann. 101 (1929), 226-237.

[11] H. Busemann, Spaces with homothetic spheres, Journal of Geometry 4-2(1974), 1-12.

[12] H. Busemann, Geometries in which the planes minimize area, Ann. Mat. Pure Appl. (IV) 55 (1961), 171-190.

[13] A.V. Pogorelov, A complete solution of Hilbert's Fourth Problem, Soviet Math. Dokl. 14 (1973), 46-49.

[14] W. Blaschke, Integralgeometrie 11, Zur Variationsrechnung, Abh. Math. Sem. Univ. Hamburg 11 (1936), 359-366.

[15] P. Finsler, Über Kurven und Flächen in allgemeinen Räumen, Dissert. Göttingen 1918 and Basel 1951.

[16] H. Rund, The Differential Geometry of Finsler Spaces, Berlin-Göttingen-Heidelberg, 1959.

[17] L. Berwald, Über die n-dimensionden Geometrien konstanter Krümmung in denen die Geraden die Kürzesten sind, Math. Zeitschr. 30 (1929), 449-469.

[18] P. Funk, Über zweidimensionale Finslersche Räume, insbesondre über solche mit geradliningen Extremalen und positiver konstanter Krümmung, Math. Zeitschr. 40 (1935), 86-93.

[19] H. Busemann, Spaces with distinguished shortest joins, A spectrum of Mathematics, Auckland, 1971, 108-120.

[20] C.M. Petty, On the geometry of the Minkowski plane, Rivista di Mat. Univ. Parma 6 (1955), 269-292.

[21] P. Kelly and L.J. Page, Symmetric perpendicularity in Hilbert geometries, Pacific J. Math. 2 (1952), 319-322.

[22] B.B. Phadke, A triangular world with hexagonal circles, Geometriae Dedicata 3(1975), 511-520.

Proceedings of Symposia in Pure Mathematics
Volume 28, 1976

HILBERT'S FIFTH PROBLEM AND RELATED PROBLEMS ON TRANSFORMATION GROUPS

C. T. YANG[1]

ABSTRACT

This article deals with Hilbert's fifth problem within the scope of transformation groups.

The study of any classical geometry such as euclidean geometry, affine geometry, projective geometry, etc. may be regarded as an investigation of a particular transformation groups. For example, if R^n is the euclidean n-space and $E(n)$ is the group of euclidean transformations of R^n (that is, the group generated by translations, rotations and reflections of R^n), then $(R^n, E(n))$ is a transformation group and the study of n-dimensional euclidean geometry is the investigation of invariants of this transformation group.

With such transformation groups in mind, Lie conceived the general concept of continuous groups of transformations of manifolds. Needless to say, Lie took differentiability for granted. Therefore a finite continuous group of transformations of a manifold in the sense of Lie is actually a differentiable transformation group in the present-day sense. Even with the assumption of differentiability, any transformation group arising from classical geometry is included in the general concept as a special case.

In 1900, Hilbert proposed the following problem:

> "How are Lie's concept of continuous groups of transformations of manifolds is approachable in our investigation without the assumption of differentiability".

This is the fifth of twenty-three problems proposed by Hilbert. See [9].

In order to ask definite questions suggested by Hilbert's fifth problem, we take for granted some basic definitions which are often used in the study of transformation groups, as seen in [15] and [3].

The following question is often regarded as Hilbert's fifth problem.

(I) Is every locally euclidean group a Lie group?

Before this question was finally answered in the affirmative in 1952, it inspired many works on topological groups and several partial answers were found. The earliest partial answer was for locally euclidean groups of dimension ≤ 2 and was given by Brouwer [5]. The most well-known partial answers are for compact locally euclidean groups and for commutative locally euclidean groups; these were given by von Neumann

AMS (MOS) subject classifications (1970). 22-03, 54H15, 57E15, 57E99.

[1]Research supported by the National Science Foundation.

[17] and Pontrjagin [19] respectively.

The final answer may be stated as:

THEOREM 1. *Every locally euclidean group is a Lie group.*

Theorem 1 is a consequence of results given in two papers: one by Gleason [7] and the other by Montgomery-Zippin [16]. These works were later simplified and extended by Yamabe [23], [24] and were further improved by Kaplansky [10]. Since the proof of Theorem 1 is very complicated and technical, it is impossible for us to sketch it here. (We refer the reader to [15] and [10] for further details.) Besides it is certainly a challenge to find a simpler proof.

The following structural theorem for locally compact groups is a by-product of efforts seeking an answer to question (I).

THEOREM 2 (Yamabe [23]). *Every locally compact group has an open subgroup which, up to an isomoprhism, is the projective limit of Lie groups.*

Since Lie was concerned with continuous groups of transformations of manifolds, it seems more appropriate to ask (I) in a more general version.

(I') *If* G *is a locally compact group which acts effectively on a manifold as a (topological) transformation group, is* G *a Lie group?*

We still do not know how to answer question (I'). However, several partial answers are known and the following are the most important ones:

THEOREM 3. *Let* G *be a locally compact group which acts effectively on a differentiable manifold* M *such that, for any* $g \in G$, $x \mapsto gx$ *is a differentiable transformation of* M. *Then* G *is a Lie group and* (G, M) *is a differentiable transformation group.*

THEOREM 4. *If* G *is a compact group which acts effectively on a manifold, and if every element of* G *is of finite order, then* G *is a finite group.*

For Theorem 3, see Montgomery [14], Bochner-Montgomery [2] and Kuranishi [13]. Theorem 4 is a consequence of Theorem 2 and a theorem of Newman [17] which says that any effective action of a finite group on a manifold cannot have uniformly small orbits unless the group is trivial.

Making use of Theorems 2 and 4, we can show that an affirmative answer to (I') is equivalent to a negative answer to the following question:

(I'') *Does there exist an effective action of a* p-*adic group of a manifold?*

So far, (I'') remains unanswered even though we know that the existence of such an action is very unnatural. The unnaturalness of such an action is indicated in Theorem 5 below which can be found in Yang [25] and Bredon-Raymond-Williams [4].

THEOREM 5. *If* X *is an* n-*dimensional separable metric space on which there is an effective action of the* p-*adic group* G, *where* p *is a prime, then the dimension of the orbit space* K/G *is either* ∞ *or* $\leqq n+3$. *If, in particular,* X *is an* n-*manifold, then the dimension of* X/G *is either* ∞ *or* $n+2$.

Next, let us interpret Hilbert's fifth problem by asking the general question below.

(II) <u>If</u> (G, M) <u>is an effective topological transformation group in which</u> G <u>is a locally euclidean group and</u> M <u>is manifold, can</u> (G, M) <u>be made into a differentiable transformation group</u>?

We first observe that G is a Lie group (Theorem 1) so that question (II) asks for a differentiable structure on M with respect to which G acts differentiably on M. Even if G is trivial, the question is by no means easy. In fact, one may find the first negative result in Kervaire [11] which shows that there is a closed PL 10-manifold which does not admit any differentiable structure. For the case in which G is the cyclic group of order 2, Bing [1] shows that there is an effective action of G on the 3-sphere such that the fixed point set is a horned sphere; the answer is again negative.

Because of these negative results, we modify question (II) by asking the following one instead.

(II') <u>Let</u> (G, M, φ) <u>be an effective topological transformation group in which</u> G <u>is a Lie group,</u> M <u>is a manifold and</u> φ <u>is an action of</u> G <u>on</u> M. <u>Find a "reasonable" condition which may be used to insure that</u> (G, M, φ) <u>can be made into a differentiable transformation group</u>.

Using recent developments in PL topology and differential topology, we find the question not only interesting but also difficult. This may be illustrated by the special case where G is the circle group, M is the $(2n + 1)$-sphere and φ is a free action. Of such a transformation group, the orbit space $M^* = M/G$ is an integral cohomology $(2n)$-manifold having the homotopy type of the complex projective n-space CP^n. Since we can recover the transformation group (G, M, φ) from the orbit space M^* and since (G, M, φ) can be made into a differentiable transformation group iff M^* can be made into a differentiable $(2n)$-manifold, our task here may be divided into three steps:

(i) Classify topological $(2n)$-manifolds M^* which have the homotopy type of CP^n.

(ii) Find those M^* which admit PL structures.

(iii) Find those PL M^* which admit differentiable structures.

For $n = 1$, all three steps are obvious. For $n = 2$, (i) cannot be done because it involves classifying certain 4-manifolds. For $n \geq 3$, (i) can be done by using the celebrated work of Kirby-Siebenman [12] in a general setting provided by Sullivan [20]. Then it is not hard to see that exactly half of them admit PL structures so that (ii) is solved. For PL $(2n)$-manifolds which have the homotopy type of CP^n, $n \geq 3$, see Sullivan [21]. As for (iii), no general result is known, but partial results may be found in Brumfiel [6].

Roughly speaking, question (II') asks when a transformation group in the topological category is also in the differentiable category. Therefore it is natural to ask questions using other categories. For the question of when a transformation group in the differentiable category is in the real analytic category or even in the real algebraic category, see Palais [18].

Before asking the question of when a transformation group in the topological category is in the PL category, we first need a "good" definition of a PL transformation group. As seen in Gluck [8], to define a PL transformation group in the obvious way is no good unless a group with a commutative identity component is used. Hence there is no natural way to give a class of transformation groups which is appropriate for the PL category and is larger than the class of differentiable transformation groups. For this purpose, we should study stratified transformation groups, where a stratified transformation group consists of a (compact) Lie group, a (locally compact) stratified space and a stratified action tried to the stratified structure of the stratified space. Of course, we must take great care in formulating the definition of a stratified space (which is slightly different from the usual definition) and the definition of a stratified transformation group so that the following hold. First, a (locally compact) space is stratified iff it is triangulable. Hence the suspension of a compact stratified space is a compact stratified space. Next, any (compact) differentiable transformation group is a stratified transformation group. Third, the orbit space of a stratified transformation group is a stratified space and the fixed point set of a stratified transformation group is a stratified space. Details of the study will be given elsewhere.

In concluding this article, we note that we interpret Hilbert's fifth problem only within the scope of transformation groups. Therefore we have not considered "infinite continuous groups" which are not groups in the present-day sense.

REFERENCES

1. Bing, A homeomorphism between the 3-sphere and the sum of two solid horned spheres. Ann. of Math. 36(1952), 354-362.
2. Bochner-Montgomery, Locally compact groups of differentiable transformations. Ann. of Math. 47(1946), 639-653.
3. Bredon, Compact Transformation Groups. Academic Press, New York, 1972.
4. Bredon-Raymond-Williams, p-adic groups of transformations. Trans. Amer. Math. Soc. 99(1961), 488-498.
5. Brouwer, Die Theorie der endlichen kontinuierlichen Gruppen unabhagin von der Axiomen von Lie. (two papers) Math. Ann. 67(1909), 246-267; 69(1910), 181-203.
6. Brumfiel, Differentiable S^1-actions on homotopy spheres. Preprint, Princeton University, 1969.
7. Gleason, Groups without small subgroups. Ann. of Math. 56(1952), 193-212.
8. Gluck, Piecewise linear groups and transformation groups. Trans. Amer. Math. Soc. 149(1970), 585-593.
9. Hilbert, Mathematical problems. Bull. Amer. Math. Soc. 8(1901-02), 437-479.
10. Kaplansky, Lie Algebras and Locally Compact Groups. Chicago University Press, Chicago, 1971.
11. Kervaire, A manifold which does not admit any differentiable structure. Comment. Math. Helv. 34(1960), 257-270.
12. Kirby-Siebenman, On the triangulation of manifolds and the Hauptvermutung.

Bull. Amer. Math. Soc. 75(1969), 742-749.

13. Kuranishi, On conditions of differentiability of locally compact groups. Nagoya Math. J. 1(1950), 71-81.

14. Montgomery, Topological groups of differentiable transformations. Ann. of Math 46(1945), 382-387.

15. Montgomery-Zippin, Tological Transformation Groups. Wiley (Interscience), New York, 1955.

16. Montgomery-Zippin, Small groups of finite-dimensional groups. Ann. of Math. 56(1952), 213-241.

17. Newman, A theorem of periodic transformations of spaces. Quart. J. Math. 2(1931), 1-8.

18. Palais, Equivariant, real algebraic differential topology. Preprint, Brandeis University, 1973.

19. Pontrjagin, Topological Groups. Princeton University Press, Princeton, 1939.

20. Sullivan, Triangulating homotopy equivalences. Thesis, Princeton University, 1965.

21. Sullivan, Triangulating and smoothing homotopy equivalences and homeomorphisms. Geometric Topology Seminar Notes. Princeton University, 1969.

22. von Neumann, Die Einfuhrung analytischer Parameter in topologischen Gruppen. Ann. of Math. 34(1933), 170-190.

23. Yamabe, On the conjecture of Iwasawa and Gleason. Ann. of Math. 58(1953), 48-54.

24. Yamabe, Generalization of a theorem of Gleason. Ann. of Math. 58(1953), 351-365.

25. Yang, p-adic transformation groups. Mich. Math. J. 7(1960), 201-218.

Department of Mathematics, University of Pennsylvania, Philadelphia, Pennsylvania 19104

Proceedings of Symposia in Pure Mathematics
Volume 28, 1976

HILBERT'S SIXTH PROBLEM:

MATHEMATICAL TREATMENT OF THE AXIOMS OF PHYSICS

A.S. Wightman[1]
Princeton, New Jersey

In enumerating his twenty-three problems, Hilbert distinguished between specific problems and general questions about the foundations of branches of mathematical knowledge. The sixth problem fell in the latter category. Since physics has gone through two revolutions in the twentieth century, one might think that the original questions posed by Hilbert would be by now primarily of archaeological interest. From this point of view, it would be natural for me to talk instead about the sixth problem that Hilbert might have posed if he were alive today. To some extent, I am going to try to follow this course in the second part of my talk. However, it is instructive first to examine exactly the question Hilbert proposed and what he and others succeeded in doing to answer it during his lifetime.

Hilbert's Statement of the Problem

"Investigations of the foundations of geometry suggest the problem: To treat in the same manner, by means of axioms, those physical sciences in which mathematics plays an important part; first of all, the theory of probability and mechanics.

As to the axioms of the theory of probability,(*) it seems to me desirable that their logical investigation should be accompanied by a rigorous and satisfactory development of the method of mean values in mathematical physics and in particular in the kinetic theory of gases.

AMS(MOS) subject classifications (1970). Primary 81-02, 81-03; Secondary 81A12, 81A17, 81A18, 82A15, 83C05.

[1]Research supported in part by the Air Force Office of Scientific Research under Contract F44620-71-C-0108

Important investigations by physicists on the foundations of mechanics are at hand; I refer to the writings of Mach,(†) Hertz,(††) Boltzmann,(**) and Volkmann.(§) It is therefore very desirable that the discussion of the foundations of mechanics be taken up by mathematicians also. Thus, Boltzmann's work on the principles of mechanics suggests the problem of developing mathematically the limiting processes, there merely indicated, that lead from the atomistic view to the laws of motion of continua. Conversely, one might try to derive the laws of motion of rigid bodies by a limiting process from a system of axioms depending upon the idea of continuously varying conditions of a material filling all space continuously, these conditions being defined by parameters. For the question of the equivalence of different systems of axioms is always of great theoretical interest.

If geometry is to serve as a model for the treatment of physical axioms, we should try first, using a small number of axioms, to include as large a class as possible of physical phenomena, and then by adjoining new axioms to arrive gradually at the more special theories. Perhaps, here a subdivision principle can be taken over from Lie's profound theory of infinite transformation groups. The mathematician will have to take into account all logically possible theories, and not merely those that approximate reality, just as he has done in geometry. He must always be careful to obtain a complete overview of the body of conclusions that follow from the assumed axiom system.

(*) Bohlmann Actuarial Mathematics second lecture in Klein and Riecke Applied Mathematics and Physics, Leipzig 1900
(†) Mach The Development of Mechanics, 2nd Edit., Leipzig 1889
(††) Hertz The Principles of Mechanics, Leipzig 1894
(**) Boltzmann Lectures on the Principles of Mechanics, Leipzig 1897
(§) Volkmann Introduction to the Study of Theoretical Physics, Leipzig 1900

Further, the mathematician has the task of testing in each instance whether a newly adjoined axiom is compatible with the previous ones. The physicist often finds himself forced by the results of his experiments to make new assumptions, even while his theory is under development, for the consistency of these new assumptions with the previous axioms, he has to rely on the new experiments or on a certain physical intuition, a practice that is not admissible in a strictly logical construction of a theory. The desired proof of the compatibility of all assumptions appears to me to be of importance, also because the effort to carry out such a demonstration always forces us most effectively toward an exact formulation of the axioms."

In singling out probability theory and its application to the kinetic theory of gases, Hilbert put his finger on one of the most profound themes of mathematical physics, one which three-quarters of a century later we can recognize in the theory of dynamical systems, ergodic theory, and statistical mechanics. Hilbert's remark on the application of Lie's theory may be regarded as employed to some extent in the modern theory of continuum mechanics, where Lie groups play an important part in the classification of the constitutive equations that describe the material properties of models [1].

For Hilbert to quote Ernst Mach, Heinrich Hertz, and Ludwig Boltzmann was only natural; what they had to say can still be read with profit by anyone interested in the foundations of mechanics or statistical mechanics[2][3][4]. Volkmann's book was a leading theoretical physics text of the day. The reference to Bohlmann was a little curious. The article referred to is a popular lecture for high school teachers on the mathematics of life insurance, but he does have a brief review of axioms for probability theory.

What Hilbert Did on Problem Six

A major period in Hilbert's scientific life was devoted to the axiomatization of physics. (It was the fifth period, according to the classification of Hermann Weyl [5]; after invariant theory, algebraic number theory, foundations of geometry, and the theory of integral equations, and before the foundations of mathematics.) He carried it forward in lectures, in his work with Ph.D. students, and in his own research. In the last category, four subjects occur which I will discuss briefly in turn.

a) Boltzmann's equation and the kinetic theory of gases [6]

The Boltzmann equation is an equation for a probability density, $F(\vec{x},\vec{v},t)$ as a function of position, $\vec{x}$, velocity, $\vec{v}$, and time, t . As is appropriate for a probability density, F satisfies

$$\iint d^3x\, d^3v\, F(\vec{x},\vec{v},t) = 1$$

$$F(\vec{x},\vec{v},t) \geqslant 0$$

From it one can calculate such quantities as the expectation value of the density, n ,

$$n(\vec{x},t) = \int d^3v\, F(\vec{x},\vec{v},t)$$

and of the velocity, $\vec{V}$,

$$n(\vec{x},t)\vec{V}(\vec{x},t) = \int d^3v\, \vec{v}\, F(\vec{x},\vec{v},t)$$

and of the thermal energy density, ε,

$$n(\vec{x},t)\varepsilon(\vec{x},t) = \frac{1}{2}\int d^3v(\vec{v} - \vec{V}(\vec{x},t)^2 F(\vec{x},\vec{v},t)$$

Explicitly, the Boltzmann equation reads

$$\frac{\partial F}{\partial t}(\vec{x},\vec{v},t) + \vec{v}\cdot\nabla_x F(\vec{x},\vec{v},t) + \vec{b}\cdot\nabla_{\vec{v}} F(\vec{x},\vec{v},t) = C(F)$$

where C is a quadratic functional of F involving integration over velocities and the function specifying the force between a pair of particles. Boltzmann's equation can be regarded as giving a statistical model of a gas, or alternatively as providing an approximate version of the description of a gas whose exact behavior would be given by the statistical mechanics of a large number of particles interacting with two-body forces.

The initial value problem for the Boltzmann equation has a natural physical interpretation: given a probability distribution $F_0(\vec{x},\vec{v})$ at time $t = 0$, find the probability distribution $F(\vec{x},\vec{v},t)$ at time t. That is not straightforward given the fact that one has a non-linear integral differential equation to deal with. In fact, the existence and uniqueness of solutions for all t is not known to this day except in special cases. [7]

However, the problem that Hilbert chose to attack was a different one and, in a sense, more subtle: the problem of normal solutions. The Boltzmann equation ought to have a distinguished class of solutions in which the probability distribution at each time can be characterized completely by a few macroscopic parameters such as the density $n(\vec{x},t)$ velocity, $\vec{V}(\vec{x},t)$, and thermal energy density $\varepsilon(\vec{x},t)$ mentioned above. The evolution of these parameters is governed by equations of a kind of macroscopic continuum mechanics. An important problem of kinetic theory is to show that general solutions of the Boltzmann equation are asymptotic for large t to normal solutions; that is the problem of the <u>approach to equilibrium</u>.

Hilbert showed how a class of normal solutions could be derived by assuming a formal Laurent series for $F(x,v,t)$ in an auxiliary parameter beginning with an inverse first power. He was very pleased when the determination of the successive terms turned out to be equivalent to solving an integral equation with symmetric kernel to which the then recently created theory of integral equations could be applied. (Actually it was only when the equation was shown by Hecke [8] to have a square integrable fifth iterate that the example was proved to fall under the then existing theory.)

Hilbert's ideas and results on the Boltzmann equation influenced D. Enskog who developed expansions for the transport coefficients (heat conductivity, viscosity, etc.) as did independently S. Chapman (see the review in [9]). These results were of great practical importance for the kinetic theory of gases, but they did not settle any of the numerous problems left open in Hilbert's work. For example, it was only in the 1960's that it was established that Hilbert's formal expansion is asymptotic to an actual solution, at least for small initial data. Even today there are difficult and interesting open problems in the subject [7].

It is notable that in choosing a problem in the application of probability theory to mechanics Hilbert chose one which is such a beast technically. There were much more attractive general questions fairly well formulated by Gibbs in his book on statistical mechanics [10]. Perhaps, Hilbert's recognition that the theory of normal solutions of Boltzmann's equation offers a non-trivial application of the theory of integral equations, determined his choice.

b) Statistical Theory of Radiation [11]

Hilbert made it clear in his papers on radiation theory that he regarded that work as an attempt to treat elementary radiation theory by the axiomatic method. He defined elementary radiation theory as "... that phenomenological part of radiation theory that rests directly on the concepts of emission and absorption and culminates in Kirchoff's laws relating emission and absorption." Even more so than for the Boltzmann equation, this is a striking choice of subject for the axiomatic method: a rather prosaic part of transport theory. Of course, this kind of elementary transport theory is absolutely basic for the theory of stellar atmospheres, but in that connection it is scarcely associated with any serious conceptual issues of physics. There was an important connection of elementary radiation theory with the discovery of quantum theory; the celebrated ultraviolet catastrophe arises from the application of the principles of statistical mechanics to black body radiation. However, the part of the subject that Hilbert studied does not bear directly on this important paradox.

One can perhaps conclude from this that Hilbert did not regard it as beneath the dignity of the axiomatic approach to carve out a modest piece of theoretical physics and reformulate it with mathematical precision, thereby obtaining greater logical coherence in its principles. Of course, it is also true that elementary radiation theory offers another opportunity for the application of the theory of integral equations. In any case, from the evidence offered by the standard references on mathematical radiation theory E. Hopf [12] and S. Chandrasekhar [13] one can conclude that Hilbert's work on radiation theory had little influence on later workers in the field.

c) General Relativity [14]

Hilbert's work on the general theory of relativity occurred at the height of his activity in mathematical physics. It was the most pretentious of his efforts in this field, as its title Die Grundlagen der Physik indicates. He again regarded it as an application of the axiomatic method to physics, but here as compared with the preceding examples, the axioms were far-reaching; together they prescribe a unified field theory of gravitation, electricity and magnetism, and the electron.

Hilbert built his proposal on an idea of Gustav Mie, according to which the existence of electrons is accounted for by alterations in the Maxwell-Lorentz equations in the presence of concentrations of electric charge. Mie derived his theory from a variational principle in which the Lagrangian depends on the vector potential $A_\mu(x)$ and its curl, $\partial_\mu A_\nu(x) - \partial_\nu A_\mu(x)$. Mie constructed his theory in accordance with the special theory of relativity.

Hilbert, who had been following the development of Einstein's ideas on general relativity quite closely, found that by letting the Lagrangian depend on the metric of space-time, and adding to it the Riemann curvature tensor, he could derive the equations of a theory of gravitation and electromagnetism. The fact that Mie's theory was formulated in terms of variational principle fitted neatly into Hilbert's axiomatic approach. Einstein had also worked on occasion with a variational principle but had imposed additional restrictions on the coordinate system that Hilbert, with his axiomatician's insistence on maximum generality strongly resisted. As a result, Einstein was handicapped in his efforts to understand how the arbitrariness in the choice of coordinates affects the initial value problem for the metric tensor. As far as the equations of gravitation are concerned, Hilbert got them out by a direct simple calculation from the variational principle, whereas Einstein arrived at them by a long route involving successively improved physical arguements. (There has been some discussion of priorities, not by Einstein and Hilbert, but by later students of their work [15] [16], since the equation

$$R_{\mu\nu} - \frac{1}{2} g_{\mu\nu} R = -\kappa T_{\mu\nu} \tag{1}$$

first appears in Hilbert's paper dated November 20, 1915, while in Einstein's papers it first appears in one dated November 25, 1915. However, the latter is quoted in the former! In any case, Hilbert always regarded the theory of gravitation as Einstein's theory. He thought his own contribution lay in the generalization of Mie's theory to give a unified theory of gravitation, electromagnetism, and the electron, in an axiomatic presentation.)

Here is a capsule version of Hilbert's theory. He assumed that space-time is a four-dimensional manifold with two distinguished differential forms, a two form

$$\sum_{\mu,\nu} g_{\mu\nu}(x) dx^{\mu} dx^{\nu}$$

with the symmetric tensor $g_{\mu\nu}(x)$, describing the gravitational field and a one form

$$\sum_{\mu} A_{\mu}(x) dx^{\mu}$$

with the vector potential, $A_{\mu}(x)$, describing the electromagnetic field.

<u>Axiom I</u> (Mie's Axiom of the World function)

$$\delta \int H \sqrt{g}\, d\omega = 0$$

where $g = -\det\{g_{\mu\nu}\}$, $d\omega = dx^0 dx^1 dx^2 dx^3$ and H is a function of $g_{\mu\nu}$ and its first and second derivatives, and of A_{μ} and its first derivatives.

The Euler Lagrange equations arising from the variational principle are 14 in number, 10 from the variation of the $g_{\mu\nu}$ and 4 from the variation of A_{μ}

Axiom II (General Invariance)

The world function, H , is an invariant under arbitrary transformations of the x^μ .

From Axiom II, Hilbert derived that four linear combinations of the Euler-Lagrange brackets are divergences. Later work by E. Noether made it clear that this is a general phenomenon with invariant variational principles (Noether's Theorem [17]). The physical significance of this result is discussed later

Axiom III (Axiom of Gravitation and Electricity)

$$H = K + L$$

where K is the curvature associated with $g_{\mu\nu}$ and L depends only on $g_{\mu\nu}$, A_μ , and $\partial_\mu A_\nu$.

With this specialization, the variation with respect to $g_{\mu\nu}$ produces the equation (1), $-\kappa T_{\mu\nu}$ being the derivative of L with respect to $g^{\mu\nu}$ The variation with respect to A_μ yields a generalized version of Mie's electromagnetic equations.

Axiom IV (Space-Time Axiom)

$$\sum g_{\mu\nu} dx^\mu dx^\nu$$

is locally Minkowskian.

From this axiom Hilbert builds up a physical interpretation of the geometry, in the manner now familiar from the general theory of relativity. He emphasized strongly the connection between the above-mentioned identities connecting the Euler-Lagrange brackets and the arbitrariness in the choice of coordinate systems: the $g_{\mu\nu}$ and A_μ in the future cannot be uniquely determined from their values and the values of their derivatives at a given time because a change in coordiantes in the future will alter the values of $g_{\mu\nu}$.

The technical parts of Hilbert's work were rapidly incorporated into the explosive developments of the general theory of relativity. Its generalized Mie theory met the fate of most unified field theories of gravitation and electricity: it led to no new results, so interest in it gradually waned. The fact that with any of the usual choices of L the theory is not gauge invariant makes it even less attractive than other unified field theories.

d) Foundations of Quantum Mechanics [18]

In the winter term of 1926-7, Hilbert gave lectures on the new quantum mechanics prepared in collaboration with his assistants, L. Nordheim and J. von Neumann. Notes were written out by Nordheim with help from von Neumann which were published later as a joint paper.

Hilbert attempted to extract a mathematical scheme covering both the matrix mechanics of Heisenberg, Born, and Jordan, and the wave mechanics of Schrödinger. One key idea came very easily, the association of operators with observables. As his other basic notion, Hilbert chose not state, but transition amplitude. Given two observables with corresponding operators A_1 and A_2 and eigenvalues α_1 and α_2, the associated transition amplitude is a complex valued function $\psi(\alpha_1,\alpha_2;A_1,A_2)$ such that

$$|\psi(\alpha_1,\alpha_2;A_1,A_2)|^2$$

is the probability of finding the value α_1 of A_1, given that A_2 has the value α_2. (The definition is patterned after the interpretation of the Schrödinger energy eigenfunctions $\psi_{E_n}(x)$ which are such that $|\psi_{E_n}(x)|^2$ is the probability density in the position, x, for an electron which is in the energy eigenstate of eigenvalue, E_n). When the basic relations of quantum mechanics are expressed in terms of transition amplitudes they involve such generalized functions as the Dirac delta function and its derivatives, concepts not rigorously defined at that time. Thus Hilbert's attempt at a mathematical formulation was, as he himself said, only partly successful. The slight shift of replacing transition amplitude by state represented by a vector in Hilbert space was made by von Neumann shortly thereafter [19], and led after the development of the theory of unbounded self-adjoint operators to von Neumann's book on mathematical foundations of quantum mechanics [20].

I do not know whether Hilbert regarded von Neumann's book as the fulfillment of the axiomatic method applied to quantum mechanics, but, viewed from afar, that is the way it looks to me. In fact, in my opinion, it is the most important axiomatization of a physical theory up to this time.

This completes the brief tour of Hilbert's achievements in mathematical physics regarded as contributions to the solution of his sixth problem It is natural to ask: Was it a good thing? Was it good for mathematics? Was it good for Hilbert? Was it good for physics?

As a partial answer to these questions, let me quote from Hermann Weyl's obituary notice for Hilbert [5]:

> "Already before Minkowski's death in 1909, Hilbert had begun a systematic study of theoretical physics,in close collaboration with his friend, who had always kept in touch with the neighboring science. Minkowski's work on relativity theory was the first fruit of these joint studies. Hilbert continued them through the years, and between 1910 and 1930 often lectured and conducted seminars on topics of physics. He greatly enjoyed this widening of his horizon and his contact with physicists, whom he could meet on their own ground. The harvest however can hardly be compared with his achievements in pure mathematics. The maze of experimental facts which the physicist has to take into account is too manifold, their expansion too fast, and their aspect and relative weight too changeable for the axiomatic method to find a firm enough foothold, except in the thoroughly consolidated parts of our physical knowledge. Men like Einstein or Niels Bohr grope their way in the dark toward their conceptions of general relativity or atomic structure by another type of experience and imagination than those of the mathematician, although no doubt mathematics is an essential ingredient. Thus Hilbert's vast plans in physics never matured."

To this I will add one point. A great physical theory is not mature until it has been put in a precise mathematical form, and it is often only in such a mature form that it admits clear answers to conceptual problems. In this sense, although quantum mechanics was discovered in 1925-6, it did not become a mature theory until the appearance of von Neumann's book. Thus, although von Neumann had nothing to do with the discovery of quantum mechanics, he had a great deal to do with the creation of quantum mechanics as the mature theory we know today. To this extent, Hilbert's axiomatic approach showed itself important for physics.

The Axiomatic Method in Physics Since Hilbert

What I have to say under this heading cannot and does not pretend to be a complete review of the applications of the axiomatic approach in mathematical physics since the 1940's. It is a miscellaneous collection of examples included because they seem to me instructive or important or both.

Before dealing with specific disciplines let me discuss two points of strategy. The first has to do with the importance of specialization. It was described by Hilbert as follows:

> "I believe that specialization plays an even more important role than generalization when one deals with mathematical problems. Perhaps in most cases in which we seek in vain the answer to a question, the cause of failure lies in the fact that we have worked out simpler and easier problems either not at all or incompletely. What is important is to locate these easier problems and to work out their solutions with tools that are as complete as possible and with concepts capable of generalization. This procedure is one of the most important levers for overcoming mathematical difficulties, and it seems to me that one usually uses it, even if unconsciously."

The strategy described by Hilbert has been followed to the letter in constructive quantum field theory as will be described later, and in certain areas of continuum mechanics.

The second point of strategy has to do with a function of axiomatization that was not so significant in the mathematical physics of his time, although it was so for its analogues in mathematical logic. It can happen that there grows up in physics a collection of principles whose internal consistency is in doubt. To settle the consistency by the construction of non-trivial examples may be very difficult mathematically. It then can be fruitful to state the physical principles as axioms in clean mathematical form and to explore their consequences. It may be possible to introduce certain abstract concepts solely in terms of the axioms and to erect a considerable theoretical structure on them without having settled the basic internal consistency questions. Such a procedure has the virtue of showing the existence of the theoretical structure as a consequence of the axioms alone. It has turned out to be fruitful in quantum field theory as will be described below. Of course, such a theory has to be regarded as incomplete until a precise characterization of the objects satisfying the axioms is obtained and in particular proofs of existence of non-trivial examples.

I now pass to the examination of examples.

A. Axiomatization of Quantum Mechanics

In the traditional physicist's treatment of classical and quantum mechanics, one begins with a classical or quantum mechanical canonical formalism consisting of coordinates and conjugate momenta finite in number and a Hamiltonian given as some function of them. Hamilton's equations for the classical system then give in differential form a flow in the phase space, a flow that describes the evolution of the state of the system with time. Analogously, for the quantum mechanical system the Schrödinger equation gives in differential form the flow in the Hilbert space of state vectors. The work of Stone and von Neumann brought the mathematical aspects of the quantum mechanical theory to a very satisfactory state. According to a theorem of Stone every continuous one-parameter group of unitary operators $U(t)$, $-\infty, < t < \infty$

$$U(t_1)U(t_2) = U(t_1+t_2)$$
$$U(t)^* = U(-t)$$

acting on a Hilbert space $\mathcal{H}$ is of the form

$$U(t) = \exp(iHt)$$

where H is a self-adjoint linear operator in $\mathcal{H}$, the Hamiltonian operator of the one-parameter group [21]. On the other hand, von Neumann had proved the spectral theorem for such possibly unbounded operators [22]. Furthermore, he had succeeded in characterizing the canonical coordinate and momentum operators $q_1 \ldots q_n$, $p_1 \ldots p_n$ by proving that the Weyl relations

$$U(\alpha_1 \ldots \alpha_n) U(\alpha_1' \ldots \alpha_n') = U(\alpha_1+\alpha_1' \ldots \alpha_n+\alpha_n')$$

$$V(\beta_1 \ldots \beta_n) V(\beta_1' \ldots \beta_n') = V(\beta_1+\beta_1' \ldots \beta_n+\beta_n')$$

$$U(\alpha_1 \ldots \alpha_n) V(\beta_1 \ldots \beta_n) = \exp(i \sum_{j=1}^{n} \alpha_j \beta_j) V(\beta_1 \ldots \beta_n) U(\alpha_1 \ldots \alpha_n)$$

have a unique continuous irreducible representation by unitary operators up to unitary equivalence[23]. The $q_1 \ldots q_n$ and $p_1 \ldots p_n$ are then defined by

$$U(\alpha_1 \ldots \alpha_n) = \exp(i \sum_{j=1}^{n} \alpha_j q_j), \quad V(\beta_1 \ldots \beta_n) = \exp(i \sum_{j=1}^{n} \beta_j p_j))$$

These results provide a clear and simple basis for discussing which wave functions are admissible in quantum mechanics, replacing the inconclusive and fumbling treatment given by the discoverers of the subject. Furthermore, they show that the usual Schrödinger mechanics gives a realization essentially unique up to unitary equivalence once the self-adjointness of the Hamiltonian has been established.

Of course, the self-adjointness of the many-body Schrödinger Hamiltonian for interesting forces remained to be shown, and it was many years before that important result was obtained [24]. However, the lack of a proof at the time does not seem to have been a significant motive in the development of the subject. What did emerge was a strong urge to find a more general and aesthetically satisfactory foundation of the quantum mechanical formalism itself, a formulation in which it was hoped the distinction between classical mechanics and quantum mechanics would be traced to some deeper source. At least two enterprises of this kind deserve mention here. The first isolated algebraic structures associated with the operators representing quantum mechanical observables. The second involved, instead, lattices.

In the first, the idea was P. Jordan's [25]. He noted that in the standard quantum mechanical formalism, if A and B are bounded observables, then $\alpha A + \alpha B$ is also assumed to be observable, α and β being any real numbers; on the other hand, AB is observable only if it is self-adjoint and that will be true only if A and B commute. The combination $A \circ B = \frac{1}{2}(AB + BA)$ is always self-adjoint if A and B are, and it is therefore a candidate for an observable. Jordan proposed the problem of finding an algebraic characterization of $A \circ B$ independent of the existence of the underlying Hilbert space. (He was actually looking for a formalism describing observables that would supersede quantum mechanics without losing its essential structure.) Jordan noted that $A \circ B$ is commutative: $A \circ B = B \circ A$, and, in general, non-associative $A \circ (B \circ C) \neq (A \circ B) \circ C$, but does satisfy

$$A^2 \circ (B \circ A) = (A^2 \circ B) \circ A$$

and he proposed the study of commutative algebras satisfying this weakened law of associativity, and the distributive law

$$A \circ (B+C) = A \circ B + A \circ C.$$

Such algebras are nowadays called commutative Jordan algebras and the theory of their structure and representations is a respectable chapter of algebra [26][27]. When a Jordan algebra has a product $A \circ B$ arising from an associative product AB via $A \circ B = \frac{1}{2}(AB+BA)$ it is said to be special, otherwise it is called exceptional.

Following Jordan's idea, Jordan, Wigner, and von Neumann developed a theory of formally real finite dimensional Jordan algebras over the real numbers and showed that with one possible exception simple algebras of this kind are special [28]; A.A. Albert showed that the exception is actually exceptional in the above sense [29]. (Formally real means that $A^2 + B^2 + \dots C^2 = 0$ implies $A = B = \dots = C = 0$; it is a natural requirement on physical grounds.) Although the exceptional algebra was a candidate for a "new form of quantum mechanics", it did not seem to be very promising in that role. However, it was natural to blame this blemish of exceptional algebras on the assumption of finite dimensionality. After all, to realize the Heisenberg commutation relation $[q,p] = i$ requires an infinite dimensional space. Von Neumann added some topological assumptions to the algebraic assumptions of Jordan, Wigner, and von Neumann and began a study of the infinite dimensional case, which he never completed [30]. The idea that exceptional infinite dimensional Jordan algebras could provide some new and more desirable form of quantum mechanics seems to have faded away. It did, however, influence one important later development, Segal's work [31] on postulates for general quantum mechanics. (Jordan's own assessment of the prospects is to be found in [32], where it is proposed to abandon Jordan algebras in favor of algebras in which the power associative law $a^\mu a^\nu = a^{\mu+\nu}$ fails.)

Segal adopted part of the algebraic assumptions of the previous work and added metric assumptions which made it possible to apply the then recently developed theory of normed algebras. More specifically, Segal assumed that the observables form a vector space, $\mathfrak{A}$, over the real numbers, that $\mathfrak{A}$ has an identity element, I , and that if $A \in \mathfrak{A}$ the powers of A, A^n , $n = 0,1,2\dots$ are well defined and such that all the usual properties of polynomials hold :

$$A^0 = I$$

and if f , g and h are real polynomials such that $f(g(\alpha)) = h(\alpha)$, $\alpha \in \mathbb{R}$, then $f(g(A)) = h(A)$ where

$$f(A) = \sum_{k=0}^{n} \beta_k A^k \quad \text{if} \quad f(\alpha) = \sum_{k=0}^{n} \beta_k \alpha^k$$

The metric postulates were

1) there exists a norm $|| \ ||$ on $\mathfrak{A}$ for which $\mathfrak{A}$ is a real Banach space

2) $||A^2 - B^2|| \leq \max(||A^2||, ||B^2||)$

3) $||A^2|| = ||A||^2$

4) A^2 is a continuous function of A.

An additional metric assumption later proved redundant by S. Sherman [ ] has been omitted. For brevity, let us call any set $\mathfrak{A}$ with the above properties a Segal system.

The first thing to note about Segal systems is that the set of all self-adjoint elements of a C* algebra is one. (Recall that a C* algebra over the complex numbers is an associative * algebra with a norm, $|| \ ||$, and an anti-involution *, which is complete in its norm regarded as Banach space and satisfies

$$||A^*A|| = ||A||^2 \quad .)$$

A Segal system is called special if it is isomorphic to the set of self-adjoint elements of some C* algebra; it is called exceptional if not. Special Segal systems are interpretable as ordinary quantum mechanics by virtue of the fundamental theorem of C* algebra theory that asserts that every C* algebra is isomorphic to an operator algebra in Hilbert space [33].

The second thing to note about Segal systems is that there exist many exceptional Segal systems, as was shown by S. Sherman [34]. D. Lowdenslager gave a necessary and sufficient condition that a Segal system be special. Unfortunately, the condition does not have any physical interpretation I know of [35].

There is a certain irony in the response to Segal's postulates. Those who are inspired by Jordan's original outlook that ordinary quantum mechanics must be superseded are encouraged by the existence of exceptional Segal systems although, it must be said, so far no one has succeeded in giving any such system an interesting physical interpretation. Those who regard Segal's postulates as an attempt to axiomatize quantum mechanics look on the existence of exceptional Segal systems as a defect of the postulates; a set of physically reasonable postulates would be satisfactory according to this view only if it implied the theorem that every Segal system is special. You can't please everybody. Segal's own assessment of the postulates from a decade and a half after is contained in the first chapter of [36].

From the vantage point of 1974 one can see that Segal's work was important not only in providing an algebraic axiomatization of quantum mechanics, but also in bringing to bear the ideas and results of modern spectral analysis (results associated with the names of Stone, Gelfand and Naimark, and Segal himself, among others). A revolution in conceptual foundations and mathematical technique resulted whose consequences are still being explored, especially in the quantum mechanics of systems of an infinite number of degrees of freedom.

Segal introduced the notion of <u>state</u> on a Segal system $\mathfrak{A}$: a state, ω, on $\mathfrak{A}$ is a linear real valued function on $\mathfrak{A}$ which is positive in the sense that

$$\omega(A^2) \geqslant 0$$

and normalized

$$\omega(I) = 1 \quad .$$

In his book on the foundations of quantum mechanics [20], J. von Neumann had already introduced this definition for the special case in which $\mathfrak{A}$ consists of all bounded self-adjoint operators on a separable Hilbert space $\mathcal{H}$. He showed that (under a somewhat vaguely formulated continuity hypothesis) every such state is of the form

$$\omega(A) = tr(\rho A) \qquad (2)$$

where ρ is a positive trace class operator of trace 1. Such an ω is called a normal (or density matrix) state. In particular, when ρ is of rank 1 with a unit vector, Φ, in its range

$$\omega(A) = (\Phi, A\Phi) \quad .$$

Such an ω is called a vector state with state vector, Φ. In general, there are many other non-normal states, states violating von Neumann's continuity hypothesis. I will have more to say about the significance of these non-normal states later.

The notions of mixture and pure state go through in Segal's general setting. A state is called a mixture if it can be written

$$\omega = \lambda\omega_1 + (1-\lambda)\omega_2$$

where $0 < \lambda < 1$, ω, and ω_2 being distinct states. Otherwise, it is called pure or extremal.

A set, S, of state is full for $\mathfrak{A}$ if for any pair $A,B \in \mathfrak{A}$

$$\omega(A) = \omega(B) \quad \text{for all} \quad \omega \in S$$

implies $A = B$. That means there are enough states in S so that any two distinct observables of the system $\mathfrak{A}$ can be distinguished by their values in some state of S.

Segal showed that every system has a full set of states and that much of the spectral theory of observables goes through, provided one makes appropriate definitions. For example, a commutative subsystem is defined to be one in which the formal product

$$A \circ B = \frac{1}{4}\left[(A+B)^2 - (A-B)^2\right]$$

(which is automatically commutative) is also associative and distributive. For special systems this coincides with the usual notion of commutativity for the underlying associative multiplication. A commutative subsystem was then shown to be isomorphic to an algebra of functions on a compact Hausdorff space, Δ ; it is reasonable to regard Δ as the spectrum of the commutative subsystem. Furthermore, there exists a probability measure on Δ such that

$$\omega(A) = \int_\Delta f_A(\delta)d\mu(\delta)$$

where f_A is the function on Δ that corresponds to the observable A . With this machinery one can define such notions as observable possessing a definite value in a state and prove such natural sounding results as : a collection of observables is simultaneously observable if and only if it forms a commutative subsystem, and a pure state on a subsystem of can always be extended to a pure state on $\mathfrak{A}$. (This last theorem was proved by Sherman [37]; it is a slightly sharpened form of what Segal had.)

These results show that to a large extent the phenomenological content of quantum mechanics (states, probability distributions for commuting observables, etc.) can be recovered from Segal's postulates. At this stage it is natural to ask whether anything is really gained by this more abstract formulation. Having attended the seminar at which Segal first put forward his results, I have memories of the answer of some physicists to this question at that time. In a polite form it can be stated: the new more general definitions may be admirable but they are a luxury. This reaction is not entirely unreasonable if one bases one's answer on the quantum mechanics of systems of a finite number of degrees of freedom. For example, the general notion of state includes eigenfunctions belonging to eigenvalues in the continuum. (In the particular case of one degree of freedom realized in the Hilbert space $L^2(R)$, $\sqrt{\delta(q-q')}$, where δ is the Dirac δ function and q' is fixed, certainly does not define a state vector but it does define a state, $\omega_{q'}$. It suffices to give $\omega_{q'}$ on operators of the form $U(\alpha)V(\beta)$

$$\omega_{q'}(U(\alpha)V(\beta)) = \begin{cases} 0 & \text{if} \quad \beta \neq 0 \\ \exp[i\alpha q'] & \text{if} \quad \beta = 0 \end{cases}$$

From these operators ω_q, extends uniquely to $\mathfrak{A}$. It is nice to be able to handle such states which would surely be regarded as singular from the point of view of the orthodox von Neumann quantum mechanics, but it is not essential.

For systems of an infinite number of degrees of freedom, the situation turned out to be completely different. After nearly three decades of further development it has become pretty evident even to the heathen that the physicists' evaluation is quite wrong. The general concept of state and its attendant cloud of definitions, theorems, and techniques is really indispensible for any general quantum mechanical theory of systems of an infinite number of degrees of freedom. This striking difference between systems of a finite number and infinite number of degrees of freedom arises in part because of the absence in the infinite case of a uniqueness theorem analogous to von Neumann's uniqueness theorm for the Weyl form of the commutation relations. In fact, for the algebras in question there are usually myriads of relevant representations and it is precisely the necessity of coping with the resulting technical problems that makes Segal's formalism indispensible.

Of course, conclusive evidence for these statements did not exist in 1948, and, in fact, it only appeared gradually over the next two decades. Here it will be discussed in more detail later on in connection with quantum field theory. There is however one general result of the theory that should be mentioned at this point: the G(elfand) N(aimark) S(egal) construction which connects states on a C* algebra with cyclic representations of the algebra [38].

Given a C* algebra, $\mathcal{B}$, and a state ω on it, it is possible to construct a cyclic representation π_ω of $\mathcal{B}$ in a Hilbert space $\mathcal{H}_\omega$ with cyclic vector Ψ_ω , such that

$$\omega(B) = (\Psi_\omega, \pi_\omega(B)\Psi_\omega)$$

The triple $\{\mathcal{H}_\omega, \pi_\omega, \Psi_\omega\}$ so constructed is unique up to unitary equivalence in the sense that any other such triple $\{\mathcal{H}_\omega', \pi_\omega', \Psi_\omega'\}$ is related to it by

$$V\,\Psi_\omega = \Psi_\omega'$$

$$V\,\pi_\omega\,V^{-1} = \pi_\omega'$$

where V is a unitary operator mapping $\mathcal{H}_\omega$ onto $\mathcal{H}_\omega'$.

The GNS construction permits one to transcribe problems about representations of a C* algebra, $\mathcal{B}$, into problems about states on $\mathcal{B}$. An important example of this is Segal's theorem that pure states yield irreducible representations. Another is the connection of automorphism groups of $\mathcal{B}$ with unitary representations. Suppose $\mathcal{G}: g \mapsto \alpha_g$ of a group $\mathcal{G}$ by automorphisms of $\mathcal{B}$:

$$\alpha_g(a_1A_1+a_2A_2) = a_1\alpha_g(A_1) + a_2\alpha_g(A_2)$$

$$\alpha_g(A_1A_2) = \alpha_g(A_1)\alpha_g(A_2)$$

$$\alpha_{g_1}(\alpha_{g_2}(A)) = \alpha_{g_1g_2}(A)$$

for all $g, g_1, g_2 \in \mathcal{G}$ and $A, A_1, A_2 \in \mathcal{B}$. Suppose further that ω is an invariant state on $\mathcal{B}$:

$$\omega(\alpha_g(A)) = \omega(A)$$

for all $g \in \mathcal{G}$ and $A \in \mathcal{B}$. Then the GNS construction yields a unitary representation of $\mathcal{G}: g \to U_\omega(g)$ such that

$$U_\omega(g)\pi_\omega(A)U_\omega(g)^* = \pi_\omega(\alpha_g(A))$$

and

$$U_\omega(g)\Psi_\omega = \Psi_\omega$$

Expressed in words, the content of the GNS construction is that any state, ω, on a C* algebra, $\mathcal{B}$, can be regarded as a vector state with state vector, Ψ_ω, for a suitable representation, π_ω, and if the state ω is invariant under a group of automorphisms the cyclic vector, Ψ_ω, will be invariant under the unitary representation, U_ω, of the group obtained from the construction. When the group in question is the Poincaré group, the invariant vector is usually called a vacuum, because of its significance in quantum field theory.

This completes my excursion on Segal's algebraic approach to the foundations of quantum mechanics. There are many related later developments; for some of them see [39]. I turn now to the second enterprise referred to above, the approach to the axiomatization of quantum mechanics through its lattice of propositions.

Birkhoff and von Neumann were the first to attempt to characterize quantum mechanics and to contrast it with classical mechanics using the lattice of propositions to which the two theories give rise [40]. They pointed out that the closed linear manifolds (≡ subspaces) of a Hilbert space $\mathcal{H}$, or equivalently in operator language the projection operators onto those subspaces, describe in quantum mechanical terms the yes or no statements (propositions) about a physical system, and that their analogues in a classical Hamiltonian system would be characteristic functions of sets in phase space or the sets themselves. In both cases there is a natural lattice structure with a join operation $\vee$ and a meet operation $\wedge$, and a partial order relation $\leq$. For classical mechanics, the join and meet of two subsets S_1 and S_2 are just the usual set theoretic union and intersection $S_1 \vee S_2 = S_1 \cup S_2$ $S_1 \wedge S_2 = S_1 \cap S_2$. For quantum mechanics, if $\mathcal{H}_1$ and $\mathcal{H}_2$ are two subspaces of $\mathcal{H}$, then their meet $\mathcal{H}_1 \vee \mathcal{H}_2$ is the subspace $\mathcal{H}_1 + \mathcal{H}_2$ they span, while their join $\mathcal{H}_1 \wedge \mathcal{H}_2$ is the set theoretic intersection $\mathcal{H}_1 \cap \mathcal{H}_2$. In both cases, the partial order relation $\leq$ means inclusion, in the first inclusion as a set of phase space, in the second inclusion as a subspace of Hilbert space. In both cases, the lattice theoretic laws

$$a \wedge a = a \qquad a \vee a = a \tag{3}$$

$$a \wedge b = b \wedge a \qquad a \vee b = b \vee a \tag{4}$$

$$a \wedge (b \wedge c) = (a \wedge b) \wedge c \qquad a \vee (b \vee c) = (a \vee b) \vee c \tag{5}$$

$$a \vee (a \wedge b) = a \wedge (a \vee b) = a \tag{6}$$

hold and the relations $a \leq b$, $a \wedge b = a$, and $a \vee b = b$ are equivalent. Furthermore, there are least and greatest elements 0 and 1 , namely, in classical mechanics the function equal to zero almost everywhere and the function equal to one almost everywhere, respectively, while in quantum mechanics it is the subspace consisting of the vector 0 , and the whole space $\mathcal{H}$, respectively.

These lattices have a physically interpretable orthocomplementation operation i.e. a map $a \mapsto a'$ of the lattice onto itself satisfying

$$(a')' = a \tag{7}$$

$$a \wedge a' = 0 \qquad a \vee a' = 1 \tag{8}$$

$$a \leq b \quad \text{implies} \quad b' \leq a' \tag{9}$$

$$(a \wedge b)' = a' \vee b' \qquad (a \vee b)' = a' \wedge b' \tag{10}$$

In classical mechanics, the orthocomplement is the operation of passing to the complementary set, in quantum mechanics the operation of passing to the orthogonal complement of a subspace.

Up to this point all listed properties of the lattice of propositions hold in both classical and quantum mechanics. Here are two valid in classical mechanics but not in quantum mechanics: the distributive laws

$$a \vee (b \wedge c) = (a \vee b) \wedge (a \vee c) \tag{11}$$

$$a \wedge (b \vee c) = (a \wedge b) \vee (a \wedge c) \quad . \tag{12}$$

An orthocomplemented lattice satisfying the distributive law is called a Boolean algebra and a basic theorem says that any Boolean algebra is isomorphic to a lattice of subsets of a set. In this sense the lattice of propositions of classical mechanics can be characterized as a Boolean algebra.

Is there some analogous way of singling out quantum mechanics? In seeking an answer to this question von Neumann and Birkhoff considered as a substitute for the distributive law the so-called modular law

$$a \vee (b \wedge c) = (a \vee b) \wedge c \tag{13}$$

They noted that the modular law is valid in quantum mechanics when the Hilbert space of states is finite dimensional, but that it fails for the lattice of all closed subspaces of an infinite dimensional Hilbert space. Now at that time von Neumann was engaged in his joint work with F.J. Murray on rings of operators and they had discovered new classes of rings of operators characterized by a dimension function taking all real positive values in an interval including the origin. Von Neumann gave a geometrical interpretation which brought out analogies between these continuous geometries and finite dimensional projective geometries [41]. With this as background, Birkhoff and von Neumann were led to conjecture that a generalized quantum mechanics exists in which the lattice of propositons is modular even though the systems have an infinite number of independent observables. The hope was that the generalized quantum mechanics would behave more like the quantum mechanics of a system with a finite dimensional Hilbert space of state vectors and that thereby some of the then existing puzzles of systems with an infinite number of degrees of freedom might be solved. This hope has remained unfulfilled as far as I know, and the problem of characterizing quantum mechanics by its lattice of propositions, unsolved by Birkhoff and von Neumann, was taken up by others, and only after a lapse of many years.

The later developments began with a reexamination of the foundations of quantum mechanics by G. Mackey [42]. Mackey's starting point was a system consisting of two notions, a set of observables, $\mathcal{O}$, and and a set of states, $\mathcal{S}$. He postulated that a state is a function defined on $\mathcal{O}$ and taking values in the set of probability measures on the real line. If α is a state and A is an observable, then $\alpha(A)$ gives the probability distribution of the values of A. It is assumed that the set of states is full i.e. if A and B are observables and $A \neq B$, then there is a state such that $\alpha(A) \neq \alpha(B)$. Furthermore, it is assumed that functions of observables are definable. That means that if $A \in \mathcal{O}$ and f is any real valued Borel function defined on the real line, there exists an observable B such that

$$\alpha(B)(E) = \alpha(A)(f^{-1}(E))$$

for any $\alpha \in \mathcal{S}$ and any Borel subset E of the real line. It is natural to call B, f(A). Using these assumptions, Mackey proceeded to an analysis of the special class of observables called questions; an observable is a question if for every state α the probability measure $\alpha(A)$ is concentrated at the real numbers 0 and 1. (In the end, what Mackey calls a question and what Birkhoff and von Neumann call a proposition may be identified, but I will preserve the original terminology since the starting points of the two analyses are different.)

If A is a question and α is a state, the probability distribution $\alpha(A)$ is completely determined by its value at the set $\{1\}$, a real positive number Mackey denotes by $m_\alpha(A)$. On the other hand, if A is any observable, E is any Borel subset of the real line and χ_E is the characteristic function of E, then $\chi_E(A)$ is a question, which he denotes Q_E^A. The family of questions Q_E^A, obtained by holding A fixed and letting E vary over the Borel subsets of the real line, uniquely determines the observable A. Since

$$m_\alpha(Q_E^A) = \alpha(A)(E)$$

the state α is completely determined by the values of m_α on the questions of $\mathcal{O}$.

Let the questions of $\mathcal{O}$ be denoted, $\mathcal{Q}$. Then $\mathcal{Q}$ has a partial order $Q_1 \leq Q_2$ iff $m_\alpha(Q_1) \leq m_\alpha(Q_2)$ for all $\alpha \in \mathcal{S}$, and one has arrived at a lattice of propositions in the sense of Birkhoff and von Neumann starting from Mackey's postulates. Mackey did not attempt to characterize the $\mathcal{Q}$ of quantum mechanics in lattice theoretic terms but he did pose a question later answered by Gleason which is an important step toward the solution of that problem. Mackey noted that the function m_α has the properties

1) If $Q_1, Q_2, \ldots$ are questions such that $Q_j \leq 1 - Q_k$ for $j \neq k$

then
$$m_\alpha(Q_1+Q_2+\ldots) = m_\alpha(Q_1) + m_\alpha(Q_2) + \ldots$$

2) $m_\alpha(1) = 1$

He calls any function from the questions to the positive real numbers satisfying 1) and 2) a measure on the questions. He asked: Is every measure on the questions the m_α of some state α ? In particular, if $\mathcal{L}$ is the lattice of closed subspaces of a separable Hilbert space, $\mathcal{H}$, as is the case in von Neumann style quantum mechanics, is every m_α of the form

$$m_\alpha(A) = \mathrm{tr}(\rho_\alpha A)$$

where ρ_α is a trace class positive operator of trace 1 on $\mathcal{H}$? Gleason showed that if $\mathcal{H}$ is of dimension three or larger the latter question has a positive answer whether $\mathcal{H}$ is understood as a Hilbert space over the real or complex numbers [43]. Gleason's result has been a guide for all later work that attempts to pass from the lattice of propositions to the Hilbert space structure of quantum mechanics. It is interesting to note that in a sense his work is a generalization of the argument of von Neumann that led to (2).

Following the lines laid down by Mackey and Gleason, the general problem was taken up again by N. Zierler [44]. Zierler took advantage of advances in lattice theory that had occurred in the meantime. Loomis had recognized that there is a weaker form of the modular law that is valid for the lattice of closed subspaces of a Hilbert space whether the Hilbert space is of finite or infinite dimension [45]. A lattice is semi-modular (or weakly modular) if

$$a \leqslant b \quad \text{implies} \quad b = a \vee (a' \wedge b) \quad .$$

There are many other equivalent conditions completely organized by S. Maeda [44]. For example, a lattice is semi-modular if

$$a \leqslant c \quad \text{and} \quad b \leqslant (1-c) \quad \text{imply} \quad a = (a \vee b) \wedge c$$

A semi-modular orthocomplemented lattice is nowadays usually called an orthomodular lattice. Zieler supplemented the assumption of semi-modularity with seven other reasonable postulates and proved that any such lattice is isomorphic to the lattice of closed subspaces of a Hilbert space over the real or complex numbers with orthocomplements corresponding.

One might think that Zierler's results would have been regarded as a satisfactory solution to the main problem of the lattice theoretic approach to the foundations of quantum mechanics, but that is not the way things happened in fact, and the story is so instructive and interesting that I will sketch it, at least in part, up to the present time. The salient points are the appearance of Mackey's book, Piron's thesis, Plymen's work connecting them, and a passionate encounter with hidden variables theorists. As you will see, all this rather highbrow mathematical activity led via the work of Bell and others to experiments that are crucial tests of the principles of quantum mechanics.

In his book [46], Mackey somewhat expanded his earlier axiomatic treatment of quantum mechanics and related it to the work of Birkhoff and von Neumann and to that of Segal. In order to make detailed comparisons with later contributions it is convenient to list Mackey's axioms explicitly. As described above the basic objects are $\mathcal{O}$, the set of observables, $\mathcal{S}$, the set of states, and a map of $\mathcal{O} \times \mathcal{S}$ into the Borel measures on $\mathbb{R}$ which he now describes in terms of a real valued function $p(A,\alpha,E)$ with $A \in \mathcal{O}$, $\alpha \in \mathcal{S}$ and $E \in \mathcal{B}$, the Borel sets of the real line $\mathbb{R}$.

<u>Axiom I</u>

$$p(A,\alpha,\emptyset) = 0 \qquad p(A,\alpha,\mathbb{R}) = 1$$

$$p(A,\alpha,E_1 \cup E_2 \cup \ldots) = \sum_{j=1}^{\infty} p(A,\alpha,E_j)$$

where $E_j \cap E_k = \emptyset$ for $j \neq k$.

<u>Axiom II</u>

$$p(A,\alpha,E) = p(A',\alpha,E) \quad \text{for all } \alpha \in \mathcal{S},\ E \in \mathcal{B}$$

implies $A = A'$

$$p(A,\alpha,E) = p(A,\alpha',E) \quad \text{for all } A \in \mathcal{O},\ E \in \mathcal{B}$$

implies $\alpha = \alpha'$

Axiom III

Let a be any member of $\mathcal{O}$ and f any real-valued Borel function on $\mathbb{R}$. Then there exists $B \in \mathcal{O}$ such that

$$p(B,\alpha,E) = p(A,\alpha,f^{-1}(E))$$

for all $\alpha \in \mathcal{S}$ and $E \in \mathcal{B}$.

Axiom IV

If $\alpha_1,\alpha_2,\ldots \in \mathcal{S}$ and $\sum_{j=1}^{\infty} t_j = 1 \quad 0 \leqslant t_j \leqslant 1$ then there exists an $\alpha \in \mathcal{S}$ such that

$$p(A,\alpha,E) = \sum_{j=1}^{\infty} t_j \, p(A,\alpha_j,E)$$

for all $E \in \mathcal{B}$ and all $\alpha \in \mathcal{O}$.

Axiom V

Let $Q_1,Q_2,$ be any sequence of questions such that $Q_j \leqslant 1 - Q_k$ for $j \neq k$. Then there exists a question Q, such that

$$m_\alpha(Q) = \sum_{j=1}^{\infty} m_\alpha(Q_j)$$

Axiom VI

If Q is any question-valued measure there exists an observable A such that

$$Q_E^A = Q_E$$

for all $E \in \mathcal{B}$.

Axiom VII

The partially ordered set of all questions in quantum mechanics is isomorphic to the partially ordered set of all closed subspaces of a separable infinite dimensional Hilbert space.

Axiom VIII

If Q is any question different from zero, there exists a state

α such that

$$m_\alpha(Q) = 1$$

Axiom IX

There exists a one-parameter semi-group $t \to V_t$ of transformations of $\mathcal{S}$ into $\mathcal{S}$ for each sequence α_1, α_2 , ... of states and each sequence γ_1, γ_2, of non-negative real numbers whose sum is 1

$$V_t(\gamma_1\alpha_1 + \gamma_2\gamma_2 + \dots) = \gamma_1 V_t(\alpha_1) + \gamma_2 V_t(\alpha_2) + \dots \quad .$$

For all $t \geqslant 0$, $a \in \mathcal{O}$, $\alpha \in \mathcal{S}$ and $B \in \mathcal{B}$, $p(A, V_t(\alpha), B)$ is a continuous function of t .

Mackey discussed the possibility of a structure theory for the lattice of questions $\mathfrak{Q}$ in the absence of Axiom VII. He pointed out that a notion of compatibility of questions can be defined which reduces when $\mathfrak{Q}$ is the lattice of subspaces of a Hilbert space to commutativity of projections. The set of all questions compatible with every question in $\mathfrak{Q}$, which he called the center of $\mathfrak{Q}$ is a Boolean algebra. Mackey suggested that it might be possible to write $\mathfrak{Q}$ as a direct sum of lattices of questions with trivial center. He noted that projective geometry and von Neumann's continuous geometry gave candidates for these constituent lattices with trivial center, so that exploration of a wide variety of possibilities would be necessary before an appropriate substitute for the somewhat ad hoc Axiom VII could be found.

Mackey noted one alteration of Axiom VII which fitted smoothly into his system; that required to take into account the existence of superselection rules. Wick, Wightman, and Wigner had used this nomenclature to describe a situation in which the set of operators representing the observables of a quantum mechanical system all commute with some operator that is not a constant multiple of the identity [47]. There is evidence that in Nature the total electric charge operator commutes with all observables, for example. In the presence of superselection rules the Hilbert space of the usual quantum mechanics splits up into a direct sum of distinguished orthogonal subspaces, and it is the subspaces of these distinguished subspaces that should appear in an adjusted Axiom VII.

Mackey did not make a detailed analysis of the relation of his axioms to those of Segal but he did note three points. From his axioms one can derive question-valued measures and from the question-valued measures self-adjoint operators representing observables. These self-adjoint operators (or rather the bounded operators among them) form the raw material for the observables of Segal's postulates. However, without the use of Mackey's Axiom VII, it would not follow that these operators form a vector space since there is no reason why the sum of two non-commuting observables should be observable. The postulates of Segal admit a wider class of states, which includes "states" that are singular from the point of view of Mackey's axioms.

Axiom IX of Mackey describes quantum dynamics, in contrast to Axioms I...VIII which describe statics. Mackey showed that from the assumed transformation law, V^t, of states, the existence of a continuous one-parameter group of unitary operators, U^t , follows such that the action of U^t on the corresponding density matrices

$$(V^t\alpha)(A) = tr(\rho_\alpha^t A)$$
$$= tr(\rho_A U^t A U^{-t})$$

In the terminology of P. Dirac, this can be regarded as a derivation of the Heisenberg picture from the Schrödinger picture.

In this connection, it is worth noting that the derivation of U^t from V^t makes heavy use of Axiom VII and weakening Axiom VII makes the problem considerably more difficult. A paper that illustrates this point is R. Kadison [48]. In it Kadison adopts the C* algebraic point of view whole hog. For him a dynamical system is a triple $\{\mathfrak{A}, \mathcal{S}, V^t\}$ consisting of an abstract C* algebra $\mathfrak{A}$, a full family of states on $\mathfrak{A}$ (positive linear normalized functions on $\mathfrak{A}$) and a one-parameter group of one-to-one affine weak continuous mappings of $\mathcal{S}$ onto $\mathcal{S}$. Kadison first shows that there is always a weakly continuous one-parameter group of automorphisms, β^t, of $\mathfrak{A}$ such that

$$\alpha(\beta^t(A)) = (V^t\alpha)(A)$$

for all t, all $A \in \mathfrak{A}$ and all $\alpha \in \mathcal{S}$. He then gives conditions under which $\beta^t(A) = U^t A U^{t*}$, $t \to U^t$ being a continuous unitary one-parameter group. The use of $\{\mathfrak{A}, \mathcal{S}\}$ as a model for the quantum mechanics of a general system is a special case of Segal's postulates which achieved widespread popularity after the appearance of the paper of Kadison and that of Haag and Kastler [49] which will be discussed later in connection with quantum field theory.

Apparently unaware of the work of Zierler, C. Piron undertook in his thesis [50], a systematic development of the lattice theoretic approach to classical and quantum mechanics. He started, as did Birkhoff and von Neumann, from a partially ordered set, L, with join, $\vee$, and meet, $\wedge$, operations satisfying the lattice theoretic laws (3)...(6). He assumed an orthocomplementation satisfying (7)...(10). He strengthened these by assuming that for any subset of elements $\{a_j; j \in J\}$ in L the meet $\bigwedge_j a_j$ and consequently the join $\bigvee_j a_j$ are defined and have the properties of greatest lower bound and least upper bound respectively i.e. he assumed that L is a complete orthocomplemented lattice. He introduced a useful lattice theoretic version of the notion of compatibility of propositions: a is compatible with b, written $a \leftrightarrow b$ iff

$$(a \wedge b') \vee b = (b \wedge a') \vee a \quad .$$

In the lattice of closed subspaces of a Hilbert space, a is compatible with b in this sense, iff the corresponding projection operators P_a and P_b commute. He then required

$$a \leq b \quad \text{implies} \quad a \leftrightarrow b \quad .$$

This turns out to be yet another equivalent form of semi-modularity.

Piron further assumed as did Zierler that the lattice of propositions, L, is atomic in the following sense. An element a of L is said to cover zero if $0 \lesssim a$ and $b \lesssim a$ implies $b = a$ or $b = 0$. An element covering zero is called an atom. (That is a traditional confusion for physicists reading papers on lattice theory.) A lattice, L, is atomic if every element $a \in L$ such that $a \neq 0$ contains an atom p, $p \lesssim a$. Finally, Piron assumed the so-called covering law: if q is an atom of L then

$$a \lesssim x \lesssim a \vee q \quad \text{implies} \quad x = a \quad \text{or} \quad x = a \vee q .$$

Thus altogether L is a complete atomic orthomodular lattice satisfying the covering law.

Let us call any lattice L satisfying these postulates a Piron lattice. Piron developed a systematic structure theory for Piron lattices. He showed that among them are lattices in which all propositions are compatible as happens in classical mechanics and also lattices with incompatibility such as occurs in quantum mechanics. One of the desirable features of Piron's general approach is that it also accommodates the possibility of quantum mechanics with superselection rules.

Piron carried out for Piron lattices the proposal that Mackey had outlined for the lattice of questions in his theory. He showed that every Piron lattice is a direct union of irreducible Piron lattices. Here L irreducible means that L has a trivial center. Every irreducible Piron lattice is in turn isomorphic to the family of subspaces of a vector space V, over some field F, with a hermitean scalar product, f. Here subspace is understood to mean any subset, $\mathcal{S}$, satisfying $(\mathcal{S}^{\perp})^{\perp} = \mathcal{S}$ where

$$\mathcal{S}^{\perp} = \{x; x \in V, f(x_1 y) = 0 \quad \text{for all} \quad y \in \mathcal{S}\}$$

Piron was unable to exclude the possibility that the field F is not the real or complex numbers nor the quaternions. For F the complex numbers,

V turns out to be automatically closed and therefore a Hilbert space, as was stated by Piron and proved by Amemiye and Araki [51]. Subspace,in the above sense, then reduces to closed subspace in the usual Hilbert space sense. In classical mechanics all these Hilbert spaces are one-dimensional while in quantum mechanics in von Neumann's original sense the family consists of a single Hilbert space.

This theory of Piron is an important step toward an alternative solution of the problem of characterizing classical mechanics and quantum mechanics in terms of their lattice of propositions. Of course, as stated it leaves open the question of the determination of the field F . It appears that definitive results have still not been obtained although there has been progress toward the goal of eliminating all possibilities except F being the complex numbers, either by further analysis of Piron's postulates themselves [52], or by the adjunction of others [53].

While Piron's analysis enables one to start from the lattice L and arrive (almost) at classical and quantum mechanics, it does not yield a detailed correspondence with Mackey's approach. When the possibility of such a correspondence was studied by Plymen he found that one of Piron's assumptions, the completeness of L , is too strong [54][55]. It would in the context of classical statistical mechanics yield the result that every subset of phase space (including each non-measurable set) corresponds to an observable. However, if completeness is weakened to σ-completeness for which the intersections $\bigwedge_j a_j$ and joins $\bigvee_j a_j$ are countable, one obtains the desired observables of classical statistical mechanics. Furthermore Plymen showed that with this and some other slight alterations in the axiom systems he could connect the C*-algebra approach of Kadison with both Mackey's and Piron's axioms. To establish the link he used the notions of σ-envelope of a C*-algebra and of Σ*-algebra developed by Davies [56]. If a C*-algebra, is given concretely as a subalgebra of $L(\mathcal{H})$ the set of all bounded operators on some Hilbert space $\mathcal{H}$ then it is a Σ*-algebra if it contains the weak limit points of all sequences formed from elements of $\mathfrak{A}$, and more generally the σ-envelope, $\mathcal{A}^{\sim}$, of $\mathcal{A}$ is the smallest Σ^*-algebra containing $\mathcal{A}$. In general the σ-envelope, $\mathcal{A}^{\sim}$, of $\mathcal{A}$ is smaller than the von Neumann algebra generated by $\mathcal{A}$ because $\mathcal{A}$ may have weak limit points that are limits of nets in $\mathcal{A}$ but not of sequences in $\mathcal{A}$. The general definition of Σ^*-algebra and σ-envelope of a C*-algebra has to get around the fact that a C*-algebra is given abstractly and not as a concrete subalgebra of an algebra of operators; suffice it say that it can be done.

Attached to a Σ^*-algebra are σ-states. These are positive linear forms on the algebra and therefore states in the sense of C*-algebra theory but also have a continuity property which reduces in the above-mentioned concrete situation to continuity under sequential weak limits.

Plymen then proved

Theorem

Let $\mathcal{A}$ be a type I separable C*-algebra and let $\mathcal{A}^{\sim}$ be its σ-envelope. Then the partially ordered set of all projections in $\mathcal{A}^{\sim}$ satisfies the essential axioms of Piron i.e. the axioms for a Piron lattice with completeness replaced by σ-completeness.

To get what he called the essential axioms of Mackey, Plymen replaced Borel function by bounded Borel function in Mackey's Axiom III and dropped Axiom VII altogether. Then he could prove

Theorem

A dynamical system $\{\mathcal{A}, \mathcal{S}, V^t\}$ in the sense of Kadison satisfies the essential axioms of Mackey if

i) $\mathcal{A}$ is a Σ^*-algebra

ii) $\mathcal{S}$ is the set of all σ-states on $\mathcal{A}$

and

Theorem

A dynamical system $\{\mathcal{A}, \mathcal{S}, V^t\}$ in the sense of Kadison satisfies the essential axioms of Mackey if

i) $\mathcal{A}$ is a W*-algebra (von Neumann algebra)

ii) $\mathcal{S}$ is the family of all normal states on $\mathcal{A}$.

These theorems, of course, give only sufficient conditions for the observables and states to satisfy the axiom systems. However, they offer a rich family of possibilities and relatively easy constructions especially if one bears in mind the result of Davies [56], that every state of a C* algebra $\mathfrak{A}$ can be extended to a σ-state of its σ-envelope $\mathfrak{A}^{\sim}$. I will say no more about other possible solutions.

With all this powerful machinery in hand, people were led to reexamine various fundamental questions of quantum mechanics. In particular, the so-called problem of hidden variables was reconsidered; Zierler and Schlesinger treated it in Zierler's formalism, Jauch and Piron in Piron's, and Kochen and Specker directly [57][58][59].

To begin with, it should be said that there has existed much disagreement over what the problem of hidden variables is and what an acceptable solution would be. For some the problem concerns an alternative interpretation of quantum mechanics; it can be formulated roughly as follows: Is it not possible that the predictions of quantum mechanics are really a result of an underlying classical deterministic theory in which hidden variables have been averaged over? Here deterministic may be interpreted as excluding the possibility of stochastic processes or not depending on the taste of the questioner. What should be emphasized is that a hidden variables theory of this kind always agrees completely with quantum mechanics in all its observable predictions. On the other hand, there are some who would only require approximate agreement with the observable predictions of quantum mechanics. Both kinds of theories will be discussed in the following, and this will not be the only respect in which conceptions of hidden variables theories will differ. That the same terms were being used for different concepts accounts in part for the passions aroused in the 1960's when the provers of no-go theorems encountered the creators or advocates of hidden variables theories.

As originally stated by von Neumann [20] the problem of hidden variables is to find out whether dispersion free states exist in quantum mechanics;

he answered the question in the negative. What von Neumann assumed of his observables and states is this:

a) The observables, $\mathcal{O}$, are all self-adjoint operators on a separable Hilbert space, $\mathcal{H}$. (von Neumann is not very explicit about the distinction between bounded and unbounded operators; here attention will be restricted to bounded operators.)

b) The states are real valued functions, α, defined on $\mathcal{O}$ and such that

$$\alpha(A^2) \geqslant 0 , \qquad \alpha(\mathbf{1}) = 1$$

$$\alpha(aA+bB) = a\,\alpha(A) + b\,\alpha(B) \qquad (\ )$$

where $A,B \in \mathcal{O}$ and a,b are real numbers.

c) If f is a real valued Borel function on the real line and $A \in \mathcal{O}$, then $f(A)$ is defined as in Mackey's Axiom III (von Neumann says all real valued functions, but he can't mean it).

Expressed in these terms a dispersion free state α is one satisfying

$$\alpha((A-\alpha(A))^2) = 0$$

i.e.

$$\alpha(A^2) - [\alpha(A)]^2 = 0 \quad .$$

As has already been remarked in connection with (2) , von Neumann deduced from a),b),c) and a not very sharply stated continuity assumption that every α is of the form $\alpha(A) = \mathrm{tr}(\rho_\alpha A)$ where ρ_α is a positive operator of trace 1. It is easy to see by considering $A = P_\Phi$ the projection operator onto a one-dimensional subspace that $\rho_\alpha = 0$ or 1 in a dispersion free state which is impossible. Hence, von Neumann's Hidden Variables Theorem: under the hypotheses a),b),c),and continuity, no dispersion free states exist.

Zierler and Schlessinger reinterpreted the hidden variables problem as follows: given a lattice of propositions, L, satisfying Zierler's axioms, can an embedding of L in some lattice L_1 be found which preserves lattice operators and is such that L_1 is a Boolean algebra. Such an embedding would be a solution to the hidden variables problem. Their conclusion: no such embedding exists unless the observables of L are all compatible.

Jauch and Piron define a hidden variables theory as one in which every state is a mixture of dispersion free states. They prove for a Piron lattice (or more accurately for a Piron lattice with one axiom slightly weakened) that in a hidden variables theory all propositons are compatible. Taking it as an empirical fact that the coordinate and momentum of a particle are incompatible observables, Jauch and Piron conclude that hidden variables theories for quantum mechanics are impossible.

Kochen and Specker work with von Neumann's quantum mechanical formalism in which the observables, $\mathcal{O}$, are all self-adjoint operators on a separable Hilbert space. However, they introduce into $\mathcal{O}$ only the structure of a partial algebra: products and sums are defined only for pairs of commuting operators. A necessary condition for the existence of a hidden variables interpretation is then the existence of an embedding of the partial algebra of observables in a commutative algebra. Restricted to the questions the partial algebra of observables is a partial Boolean algebra of questions. For such a partial Boolean algebra Kochen and Specker prove

Theorem

A necessary and sufficient condition that a partial Boolean algebra, $\mathcal{N}$, be embeddable in a Boolean algebra, $\mathcal{B}$, is that for every pair of distinct elements a and b in $\mathcal{N}$ there is a homomorphism $h : \mathcal{N} \to Z_2$ such that $h(a) \neq h(b)$.

Kochen and Specker apply this criterion to the partial Boolean algebra of propositons of quantum mechanics and show no such homomorphism h exists. The method is to construct a finite partial Boolean subalgebra of the partial Boolean algebra of linear subspaces of three-dimensional Euclidean space which has no such homomorphism. They do this by a direct and ingenious geometrical construction. Then they prove that the partial Boolean algebra of linear subspaces of three-dimensional Euclidean space can be embedded in the partial Boolean algebra of quantum mechanical propositions of any system that has states of angular momentum 1. The crucial propositions are those that say: the component of angular momentum along direction $\vec{n}$ is not zero. Elementary calculations show that these observables satisfy the constraint: if $\vec{n}_1, \vec{n}_2, \vec{n}_3$ are three orthogonal directions, two of the three observables $S_{n_1}^2, S_{n_2}^2, S_{n_3}^2$ are equal to 1 and one is 0. The preceding proof of non-existence for the homomorphism then offers 117 directions such that the predictions of a hidden variables theory must fail on at least one triple of orthogonal directions. Kochen and Specker's proof of the non-existence of a hidden variables interpretation of quantum mechanics applies for any system where Hilbert space has dimension ≥ 3. For dimension two they actually constructed a hidden variables theory. satisfying their hypotheses but not von Neumann's. A similar two-dimensional theory was earlier constructed by J. Bell [60] and will be discussed in connection with his work.

All three of these no-go theorems are notable because they do not explicitly assume the additivity of states on sums of non-commuting observables (14), postulated by von Neumann. However, the new theorems evoked protests from some of the advocates of hidden variables theories, who felt that physically objectionable hypotheses were merely better hidden in the new theorems. (A good sampling of the reaction is contained in Reviews of Modern Physics 40 (1966).) To understand the vehemence of the response, it is important to recognize the existence of another tradition in hidden variables theories associated with such names as de Broglie, Einstein, Rosen, and Podolsky, and Bohm, going back to the earliest days of quantum mechanics. (For a review of these ideas, sometimes referred to as the causal interpretation of quantum mechanics, up to 1957, see [61].) In my opinion, the outcome of the collision was a striking clarification of the subject. The believers in hidden variables were led to make their ideas much more specific and precise and the lovers of lattice theory were forced to recognize that some of their postulates have a non-trivial physical content, which had still to be tested empirically.

Notable for its clarifying influence was a contribution of J.Bell[60][61]. Bell constructed a hidden variables interpretation for the quantum mechanics of a two-dimensional Hilbert space and used it to clarify the significance of von Neumann's and Jauch and Piron's assumptions. One of the beauties of this example is its totally elementary character, so I will give it explicitly.

In two dimensions, the hermitean matrices may be written

$$A_{\alpha,\vec{\beta}} = \alpha 1 + \vec{\beta}\cdot\vec{\tau} \tag{15}$$

where the notation stands for

$$\vec{\beta}\cdot\vec{\tau} = \sum_{j=1}^{3} \beta_j\tau_j \;; \qquad \tau_1 = \begin{pmatrix} 0 & 1 \\ 1 & 0 \end{pmatrix} \tau_2 = \begin{pmatrix} 0 & -i \\ i & 0 \end{pmatrix} \quad \tau_3 = \begin{pmatrix} 1 & 0 \\ 0 & -1 \end{pmatrix}$$

and α and $\vec{\beta}$ are real. Two, say $\alpha_1 + \vec{\beta}_1\cdot\tau$ and $\alpha_2 + \vec{\beta}_2\cdot\vec{\tau}$, such commute iff the exterior product $\vec{\beta}_1 \times \vec{\beta}_2 = 0$, and that holds only when they are collinear. The eigenvalues of the hermitean operator (15) are

$$\alpha \pm |\vec{\beta}|$$

A density matrix describing a pure state in the two-dimensional Hilbert space is given by

$$\rho_{\vec{\gamma}} = \frac{1}{2}(1 + \vec{\gamma}\cdot\vec{\tau}) \tag{16}$$

where $\vec{\gamma}$ is real and $\vec{\gamma}^2 = 1$. Thus the expectation value of (15) in the state (16) is

$$\alpha + \vec{\beta}\cdot\vec{\gamma}$$

A hidden variables interpretation is given by a space Γ and two mappings, one of observables into functions on Γ

$$\{\alpha,\vec{\beta}\} \rightarrow F_{\alpha,\vec{\beta}} \tag{17}$$

and the other of states into measures on Γ

$$\vec{\gamma} \to d\mu_\gamma \tag{18}$$

such that

$$\mathrm{tr}[\rho_\gamma A_{\alpha,\vec{\beta}}] = \alpha + \vec{\beta}\cdot\vec{\gamma} = \int_\Gamma F_{\alpha,\vec{\beta}}(\xi)\, d\mu_\gamma(\xi) \tag{19}$$

Bell takes $\xi = \{\vec{\gamma}',\lambda\}$ where γ' runs over the unit sphere and λ over the interval $-\frac{1}{2} \leq \lambda \leq \frac{1}{2}$, and writes

$$d\mu_{\vec{\gamma}}(\vec{\gamma}',\lambda) = \delta(\vec{\gamma}-\vec{\gamma}')\, d\vec{\gamma}'\, d\lambda$$

$$F_{\alpha,\vec{\beta}}(\vec{\gamma}',\lambda) = \alpha + \tfrac{1}{2}|\vec{\beta}|\,\mathrm{sgn}(\lambda|\vec{\beta}| + |\vec{\beta}\cdot\vec{\gamma}'|)\,\mathrm{sgn}(\vec{\beta}\cdot\vec{\gamma}') \tag{20}$$

Elementary calculations show that (19) holds as does

$$aF_{\alpha_1,\vec{\beta}_1} + bF_{\alpha_2,\vec{\beta}_2} = F_{a\alpha_1+b\alpha_2,\, a\vec{\beta}_1+b\vec{\beta}_2}$$

whenever $\vec{\beta}_1 \times \vec{\beta}_2 = 0$, but not in general.

How does this hidden variables theory manage to escape von Neumann's proscription? Bell's answer was that von Neumann asked that his dispersion free states be additive on the sum of any two hermitean operators whether those operators commute or not. Here that requirement would mean that F $F_{\alpha,\vec{\beta}}(\xi)$ for ξ fixed would be linear in α and $\vec{\beta}$, since ξ labels the dispersion free states. That is not an assumption with an immediate physical interpretation. It is violated here because the function $F_{\alpha,\vec{\beta}}$ has values that are always eigenvalues of $A_{\alpha,\vec{\beta}}$ and those eigenvalues do not depend linearly on $\vec{\beta}$. Conclusion: to be a serious objection against hidden variables interpretations of quantum mechanics a no-go theorem should not assume that the dispersion free states are additive on non-commuting observables. (Kochen and Specker independently obtained a hidden variables theory for the quantum mechanics of a two-dimensional Hilbert space. In their case the space Γ is the two sphere. If $A = \alpha + \vec{\beta}\cdot\vec{\tau}$ with eigenvalues $\lambda_{\substack{1\\2}} = \alpha \pm |\vec{\beta}|$, they choose for $|\vec{\beta}| \neq 0$

$$F(A),\vec{\xi}) = \begin{cases} \lambda_1 & \text{if } \vec{\xi} \in \text{hemisphere of pole } \vec{\beta}/|\vec{\beta}| \\ \lambda_2 & \text{if } \vec{\xi} \notin \quad " \quad " \quad " \quad " \end{cases}$$

If $\vec{\beta} = 0$, $F = \alpha$. The measure for a ψ whose density matrix is $\frac{1}{2}(1+\vec{\gamma}\cdot\vec{\xi})$ is

$$d\mu_{\vec{\gamma}}(\xi) = \begin{cases} \frac{1}{\pi}\,\vec{\gamma}\cdot\vec{\xi}\; d\omega(\vec{\xi}) & \vec{\gamma}\cdot\vec{\xi} \geqslant 0 \\ 0 & \gamma\cdot\vec{\xi} < 0 \end{cases}$$

where $d\omega(\vec{\xi})$ is the usual invariant area on the sphere of radius 1 . An elementary calculation shows

$$(\psi,A\psi) = \alpha + \vec{\gamma}\cdot\vec{\beta} = \int F(A,\xi)d\mu_{\gamma}(\xi).)$$

The Jauch Piron no-go theorem apparently avoids this pitfall. How does the two-dimensional model manage to survive the Jauch-Piron prohibitions? Bell's answer was that another property assumed by Jauch and Piron while true for propositions in quantum mechanics is not of direct phenomenological significance. It is: if for some state α and two propositions a and b

$$\alpha(a) = 1 = \alpha(b) \tag{21}$$

then

$$\alpha(a \wedge b) = 1 \quad . \tag{22}$$

This hypothesis is violated in the model, as the following elementary calculation shows.

The propositions of the two-dimensional Hilbert space are given by

$$0,\ 1,\ \text{and}\ \ \frac{1}{2}(1+\vec{\beta}\cdot\vec{\tau}) \quad \text{where } \vec{\beta} \text{ is real and } \vec{\beta}^2 = 1 \ .$$

Let α be a dispersion free state. Since $\frac{1}{2}(1+\vec{\beta}\cdot\vec{\tau})$ and $\frac{1}{2}(1-\vec{\beta}\cdot\vec{\tau})$ commute, we are allowed to use additivity to obtain

$$\alpha(\tfrac{1}{2}(1+\vec{\beta}\cdot\vec{\tau})) + \alpha(\tfrac{1}{2}(1-\vec{\beta}\cdot\vec{\tau})) = \alpha(1) = 1$$

so we conclude that in a dispersion free state α

$$\text{either}\quad \alpha(\tfrac{1}{2}(1+\vec{\beta}\cdot\vec{\tau})) = 1 \quad \text{or} \quad \alpha(\tfrac{1}{2}(1-\vec{\beta}\cdot\vec{\tau})) = 1 \ .$$

If $\vec{\beta}_1$ and $\vec{\beta}_2$ are non-collinear unit vectors choose signs in

$$a = \frac{1}{2}(1 \pm \vec{\beta}_1 \cdot \vec{\tau}) \qquad b = \frac{1}{2}(1 \pm \vec{\beta}_2 \cdot \vec{\tau})$$

so that $\alpha(a) = \alpha(b) = 1$. Now a and b project onto distinct one-dimensional subspaces of the two-dimensional Hilbert space. Those subspaces therefore intersect only in the vector 0 and therefore $a \wedge b = 0$ and

$$\alpha(a \wedge b) = 0$$

contradicting (22) .

Gleason's work makes neither of the two preceding objectionable hypotheses, but as Bell (and independently Kochen and Specker) pointed out it can be used to give another hidden variables no-go theorem. Of course, in his work Bell could not use the two-dimensional model as an example to test for other objectionable hypotheses since Gleason's theorem assumes dimension $\geqslant 3$. What he did was to introduce a classification of hidden variables theories which A. Shimony has denoted contextual vs. non-contextual [62]. The correspondence between observables and functions on Γ

$$\mathcal{O} : A \longmapsto F(A, \xi, \mathcal{B})$$

may depend not only on the observable A but also on whatever other observables are measured, say some set of compatible observables $\mathcal{B}$. If $F(A, \cdot, \mathcal{B})$ is independent of $\mathcal{B}$, the hidden variables theory is said to be non-contextual; if it depends on $\mathcal{B}$, it is contextual. Of course, the expectation value

$$\alpha(A) = \int_\Gamma F(A, \xi, \mathcal{B}) d\omega_\alpha(\xi)$$

is always assumed independent of $\mathcal{B}$.

In this terminology, what the no-go theorem of Kochen and Specker offers is the statement: no non-contextual hidden variables interpretation exists for quantum mechanics in Hilbert spaces of dimension $\geqslant 3$. Bell offered a proof in the spirit of Gleason's theorem that no dispersion free state exists in a Hilbert space of dimension $\geqslant 3$, for a non-contextual hidden variables theory.

The moral to be drawn from the no-go theorems was therefore that apart from the two-dimensional case (where the above hidden variables theory is evidently non-contextual), the possible hidden variables theories must be contextual. Now that is what some of the advocates of hidden variables theories felt they had been saying all along. It was not even difficult to find statements to support them, sometimes from unexpected sources. Bell quoted Niels Bohr on "...the impossibility of any sharp distinction between the behavior of atomic objects and the interaction with the measuring instruments which serve to define the conditions under which the phenomena appear."

Gudder supplemented this general moral with a sharp theorem [63]. Starting from axioms for the lattice of propositions differing slightly from those for a Piron lattice, he gave a definition of the notion that a lattice of propositions admits a hidden variables theory , this theory being in general contextual. He then showed that every lattice of propositions in his sense, admits a hidden variables theory in his sense.

As for contextual hidden variables theories, Bell's paper [60] introduced a distinguished class, local theories, and conjectured that they could never agree in all respects with quantum mechanics. In [61], Bell proved this conjecture. He considered all local hidden variables theories of a certain simple system and proved that their predictions had to satisfy certain inequalities violated by the quantum mechanical predictions. This opened a new chapter in the foundations of quantum mechanics, both because the experiments turned out, after some further developments, to be practical, and because they involve one of the touchstones of all discussions of hidden variables, the Einstein Podolsky Rosen Paradox [64].

The EPR Paradox arises in any experiment in which a composite system, initially localized in some region in space,decays into two fragments that fly apart. For simplicity, suppose the fragments carry some two-valued state variable say $\sigma_1 = \pm 1$ and $\sigma_2 = \pm 1$ respectively. (The fragments might be photons and the state variable polarization, for example.) A typical wave function of the composite system might be

$$\Phi(\vec{p}_1,\vec{p}_2) = \frac{1}{\sqrt{2}}[\psi_+(\vec{p}_1)\otimes\psi_-(\vec{p}_2) - \psi_-(\vec{p}_1)\otimes\psi_2(\vec{p}_2)] \qquad (23)$$

where $\vec{p}_1$ and $\vec{p}_2$ are the momenta of the two fragments, ψ_+ is the eigenvector of eigenvalue +1 of the two-valued variable and the orthogonal wave function ψ_- is the eigenvector of the eigenvalue −1. This wave function predicts that if one of the fragments has the eigenvalue +1 of the two-valued variable, the other is guaranteed to have the value −1. On the other hand, if we introduce two other functions

$$\psi_\perp = \frac{1}{i\sqrt{2}}(\Psi_+ - \Psi_-) \qquad \Psi_{\shortparallel} = \frac{1}{\sqrt{2}}(\psi_+ + \psi_-)$$

they will also be orthogonal and a little algebra shows

$$\Phi(\vec{p}_1,\vec{p}_2) = \frac{i}{\sqrt{2}}[\psi_{\shortparallel}(\vec{p}_1)\otimes\psi\ (\vec{p}_2) - \psi_\perp(\vec{p}_1)\otimes\psi_{\shortparallel}(\vec{p}_2)] \ .$$

Written in this form the wave function is seen to have the property that if one fragment is in the state $\psi_{\shortparallel}$ the other is in the state $\psi_\perp$ and vice versa. The ERP Paradox arises from these statements as the assertion: the state of one fragment depends on what experiment is chosen to be done on the other, even though it may happen that there is no time for a light signal to travel from one fragment to the other to communiate the choice.

Einstein found the EPR Paradox serious evidence for the incompleteness of quantum mechanics and tried to devise alternatives in which it did not occur. He made his position very clear in [65]: "But on one supposition we should, in my opinion, absolutely hold fast: the real factual situation of the system S_1 is independent of what is done with the system S_2, which is spatially separated from the former."

A <u>local</u> hidden variables theory is one which follows Einstein's injunction. For it, the mapping of observables into functions on Γ

$$\mathcal{O}: \ A \to F(A,\cdot,\mathcal{B})$$

is restricted by the requirement that if A corresponds to a measurement in one space-time region R_1 then F has no dependence on observables in $\mathcal{B}$ corresponding to space-time regions, R_2, separated from R_1 by a space-like interval. Bell took the example of two spin 1/2 particles, and computed the expectation value of $\vec{\sigma}_1 \cdot \vec{a}\, \vec{\sigma}_2 \cdot \vec{b}$, where $\vec{a}$ and $\vec{b}$ are any two unit vectors, in a singlet state which is the analogue of (23) for spin 1/2, $\vec{\sigma}_1 \cdot \vec{a}$ being measured in R_1 and $\vec{\sigma}_2 \cdot \vec{b}$ in R_2. In quantum mechanics the answer is

$$P(\vec{a},\vec{b}) = - \vec{a} \cdot \vec{b} \quad . \tag{24}$$

In the hidden variable theory, it is

$$P(\vec{a},\vec{b}) = \int F(\vec{\sigma}_1 \cdot \vec{a}, \xi, \mathcal{B}_1) F(\vec{\sigma}_2 \cdot \vec{b}, \xi, \mathcal{B}_2) d\mu(\xi) \tag{25}$$

where $\mathcal{B}_1$ and $\mathcal{B}_2$ refer to the observables of regions R_1 and R_2. From (25), Bell derived the inequality

$$1 + P(\vec{b},\vec{c}) \geq |P(\vec{a},\vec{b}) - P(\vec{a},\vec{c})|$$

which is not satisfied by the quantum mechanical answer (24). The derivation uses nothing but Schwarz's inequality and the fact that the functions F are of absolute value 1, since $\vec{\sigma}_1 \cdot \vec{a}$ has eigenvalues ± 1.

Bell's inequality shows that no local hidden variables theory can ever explain the spin correlations, if those correlations agree with the quantum mechanical predictions. Unfortunately, the experiments for this case are very difficult. However, for suitably chosen pairs of optical photons slightly different inequalities can be derived in the spirit of Bell, and those are accessible to test [66]. The first successful experiment was made by Freedman and Clauser [67] and it gave agreement with quantum mechanics and a clear-cut violation of the inequalities of Clauser, Home, Shimony, and Holt. If that were the end of the tale one could say goodbye to local hidden variables theories and let the matter rest. However, there is another completed experiment of which an account is circulating in preprint form. It studies a different transition and its results disagree with quantum mechanics and are just at the limit of the range allowed by local hidden variables theories. When such experimental discrepancies occur, the solution is always more and, if possible, better experiments. Theoretical physicists usually have little to contribute. However, I do have one constructive suggestion, if quantum mechanics turns out to be wrong I would propose a six months cooling-off period during which Physical Review Letters and the Physical Review would shut down while people think things over.

My first example of axiomatization of physics has turned out to involve a rather long story even though many interesting works on the axiomatization of quantum mechanics have not even been mentioned [68]. It seems to me that Hilbert's precepts have been followed to a reasonable extent in this subject and that as a result we have a deepened understanding of the foundations

B. Axiomatization in Quantum Field Theory

As has already been remarked at the beginning of the preceding section on axiomatization in general quantum mechanics, systems of a finite number of degrees of freedom, the internal consistency of the quantum mechanical description was not an issue after the work of the late twenties and early thirties. In part, the generally accepted opinion that the theory is sound was based on the von Neumann uniqueness theoresm for the Weyl relations and on Stone's theorem; in part it was based on the (then still unproved) self-adjointness of the n-body Schrödinger Hamiltonians with reasonable interactions among the particles. On the other hand, for systems of an infinite number of degrees of freedom such as electrons interacting with electromagnetic radiation the situation was very different. The occurrence of ambiguities or infinities both in the formulation of the basic equations of those theories and in the approximate solutions then constructed was a strong motivation to modify the foundations of the theory including, possibly, quantum mechanics itself. Everything from the non-conservation of energy to the quantization of space-time was advocated as a cure and tried at one time or another during the thirties. There were few who supported the view that by taking the equations as they stood seriously but interpreting them with more physical mathematical sophistication, it might be possible to arrive at a consistent theory. Nevertheless, that is the lesson which seems to be emerging after four decades of further development. Step by step, apparent difficulties of internal consistency have been overcome, I do not want to give the impression that the main problems of the subject have been solved. In fact, many remain open. Nevertheless, much has been learned, and nowhere has any indication been found of basic inconsistencies when quantum mechanics is properly applied to systems of an infinite number of degrees of freedom. What seems to be called for is not radical alteration of the foundations but more of the subtle and hard analysis that got us this far. My purpose here is to explain the role that axiomatization has played in these affairs up to this point and its possible usefulness in the future.

The most important development during the late 1930's and 1940's in the process of clarification of the foundations of quantum field theory was the theory of renormalization which we owe to P. Dirac, H. Kramers, J. Schwinger, S. Tomonaga, R. Feynman, F. Dyson, A. Salam, and others[69]. The theory of renormalization in the form they developed dealt with the expansion of the matrix elements of the scattering matrix in electrodynamics in powers of the electric charge. The analysis showed that the ambiguities in the coefficients of these expansions can be classified and collected into power series for a few basic parameters of the theory. When these parameters are fixed from experiment, a process referred to as renormalization, one has renormalized perturbation series in powers of the renormalized charge which are completely well defined as formal power series. That is a remarkable result. What is even more remarkable is that when the theory is applied to electrons and muons interacting with electromagnetic radiation, the first few terms of the series give predictions that agree fantastically accurately with experiment.

The theory of renormalization starts with formal power series for the scattering matrix elements in powers of the unrenormalized charge and produces by the process of renormalization formal power series in the renormalized charge, power series where coefficients are free of ambiguities and infinities. A natural question is then: are the renormalized series convergent or divergent? Although there is even yet no proof in quantum electrodynamics, it is generally believed on the basis of a study of simpler cases that they are divergent. Of course, the sum of the first few terms of a divergent series can be an accurate numerical representation of a function if the series is an asymptotic expansion of the function and the expansion parameter is sufficiently small. The expansion parameter in quantum electrodynamics when put in dimensionless form is $e^2/\hbar c \approx 1/137$ and the first few coefficients of the expansions seem to be of order of magnitude one, so it is reasonable to explain the success of quantum electrodynamics in this way. For typical theories of strong interactions, the analogous expansion parameter can be $g^2/\hbar c \approx 15$ and again the first few coefficients are of order of magnitude 1 so the expansions are not of practical interest.

Thus at this stage in the development, which occured in the early 1950's, the key problem was to find some non-perturbative approach to the solutions of quantum field theories. What made the situation confusing was the welter of heuristic material formally related to such solutions, most of which it was either impossible or inadvisable to put in precise mathematical form. The pioneering work of K. Friedrichs [70] and J. Cook [71] on the mathematical objects appearing in simple field theory models had only scratched the surface. An important preliminary in the problem was therefore to decide what concepts should be used to state it and what should be used to describe an admissible solution.

In order to understand how these concepts gradually emerged it is useful to recall the traditional approach to the subject. In it one starts from a classical theory of fields satisfying some partial differential equations. One puts the equations in Hamiltonian form by finding a suitable set of canonical variables and a Hamiltonian function of them, such that the Hamiltonian equations for the time development of the canonical variables are equivalent to the original partial differential equation. For example, if $\phi(x)$ is a real valued function defined on space-time satisfying

$$(\square + m^2)\,\phi(x) = -\,4\lambda\,\phi(x)^3 \tag{26}$$

where $\lambda > 0$ and $\square = \left(\frac{\partial}{\partial ct}\right)^2 - \Delta_3$, one takes as canonical variables $\phi(t,\vec{x})$ and $\pi(t,\vec{x}) = \frac{\partial}{\partial ct}\,\phi(t,\vec{x})$

If one chooses

$$H = \int d^3x\,[\pi(t,\vec{x})^2 + (\Delta\phi(t,\vec{x}))^2 + m^2\,\phi(t,\vec{x})^2 + \lambda\,\phi(t,\vec{x})^4] \tag{27}$$

and takes as Poisson brackets

$$\{\phi(t,\vec{x})\,,\,\phi(t,\vec{y})\} = 0\,,\ \{\pi(t,\vec{x})\,,\,\pi(t,\vec{y})\} = 0\,,\ \{\phi(t,\vec{x})\,,\,\pi(t,\vec{y})\} = \delta(\vec{x}-\vec{y})$$

then Hamilton's equations

$$\frac{\partial}{\partial ct}\,\pi(t,\vec{x}) = -\,\frac{\delta H}{\delta\phi(t,\vec{x})} = \{\pi(t,x)\,,\,H\}$$

$$\frac{\partial}{\partial ct}\,\phi(t,\vec{x}) = \frac{\delta H}{\delta\pi(t,\vec{x})} = \{\phi(t,\vec{x})\,,\,H\}$$

are equivalent to (26). Of course, this reduction to Hamiltonian form could only be formal at that time since the appropriate theory of infinite dimensional Hamiltonian systems was still to be created. However, the main point of reducing this classical field theory to this Hamiltonian form is that there is a standard quantization procedure for it: regard the canonical variables as operators in some Hilbert space and replace all Poisson brackets by $\frac{i}{\hbar}$ times commutators. Then $\phi(t,\vec{x})$ and $\pi(t,\vec{x})$ are promoted to be operators satisfying the canonical commutation relations

$$[\phi(t,\vec{x}) , \phi(t,\vec{y})] = 0 = [\pi(t,\vec{x}) , \pi(t,\vec{y})]$$

$$[\phi(t,\vec{x}) , \pi(t,\vec{y})] = i\hbar\delta(\vec{x}-\vec{y}) \tag{28}$$

and the H defined in terms of them by (27) becomes an operator. (Since I will not have occasion to say more about the classical field theory, I will not introduce a notation to distinguish the operators $\phi(t,\vec{x})$ and $\pi(t,\vec{x})$ from the real valued functions; from this point on they will be operators.)

The commutation relations (28) can be restated in Weyl's form if one introduces the exponentials of the smeared fields

$$U(f) = \exp i\int d\vec{x}\ f(\vec{x})\ \phi(t,\vec{x})$$

$$V(g) = \exp i\int d\vec{x}\ g(\vec{x})\ \pi(t,\vec{x})$$

The analogues of Weyl's relations are then

$$U(f_1)\ U(f_2) = U(f_1+f_2)$$

$$V(g_1)\ V(g_2) = V(g_1+g_2) \tag{29}$$

$$U(f)\ V(g) = \exp\text{-}i(f,g)\ V(g)\ U(f)$$

where

$$(f,g) = \int d\vec{x}\ f(\vec{x})\ g(\vec{x})$$

The choice of the classes of test functions f and g is arbitrary. One might take for example f, and g real, C^∞ and of compact support. U(f) and U(g) are assumed unitary. It is reasonable by analogy with the finite dimensional case to ask for some continuity of U(f) as a function of f and $V(g)$ as a function of g. The weakest such requirement seems to be $U(\lambda f)$ and $V(\lambda g)$ weakly continuous in λ at $\lambda = 0$ for each fixed f and g.

It would seem at first sight that here are the elements of a non-perturbative formulation of a quantum field theory. One would only have to investigate the representations of this Weyl form of the commutation relations. Then, having chosen an appropriate representation one would insert the resulting $\phi(t,\vec{x})$ and $\pi(t,\vec{x})$ in (27), compute the corresponding Hamiltonian operator H and prove H self adjoint on an appropriate domain. It's a very appealing picture. What goes wrong?

The first difficulty is that the Weyl relations (29) do not have a unique irreducible representation since they describe an infinite number of degrees of freedom. That would not be an insuperable difficulty if a convenient method existed for singling out an appropriate representation from all others. However, as was gradually realized over nearly a decade, there is no reasonable parametrization of the irreducible representations. In general, two different Hamiltonians, even though they differ only in the value of a parameter such as λ, will require two inequivalent irreducible representations of the commutation relations. In fact, it is probably more fruitful to think of using the classification of dynamics (via the Hamiltonian operator) to parametize irreducible representations of the commutation relations, than to try to obtain an a priori classification of the representations of the commutation relations and then to single out an appropriate one for a given Hamiltonian. A mathematically more precise version of the statement that the canonical commutation relations possess a maze of irreducible representations is a by product of the work of Mackey [72] and Glimm [72] on group and algebras possessing type II and type III representations: the space of unitary equivalence classes of irreducible representations of the commutation relations does not possess a separable Borel structure.

As will be described later, some extraordinary work has recently led to the construction of the appropriate representations for the theory (27), but in two dimensional space-time rather than the more physical four dimensional space-time. So this Hamiltonian quantization scheme is not empty. However, let me continue with my enumeration of general difficulties.

To describe the next difficulty I need the notion of ΦΟΚ representation or more accurately family of representations plays an important role in the non-perturbative approach to quantum field theory via canonical quantization. It makes possible the solution of the wave equation (26) when $\lambda = 0$; the field $\phi(t,\vec{x})$ then satisfies the linear differential equation

$$(\Box + m^2)\ \phi(t,\vec{x}) = 0$$

and the canonical commutation relations (28) can be regarded as initial conditions for the commutator $[\phi(t,\vec{x}) \, , \ \phi(t',\vec{y})]$. The integration yields

$$[\phi(x) \ , \ \phi(y)] = \frac{1}{i}\,\Delta(m,x-y)$$

where $\Delta(m,x)$ is odd

$$\Delta(m,x) = -\ \Delta(m,x) \ ,$$

and satisfies the homogeneous wave equation

$$(\Box + m^2)\ \Delta(m,x) = 0$$

as well as the initial condition

$$\frac{\partial}{\partial x^0}\,\Delta(m,x)\,\Big|_{x^0=\,0} = \delta(\vec{x})$$

with $x^0 = ct$.

To construct $\phi(t,x)$, one begins with ΦΟΚ space, the Hilbert space, $\mathfrak{F}_s(\mathcal{H}^{(1)})$, defined as the direct sum

$$\mathcal{F}_s(\mathcal{H}^{(1)}) = \bigoplus_{n=0}^{\infty} (\mathcal{H}^{(1)\otimes n})_s$$

where by convention for $n = 0$ the direct summand is the complex numbers regarded as a Hilbert space with the scalar product $\bar{z}w$. The subscript s stands for symmetrized (under permutations). $\mathcal{H}^{(1)} = L^2(\mathbb{R}^3)$, so $\mathcal{H}^{(1)\otimes n} = L^2(\mathbb{R}^{3n})$ and $(\mathcal{H}^{(1)\otimes n})_s$ is the subspace of $L^2(\mathbb{R}^{3n})$ symmetric under permutations of $\vec{x}_1 \ldots \vec{x}_n$. One then defines the number operator N

$$(N\Phi)^{(n)}(\vec{x}_1 \ldots \vec{x}_n) = n\Phi^{(n)}(\vec{x}_1 \ldots \vec{x}_n)$$

and the annihilation and creation operators

$$(a(f)\Phi)^{(n)}(\vec{x}_1 \ldots \vec{x}_n) = \sqrt{n+1} \int f(\vec{x})\, d\vec{x}\, \Phi^{(n+1)}(\vec{x}\vec{x}_1 \ldots \vec{x}_n)$$

$$(a^+(f)\Phi)^{(n)}(\vec{x}_1 \ldots \vec{x}_n) = \frac{1}{\sqrt{n}} \sum_{j=1}^{n} f(\vec{x}_j)\, \Phi^{(n-1)}(x_1 \ldots \hat{x}_j \ldots x_n)$$

for each $f \in \mathcal{H}^{(1)}$. N, $a(f)$, and $a^+(f)$ are bounded operators on the subspaces $\bigoplus_{n=0}^{P} (\mathcal{H}^{(1)\otimes n})_s$ and are closed densely defined operators on $\mathcal{F}_s(\mathcal{H}^{(1)})$.

They satisfy

$$[a(f), a(g)] = 0 = [a^*(f), a^*(g)]$$

$$[a(f), a^*(g)] = \int d\vec{x}\, f(\vec{x})\, g(\vec{x})$$

$$a(f)^* = a^*(\bar{f})$$

$$[a(f), N] = a(f)$$

From the a's and a^*'s one obtains a representation of the Weyl form of the canonical commutation relations by introducing the self-adjoint operators

$$\phi_{\Phi OK}(f) = (a(f) + a^*(f))$$

$$\pi_{\Phi OK}(f) = \frac{1}{i}(a(f) - a^*(f))$$

and defining

$$U_{\Phi OK}(f) = \exp i\, \phi_{\Phi OK}(f)$$

$$V_{\Phi OK}(f) = \exp i\, \pi_{\Phi OK}(f)$$

Now the field $\phi(t,f) = \int d\vec{x}\, \phi(t,\vec{x})\, f(\vec{x})$ is defined in terms of this ΦOK representation by

$$\phi(t,f) = a(e^{-i\omega t}\, \omega^{-\frac{1}{2}} f) + a^*(e^{i\omega t}\, \omega^{-\frac{1}{2}} f)$$

Alternatively, we can define a representation of the Weyl relations by

$$U_\omega(f) = U_{\Phi OK}(\omega^{-\frac{1}{2}} f)$$

$$V_\omega(f) = V_{\Phi OK}(\omega^{\frac{1}{2}} f)$$

The representations $\{U_\omega, V_\omega\}$ of the Weyl relations are unitarily inequivalent for distinct values of m. However all such representations possess a <u>vacuum state</u> Φ_0, satisfying

$$[\phi_\omega(\omega f) + i\, \pi_\omega(f)]\, \Phi_0 = 0$$

for all $f \in \mathcal{H}^{(1)}$, an equation which is derived from the fact that the vector

$$\Phi_0 = \{1, 0, 0, \ldots\}$$

of $\mathfrak{F}_s(\mathcal{H}^{(1)})$ is annihilated by all the annihilation operators $a(f)$.

The vacuum state contains no particles in the sense that

$$N\,\Phi_0 = 0$$

This explicit construction yields an object, $\phi(t,\vec{x})$, called <u>the free scalar field of mass m</u>. Given the field operators so constructed one can ask what sense can be given to the expression (27) for the Hamiltonian. It has been known since the early days of field theory that (27) as it stands is meaningless but that the subtraction of a constant, the zero point energy of the vacuum yields a well defined expression. This procedure can be expressed in the following well defined mathematical form. Consider the expression

$$(\Phi, \frac{1}{2}\,(\pi(t,\vec{x})\;\pi(t,\vec{y}) + (\nabla\phi)(t,\vec{x})\cdot(\nabla\phi)(t,\vec{y})\;m^2\;\phi(t,\vec{x})\;\phi(t,\vec{y}))\;\Psi)$$

$$-\,(\Phi_0\,,\,\frac{1}{2}\,(\pi(t,\vec{x})\;\pi(t,\vec{y}) + (\nabla\phi)(t,\vec{x})\cdot(\nabla\phi)(t,\vec{y})\;m^2\;\phi(t,\vec{x})\;\phi(t,\vec{y}))\;\Phi_0)(\Phi,\Psi)$$

For $\Phi\,,\,\Psi \in \bigoplus_{n=0}^{P} (\mathcal{H}^{(1)} \otimes n)_s$ and such that $\Phi^{(n)}$ and $\Psi^{(n)}$ are, say C^∞ and of compact support, the expression is a continuous function of $\vec{x}$ and $\vec{y}$, the singularity in $(\Phi,\,\pi)t,\vec{x})\;\pi(t,\vec{y})\;\Psi)$ being cancelled by that in the vacuum expectation value $(\Phi_0\,,\,\pi(t,\vec{x})\;\pi(t,\vec{y})\;\Phi_0)$ and similarly for the other two terms. The limit as $\vec{x}\to\vec{y}$ is denoted

$$(\,\Phi,\,\frac{1}{2}\,(:\pi^2:(t,\vec{x}) + :(\nabla\phi)\cdot(\nabla\phi):(t,\vec{x}) + m^2{:}\phi^2{:}(t,\vec{x}))\Psi)$$

It is integrable in $\vec{x}$ and the form defined by the integral defines an operator

$$H = \frac{1}{2}\int dx\,\frac{1}{2}\,(:\pi^2:\;(t,\vec{x}) + :(\nabla\phi)\cdot(\nabla\phi):\;(t,\vec{x}) + m^2\;:\phi^2:\;(t,\vec{x}))$$

which in fact governs the time development of the fields in the sense that

$$\phi(t,\vec{x}) = \exp(iHt)\;\phi(0,\vec{x})\;\exp(-iHt)$$

With these definitions, the free field of mass m becomes a well defined mathematical object. It has some features worth noting. $\phi(t,\vec{x})$ is not an operator valued function of t and $\vec{x}$, but $\phi(t,f)$ the field smeared in $\vec{x}$ with a test function, f, is an (unbounded) operator-valued function of t for each fixed f. Even for $\lambda = 0$, the field, $\phi(t,\vec{x})$, is an operator valued distribution in $\vec{x}$ for fixed t. For $:\phi^2:(t,\vec{x})$ the situation turns out to be worse. It is not an operator valued function of t and $\vec{x}$, but neither is $:\phi^2:(t,f)$ an operator valued function of t for space-time dimension > 2. It is only when one smears in <u>both</u> t and $\vec{x}$ to obtain $:\phi^2:(g)$ with g a test function on space time that one gets an operator. The field in two dimensional space time is exceptional; for it, $:\phi^2:(t,f)$ is an operator. (We begin to see emerging technical simplifications for two dimensions of space time, which have played an important role in constructive field theory.)

Since $\phi(t,x)^2$ can be made well defined by reinterpreting it as $:\phi^2:(t,x)$, it is natural to hope than an analogous procedure will work for higher powers. In fact, that is the case. There is a well defined quantity $:\phi^n:(t,\vec{x})$ obtained by passing to the limit in an expression with points separated and suitable vacuum expectation values subtracted. For example,

$$:\phi^3:(t,\vec{x}) = \lim_{\vec{x}_1,\vec{x}_2,\vec{x}_3\to x} :\phi(t,\vec{x}_1)\,\phi(t,\vec{x}_2)\,\phi(t,\vec{x}_3):$$

where

$$\begin{aligned}:\phi(t,\vec{x}_1)\phi(t,\vec{x}_2)\phi(t,\vec{x}_3): &= \phi(t,\vec{x}_1)\phi(t,\vec{x}_2)\phi(t,\vec{x}_3)\\ &- (\Phi_0,\phi(t,\vec{x}_1)\phi(t\ \vec{x}_2)\Phi_0)\phi(t,\vec{x}_3)\\ &- (\Phi_0,\phi(t,\vec{x}_1)\phi(t,\vec{x}_3)\Phi_0)\phi(t,\vec{x}_2)\\ &- (\Phi_0,\phi(t,\vec{x}_2)\phi(t,\vec{x}_3)\Phi_0)\phi(t,\vec{x}_1)\end{aligned}$$

Just as for $n = 2$, $:\phi^n:(t,\vec{x})$ for $n > 2$ is an operator-valued distribution in space time but $:\phi^n:(t,f)$ is an operator-valued function only in two-dimensional space time.

With this somewhat lengthy discussion of the free field out of the way, we are ready for interacting fields. The natural course is to try to define a Hamiltonian operator, H , for $\lambda > 0$ as the operator associated with the form

$$\int dx \, \{(\Phi,[\tfrac{1}{2}\,(:\pi^2:(t,\vec{x}) + :(\nabla\phi)\cdot(\nabla\phi)(t,\vec{x}) + m^2:\phi^2:(t,\vec{x}) + \lambda:\phi^4:(t,\vec{x})]\Psi)\} \tag{30}$$

What happens? The answer is that when $\lambda > 0$ the integral diverges for so large a set of Φ and Ψ that the form cannot be used to define an operator.

These are two quite different causes for this divergence which operate simultaneously in four-dimensional space-time and in the early 1950's it was quite a puzzle to disentangle them. A decisive role in the process of clarification was played by Haag's theorem, which in one of its weaker variants goes as follows.

Theorem

Suppose that in a Hilbert space, $\mathcal{H}$, there exists an irreducible representation $\{U(f),V(g)\}$ of the Weyl form of the commutation relations (29), which is a Fock representation in the sense that there is a vector, Φ_0, such that

$$\frac{d}{d\lambda}\,(U(\lambda f) + V(\lambda f))\Big|_{\lambda=0}\Phi_0 = 0$$

Suppose, in addition, that there is in $\mathcal{H}$ a unitary representation of the Euclidean group of R^3 : $\{\vec{a},R\} \to W(\vec{a},R)$ which acts covariantly on the $\{U(f),V(g)\}$ in the sense that

$$W(\vec{a},R)U(f)W(\vec{a},R)^* = U(\{a,R\}f)$$

$$W(\vec{a},R)V(g)W(\vec{a},R)^* = V(\{a,R\}g)$$

where

$$(\{\vec{a},R\}f)(x) = f(R^{-1}(\vec{x}-\vec{a}))$$

and similarly for g .

If $W(\vec{a},R)$ has an invariant vector , Ψ_0 :

$$W(\vec{a},R)\Psi_0 = \Psi_0$$

unique up to normalization, then up to normalization

$$\Phi_0 = \Psi_0$$

Expressed in words this theorem says, roughly, that in a Euclidean invariant theory using a Fock representation of the Weyl relations, the vacuum is the no-particle state. To apply this to the analysis of the Hamiltonian (30), note that H is expected to have an eigenvector invariant under translation, unique up to normalization, namely the vacuum Ψ_0. A direct elementary calculation shows that Φ_0 is not an eigenvector of H . Conclusion: if it is possible to define a respectable H by inserting an appropriate representation of the Weyl relations in (30) , that representation cannot be a Fock representation.

Nowadays, we accept this conclusion rather calmly, but in the 1950's it was a bit of a shock because the books on field theory, to the extent that they discussed such matters at all, treated only Fock representations. (An exception was K. **Friedrichs** [70].)

Not all divergence phenomena can be blamed on the use of the wrong representation of the Weyl relations; in general, there are ultra-violet divergences in addition. These divergences are weakened in space times of lower dimension, and for the particular theory (30), no ultra-violet divergences occur for two-dimensional space time. Apart from an additive constant in the square bracket in (30) that has to be adjusted to make the lower bound of H equal zero, the removal of the divergences in H is accomplished solely by using the correct representation of the Weyl relations. On the other hand, for space-time dimension 3, this procedure does not suffice. What is even more important for the evaluation of the prospects of the canonical formalism is the fact that when the renormalized perturbation series for the operator $\phi(t,\vec{x})$ and $\pi(t',\vec{y})$ are obtained in four dimensions they are so singular that the fields at fixed time $\phi(t,f)$ and $\pi(t',g)$ do not exist as operators, so the commutator $[\phi(t,f),\pi(t',g)]$, while it can be defined as a distribution in t and t' by smearing with test functions is too singular at $t = t'$ for the canonical commutation relations to hold.

Although this evidence for the inadequacy of a field theory formalism based on the Weyl relations is not a proof because it relies on a (presumably divergent) perturbation expansion, it is rather convincing evidence that the paradigm of quantization under discussion is not general enough to cover all important cases in four dimensions of space time. Thus some objects other than canonical pairs of fields at fixed time have to be used in the description of relativistic field theories if all the reasonable examples are to be included. Historically, the S-matrix elements and Green's functions of the theory were very important in the development of such a description and it is instructive to see how this came about.

In his work on the renormalization of quantum electrodynamics in 1948, Dyson had chosen to work with the S-matrix (collision matrix), a unitary matrix which collects all the probability amplitudes for collision processes [74]. To work with the S-matrix is natural because all physical predictions about collisions which the theory makes are expressed directly in terms of it. If one has shown how renormalization can be used to give a finite S-matrix to all orders in an expansion in the coupling constant, one can claim to have removed the infinities from all physical predictions of the theory. The importance of the S-matrix as a natural description for elementary particles had been emphasized by Heisenberg in a famous series of papers [75]. Heisenberg went farther and advocated abandoning field equations as a description of dynamics in favor of direct assumptions about the S-matrix but that idea is irrelevant for our present purposes. Suffice it to say that the S-matrix had been impressed on the minds of those who followed developments in elementary particle physics as a convenient description of the predictions of relativistic theories of elementary particles. The work of Feynman, Schwinger, and Dyson systematically simplified the calculation of the perturbation series for S-matrix elements.

It was soon recognized that S-matrix elements themselves can be expressed in terms of another set of objects which came to be known (in partial conflict with standard mathematical terminology) as Green's functions. Green's functions may be defined as time-ordered expectation values of products of fields in the physical vacuum. Explicitly

$$G^{(n)}(x_1 \dots x_n) = (\Psi_0 (\phi(x_1), \dots \phi(x_n))_+ \Psi_0) \tag{31}$$

where Ψ_0 the physical vacuum is supposed to be a vector invariant under space and time translations, and the symbol $(\ ,\dots)_+$ stands for the product of the field operators in the order of the times $x_1^0 \geq \dots \geq x_n^0$. Of course, this definition of the Green's functions begs the question of the existence of the theory and the physical vacuum, and at that time it was a formal expression for the Green's functions as the ratio of two-power series in λ, which was the definition of the Green's functions for practical purposes.

$$G^{(n)}(x_1 \dots x_n) = \frac{\displaystyle\sum_{\ell=0}^{\infty} \frac{(-i)^\ell}{\ell!} \int \dots \int d\xi_1 \dots d\xi_n (\Phi_0, (\mathcal{L}_I(\xi_1), \dots \mathcal{L}_I(\xi_\ell), \phi^{(0)}(x_1) \dots \phi^{(0)}(x_n))_+ \Phi_0)}{\displaystyle\sum_{\ell=0}^{\infty} \frac{(-i)^\ell}{\ell!} \int \dots \int d\xi_1 \dots d\xi_n (\Phi_0, (\mathcal{L}_I(\xi_1), \dots \mathcal{L}_I(\xi_n))_+ \Phi_0)} \tag{32}$$

Here $\mathcal{L}_I(\xi) = \lambda : \phi^{(0)4} : (\xi)$, $\phi^{(0)}$ is the free field just described, and Φ_0 is the vacuum of that free field, customarily called the <u>bare</u>

vacuum to emphasize that it is different from the physical vacuum, Ψ_0 . These expressions look suspiciously like those appearing in the Fredholm theory of integral equations and they are indeed related, but I won't go into that. It should be emphasized that the series in (32) are unrenormalized and it is the art of renormalization theory to separate infinite parts from the integrals and resum to display the series in renormalized form. It should also be noted that in addition to the infinities which arise as a result of ultra-violet divergences in the terms of (32) there are infinities that result from the infinite volume of space-time. This is the form taken by the infinities which earlier forced the use of a non-Fock representation.

The formulae (31) and (32) were first given by Gell-Mann and Low[76]. Variants will turn up in various guises in the following. In one form or another they are fundamental for field theory.

The connection between Green's functions and S-matrix elements can be made explicit as follows. Let $\hat{G}^{(n)}(p_1 \ldots p_n)$ be the Fourier transform of $G^{(n)}(x_1 \ldots x_n)$. Because of the translation invariance of $G^{(n)}$, $\hat{G}^{(n)}$ is of the form

$$\hat{G}^{(n)}(p_1 \ldots p_n) = \delta(\sum_{j=1}^{n} p_j) F^{(n)}(p_1 \ldots p_n)$$

where δ is the Dirac delta function. If for simplicity, it is assumed that the theory predicts just one kind of particle of mass m , then $F^{(n)}$ will have simple poles at $p_j^2 = m^2$, $j = 1, \ldots n$, and the residue is up to inessential factors , the S-matrix element for reactions in which the total number of incoming and outgoing particles is n . To obtain the amplitude for, say , two particles incoming of momentum p_1 and p_2 and $n-2$ outgoing of momenta $p_3 \ldots p_n$, the residue has to be evaluated at $\{-p_1, -p_2, p_3 \ldots p_n\}$. The prescription has an easy generalization to the case of several kinds of particles of different masses and spins which will not be described here.

This connection between Green's functions and S-matrix elements, conventionally called a reduction formula, was initially discussed in the context of specific field theories, but it was soon recognized that it is a general feature of quantum field theories that have certain general properties. There is a striking contrast between the basic paper of Lehmann, Symanzik, and Zimmermann [77] in which reduction formulae were first established from general principles and the papers on renormalization theory of the S-matrix that preceded it. The latter are studded with technicalities heavily dependent on the detailed form of the dynamics. The former uses only general properties, the decay of solutions of $(\Box + m^2)u(x) = 0$ for large times, relativistic invariance, etc. Of course, LSZ do assume that the dynamics is such as to produce the particle states whose scattering theory is the subject of their results, and that is an implicit restriction on the dynamics.

During this same period, another family of general ideas was brought into contact with quantum field theory and led to a further clarification of some aspects of the subject. This was Wigner's analysis of relativistic invariance in quantum mechanics in terms of the representations of the Poincaré (≡ inhomogeneous Lorentz) group.

Wigner showed under quite general assumptions that if a quantum mechanical theory whose pure states are described by vectors in a Hilbert space, $\mathcal{H}$, is invariant under the connected component of the identity, $\mathcal{P}_+^\uparrow$, in the Poincaré group, there is a uniquely determined continuous unitary representation, $U:\{a,A\} \mapsto U(a,A)$, of its covering group ISL(2,C) (≡ inhomogeneous SL(2,C)) in $\mathcal{H}$, which gives the transformation law of states under space-time translation and Lorentz transformation [78] [79]. Since the structure of the continuous unitary representations of the translation sub-group is known to be

$$U(a,1) = \exp(iP\cdot a) = \int \exp(ip\cdot a)dE(p) \tag{33}$$

where E is a projection-valued measure on $\hat{\mathbb{R}}^4$ (the dual of $\mathbb{R}^4$) and p in this formula has an immediate physical interpretation as a possible value of the energy momemtum operator, P, it makes sense to define the energy-momentum spectrum of a Poincaré invariant quantum mechanical theory as the support of the measure E. Because the group multiplication law dictates

$$U(0,A)U(a,1)U(0,A)^{-1} = U(\Lambda(A)a,1) \tag{34}$$

where $\Lambda(A)$ is the element of the Lorentz group determined by $A \in SL(2,C)$, it follows that the energy momentum spectrum is invariant under restricted Lorentz transformations. Thus, the spectrum splits into parts

$p^0 > 0,\ p^2 > 0$	positive energy positive mass hyperboloids
$p^0 > 0,\ p^2 = 0$	positive energy light cone
$p = 0$	zero energy -momentum
$p^0 < 0,\ p^2 = 0$	negative energy light cone
$p^0 < 0,\ p^2 > 0$	negative energy positive mass hyperboloids
$p^2 < 0$	space-like hyperboloids

Since isolated systems with energy momentum in the last three categories have never been observed, it is a natural physical requirement to impose that the energy momentum spectrum be empty there; that is the spectral condition. It is absolutely fundamental for later developments in quantum field theory.

Wigner made a complete analysis of the irreducible representations of $ISL(2,\mathbb{C})$ satisfying the spectral condition and found they can be labeled by a mass $P^2 = M^2$ and, when $M > 0$, a spin, s, which takes one of the values $0, \frac{1}{2}, 1, \frac{3}{2}, \ldots$. For $M = 0$, the irreducibles are labeled by a helicity $s = 0, \pm\frac{1}{2}, \pm 1, \pm\frac{3}{2}, \ldots$ if the spectrum is the light cone. For $P = 0$, one has the trivial one-dimensional identity representation. (There are two other one-parameter families of representations of mass zero which won't be described since they have turned out to be of no physical interest.) Wigner interpreted these irreducible representations as the transformation laws of what he called elementary systems.

This notation is not intended to imply that an elementary system cannot be a composite particle. Any system whose state space cannot be decomposed into orthogonal subspaces in an invariant way is an elementary system in this sense, e.g. an isolated nitrogen ^{14}N nucleus as well as an isolated electron.

The application of Wigner's analysis to relativistic quantum field theory required two things: an interpretation of the physical meaning of reducible representations, $U(a,A)$ of $ISL(2,\mathbb{C})$ and an analysis of the relation of $U(a,A)$ to the field operators. To understand the first, it is useful to examine the simplest case of interaction in a composite system: the collision of two identical mass m, spin zero particles. If the particles do not interact at all, the transformation law of states should be given by the symmetrized tensor product

$$U^{(2)} = ([m,0] \otimes [m,0])_s$$

where for brevity $[m,0]$ denotes the unitary representation for a mass m spin zero particle. Does the introduction of an interaction between the particles change this representation? If the interaction is such that the particles can produce a bound state of mass M and spin zero, one would expect to have to add $[M,0]$ as a direct summand to the representation. If the collision can also produce a third particle, one would expect to have the direct summand

$$([m,0] \otimes [m,0] \otimes [m,0])_s$$

and so on for the production of more particles. However, there appears to be no other way that the representation of $ISL(2,\mathbb{C})$ can display the existence of interaction. Thus, one comes to the following procedure for determining a reducible representation to describe the relativistic transformation law of a field theory.

1) List all stable particles.

2) Construct appropriately symmetrized (or in the case of fermions

anti-symmetrized) tensor products for all possible numbers of these particles.

3) Take the direct sum of all these possibilities.

For example, if there is a single kind of mass m, spin zero boson, the representation is

$$\mathbb{1} \oplus [\bigoplus_{n=1}^{\infty} ([m,0]^{\otimes n})_s] \quad .$$

Here the identity representation has been added to those mentioned above in order that the theory shall have a unique vacuum state. That the unitary equivalence class of the representation of ISL(2,$\mathbb{C}$) for an interacting field theory is the same as that of a free field theory for particles of the same mass is quite intuitive, once one has developed a little feeling for scattering theory. In the early 1950's the failure to recognize it was a temporary barrier to the application of Wigner's analysis to field theory. The statement is a theorem in any relativistic quantum field theory in which the collision states are asymptotically complete as will be explained in due course.

The transformation law of fields under translations given in (34) has an obvious extension to a transformation law under ISL(2,C)

$$U(a,A)\phi(f)U(a,A)^{-1} = \phi(\{a,A\}f) \tag{35}$$

where

$$(\{a,A\}f)(x) = f(\Lambda(A^{-1})(x-a)) \tag{36}$$

and f is a test function on sapce-time. This transformation law of fields has an easy generalization to several component fields transforming under an $N \times N$ matrix representation $S: A \to S(A)$ with $A \in SL(2,C)$. Then one introduces N-tuples of test functions and defines $\phi(f) = \sum_{\alpha=1}^{N} \int \phi_\alpha(x) f_\alpha(x) d^4x$. The transformation law (35) again holds but (36) is replaced by

$$(\{a,A\}f)(x) = S(A^{-1})^T f(\Lambda(A^{-1})(x-a) \quad . \tag{37}$$

The transformation law (35) and the characterization of the vacuum as the unique invariant state

$$U(a,A)\Psi_0 = \Psi_0 \tag{38}$$

are nowadays taken so much for granted that it is difficult to recover the frame of mind of the early 1950's when the vacuum was thought of as expressed in terms of the bare vacuum and therefore as a very complicated state boiling with virtual particles.

The preceding thumbnail sketch of the development of ideas about Green's functions should make it plausible why it was natural to abandon the canonical formalism based on the Weyl relations (29) for one based on vacuum expectation values of products of fields. The Green's functions themselves provided one such natural family, because of their close connection with S-matrix elements, and such a formalism was extensively developed [80]. However, there is an alternative approach which is somewhat easier to make mathematically rigorous, which uses ordinary products rather than time-ordered products of fields [81][82][83]. With such quantities in mind it is natural to look for a set of axioms for fields that permits the construction of vacuum expectation values and the proof of their main properties. Using ordinary products rather than time-ordered products one can also get a simple reconstruction theorem which, under a few general hypotheses on vacuum expectation values, permits one to reconstruct a Hilbert space of states, $\mathcal{H}$, continuous unitary representation U of $SL(2,C)$, a vacuum vector, Ψ_0, and fields such that the vacuum expectation value of the reconstructed fields coincide with the initial vacuum expectation values. By admitting from the outset that fields may become operators only when smeared with test functions on space-time rather than on space at a fixed time, one secures the required generality to treat renormalizable theories in four-dimensional space-time, at least in perturbation theory.

To cut a long story short, let me give the axioms for a quantum field theory of a single hermitean scalar field.

<u>Definition</u>

Let $\mathcal{H}$ be a Hilbert space and U be a continuous unitary representation of ISL(2,C) in $\mathcal{H}$ satisfying the spectral condition. Suppose U has a unique invariant vector, Ψ_0 , (the vacuum). Then a hermitean scalar field, ϕ , is an operator-valued distribution on space-time that is relativistically invariant and local. Explicitly, for each test function, f , there is a linear (possibly unbounded) operator, $\phi(f)$, defined on a dense linear set, $D \subset \mathcal{H}$ and such that

$$U(a,A)D \subset D \tag{39}$$

$$\Psi_0 \in D \tag{40}$$

$$\phi(f)D \subset D \tag{41}$$

$(\Phi,\phi(f)\Psi)$ is a continuous linear functional of f for each fixed $\Phi,\Psi \in D$ and on D

$$\phi(f)^* = \phi(\bar{f}) \tag{42}$$

$$U(a,A)\phi(f)U(a,A)^{-1} = \phi(\{a,A\}f) \tag{43}$$

$$[\phi(f),\phi(g)] = 0 \quad \text{if} \quad f(x)g(y) = 0 \quad \text{for all } x,y \text{ such that } (x-y)^2 \geqslant 0 . \tag{44}$$

The property (44) of local commutativity is an expression of the requirement that influence should not propagate at speeds faster than that of light. When combined with the relativistic transformation law (43) and the spectral condition, it gives relativistic quantum field theory much of its special flavor.

The axioms as given above do not specify the space of test functions. Although Schwartz's space, $\mathcal{S}$, the space of infinitely differentiable functions of rapid decrease fitted naturally with the ideas then current on field theory, it was emphasized in [83], that much of theory goes through also for Schwartz's space, $\mathcal{D}$, and that other spaces might be more natural for special problems. Further developments have borne this out. In [85] Jaffe introduced a family of test function spaces which permit the fields n to be appreciably more singular than those of conventional renormalizable field theory and there is good evidence that just such spaces are called for in certain interesting classes of models.

One feature of the axioms that appears a bit awkward is the domain D. Is it not possible to assume simply that $D = \mathcal{H}$? The answer is no; in general $\phi(f)$ can at best be defined, consistent with (42), on a dense domain that is not the whole space,because it is unbounded. There is a particular subset, D_0, of D which is easy to work with. It consists of all vectors of the form $P\,\Psi_0$ where P is a polynomial in the smeared fields. Clearly $D_0 \subset D$ so one can always form the quantities

$$(\Psi_0, \phi(f_1)\ldots\phi(f_n)\Psi_0) \tag{45}$$

multilinear functionals separately continuous in the variables $f_1 \ldots f_n$, the vacuum expectation values of products of fields referred to above. If D_0 is dense in $\mathcal{H}$ it is reasonable to regard the scalar field as a complete description of the system because every state can be approximated by states of the form $P\,\Psi_0$. One then says that $\mathcal{H}$, U, and their attendant machinery Ψ_0, D_0, and ϕ are a quantum field theory of the scalar field ϕ.

In order to state the reconstruction theorem that permits a quantum field theory in this sense to be reconstructed from its vacuum expectation values, it is necessary to develop the theory of the vacuum expectation values a little farther. The first step is to extend the multilinear functional to become an invariant distribution on R^{4n}. It is a corollary of the nuclear theorem of L. Schwartz that there is a unique distribution on R^{4n} giving meaning to the symbol

$$(\Psi_0, \phi(x_1)\ldots\phi(x_n)\Psi_0) \tag{46}$$

which when tested with a product $f(x_1 \ldots x_n) = \prod_{j=1}^{n} f_j(x_j)$ agrees with (45). The invariance follows from the elementary calculation

$$\begin{aligned}(\Psi_0, \phi(f_1)\ldots\phi(f_n)\Psi_0) &= (U(a,A)\Psi_0, U(a,A)\phi(f_1)\ldots\phi(f_n)\Psi_0) \\ &= (\Psi_0, \phi(\{a,A\}f_1)\ldots\phi(\{a,A\}f_n)\Psi_0)\end{aligned} \tag{47}$$

and the fact that the resulting invariance extends immediately to (46). From the translation invariance it follows that there exists a distribution $F^{(n)}$ on $R^{4(n-1)}$ such that

$$F^{(n)}(x_1-x_2, \ldots x_{n-1}-x_n) = (\Psi_0, \phi(x_1)\ldots\phi(x_n)\Psi_0) \quad .$$

The identity

$$\begin{aligned}&(\Psi_0, \phi(x_1)\ldots\phi(x_j)U(a,1)\phi(x_{j+1})\ldots\phi(x_n)\Psi_0) \\ &\qquad = F^{(n)}(x_1-x_2, \ldots, x_j-x_{j+1}-a, \ldots x_{n-1}-x_n)\end{aligned}$$

which follows from the transformation law (43) of the field and the invariance of the vacuum (38), yields as a consequence of the spectral condition that $F^{(n)}$ is the Fourier transform of a distribution $\hat{F}^{(n)}(p_1 \ldots p_{n-1})$ whose support lies in the cone

$$p_j^{\,2} \geq 0 \quad p_j^{\,0} \geq 0 \qquad j = 1, \ldots n-1 \quad . \tag{48}$$

From this and the uniqueness of the vacuum it follows that the $F^{(n)}$ have the <u>cluster property</u>

$$\begin{aligned}&\lim_{\lambda\to\infty} F^{(n)}(\xi_1 \ldots, \xi_j+\lambda a, \xi_{j+1} \ldots \xi_{n-1}) \\ &\qquad = F^{(j)}(\xi_1 \ldots \xi_{j-1})F^{(n-j)}(\xi_j \ldots \xi_{n-1})\end{aligned} \tag{49}$$

where a is any space-like vector. Local commutativity implies that the $F^{(n)}$ satisfy

$$F^{(n)}(\xi_1 \ldots \xi_{j-1} \xi_j \xi_{j+1} \ldots \xi_n) = F^{(n)}(\xi_1 \ldots \xi_{j-1}+\xi_j, -\xi_j, \xi_{j+1}+\xi_j \ldots \xi_n) \quad (50)$$

for all space-like ξ_j. Finally, the $F^{(n)}$ are Lorentz invariant as a consequence of (47)

$$F^{(n)}(\Lambda\xi_1, \ldots \Lambda\xi_{n-1}) = F^{(n)}(\xi_1, \ldots \xi_{n-1}) \quad (51)$$

and satisfy the positivity conditions

$$\sum_{j,k} \iint \overline{f_j(x_1 \ldots x_j)}\, F^{(j+k)}(x_j \ldots x_1 y_1 \ldots y_k) f_k(y_1 \ldots y_k) \geq 0$$
$$dx_1 \ldots dx_j dy_1 \ldots dy_k \quad (52)$$

for any finite sequence of test functions $f_1 \ldots f_2 \ldots$. The positivity condition simply expresses the fact that $||P\Psi_0||^2 \geq 0$ for any polynomial, P, in the smeared fields.

The reconstruction theorem now states

Theorem

Let $F^{(0)} = 1$, $F^{(1)} = \text{const}$, $F^{(2)}(\xi_1)$, $F^{(3)}(\xi_1, \xi_2), \ldots$ be any sequence of tempered distributions that are Lorentz invariant (51), have Fourier transforms with support in the cone (48), and satisfy the local commutativity conditions (50), the cluster property (49) and the positivity conditions (52).

Then there exists a Hilbert space $\mathcal{H}$ and in it a continuous unitary representation, U, of the Poincaré group, $\mathcal{P}_+^\uparrow$, satisfying the spectral condition, and having a unique vacuum, Ψ_0, a field ϕ with domain D_0 satisfying the transformation law (43) and local commutativity (44). Polynomials in the smeared field applied to the vacuum Ψ_0 are dense in $\mathcal{H}$ and

$$(\Psi_0, \phi(x_1)\ldots\phi(x_n)\Psi_0) = F^{(n)}(x_1-x_2,\ldots x_{n-1}-x_n) .$$

The reconstruction theorem makes it possible to translate problems originally expressed in terms of fields into equivalent problems for vacuum expectation values. It is well adapted for the study of general questions of the quantum theory of fields in which the detailed form of the dynamics plays no role. To convince you of that let me give some examples of the results achieved in the first few years after the introduction of the formalism.

As a byproduct of the proof that the Fourier transform $\hat{F}^{(n)}$ of $F^{(n)}$ has its support in the cone (48), we have that $F^{(n)}$ is the boundary value of a holomorphic function in the tube of $\mathbb{C}^{4n}$ given by the points

$$\xi_1 - i\eta_1 \qquad \xi_2 - i\eta_2 \quad \ldots \quad \xi_n - i\eta_n$$

with $\xi_1\ldots\xi_n$ $\eta_1\ldots\eta_n$ real and $\eta_1\ldots\eta_n$ in the future cone: $\eta_j^{\,2} > 0$, $\eta_j^{\,0} > 0$. As was shown in [85], the Lorentz invariance (51) then implies that the holomorphic function in question can be continued analytically to become a single-valued analytic function on the so-called extended tube obtained from the tube by letting the complex Lorentz group act via

$$\zeta_1\ldots\zeta_{n-1} \rightarrow \Lambda\zeta_1,\ldots\Lambda\zeta_{n-1}$$

Λ any element of the proper complex Lorentz group. Because the extended tube contains real points for which the $\xi_1\ldots\xi_{n-1}$ are space-like, the local commutativity relation (50) is equivalent to the invariance of the holomorphic functions under the action of the permutation group. More explicitly, if we use the same symbol $F^{(n)}$ to denote the holomorphic function and its boundary value distribution as well

$$F^{(n)}(\zeta_1,\ldots\zeta_{j-1},\zeta_j,\zeta_{j+1}\ldots\zeta_{n-1}) = \tag{53}$$
$$F^{(n)}(\zeta_1\ldots\zeta_{j-1}+\zeta_j,-\zeta_j,\zeta_{j+1}+\zeta_j\ldots\zeta_{n-1})$$

The recognition of the relationship of the vacuum expectation values with holomorphic functions led to extraordinarily simple and elegant proofs of two fundamental theorems of the general theory of quantized fields: the PCT Theorem and the Spin-Statistics Theorem [86][87]. The PCT Theorem says that every relativistic quantum field theory possesses PCT symmetry. That means, in the context of the theory of a scalar field described above, that there exists an anti-unitary operator Θ such that

$$\Theta U(a,A)\Theta^{-1} = U(-a,A) \tag{54}$$

and

$$\Theta\phi(f)\Theta^{-1} = \phi(\overline{f}) \ . \tag{55}$$

The Spin-Statistics Theorem says that particles whose fields transform according to a representation of ISL(2,C) that is single-valued on $\mathcal{P}_+^\uparrow$ cannot have Fermi-Dirac statistics while those for which the representation is double-valued cannot have Bose-Einstein statistics. (For a more detailed description see [88].) These two theorems have very important practical consequences and were evidence, convincing to some at least, that it was worthwhile pursuing further the general consequences of the axioms. Another family of results, even more convincing in this respect is the Haag-Ruelle collision theory [89][90]. This theory takes as its starting point the above axioms together with the assumption that the energy momentum spectrum has a gap above the vacuum followed by a number of isolated hyperboloids of definite mass. It shows that accompanying these single-particle states there must be many particle states with an asymptotic behavior at times $\pm\infty$, consistent with the general theory of collisions. Thus implicit in the axioms of field theory is a scattering theory for all particles indicated by the discrete masses and spins of the single-particle spectrum. This does not show that the collision states span the whole Hilbert space; that is the property of <u>asymptotic completeness</u> which remains an open problem to this day (see below).

These results together with others such as the Reeh-Schlieder Theorem [91], and the theory of Borchers classes [92], do not settle the fundamental question of the existence of non-trivial local relativistic quantum field theories, but they do give a strong impression of coherence. For a time there was some hope that the general theory could be pushed farther, far enough so that the existence problem could be settled on the basis of a general structure theory of local fields. In one form the idea involved was as follows. First find a parametrization of the family of all holomorphic functions $F^{(n)}$ analytic in the extended tube invariant under the Lorentz group and satisfying the relations (53) guaranteeing local commutativity. (This part of the proposal is one form of what is sometimes called the linear programme; the conditions to be satisfied only involve one $F^{(n)}$ at a time.) Second, take a sequence of these $F^{(n)}$ and impose the cluster property (49) and the positivity condition (52) ; these provide connections between the different $F^{(n)}$ for different values of n sufficient for the reconstruction of a theory. Some progress was made on this program in the early 1960's [93][94][95] but the technical problems turned out to be too difficult to be handled easily with the machinery then available, and most of the workers in the field turned to other approaches.

Before I describe these and more recent developments let me comment on the general significance of the axiomatic method in the historical evolution just described. I think it accords with the attitudes expressed by Hilbert although the role played was in some respects different from anything he envisioned. Before the advent of axiomatic field theory, there was no clear answer to the questions: what is a quantized field and what is a quantum field theory? Axiomatic quantum field theory crystallized simple general mathematical definitions answering these questions which not only incorporated many of the heuristic insights based on renormalized perturbation theory but, as has just been described, produced a significant body of general theorems valid in any relativistic quantum field theory. However, the axiomatic theory did not solve the equations of any of the famous

Lagrangean field theories such as the quantum electrodynamics of spin one-half particles. Further, it did not establish the existence of any single non-trivial theory. (The free field theories described above are regarded as trivial in this context, although, of course, they do show the consistency of the axioms.) It is this absence of non-trivial examples which made the situation somewhat anomalous. Normally in applications of the axiomatic approach, one has at least a core of interesting and non-trivial examples and part of the point of the investigation of axioms is to characterize what else can satisfy the axioms. Think for example of Hilbert's fifth problem where one had the Lie groups in hand and the problem was to prove that there is nothing else satisfying the assumptions.
In spite of this somewhat anomalous situation, by the early 1960's axiomatic quantum field theory had made quite a decent contribution to the progress of relativistic quantum field theory. It had at least made a serious mathematical treatment of relativistic quantum field theory into a respectable problem. It offered a conceptually simple statement of what was wanted without obvious internal contradictions. It was a conservative alternative to the radical proposals for alterations in the foundations of relativistic quantum theory.

In the above discussion, mainly for reasons of brevity, no mention has been made of the developments in axiomatic quantum field theory growing out of the systematic use of Green's functions rather than vacuum expectation values of products of fields. A priori it would seem reasonable that the two approaches should be equivalent. In particular, starting from the vacuum expectation values of products of fields it should be possible to define the Green's functions as vacuum expectation values of time-ordered products. In practice, however, the ambiguity of the definition associated with the arbitrariness of the time-ordered product at points where two arguments coincide was an impediment in the smooth passage from one version of the theory to the other. In general, work on the Green's functions simply assumed they were defined somehow, and had the properties one would expect if the singularities appearing are the same as those which show up in the perturbation series of typical model field theories. Within these assumptions, it was possible to obtain an analogue of the reconstruction theorem [96]. The disadvantage of this set up for proving general theorems about fields is the somewhat implicit character of the assumptions about the singularities of Green's functions at points of coincidence of their arguments. The considerable advantages for applications to scattering theory became evident on the systematic introduction of retarded and advanced functions into the theory [97]. These are distributions which have their supports in cones in the differences of space-time coordinates and so their Fourier transforms are boundary values of holomorphic functions in the dual (energy-momentum) variables. For example, the two-point retarded function for the scalar field $\phi(x)$ is

$$\frac{1}{2}\,[1 + \mathrm{sgn}(x^0-y^0)]\;(\Psi_0,[\phi(x),\phi(y)]\Psi_0).$$

It vanishes except in the cone $x^0-y^0 \geqslant 0$, $(x-y)\cdot(x-y) \geqslant 0$. There are analogues for the n-point functions. By relating the retarded functions to Green's functions and using the reduction formula to connect these in turn to S-matrix elements, one can get from the general properties of the theory to holomorphy properties of S-matrix elements. The relations between real and imaginary parts of matrix elements resulting from the holomorphy of the scattering amplitude are known as <u>dispersion relations</u>. Their discovery in the mid-1950's led to a whole industry in the application of dispersion relations to elementary particle reactions. They have been of particular importance in studies of strong dynamics where perturbation theory is not a practical technique. In optimal cases, dispersion relations provide connections between observable cross-sections which are consequences only of the axioms of relativistic quantum field theory and some assumptions about high energy behavior of the S-matrix elements in question. Such relations are so far in agreement with experiment and provide an important test of the validity of relativistic quantum field theory. Their practical application in the hands of A. Martin and others [98] is of such evident physical importance that it has partly counteracted the reputation for abstractness, austerity, and barrenness axiomatic field theory had acquired among physicists.

The second main development in the axiomatic approach to field theory in the 1960's is what has been called local quantum theory. It is a child of the marriage between Segal's algebraic approach to quantum mechanics through C*-algebras and the ideas of relativistic quantum field theory. It grew out of the investigations of scattering theory that led to the Haag-Ruelle theory. It was found convenient there to work with bounded operators representing observables rather than with fields. There is a straightforward formal procedure for getting from quantum field theory to such a formalism; it is just what was done in passing from the commutation relations (28) to the Weyl relations (29). If $\phi(f)$ is self-adjoint and unbounded, $\exp i\,\phi(f)$ is unitary and bounded. The dependence of the field $\phi(f)$ on space-time is explicit through its linear dependence on the test function f. For the bounded operators the analogous information is kept track of by giving the algebras $\mathcal{A}(\mathcal{O})$ generated the bounded operators $\exp i\,\phi(f)$ with the support of f contained in the bounded open set $\mathcal{O}$ of space-time.

Clearly one has

$$\mathcal{A}(\mathcal{O}_1) \subset \mathcal{A}(\mathcal{O}_2) \quad \text{if} \quad \mathcal{O}_1 \subset \mathcal{O}_2 \tag{56}$$

and as an expression of local commutativity

$$\mathcal{A}(\mathcal{O}_1) \subset \mathcal{A}(\mathcal{O}_2)' \tag{57}$$

if $\mathcal{O}_1$ is space-like separated from $\mathcal{O}_2$, where the prime indicates the commutant. The action of the Poincaré group on the fields via (43) implies that the local algebras are moved around in an analogous way

$$U(a,A)\,\mathcal{A}(\mathcal{O})\,U(a,A)^{-1} = \mathcal{A}(\{a,\Lambda(A)\}\mathcal{O}) \ . \tag{58}$$

The norm closure, $\mathcal{A}$, of the union $\bigcup_{\mathcal{O}} \mathcal{A}(\mathcal{O})$ of all the local algebras $\mathcal{A}(\mathcal{O})$ is called the quasi-local algebra. $\mathcal{A}$ is a C*-algebra by definition so one can make contact with Segal's C*-algebra approach to quantum mechanics. Here $\mathfrak{A}$ is equipped with an additional structure, the distinguished family of subalgebras $\mathfrak{A}(\mathcal{O})$ satisfying (56) (57) and (58). It is extraordinary how rich in implications this structure of local quantum theory turns out to be. Haag explains this by the remark that in essence most physical predictions are geometrical in the sense that they can be expressed in terms of the correlations between occurrence or non-occurrence of events in space-time regions. Araki gave an illustration of this point when he showed that with additional assumptions including the requirement that each $\mathfrak{A}(\mathcal{O})$ be a von Neumann algebra, an analogue of the Haag-Ruelle scattering theory exists for local quantum theory [99].

Haag and Kastler gave a critical general discussion of local quantum theory as an alternative formulation of the general ideas of relativistic quantum theory [49]. They worked throughout with C*-algebras of bounded operators and exploited the highly developed theory of those algebras. They found some of the mathematical ideas of the representation theory tailor-made for the physical problems they wished to discuss. I will give an example since it is mathematically very appealing and physically important.

In general quantum mechanics one traditionally introduces a strong notion of physical equivalence of quantum theories as follows. A necessary condition for two theories to be physically equivalent is surely that they have the same set of observables. However, more is involved; the two theories will attribute operators $\pi_1(A)$ and $\pi_2(A)$ respectively to an observable A. For the two theories to be physically equivalent in this strong sense, there must exist a one-to-one mapping of their state vectors such that if Ψ_1 corresponds to Ψ_2

$$(\Psi_1, \pi_1(A)\Psi_1) = (\Psi_2, \pi_2(A)\Psi_2)$$

for all observables A. This notion of physical equivalence is thus equivalent to the unitary equivalence of the representations π_1 and π_2. It lies at the bottom of Wigner's analysis of relativistic invariance in quantum mechanics.

A much weaker notion of physical equivalence of two theories is the following. For every $\varepsilon > 0$, every finite set of observables $A_1 \ldots A_n$ and state vector Ψ_1 of the first theory there is a state vector Ψ_2 of the second such that

$$|(\Psi_1, \pi_1(A_j)\Psi_1) - (\Psi_2, \pi_2(A_j)\Psi_2)| < \varepsilon \qquad j=1,\ldots n$$

and similarly with theories 1 and 2 interchanged. This weaker notion of equivalence is actually much closer to what one can test in the laboratory than the stronger notion. To the delight of all Pythogoreans, the weak notion of equivalence turned out to be exactly that introduced earlier by Fell to discuss certain aspects of the representation theory of C*-algebras [100].

In the context of local quantum theory in which only faithful representations are of interest, one of Fell's theorems says that <u>any</u> two representations of the quasi-local algebra are equivalent in this sense. Haag and Kastler recognized in this theorem of Fell a possible solution of one of the puzzles of general quantum theory, the origin of superselection rules. Since the early 1950's, it had been recognized that there are some operators in quantum mechanics that commute with all observables [47]. For example, the operator of total electric charge, Q, or the operator $(-1)^{2J}$ that is one on the states of integer angular momentum, J, and -1 on states of half-odd integer angular momentum. The existence of these operators has as a consequence that the pure states that one can produce in the laboratory are always characterized by a definite value of the charge Q or of $(-1)^{2J}$. The state space is then a direct sum of subspaces, called <u>superselection sectors</u> that have definite values of these quantum numbers. The representations of the quasi-local algebra of observables $\mathfrak{A}$ have to leave the superselection sectors invariant and to be distinct in different sectors: the restrictions of the representations of $\mathfrak{A}$ to distinct sectors are weakly equivalent but unitary inequivalent. Haag and Kastler interpreted this situation in terms of what is called the "particles behind the moon" argument. To see the equivalence of , say, the representation in the vacuum sector $(Q=0,(-1)^{2J}=1)$ with the $(Q=e,(-1)^{2J}=1)$ sector, one considers states of the vacuum sector in which a pair of charges $Q = \pm e$ are present and one of them is far away ("behind the moon") from the region in which the measurement is being made. Then the results of measurements made in such states differ inappreciably from those in which the particle behind the moon is absent altogether.

According to these ideas of Haag and Kastler the superselection sectors of a local quantum theory are predicted by finding the unitary equivalence classes of representations of the quasi-local algebra that satisfy appropriate physical conditions. This general point of view has been made into a quantitative theory under further assumptions by Doplicher, Haag, and Roberts [101][102].

Local quantum theory in the sense of Haag and Kastler has continued to show itself a flexible and elegant formalism for the investigation of general questions in relativistic quantum theory. However, its role has been much more important in the recent development of statistical mechanics. (For this application one replaces the local commutativity condition (44) by a corresponding assumption appropriate to non-relativistic Schrödinger particles if they constitute the system under consideration.) The point of view is that the formalism of local algebras of observables provides just the right language for the description of the quantum mechanics of actually infinite systems. It is worth remarking that it is far from obvious that such a quantum mechanics of actually infinite systems should exist. What is plausible on physical grounds is that various quantities of physical significance should have limits as the system in question becomes infinitely large. More precisely, the quantities in question should have limits as the number of particles $N \to \infty$ and the volume $V \to \infty$ provided $\frac{N}{V}$, the density of particles, has a limit; this is called the <u>thermodynamic limit</u>.

In equilibrium statistical mechanics, equilibrium states of finite systems are assumed to be Gibbs states. Such a Gibbs state is given by

$$\omega_\beta(A) = \mathrm{tr}(\rho_\beta A)$$

where

$$\rho_\beta = \exp(-\beta H)/\mathrm{tr}(\exp(-\beta H))$$

H being the Hamiltonian of the finite system. The equilibrium states of the corresponding infinite system are then thermodynamic limits of these. A fundamental problem is to obtain an intrinsic characterization of such equilibrium states. A fundamental paper of Haag, Hugenholtz, and Winnink singled out the class of KMS states showing that every equilibrium state is a KMS state [102]. Furthermore, they showed that the representations of the quasi-local algebra arising from a KMS state have a remarkable structure connected with some of the deepest general results in the theory of operator algebras, the Tomita-Takesaki theory [103].

The work on KMS states also settles a question that was left open in the earlier discussion of Segal's algebraic approach to quantum mechanics: KMS states are in general non-normal and, nevertheless, the representations they generate are physically reasonable. Thus the full generality of Segal's formalism is required to handle the practical requirements of statistical mechanics.

It is perhaps symbolic of the impact that mathematical physics has had on the theory of operator algebras in the last decade that the letters KMS should be the initials of Kubo, Martin, and Schwinger, three distinguished theoretical physicists not noted for their interest in von Neumann and C*-algebras. It is worth noting in addition that in the fruitful interaction between statistical mechanics and the theory of operator algebras that has occurred there the level of picayune and juvenile neurosis is conspicuously lower than normally visible when physicists and mathematicians discuss a matter they both care about.

At the moment there is no satisfactory general theory connecting local quantum theory with axiomatic quantum field theory. There is a class of special cases in which the vacuum is an analytic vector for the field and so $\phi(f)|D_0$ is essentially self-adjoint. This permits the passage from $\phi(f)$ to $\exp i\,\phi(f)$ used to arrive at $\mathfrak{A}(\mathcal{O})$ in the beginning of our discussion. Furthermore the properties (56) (57) (58) of $\mathfrak{A}(\mathcal{O})$ follow from the axioms for $\phi(f)$ [104]. However, in general, this treatment is not possible, $\phi(f)$ not being essentially self-adjoint on D_0. What is worse, even if one could pick a self-adjoint extension of $\phi(f)|D_0$ one would not be guaranteed without further argument that local commutativity (44) for $\phi(f)$ implies local commutativity (57) for the corresponding bounded operators. To establish the connection between local quantum theory and relativistic quantum field theory in general, it appears that one will have to dig more deeply into the structure of the theory. In the context of some of the specific models treated in constructive field theory the connection is more accessible and has indeed been established [105].

For this reason during the late 1960's and early 1970's local quantum theory and relativistic quantum field theory to a considerable extent developed in parallel influencing each other by analogy rather than providing direct aid to each other. It is one of the basic problems of the general theory of quantized fields to forge a closer link.

It remains to discuss the subject that has provided the most significant progress in the mathematical foundations of relativistic quantum field theory in the last decade: <u>constructive quantum field theory</u>. This name nowadays connotes two things. First, the objective of this kind of theory is to construct concrete examples of field theories. Second, and less obvious from the terminology, the examples are chosen from the traditional repertoire provided by heuristic Lagrangian quantum field theory. In the high moments of axiomatic quantum field thoery at the beginning of the 1960's when the possibility of a general structure theory was being taken seriously, it was regarded by some as a virtue of axiomatic quantum field theory that it did not stoop to discuss the theories arising from specific Lagrangians. When attempts at a general structure theory petered out without giving a single non-trivial example, it was natural to turn to such Lagrangian theories for special cases for frontal attack. As I have already remarked earlier in this talk what happened then provides clear examples of the phenomena touched on in Hilbert's remarks about specialization as a technique for making progress in mathematics. To make this point in detail requires a brief recapitulation of the history of the subject.

The origins of constructive quantum field theory can be traced to the study of certain field theory models of the early 1960's. These models are all Lagrangian quantum field theories obtained by eliminating one or another or several of the essential difficulties of a full relativistic field theory. For example, the model of self-coupled scalar field, described in (30) can be simplified by introducing two cutoffs, a space cutoff and an ultra-violet cutoff. This amounts in one version to replacing the interaction term $\lambda \int d^3x :\phi^4:(\vec{x})$ in (30) by

$$\lambda \int d^3x\, g(x):\phi_K{}^4:(x)$$

where $g(x)$ is a positive C^∞ function of compact support on R^3 (describing the box in which interaction is allowed to take place), and ϕ_K is obtained from ϕ by truncating its Fourier transform, by multiplying it with the characteristic function of the interval $[-K,K]$ (K is called an ultra-violet cutoff). The use of such double cutoffs goes back to the earliest days of quantum field theory. They have the effect that the cutoff Hamiltonian is mathematically well defined for ϕ and π a ΦOK representation of mass m_0.

In the unpublished but widely circulated thesis of A. Jaffe [106] such cutoff models were shown to have solutions with unique vacuum states and Green's functions. Similar cutoffs were introduced into the Yukawa interaction between a fermion field and a boson field, and the Green's functions shown to exist by O. Lanford [107]. (The non-degeneracy of the vacuum could not be established for all coupling strengths. The self-adjointness of the Hamiltonian had already been established earlier by Y. Kato for a similar cutoff model [108].

A simplified model of a somewhat different kind was studied by E. Nelson. He considered a non-relativistic Schrödinger particle coupled to a relativistic meson field [109]. A significant feature of this model is that it requires an infinite renormalization which Nelson carried out without resort to perturbation theory. Nelson's method used a so-called dressing transformation on a cutoff version of the Hamiltonian. This is an equivalence transformation which displays the cutoff Hamiltonian in a new form in which the singular dependence on the cutoff is explicitly displayed. The use of this technique in the context of relativistic field theory goes back to J. Schwinger in his initial work on renormalization theory [110]. It was developed at some length by K. Friedrichs in his Boulder lectures of 1960, although no proof of the effectiveness of the method was given [111].

It was about at this point that the possibility of defining a field theory without cutoffs really began to be taken seriously, and Hilbert's principle of specialization began to be applied in earnest. The first point was to pass to the Lagrangian field theory model with the least possible technical difficulties. This was unquestionably the so-called $(\lambda\phi^4)_2$ model, precisely the one that has been described by the Hamiltonian (30), except that the number of space-time dimensions is 2, as is indicated by the subscript 2. In perturbation theory this model has no ultra-violet divergences at all so in principle the only obstacle to a solution is that arising from Haag's theorem: one does not know which representation of the ϕ and π to insert in (30) to define the Hamiltonian of the theory. A method for circumventing this difficulty was proposed by Guenin [112]. In it one proposes to use the ΦOK representation of ϕ in π in

$$H(g) = H + \lambda \int dx\, g(x):\phi^4:(x)$$

with H_0 the free field Hamiltonian and g a positive smooth function of compact support equal to 1 on some large interval. Because the theory should predict that influence moves slower than light the time development

$$\exp[i\phi(t,f)] = \exp(itH(g))\exp(i\phi(f))\exp(-itH(g))$$

$$\exp[i\pi(t,f)] = \exp(itH(g))\exp(i\pi(f))\exp(-itH(g))$$

ought to be independent of g provided that the support of f is in the interval where $g = 1$ and the points (t,x) with $x \in \text{supp } f$ lie in the dependence domain of that interval. $H(g)$ is said to be a locally correct Hamiltonian. If one patches together the time developments from all the locally correct Hamiltonians one gets an automorphism of the algebra of the commutation relations. However, by Haag's theorem, it cannot be an automorphism implemented by a Hamiltonian. Thus Guenin's proposal gives the right time development and therefore the right quasi-local algebra on space-time but the wrong representation of it (the representation does not have a unitarily implemented time translation, nor a vacuum Ψ_0). In [113] Segal announced the result that if the operators $H(g)$ and $\lambda \int dx\, g(x){:}\phi^4{:}(x)$ are self-adjoint that Guenin's program really works for the $(\lambda\phi^4)_2$ model. Glimm and Jaffe supplied the necessary proof of self-adjointness [114][115].

To obtain the physical representation of the quasi-local algebra, one has to construct a (non-normal!) state invariant under the time translation automorphisms. That state will then by the GNS construction determine a representation of the quasi-local algebra in which there is a Hamiltonian, H, and a vacuum Ψ_0.

The construction of the vacuum state is obtained by using refined information on the $H(g)$. Glimm and Jaffe showed that $H(g)$ has a simple eigenvalue at the lower end of its spectrum, whose eigenvector could be used as an approximate vacuum [116]. By a compactness argument these approximate vacuum states have limit points as $g \to 1$. In a remarkable piece of analysis, they showed that suitably averaged every one of these limit points provides a physical representation of the theory with positive Hamiltonian and a physical vacuum vector Ψ_0 [117].

These results were soon extended to other technically more difficult models: first to $P(\phi)_2$ the theory in which $\lambda\phi^4$ is replaced by an arbitrary polynomial bounded below, then to Y_2 the interaction between a scalar field and a two-compoinent field satisfying the anti-commutation relations. Y_2 is more difficult because it involves two renormalizations [118][119].

At this point several new families of ideas were introduced which not only led to new results but also had the effect of making some previous results easier to derive. The first of these was the connection of quantum field theory with probability theory via Markoff processes defined on Euclidean spaces [120][121]. Second was a more general connection between quantum field theory in Minkowski space and what is called Eulidean field theory in Euclidean space [122]. Finally, came the connection of Euclidean field theory with the ideas and methods of statistical mechanics [123][124]. With this new armory of ideas, problems became accessible which previously were out of reach. The existence of the Green's functions and the single particle states for the $P(\phi)_2$ model was established, the boundedness below of the $(\phi^4)_3$ Hamiltonian proved [125][126]. Conditions for the uniqueness of the solution of $P(\phi)_2$ were found [127][128]. The $P(\phi)_2$ model was shown to define a local quantum theory in the sense of Haag and Kastler and also to satisfy the axioms of relativistic quantum field theory [125]. Thus, in a sense the results on the $P(\phi)_2$ model summarized in [129], settle once and for all the existence of non-trivial relativistic quantum field theories, <u>if</u> one is content to accept evidence from two dimensions.

During all of this period the mathematical physicists who persisted in the study of these models in two- and three-dimensional space-time expected and received some derision from their physicist friends who always asked: when will you get to four-dimensional space-time? That is a good question but the implied view that all the effort that has gone into lower dimensions has been wasted, is not supported by the evidence. In my opinion, Hilbert's principles of specialization have been applied effectively and with taste in the development of constructive field theory up to this point. Whether all this superbly plotted preliminary work on simplified problems will help with the four-dimensional problems remains to be seen.

REFERENCES

[1] C. Truesdell Six Lectures on Modern Natural Philosophy Springer-Verlag, New York, 1966

[2] E. Mach The Science of Mechanics Fifth Edit. Open Court Lasalle, Ill. London 1942

[3] H. Hertz The Principles of Mechanics MacMillan, New York, 1899

[4] L. Boltzmann Lectures on Gas Theory U. Cal. Press, Berkeley, 1964

[5] H. Weyl David Hilbert and His Mathematical Work Bull. Amer. Math. Soc. 50 (1944) 612-654

[6] D. Hilbert Begründung der kinetischen Gastheorie Math. Ann. 71 (1912) 562-577

[7] H. Grad Singular and Non-uniform Limits of Solutions of the Boltzmann Equation p 269-308 in Transport Theory AMS Proceedings Vol. I Providence 1969

[8] E. Hecke Über die Integralgleichung der kinetischen Gastheorie Math. Zeits. 12 (1922) 274-286

[9] H. Grad Principles of the Kinetic Theory of Gases pp 205-294 Vol. XII in Handbuch der Physik Ed. S. Flügge

[10] J.W. Gibbs Elementary Principles in Statistical Mechanics Vol II of Collected Works of J. Willard Gibbs Longmans Green, New York, 1928

[11] D. Hilbert Begründung der Elementaren Strahlungstheorie Gesam Abhl. Vol. III Springer, Berlin, 1970 pp 217-258

[12] E. Hopf Mathematical Problems of Radiative Equilibrium Cambridge Tracts No. 31 Cambridge Univ. Press 1934

[13] S. Chandrasekhar Radiative Transfer Dover, New York, 1960

[14] D. Hilbert Die Grundlagen der Physik Gesam. Abhl. Vol. III pp 258-289

[15] J. Mehra Einstein, Hilbert, and the Theory of Gravitation in The Physicists Conception of Nature Reidel, Boston, 1973

[16] E. Guth Contribution to the History of Einstein's Geometry as a Branch of Physics in Relativity Plenum Press 1970

[17] E. Noether Invariante Variationsprobleme Gött. Nach. (1918) 235-257

[18] D. Hilbert, J. v. Neumann, L. Nordheim Über die Grundlagen der Quantenmechanik Math. Ann. 98 (1927) 1-30

[19] J. von Neumann Mathematische Begrundung der Quantenmechanik Gött. Nach. (1927) 1-57

[20] J. von Neumann Mathematical Foundations of Quantum Mechanics Princeton Press 1955

[21] M.H. Stone Linear Transformations in Hilbert Space III Operational Methods and Group Theory Proc. Nat. Acad. Sci. USA 10 (1930) 172-5

[22] J. von Neumann Allgemeine Eigenwerttheorie Hermitischen Funktional Operatoren Math. Ann. 102 (1929) 49-131

[23] J. von Neumann Die Eindeutigkeit der Schrödingerschen Operatoren Math. Ann. 104 (1931) 570-8

[24] T. Kato Fundamental Properties of Schrödinger Operators of Hamiltonian Type Trans. Amer. Math. Soc. 70 (1951) 195-211

[25] P. Jordan Über die Multiplikation quantenmechanischen Grössen Zeits für Phys. 80 (1933) 285-91

[26] H. Braun, M. Koecher Jordan-Algebren Grundlehren der Math. Wiss. Bd. 128 Springer, Berlin, 1966

[27] N. Jacobson Structure and Representations of Jordan Algebras Amer. Math. Soc. Prov. 1968

[28] P. Jordan, J. von Neumann, E. Wigner On an Algebraic Generalization of the Quantum Mechanical Formalism Ann. of Math. 35 (1934) 29-64

[29] A.A. Albert On a Certain Algebra of Quantum Mechanics Ann. of Math. 35 (1934) 65-73

[30] J. von Neumann On an Algebraic Generalization of the Quantum Mechanical Formalism (Part I) Mat. Sbornik 1 (1936) 415-984

[31] I. E. Segal Postulates for General Quantum Mechanics Ann. of Math. (2) 48 (1947) 930-48

[32] P. Jordan Über das Verhältnis der Theorie der Elementarlänge zur Quantentheorie Comm. in Math. Phys. 9 (1968) 279-292

[33] I. M. Gelfand and M. Neumark On the Embedding of Normed Rings into the Ring of Operators in Hilbert Space Mat. Sbornik 12 (1943) 197-213

[34] S. Sherman On Segal's Postulates for General Quantum Mechanics Ann. of Math 64 (1956) 593-61

[35] D. Lowdenslager On Postulates for General Quantum Mechanics Proc. Amer. Math. Soc. 8 (1957) 88-91

[36] I.E. Segal Mathematical Problems of Relativistic Physics Amer. Math. Soc. Providence 1963

[37] S. Sherman Non-negative Observables Are Squares Proc. Amer. Math. Soc. 2 (1951) 31-3

[38] I. Gelfand Normierte Ringe Mat. Sbornik 9 (1941) 3-24

[39] G. Emch Algebraic Methods in Statistical Mechanics and Quantum Field Theory Wiley Interscience, New York, 1972

[40] G. Birkhoff and J. von Neumann The Logic of Quantum Mechanics Annals of Math. (1936) 823-843

[41] J. von Neumann Continuous Geometry Princeton Press 1960

[42] G. Mackey Quantum Mechanics and Hilbert Space Amer. Math. Monthly 64 (1957) 45-57

[43] A. Gleason Measures on the Close Subspaces of a Hilbert Space J. Rat. Mech. and Anal. 6 (1957) 885-94

[44] N. Zierler Axioms for Non-relativistic Quantum Mechanics Pacific Jour. Math. 11 (1961) 1151-69

[45] L. Loomis Lattice Theoretic Background of the Dimension Theory of Operator Algebras Memoirs of Amer. Math. Soc. No 18 (1955)

[46] G. W. Mackey Mathematical Foundations of Quantum Mechanics Benjamin, New York, 1963

[47] G.C. Wick, A.S. Wightman, E.P. Wigner The Intrinsic Parity of Elementary Particles Phys.Rev. 88 (1952) 101-5

[48] R. Kadison Transfo... .cions of States in Operator Theory and Dynamics Topology 3 (1965) 177-98

[49] R. Haag and D. Kastler An Algebraic Approach to Quantum Field Theory Jour. Math. Phys. 5 (1969) 848-61

[50] C. Piron Thesis University of Geneva Axiomatique Quantique Helv. Phys. Acta 37 (1964) 439-468

[51] Amemiye and H. Araki A Remark on Piron's Paper Publ. of Res. Inst. for Math.Sci., Kyoto, Ser. A 2 (1967) 423-7

[52] J. P. Eckmann and Ph. Zabey Impossibility of Quantum Mechanics in a Hilbert Space over a Finite Field Helv. Phys. Acta 42 (1969) 420-4

[53] S. Gudder and C. Piron Observables and the Field in Quantum Mechanics J. Math. Phys. 12 (1971) 1583-8

[54] R. Plymen A Modification of Piron's Axioms Helv. Phys. Acta 41 (1968) 69-74

[55] R. Plymen C* Algebras and Mackey's Axioms Comm. Math. Phys. 8 (1968) 132-146

[56] E. Davies The Borel Structure of C* Algebras Comm. Math. Phys. 8 (1968) 147-163

[57] N. Zierler and M. Schlesinger Boolean Embeddings of Orthomodular Sets and Quantum Logic Duke Jour. 32 (1965) 251-62

[58] J. Jauch, C. Piron Can Hidden Variables be Excluded in Quantum Mechanics Helv. Phys. Acta 36 (1963) 827-37

[59] S. Kochen, E. Specker The Problem of Hidden Variables in Quantum Mechanics Jour. Math. and Mech. 17 (1967) 59-87

[60] J. Bell On the Problem of Hidden Variables in Quantum Mechanics Rev. Mod. Phys. 38 (1966) 447-52

[61] J. Bell On the Einstein Podolsky Rosen Paradox Physics 1 (1965) 195-200

[62] A. Shimony Experimental Test of Local Hidden Variables Theories pp 182-194 in Foundations of Quantum Mechanics Academic, New York, 1971

[63] S. P. Gudder On Hidden Variables Theories Jour. Math. Phys. 11 (1970) 431-6

[64] A. Einstein, B. Podolsky, N. Rosen Can Quantum-Mechanical Description of Reality Be Considered Complete Phys. Rev. 47 (1935) 777-80

[65] A. Einstein p 85 in Albert Einstein: Philosopher-Scientist Ed. P. Schilpp Library of Living Philosophers, Evanston, Ill. 1949

[66] J. Clauser, M. Horne, A. Shimony, R. Holt Proposed Experiment to Test Hidden Variables Theory Phys. Rev. Lett. 23 (1969) 880-4

[67] S. Freedman and J. Clauser Experimental Test of Local Hidden-Variables Theories Phys. Rev. Lett. 28 (1972) 938-41

[68] G. Ludwig Deutung des Begriffs "physikalische theorie"... Springer, Berlin, 1970 B. Mielnik Theory of Filters Comm. Math. Phys. 15 (1969) 1-46

[69] All the relevant papers are reprinted in Selected Papers on Quantum Electrodynamics Ed. J. Schwinger, Dover, New York, 1958 , except H.A. Kramers Non-relativistic Quantum-Electrodynamics and Correspondence Principle pp 241-68 Rapports due 8^{e} Conseil Solvay 1948, and

A. Salam Renormalized S-Matrix for Scalar Electrodynamics Phys.Rev. 86 (1952) 731-44

[70] K. Friedrichs Mathematical Aspects of the Quantum Theory of Fields Interscience Publishers, New York, 1953

[71] J. Cook The Mathematics of Second Quantization Trans. Amer. Math. Soc. 74 (1953) 222-45

[72] G. Mackey Borel Structures in Groups and Their Duals Trans. Amer. Math. Soc. 85 (1957) 134-165

[73] J. Glimm Type I C* Algebras Ann. Math. 73 (1961) 572-612

[74] F.J. Dyson The S-Matrix in Quantum Electrodynamics Phys Rev. 75 (1949) 1736-55

[75] W. Heisenberg Die "beobachtbaren Grössen" in der Theorie der Elementarteilchen Zeits. für Phys. 120 (1943) 513-38, II 120 (1943) 673-702

[76] M. Gell-Mann and F. Low Bound States in Quantum Field Theory Phys. Rev. 84 (1951) 350-4

[77] H. Lehmann, K. Symanzik, W. Zimmermann Zur Formulierung quantisierter Feldtheorien Nuovo Cim. 1 (1955) 205-225

[78] E. Wigner Unitary Representations of the Inhomogeneous Lorentz Group Ann. Math. 40 (1939) 149-204

[79] V. Bargmann and E. Wigner Group Theoretical Discussion of Relativistic Wave Equations Proc. Nat. Acad. Sci. U.S.A. 34 (1946) 211-223

[80] K. Hepp On the Connection Between Wightman and LSZ Quantum Field Theory pp 135-246 in Axiomatic Field Theory 1965 Brandeis Lectures Gordon and Breach, N.Y. 1966

[81] A.S. Wightman Quantum Field Theory in Terms of Vacuum Expectation Values Phys. Rev. 101 (1956) 860-6

[82] A.S. Wightman Quelques Problèmes Mathématiques de la Théories Quantique Relativiste pp 1-38 in Les Problèmes Mathématiques de la Théorie Quantique des Champs CNRS 1959

[83] A.S. Wightman and L. Gårding Fields as Operator-Valued Distributions in Relativistic Quantum Field Theory Arkiv for Fysik 28 (1964) 129-184

[84] A. Jaffe Form Factors at Large Momentum Transfer Phys Rev. Lett. 17 (1966) 661-3

[85] D. Hall and A.S. Wightman A Theorem on Invariant Analytic Functions With Applications to Relativistic Quantum Field Theory Mat. Fys. Medd. Danish Vid. Selsk. 31 No 5 (1957)

[86] R. Jost Eine Bemerkung zum CTP Theorem Helf. Phys. Acta 30 (1957) 409-16

[87] N. Burgoyne On the Connection of Spin with Statistics Nuovo Cim. 8 (1958) 607-9

[88] R. Jost The General Theory of Quantized Fields Amer. Math. Soc., Providence 1965

[89] R. Haag Discussion des 'Axiomes' et des Propriétés asymptotiques d'une Théorie des Champs locale avec particules composées pp 151-62 in Les Problèmes Mathématiques de la Théorie Quantique des Champs CNRS 1959

[90] D. Ruelle On the Asymptotic Condition in Quantum Field Theory Helv. Phys. Acta 35 (1962) 147-163

[91] H. Reeh and S. Schlieder Bemerkungen zur Unitäräquivalenz von Lorentz invarianten Feldern Nuovo Cimento 22 (1961) 1051-68

[92] H.J. Borchers Über die Mannigfaltigkeit der interpolierenden Felder zer eimer kausalen S-Matrix Nuovo Cimento 15 (1960) 784-94

[93] G. Källén and J. Toll Integral Representation for the Vacuum Expectation Value of Three Scalar Local Fields Helv. Phys. Acta 33 (1960) 753-72

[94] H. Araki Wightman Functions, Retarded Functions, and Their Analytic Continuations Prog. Theoret. Phys. Supp. 18 (1961) 83-125

[95] H. Epstein Some Analytic Properties of Scattering Amplitudes in Quantum Field Theory pp 1-133 in Axiomatic Field Theory 1965 Brandeis Lectures Gordon and Breach, New York, 1966

[96] V. Glaser, H. Lehmann, W. Zimmermann Field Operators and Retarded Functions Nuovo Cim. 6 (1957) 1122-8

[97] H. Lehmann, K. Symanzik, W. Zimmermann On the Formulation of Quantized Field Theories II Nuovo Cim. 6 (1957) 319-33

[98] A. Martin Scattering Theory: Unitarity, Analyticity, and Crossing Lecture Notes in Physics 3 Springer, Berlin, 1969

[99] H. Araki Local Quantum Theory I pp 65-96 in Local Quantum Theory Proceedings of the Int. School of Phys. Enrico Fermi Course 45 Ed. R. Jost, Academic, New York, 1969

[100] J. Fell The Dual Spaces of C* Algebras Trans. Amer. Math. Soc. 94 (1960) 365-403

[101] S. Doplicher, R. Haag, and J. Roberts Fields, Observables, and Gauge Transformations I Comm. Math. Phys. 13 (1969) II ibid 15 (1969) 173-200

[102] R. Haag, N. Hugenholtz, M. Winnink On the Equilibrium States in Quantum Statistical Mechanics Comm Math. Phys. 5 (1967) 215-36

[103] M. Takesaki Tomita's Theory of Modular Hilbert Algebras and Its Applications Springer, Berlin, 1970

[104] H. Borchers, W. Zimmermann On the Self-Adjointness of Field Operators Nuovo Cim. 31 (1964) 1047-59

[105] J. Glimm, A. Jaffe, and T. Spencer The Particle Structure of the Weakly Coupled $P(\phi)_2$ Model and Other Applications of High Temperature Expansions pp 132-142 in Constructive Quantum Field Theory Vol 25 of Lecture Notes in Physics Springer, Berlin, 1973

[106] A. Jaffe Dynamics of a Cutoff $\lambda\phi^4$ Field Theory Princeton Thesis 1965, unpublished

For a more complete bibliography on the early development of constructive field theory, see A.S. Wightman Constructive Field Theory; Introduction to the Problems pp 1-85 in Fundamental Interactions in Physics and Astrophysics Ed. G. Iverson, A. Perlmutter, S. Mintz, Plenum 1972

[107] O. Lanford Construction of Quantum Fields Interacting by a Cutoff Yukawa Coupling Princeton Thesis 1966, unpublished

[108] Y. Kato Some Converging Examples of Perturbation Series in Quantum Field Theory Prog. Theoret. Phys. 26 (1961) 99-122

[109] E. Nelson Interaction of Non-relativistic Particles with a Quantized Scalar Field Jour. Math. Phys. 5 (1964) 1190-7

[110] J. Schwinger (unpublished lectures) The results of the application of the method are summarized in On Quantum-Electrodynamics and the Magnetic Moment of the Electron Phys. Rev. 73 (1948) 416 .

[111] K. Friedrichs Perturbation of Spectra in Hilbert Space Amer. Math. Soc., Providence, 1965

[112] M.Guenin On the Interaction Picture Comm.Math. Phys. 3 (1966) 120-31

[113] I. Segal Notes Toward the Construction of Non-linear Relativistic Quantum Fields I The Hamiltonian in Two Space Time Dimensions as the Generator of a C*-automorphism Group Proc. Nat. Acad. Sci. USA 57 (1967) 1178-1183

[114] J. Glimm and A. Jaffe A $\lambda\phi^4$ Quantum Field Theory Without Cutoffs Phys. Rev. 176 (1968) 1945

[115] I.E. Segal Non-linear Functions of Weak Processes I, II Jour. Func. Anal. 4 (1969) 404-56 and 6 (1970) 29-75

[116] J. Glimm and A. Jaffe The $\lambda(\phi^4)_2$ Quantum Field Theory Without Cutoffs: II The Field Operators and the Approximate Vacuum Ann. Math. 91 (1970) 362-401

[117] J. Glimm and A. Jaffe The $(\lambda\phi^4)_2$ Quantum Field Theory Without Cutoffs III The Physical Vacuum Acta Math. 125 (1970) 203-67

[118] J. Glimm and A. Jaffe The Yukawa_2 Quantum Field Theory Without Cutoffs

Jour. Fcnal.Anal. 7 (1971) 323-57 and earlier articles cited there

[119] R. Schrader Yukawa Field Theory in Two-Dimensional Space Time Without Cutoffs Annals Phys. 70 (1972) 412-57

[120] E. Nelson The Free Markoff Field Jour. Fcnal. Anal. 12 (1973) 211-227 Construction of Quantum Fields from Markoff Fields ibid 12 (1973) 97-112

[121] E. Nelson Probability Theory and Euclidean Field Theory pp 94-124 in Constructive Field Theory [129]

[122] K. Osterwalder Euclidean Green's Functions and Wightman Distributions pp 71-93 in Constructive Field Theory [129]

[123] F. Guerra, L. Rosen, B. Simon The $P(\phi)_2$ Euclidean Quantum Field Theory as Classical Statistical Mechanics Ann. Math. to appear 1974

[124] B. Simon Correlation Inequalities and the Mass Gap in $P(\phi)_2$ I Comm Math. Phys. 31 (1973) 127-36, II Ann. of Math., to appear

[125] J. Glimm, A. Jaffe, T. Spencer The Wightman Axioms A Particle Structure in the $P(\phi)_2$ Quantum Field Theory Model Ann. of Math.,to appear

[126] J. Glimm, A. Jaffe Positivity of the ϕ_3^4 Hamiltonian Fortschutte der Physik 21 (1973) 327-76

[127] J. Glimm, A. Jaffe, T. Spencer [125] and T. Spencer The Mass Gap for the $P(\phi)_2$ Quantum Field Theory Model with a Strong External Field to appear

[128] B. Simon Correlation Inequalities and the Mass Gap in $P(\phi)_2$ II Ann of Math, to appear

[129] Constructive Quantum Field Theory Ed. G. Velo and A. Wightman Lecture Notes in Physics No. 25 Springer, Berlin, 1973

[130] V. Arnold and A. Avez Ergodic Problems of Classical Mechanics Benjamin 1968

[131] J. Moser Stable and Random Motions in Dynamical Systems Annals of Math. Studies No. 77 Princeton 1973

[132] D. Ruelle Statistical Mechanics Rigorous Results Chapter 7, Benjamin, New York, 1969

[133] ibid Chapter 4

[134] F.J. Dyson Missed Opportunities Bull. Amer. Math. Soc. 78 (1972) 635-52

[135] K. Symanzik On the Many Particle Structure of Green's Functions in Quantum Field Theory Jour. Math. Phys. 1 (1960) 249-73

[136] K. Symanzik Grundlagen und gegenwärtiger Stand der feldgleichungsfreien Feldtheorie pp 275-298 in Werner Heisenberg und die Physik unserer Zeit Vieweg Braunschweig 1961

Proceedings of Symposia in Pure Mathematics
Volume 28, 1976

HILBERT'S SEVENTH PROBLEM

ON THE GEL'FOND-BAKER METHOD AND ITS APPLICATIONS

R. TIJDEMAN

This survey consists of fourteen sections. The section headings describe the scope of the paper.

AMS (MOS) Subject classifications (1970).
Primary 10F, 10B, 10H15, 12A50.
Secondary 10C10, 10G05, 30A08.

1. *Early history.*

Long before the first rigorous proof of the existence of transcendental numbers was given, conjectures were made on the transcendence of specific numbers. Around 1750 it was conjectured that e, π and ${}^a\log b \ (= \log b / \log a)$ with $a, b \in \mathbb{Q}, a > 1, b > 1, {}^a\log b \notin \mathbb{Q}$ are transcendental. (See Euler (1748), Lambert (1768).) The transcendence of π was of particular interest, since it was known that it would imply the negative solution of the classical problem of squaring the circle. (See for example Lambert (1768) Op.Math. II p. 159.) The first proof that transcendental numbers do exist was given by Liouville (1844). He showed that irrational algebraic numbers cannot be "too well" approximated by rational numbers, and that numbers like $\sum_{n=1}^{\infty} 2^{-n!}$ are therefore transcendental. Cantor's proof that almost all numbers are transcendental was not given until 1874. However, neither Liouville's nor Cantor's method is of any use for numbers like e or π.

The proof of the transcendence of e by Hermite (1873) was the beginning of a prosperous period for transcendental number theory. Further developments enabled Lindemann (1882) to prove the transcendence of π. More generally the following result was established. (Weierstrass (1885).)

THEOREM 1A. *If $\alpha_1, \dots, \alpha_n$ are distinct algebraic numbers, then $e^{\alpha_1}, \dots, e^{\alpha_n}$ are linearly independent over the field of all algebraic numbers.*

This theorem includes the transcendence of both e^{α} $(\alpha \in \mathbb{A}, \alpha \neq 0)$ and π and $\log \alpha$ $(\alpha \in \mathbb{A}, \alpha \neq 0, 1)$. (Here we denote the field of algebraic numbers by $\mathbb{A}$). In the last part of the nineteenth century Hilbert and several other authors devoted papers to this method, but Euler's problem on the transcendence of ${}^a\log b$ remained open. Hilbert felt the importance of this conjecture and listed the following natural extension as seventh problem in his address to the International Mathematical Congress in Paris in 1900. (Hilbert (1900).)

The expression α^{β} for an algebraic base α and an irrational algebraic exponent β, e.g., the number $2^{\sqrt{2}}$ or $e^{\pi} = i^{-2i}$, always represents a transcendental or at least an irrational number.

(By taking $\alpha = a, \beta = {}^a\log b$ one has Euler's conjecture.) Hilbert added that the solution of this problem must lead to entirely new methods and to a new insight into the nature of special irrational and transcendental numbers. Siegel, who came to Göttingen as a student in 1919, was always to remember a lecture which he heard from Hilbert at this time (See Reid (1970) p. 164).

Hilbert mentioned Riemann's hypothesis and said that there had recently been much progress and that he was very hopeful that he himself would live to see it proved. Fermat's problem had been around for a long time and apparently demanded entirely new methods for its solution; perhaps the youngest members of the audience would live to see it solved. However, as for establishing the transcendence of $2^{\sqrt{2}}$, no one present in the lecture hall would live to see that.

2. *The theorem of Gel'fond-Schneider.*

The first partial solution of the Euler-Hilbert problem, namely in case β is an imaginary quadratic irrational, was given by Gel'fond (1929) hardly ten years later. It includes the transcendence of e^{π}. Kuz'min (1930) and Siegel (unpublished, see for example Reid (1970) p. 164) proved independently that Gel'fond's method can be extended to the case that β is a real quadratic irrational number. This proved the transcendence of $2^{\sqrt{2}}$! Siegel (1949, p. 84) remarked that this shows that one cannot guess the real difficulties of a problem before having solved it.

Just at the same time Siegel (1929) made an important contribution to the method of Hermite-Lindemann. We shall not follow later developments of this method here, but we note that Siegel's paper contains an extremely useful lemma. A simple application of the box principle enables one to construct auxiliary functions which have many zeros at certain prescribed points. On combining this and other new ideas with the method of the earlier partial solutions Gel'fond (1934) and independently Schneider (1934 I) obtained the complete solution of the strongest form of Hilbert's seventh problem. Although there is one important difference between both proofs, which we shall indicate later, the main principles of both proofs are the same.

We give three equivalent formulations of the theorem of Gel'fond-Schneider:

THEOREM 2A. a) *If* $\alpha, \beta \in \mathbb{A}$, $\alpha \neq 0,1$, $\beta \notin \mathbb{Q}$, *then* $\alpha^{\beta} \notin \mathbb{A}$.

b) *If* $\alpha, \beta \in \mathbb{A}$, $\alpha \neq 0,1$, $\beta \neq 0,1$, *then* $\log\alpha / \log\beta$ *is either rational or transcendental.*

c) *Let* $\alpha, \beta \in \mathbb{A}$, $\alpha \neq 0$, $\beta \neq 0$. *If* $\log\alpha$ *and* $\log\beta$ *are linearly independent over* $\mathbb{Q}$, *then they are linearly independent over* $\mathbb{A}$.

(Here and elsewhere it does not matter which branch of the logarithm is taken, but one should not take different determinations for the logarithm of the same number on different occasions).

One year later Gel'fond (1935) derived so-called approximation measures for α^β and $\log\alpha/\log\beta$. The height of an algebraic number ξ is defined as the maximal absolute value of the coefficients of its minimal defining polynomial. One of Gel'fond's results is that under the conditions of b) for every $\varepsilon > 0$ and every positive integer N there exists a constant $H_0 = H_0(\alpha, \beta, \varepsilon, N)$ such that

$$\left| \frac{\log\alpha}{\log\beta} - \xi \right| > \exp\left(-(\log H)^{5+\varepsilon}\right) \tag{2.1}$$

for every algebraic number ξ of degree N and height at most H with $H \geq H_0$. In 1939 and 1949 Gel'fond found improvements which imply that the constant 5 in the exponent of $\log H$ can be replaced by 2. (See Gel'fond (1939),(1949),(1952).) So Gel'fond had proved much more than Hilbert had asked for. His ingenious method is fundamental for a very important part of transcendental number theory.

3. *Gel'fond's method.*

In this sectionthe main principles of Gel'fond's method are explained. Suppose that α and β are algebraic numbers, $\alpha \neq 0,1$, $\beta \neq 0,1$, such that

$$\left| \frac{\log\alpha}{\log\beta} - \xi \right|$$

is extremely small, say smaller than $\exp\left(-(\log H)^{2+\varepsilon}\right)$ where H is the height of the algebraic number ξ. We distinguish four steps. A detailed account of proofs along these lines can be found in Cijsouw (1972, Ch V) and in Cijsouw (1974 a).

Step 1. *Construction of an auxiliary function* $F(z)$ *such that* $|F^{(t)}(p)|$ *is very small for many values of* p *and* t.

Let

$$F(z) = \sum_{k=0}^{K-1} \sum_{\ell=0}^{L-1} C_{k\ell}\, e^{(k\log\alpha + \ell\log\beta)z}, \tag{3.1}$$

where the coefficients $C_{k\ell}$ are integers to be chosen below. It follows that for $p = 1,2,\cdots$

$$\frac{1}{(\log\beta)^t} F^{(t)}(p) = \sum_k \sum_\ell C_{k\ell} \left(k\frac{\log\alpha}{\log\beta} + \ell\right)^t \alpha^{kp} \beta^{\ell p}.$$

We define

$$\Phi_{tp} = \sum_k \sum_\ell C_{k\ell} (k\xi + \ell)^t \alpha^{kp} \beta^{\ell p}.$$

It is an immediate consequence of our assumption that

$$(3.2) \qquad \left| \frac{F^{(t)}(p)}{(\log\beta)^t} - \Phi_{tp} \right|$$

is very small for very many values of p and t, if the coefficients are not too large, say

$$(3.3) \qquad |C_{k\ell}| \leq \exp\left((\log H)^{2+\varepsilon/2}\right).$$

We now use Siegel's lemma (see for its simplest form Baker (1966, Lemma 1) and for its most ready-made form Cijsouw (1974 a, Lemma 6)) to choose the integers $C_{k\ell}$ subject to (3.3) and such that $\Phi_{tp} = 0$ for many values of p and t, say $p = 0, 1, \ldots, P-1$ and $t = 0, 1, \ldots, T-1$. Hence, $|F^{(t)}(p)|$ is very small for these values of p and t.

Step 2. *Proof that* $|F^{(t)}(p)|$ *is small for very many values of* p *and* t. Since $|F^{(t)}(p)|$ is very small for $p = 0, 1, \ldots, P-1$ and $t = 0, 1, \ldots, T-1$ it is easy to construct an analytic function $F^*(z)$, not differing much from $F(z)$, such that $F^{*(t)}(p) = 0$ for these numbers p and t. Hence,

$$F^*(z) \Big/ \prod_{p=0}^{P-1} (z-p)^T$$

is an analytic function. Let $2P \leq R_2$ and $5R_2 \leq R_1$. Then, by Cauchy's residue theorem, for $|z| = R_2$

$$F^*(z) = \frac{1}{2\pi i} \int_{|\zeta|=R_1} \frac{F^*(\zeta)}{\zeta - z} \prod_{k=0}^{P-1} \left(\frac{z-p}{\zeta-p}\right)^T d\zeta.$$

Hence,

$$\max_{|z|=R_2} |F^*(z)| \le 2\left(\tfrac{1}{3}\right)^{PT} \max_{|z|=R_1} |F^*(z)| .$$

On using (3.1) and (3.3) the right-hand side can easily be estimated. By an appropriate choice of the parameters P, T, R_1, R_2 this upper bound will be rather small. Hence, $|F^*(z)|$ and therefore also $|F(z)|$ is rather small for $|z| = R_2$. By a second application of Cauchy's theorem we have

$$F^{(t)}(p) = \frac{t!}{2\pi i} \int_{|z|=R_2} \frac{F(z)\,dz}{(z-p)^{t+1}} .$$

So we find that $|F^{(t)}(p)|$ is small for $p = 0,1,\ldots,P-1$ and $t = 0,1,\ldots,T'-1$, where T' is considerably larger than T. A ready-made lemma for this step is given by Cijsouw (1974, Lemma 7).

Step 3. *Proof that* $|F^{(t)}(p)|$ *is very small for very many values of* p *and* t.
Since $|F^{(t)}(p)|$ is small and (3.2) is very small, we obtain that $|\Phi_{tp}|$ is small for $p = 0,1,\ldots,P-1$ and $t = 0,1,\ldots,T'-1$. We see from its definition that $\Phi_{tp} \in \mathbb{A}$. Assume $\Phi_{tp} \neq 0$. Let M be the smallest rational integer such that $M\Phi_{tp}$ is an algebraic integer. Then the product of $M\Phi_{tp}$ and its conjugates is in absolute value at least 1. It is not difficult to compute upper bounds for M and the absolute values of the conjugates of $M\Phi_{tp}$, and this gives immediately a non-trivial lower bound for $|\Phi_{tp}|$. By an appropriate choice of our parameters this lower bound is incompatible with the upper bound for $|\Phi_{tp}|$ derived in step 2. We conclude that $\Phi_{tp} = 0$ for $p = 0,1,\ldots,P-1$ and $t = 0,1,\ldots,T'-1$. Since (3.2) is very small, we obtain that $|F^{(t)}(p)|$ is very small for $p = 0,1,\ldots,P-1$ and $t = 0,1,\ldots,T'-1$. A ready-made lemma for this step is given by Fel'dman (1968 a, Lemma 2).

Step 4. *Proof that* $|F^{(t)}(p)|$ *cannot be very small for very many values of* p *and* t.
(An alternative step is to prove by the use of determinants that Φ_{tp} cannot be zero for very many values of p and t.) It is a well known fact that an

analytic function cannot have infinitely many zeros in a finite disk. In the case of exponential polynomials there is a simple quantitative upper bound for the number of zeros. In particular, $F(z)$ cannot have more than $3K^2 + 4KR\,(|\log\alpha| + |\log\beta|)$ zeros in a disk of radius R. (See Tijdeman (1971 a).) By a modification of this theorem one can use the fact that not all integers $C_{k\ell}$ vanish to prove that $|F(z)|$ cannot have more than this number of very small values in well spaced points in a disk of radius R. Such a result can be found in Tijdeman (1973 a). The application of this lemma gives immediately a contradiction with the final result of step 3. The contradiction completes the proof of the original assertion.

The main difference with Schneider's proof is that Schneider uses the auxiliary function

$$F(z) = \sum_{p=0}^{P-1} \sum_{t=0}^{T-1} C_{pt}\, z^t e^{pz}$$

in the points $z = k\log\alpha + \ell\log\beta$ and applies the addition formula $e^{az}.e^{bz} = e^{(a+b)z}$ instead of differentiating. Therefore Schneider's method is more appropriate for generalizations to functions satisfying an addition theorem, while Gel'fond's method is advantageous in generalizations to functions satisfying differential equations.

4. *Baker's method.*

The following statement is a natural extension of the Gel'fond-Schneider theorem. (Theorem 2.1.c)

Let $\alpha_1,\dots,\alpha_n$ be non-zero algebraic numbers. If $\log\alpha_1,\dots,\log\alpha_n$ are linearly independent over $\mathbb{Q}$, then they are linearly independent over $\mathbb{A}$.

Such a result gives an analogue for logarithms of Theorem 1 A for exponentials. Gel'fond and Linnik (1948) proved that a quantitative form of this result for $n = 3$ would have an interesting application to the determination of all imaginary quadratic fields with class number 1. This last problem was completely settled by Stark (1967) using another method. At the same time Baker developed a method which might be considered as an extension of Gel'fond's method. From 1966 on he published the above and several related results. His ideas made it also possible to give better approximation measures for $\log\alpha/\log\beta$. To indicate Baker's method we shall sketch a proof of the following improvement of (2.1).

$$\left|\frac{\log\alpha}{\log\beta} - \xi\right| > \exp\left(-(\log H)^{1+\varepsilon}\right). \tag{4.1}$$

We follow the same steps as in section 3. We remark that the proofs in Baker's papers have a rather different form and that a complete account of proofs along our lines can be found in Cijsouw (1972, Ch VI) and Cijsouw (1974 b).

We assume throughout that α and β are algebraic numbers, $\alpha \neq 0,1$, $\beta \neq 0,1$, such that (3.1) is extremely small, say smaller than $\exp\left(-(\log H)^{1+\varepsilon}\right)$, where H is the height of ξ.

Step 1. *Construction of an auxiliary function* $F(z)$ *and some derived functions* $F_{\tau t}(z)$ *such that* $|F_{\tau t}(p)|$ *is very small for many values of* τ, t *and* p.

Let

$$F(z) = \sum_{k=0}^{K-1}\sum_{l=0}^{K-1}\sum_{m=0}^{M-1} C_{klm}\, z^m e^{(k\log\alpha + l\log\beta)z}.$$

Note that powers of z appear in the auxiliary function; usually M is much greater than K. It follows that

$$\frac{F^{(t)}(z)}{(\log\beta)^t} = \sum_{\tau=0}^{t}\binom{t}{\tau} F_{\tau t}(z),$$

where

$$F_{\tau t}(z) = \sum_k\sum_l\sum_m C_{klm}\frac{m!}{(m-\tau)!} z^{m-\tau}\left(k\frac{\log\alpha}{\log\beta}+l\right)^{t-\tau}\alpha^{kz}\beta^{lz}.$$

We define for $p = 0,1,2,\ldots$

$$\Phi_{\tau t p} = \sum_k\sum_l\sum_m C_{klm}\frac{m!}{(m-\tau)!} p^{m-\tau}(k\xi + l)^{t-\tau}\alpha^{kp}\beta^{lp}$$

Hence, $F_{\tau t}(z)$ is an exponential polynomial, $\Phi_{\tau t p}$ is an algebraic number and

(4.2) $$\left| F_{\tau t}(p) - \Phi_{\tau t p} \right|$$

is very small for a great many values of τ, t and p, if the coefficients $C_{k\ell m}$ are not too large. We use Siegel's lemma in such a way that $\Phi_{\tau t p} = 0$ for many values of τ, t and p. Hence, $|F_{\tau t}(p)|$ is very small for these values of τ, t and p.

Steps 2 and 3. *Proof that* $|F_{\tau t}(p)|$ *is very small for a great many values of* τ, t *and* p, *and hence* $|F^{(t)}(p)|$ *is very small for very many values of* t *and* p.

By applying to $F_{\tau t}(z)$ the same arguments as were applied in steps 2 and 3 of section 3 to $F(z)$, we find that $|F_{\tau t}(p)|$ is very small for very many values of τ, t and p.

However, this is not sufficient to obtain a contradiction and a new element is introduced, namely induction. Steps 2 and 3 are applied again and again and each time we obtain more values of τ, t and p such that $|F_{\tau t}(p)|$ is small. After a finite number of induction steps we have a great many such values. More precisely, $|F_{\tau t}(p)|$ is very small for very many values of t and p, and for each such choice of p and t is very small for $\tau = 0, 1, \ldots, t$. By the definition of $F_{\tau t}(z)$ this implies that $|F^{(t)}(p)|$ is very small for very many values of t and p. (For convenience we have omitted quantitive forms of these assertions.)

Step 4. The further proof is completely similar to that in section 3. In Baker's papers determinants and induction are used in the final arguments.

5. *Linear forms in the logarithms of algebraic numbers.*

In this section we formulate a few more results which can be proved by the method of the previous section and its later refinements.

We denote by $\beta_0, \beta_1, \ldots, \beta_n$ non-zero algebraic numbers of degrees at most d and with heights at most B. By $\alpha_1, \alpha_2, \ldots, \alpha_n$ we denote non-zero algebraic numbers of degrees at most d and heights at most $A_1, A_2, \ldots, A_n$ respectively. We assume $d \geq 4$, $A_1 \geq 4, \ldots, A_n \geq 4$ throughout.

Further, we put $A = \max_{1 \le j \le n} A_j$, $A' = \max_{1 \le j < n} A_j$, $Q = \prod_j \log A_j$.

The following results generalize Theorem 2 A b).

THEOREM 5 A. (Baker 1967a)).
$\beta_1 \log \alpha_1 + \dots + \beta_n \log \alpha_n$ *is transcendental if at least one of the following conditions holds.*

a) $\log \alpha_1, \dots, \log \alpha_n$ *are linearly independent over* $\mathbb{Q}$.

b) $\beta_1, \dots, \beta_n$ *are linearly independent over* $\mathbb{Q}$.

Baker's original approximation measures for $\beta_1 \log \alpha_1 + \dots + \beta_n \log \alpha_n$ were considerably improved by Fel'dman (1968a). He proved, using a more complicated auxiliary function:

THEOREM 5 B. *Let* $\log \alpha_1, \dots, \log \alpha_n$ *be fixed values of the logarithms which are linearly independent over* $\mathbb{Q}$. *Then there exists an effectively computable constant* $C = C(n,d)$ *such that*

$$|\beta_0 + \beta_1 \log \alpha_1 + \dots + \beta_n \log \alpha_n| > \exp\left(-C (\log A)^{12n^2+4n} \log B\right).$$

In fact Fel'dman gave the constant C almost explicitly. The lower bound is the best possible with respect to B.

In more recent papers the dependence on the heights of $\alpha_1, \dots, \alpha_n$ became a subject of research. First we quote some results for the special case that all coefficients $\beta_1, \dots, \beta_n$ are integers.

THEOREM 5 C. (Baker 197?)) *There exists an effectively computable constant* $C = C(n,d)$ *such that the inequalities*

$$0 < |\beta_1 \log \alpha_1 + \dots + \beta_n \log \alpha_n| < \exp(-C\, Q \log Q \log B)$$

have no solutions in integers $\beta_1, \dots, \beta_n$ *of absolute values at most* B.

In case one of the α's has a height which is much larger than those of the others, the following result gives more information. (Tijdeman (197?), see also Baker (1972)):

THEOREM 5 D. *There exists an effectively computable constant* $C = C(n,d)$ *such that the inequalities*

$$0 < |\beta_1 \log \alpha_1 + \ldots + \beta_n \log \alpha_n| < \exp\left(-C(\log A')^{2n^2+7n} \log A_n \log B\right)$$

have no solutions in integers $\beta_1, \ldots \beta_n$ *of absolute values at most* B.

It is an immediate consequence of these results that if $|\beta_1 \log \alpha_1 + \ldots + \beta_n \log \alpha_n|$ is very small, but not zero, then B cannot be very large. A first result of this type was given by Baker (1968 a):

THEOREM 5 E. *Let* $0 < \delta \leq 1$. *Denote by* $\log \alpha_1, \ldots, \log \alpha_n$ *the principal values of the logarithms. If rational integers* $\beta_1, \ldots, \beta_n$ *exist, with absolute values at most* B *such that*

$$0 < |\beta_1 \log \alpha_1 + \ldots + \beta_n \log \alpha_n| < e^{-\delta B}, \tag{5.1}$$

then

$$B < \left(4^{n^2} \delta^{-1} d^{2n} \log A\right)^{(2n+1)^2}.$$

Very recently Fel'dman (1972) and Baker (1973) have given upper bounds for B which behave much better with respect to A. They proved, for example:

THEOREM 5 F. *If, for some* $\delta > 0$ *there exist rational integers* $\beta_1, \ldots, \beta_{n-1}, \beta_n = -1$ *with absolute values at most* B *such that (5.1) holds, then* $B < C \log A_n$ *for some effectively computable constant* C *depending only on* n, d, A' *and* δ.

In the general case of algebraic coefficients similar results are known. The following theorem is due to Stark (1973 a):

THEOREM 5 G. *Suppose that* $B < H^{\log H}$, $\varepsilon > 0$, $\delta > 0$.

If

$$0 < |\beta_1 \log \alpha_1 + \ldots + \beta_n \log \alpha_n| < e^{-\delta H},$$

then

$$H < C\,\mathcal{Q}^{1+\varepsilon},$$

where $C = C(n, d, \varepsilon, \delta)$ *is an effectively computable constant.*

For estimates concerning the behaviour of C as a function of n see Ramachandra (197?) and Shorey (1974).

Several authors, for example Gel'fond (1940), Adams (1966), Brumer (1967), Coates (1969), Sprindžuk (1972), have given $\mathfrak{p}$-adic analogues of the above theorems. We give two examples. Gel'fond (1940) published a $\mathfrak{p}$-adic analogue of (2.1) with $3+\varepsilon$ instead of $5+\varepsilon$. Kaufman (1971) proved the following analogue of Fel'dman's Theorem 5 B:

THEOREM 5 G. *Let* $\alpha_1, \dots, \alpha_n$ *be algebraic numbers which are multiplicatively independent over* $\mathbb{Q}$. *Let* $\beta_0, \dots, \beta_n$ *be algebraic numbers of heights at most* B. *Denote the degree of* $\mathbb{Q}(\alpha_1, \dots \alpha_n, \beta_0, \beta_1, \dots, \beta_n)$ *by* d *and let* $\mathfrak{P}$ *be a prime ideal of this field. Assume* $|\alpha_i - 1|_{\mathfrak{P}} \leq p^{-1}$ *for* $i = 1, \dots, n$. *Then there exists an effectively computable constant* C *independent of* B, *such that*

$$|\beta_0 + \beta_1 \log \alpha_1 + \dots + \beta_n \log \alpha_n|_{\mathfrak{P}} > p^{-C \log B}.$$

A central problem of future research in this field will be to establish, under suitable conditions, the algebraic independence of the logarithms of algebraic numbers.

6. *Imaginary quadratic fields with small class numbers.*

As mentioned before Stark (1967) determined all imaginary quadratic fields with class number 1. Several proofs have been given since, some of which are based on Baker's method. Baker (1971 a) and Stark (1971 a) gave independently of each other an effective algorithm to determine all imaginary quadratic fields of class number 2. In both proofs estimates of sums of logarithms of algebraic numbers play an important role. Sketches of the main ideas of the proofs of these results can be found in several survey papers. See for example Stark (1969, 1971 b) and Baker (1971 b).

The above mentioned results are partial confirmations of a conjecture made by Gauss in 1802. He conjectured that there are only finitely many negative discriminants associated with any given class number and moreover that the tables of discriminants which he had drawn up were in fact complete in the cases of relatively small class numbers. (See Gauss (1802) § 303.)

The first conjecture was verified by Heilbronn in 1934. It follows from extensions of his work by Siegel and Brauer that each imaginary quadratic field of class number h has discriminant at least $ch^{2-\varepsilon}$ for some constant $c = c(\varepsilon)$, $\varepsilon > 0$. However, their arguments are not effective. New methods are required to find effective algorithms to determine all imaginary quadratic fields of a fixed class number greater than 2, but it seems not unlikely that the method of Gel'fond-Baker will be used in such algorithms.

7. *Upper bounds for solutions of Diophantine equations.*

Many applications of the Gel'fond-Baker method concern the computation of effective upper bounds for solutions of Diophantine equations with only finitely many solutions. In some cases the complete set of solutions has been obtained by additional arguments and by the use of computers. See for example Baker and Davenport (1969), Ellison *et al.* (1972).

Gel'fond (1940) used the p-adic analogue mentioned at the end of section 5 to derive such a result for the equation in rational integers x, y, z,

$$\alpha^x + \beta^y = \gamma^z,$$

where α, β and γ are real algebraic numbers. Similar applications of Gel'fond's method were given by Schinzel (1967) for equations of the form $x^2 + d = p^m$ in positive integers x, m, where d is a given positive integer and p is a prime of certain type. Schinzel's results follow from estimates for the solutions of linear recurrences of the second order with integer coefficents.

Baker's extension of Gel'fond's method led to very important new applications. Let f be an irreducible binary form with integer coefficients and of degree at least 3. Let m be a non-zero integer. The equation $f(x,y) = m$ in integers x, y is named after Thue, who proved in 1909 that there are only finitely many solutions. Baker (1968 b), Fel'dman (1971), Sprindžuk (1972) and Stark (1973 b) have given upper bounds for these solutions. Baker used his estimate to prove that every solution (x,y) of the equation

(7.1) $$y^2 = x^3 + k$$

where k is a non-zero integer, satisfies

(7.2) $$\max(|x|, |y|) \leq \exp\left(\left(10^{10}|k|\right)^{10^4}\right).$$

Stark (1973 b) improved this estimate by

(7.3) $$\max(|x|,|y|) \leq \exp\left(c\,|k|^{1+\varepsilon}\right),$$

for every $\varepsilon > 0$, where $c = c(\varepsilon)$ is effectively computable. Effective estimates have also been given for the solutions of the more general equation

(7.4) $$y^m = P(x),$$

where P is a polynomial with integer coefficients and at least two simple zeros in case $m \geq 3$, and at least three simple zeros in case $m = 2$. See Baker (1968 c, 1969 a), Sprindžuk (1973).

Coates (1970 a) and Sprindžuk (1972) generalized Baker's work on Thue's equation (7.1) in another direction. They determined upper bounds for the integer solutions $(x, y, z_1, \ldots, z_s)$ of the equations $f(x,y) = m\,p_1^{z_1} \cdots p_s^{z_s}$ and $y^2 = x^3 + k\,p_1^{z_1} \cdots p_s^{z_s}$, where $p_1, \ldots, p_s$ are given primes. For a sketch of the proofs of these results see Baker (1971 c).

A very recent application of Baker's method deals with another generalization of (7.1), namely the equation in integers $x > 1$, $y > 1$, $m > 1$, $n > 1$,

(7.5) $$x^m - y^n = k,$$

where k is a fixed non-zero integer. The case $k = 1$ was already considered by Catalan (1844) who conjectured that $3^2 - 2^3 = 1$ would be the only solution. Cassels (1953) made the weaker conjecture that in this case there are only finitely many solutions. The latter conjecture has now been proved by Choodnovski (written communication) and Tijdeman (197?) independently. Using Theorem 5 D it is first proved that there exist effectively computable upper bounds for m and n. Baker's estimate for the solutions of (7.4) then gives effective upper bounds for x and y. Choodnovski has even given the same result for the general equation (7.5). His upper bound for the solutions of (7.5) is of the form $\exp(\exp(e^{Ck}))$ for some absolute constant C. It is very probable that Choodnovski's powerful method will be of great value for future research.

In some cases improvements upon the upper bounds for the solutions would have important consequences. For example, one would like to prove that

$$\max(|x|,|y|) \leq \exp\left((\log k)^C\right)$$

for every solution (x,y) of (7.1). M.Hall Jr. has conjectured that even $\max(|x|,|y|) \leq Ck^3$, but nothing better than (7.3) is known.

8. *Integers with many small prime factors.*

Let $p_1, \ldots, p_s$ be fixed primes and let $n_1 = 1 < n_2 < \ldots$ be the set of all positive integers which are composed of these primes. It follows from Thue's result mentioned in the preceding section that $n_{i+1} - n_i \to \infty$ as $i \to \infty$. However, his proof was ineffective. Cassels (1963) used Gel'fond's method to prove the effective analogue: For every integer k the solutions of $n_{i+1} - n_i = k$ can be determined in a finite number of steps. On combining Theorem 5 B with some simple facts about continued fractions it can be proved that there exist effectively computable constants C_1 and C_2 such that

$$\frac{n_i}{(\log n_i)^{C_2}} < n_{i+1} - n_i < \frac{n_i}{(\log n_i)^{C_1}}, \qquad \text{for } i \geq 2.$$

See Tijdeman (1973 b, 1974). Schinzel (1967) investigated the related problem of finding non-trivial lower bounds for the absolute value and the greatest prime factor of the difference $x^\nu - \varepsilon P_1^{z_1} \ldots P_s^{z_s}$, where $\nu = 2$ or 3, $\varepsilon = \pm 1$ and $P_1, \ldots, P_s$ are fixed positive integers. The following result answers an old problem of Wintner.

THEOREM 8A. *For every* $\varepsilon > 0$ *there exists an infinite sequence of primes* $P_1 < P_2 < \cdots$ *such that if* $n_1 < n_2 < \ldots$ *is the set of all integers composed of these primes, then*

$$n_{i+1} - n_i > n_i^{1-\varepsilon} \qquad \text{for all } i.$$

The proof is based on a theorem of A. Baker (1973 a) which generalizes Theorem 5 C. See Tijdeman (1973 b).

Several results give estimates for the greatest prime divisor of an integer. Let P be a polynomial with integer coefficients, and let n run through the positive integers. Schinzel (1967) proved that the greatest prime factor p_m of $P(m)$ satisfies $p_m > C \log\log m$ if P is of degree 2. The general case was treated by Keates (1969), Coates (1970 a) and Sprindžuk (1971 a).

The best general result known is due to Sprindžuk: $p_m > C \log\log m / \log\log\log m$.

Let p be the greatest prime factor of $(n+1)\ldots(n+k)$. It is easy to see that k is bounded if p is fixed. We denote the maximal possible value of k by $f(p)$. Erdős (1955) conjectured $f(p) = O((\log p)^2)$ and remarked that the proof of $f(p) < \pi(p)$ for all large p would already be difficult. This last inequality has been proved by Ramachandra (1971) and Tijdeman (1972). The best result at this moment is given by Shorey (1974).

$$f(k) = O\left(\frac{k}{\log k} \cdot \frac{\log\log\log k}{\log\log k}\right).$$

The proof is based on work of Jutila, Ramachandra and Shorey, and uses Baker's method, sieve methods and the Hoheisel-Ingham theorem on the difference of consecutive primes.

Let a and b be integers with the same greatest prime factor p with $a < b$. Erdős and Selfridge (1971) conjectured $b - a > (\log a)^C$ for some constant C. Baker's estimate (7.2) leads to $b - a > 10^{-6} \log\log a$ and, more precisely, to

$$\log(b-a) + p > 10^{-6} \log\log a. \qquad \text{(Tijdeman (1973 b).)}$$

There are numerous related results in the literature, many of which are due to Erdős. Most of the proofs are elementary and the Gel'fond-Baker method is one of the first non-elementary methods which have proved to be useful in this area. It is very possible that these result will grow to be part of a coherent theory, in which the Gel'fond-Baker method and its future developments will be one of the main tools.

9. *Characterization of arithmetical functions.*

I mention two applications of the Gel'fond-Baker method to arithmetical functions: Linnik and Chudakov (1950) gave a characterization of all completely multiplicative functions f such that $f(p) = 0$ for all but finitely many primes p and such that $\sum_{n=1}^{N} f(n)$ is bounded. Baker, Birch and Wirsing (1973) proved the conjecture of Chowla that there does not exist a rational-valued function $f(n)$, periodic with prime period p such that $\sum_{n=1}^{\infty} f(n)/n = 0$.

10. *Approximations of algebraic numbers by rationals.*

The exact formulation of Liouville's theorem metnioned in 1., is that for every algebraic number α of degree $n \geq 2$ there exists an effectively computable constant $C = C(\alpha) > 0$ such that

(10.1) $$\left|\alpha - \frac{p}{q}\right| > \frac{C}{q^n}$$

for all integers p and q. If $n=2$, this result is best possible. For $n \geq 3$ there are ineffective improvements by Thue, Siegel, Dyson and finally by Roth (1955). The final Thue-Siegel-Roth theorem states that in (10.1) the exponent n can be replaced by $2+\varepsilon$ for any $\varepsilon > 0$. Baker (1968 b) has given the first general effective improvement upon Liouville's theorem. The best effective result up to now due to Fel'dman (1971) reads as follows. There exist effectively computable constants $C > 0$ and $\varepsilon > 0$ such that

(10.2) $$\left|\alpha - \frac{p}{q}\right| > \frac{C}{q^{n-\varepsilon}}$$

for all integers p and q. Several authors have proved comparable results for special values of α, for example Baker (1964) proved (10.2) for $\alpha = \sqrt[3]{2}$, $\varepsilon = 0.045$ and $C = 10^{-6}$. A p-adic analogue of the general result was proved by Sprindžuk (1971 b). Also in this case there remains a big gap between ineffective and effective results and further improvements of the latter ones are very desirable. More information on diophantine approximations and the place of Baker's results in this theory can be found in a survey paper by W.M. Schmidt (1971).

11. *Transcendence and algebraic independence.*

The method of Gel'fond-Baker is so important in transcendental number theory that we have to restrict ourselves to only some developments in the theory. A complete account to 1966 has been given in an excellent survey paper by Fel'dman and Shidlovskii (1967). I shall therefore pay relatively more attention to later developments.

In **1.** and **5.** we mentioned the transcendence of $\beta_1 e^{\alpha_1} + \dots + \beta_n e^{\alpha_n}$ and $\beta_1 \log\alpha_1 + \dots + \beta_n \log\alpha_n$ for algebraic α's and β's satisfying certain natural conditions. Baker (1967) also proved the transcendence of

$$e^{\beta_0}\alpha_1^{\beta_1} \dots \alpha_n^{\beta_n}$$

for algebraic numbers $\beta_0, \beta_1, \dots, \beta_n$ and non-zero algebraic numbers $\alpha_1, \dots, \alpha_n$, if at least one of the following condition holds:

a) $\beta_0 \neq 0$.

b) $1, \beta_1, \beta_2, \ldots, \beta_n$ are linearly independent over $\mathbb{Q}$.

For all these numbers approximation and transcendence measures have been given. (See Fel'dman and Shidlovskii (1967), Cijsouw (1972).) Moreover, there are many similar results for α's and β's which are "almost algebraic", in the sense that the numbers can be closely approximated by algebraic numbers. (See for example Šmelev (1972 a), Bundschuh (1973), Wallisser (1973).)

Many transcendence results are of the following type: At least one of the numbers $\exp(e)$ and $\exp(e^2)$ is transcendental (Waldschmidt (1973 b), Brownawell (1974)). This particular result answers a question posed by Schneider (1957, p. 138). We define the transcendence degree of a finite set of numbers as the maximal number of elements of that set which are algebraically independent over $\mathbb{Q}$. Brownawell noticed that it follows from his general result that for any complex number t the transcendence degree of $\exp(t), \exp(t^2), \exp(t^3)$ is at least 1. Ramachandra (1968) proved that for every algebraic number α with $\alpha \neq 0$, $\log \alpha \neq 0$ and every $t \notin \mathbb{A}$ the transcence degree of $\alpha^t, \alpha^{t^2}, \alpha^{t^3}$ is at least 1.

More generally, one has the following theorem of Schneider-Siegel (unpublished), Lang (1966), Ramachandra (1968):

THEOREM 11 A. *Let the sets of complex numbers* $\alpha_1, \alpha_2, \alpha_3$ *and* β_1, β_2 *each be linearly independent over* $\mathbb{Q}$. *Then the transcendence degree of the six numbers* $e^{\alpha_i \beta_j}$ *is at least* 1.

Gel'fond extended his method to show that certain sets have transcendence degree at least 2; see Gel'fond (1952). He proved this, for example, for the set $\alpha^{\beta}, \alpha^{\beta^2}, \alpha^{\beta^3}, \alpha^{\beta^4}$ where α and β are algebraic numbers with $\alpha \neq 0, 1$ and β of degree at least 3. His work was generalized and improved by Šmelev (1968, 1971 a), Tijdeman (1971 b), Waldschmidt (1971, 1973a), Brownawell (197?) and others. We give a typical result.

THEOREM 11 B. *Let the sets of complex numbers* $\alpha_1, \ldots, \alpha_k$ *and* $\beta_1, \ldots, \beta_\ell$ *each be linearly independent over* $\mathbb{Q}$. *If* $2(k+\ell) \leq k(\ell+1)$, *then the transcendence degree of the* $k(\ell+1)$ *numbers* $\alpha_i, e^{\alpha_i \beta_j}$ *is at least* 2.

(Gel'fond's result is obtained by taking $k = \ell = 3$, $\alpha_i = \beta^{i-1}$, $\beta_j = \beta^{j-1} \log \alpha$.) For the most general formulations see Waldschmidt (1973 a) and Brownawell (197?). Some p-adic analogues were given by Shorey (1972 a,b).

Very recently, Šmelev (1972 b) and Choodnovski (197?) have proved similar results for sets of higher transcendence degrees. We give a result announced by Choodnovski:

THEOREM 11C. *Let* α *and* β *be algebraic numbers,* $\alpha \neq 0, 1$. *If the degree of* β *is at least* $2^n - 1$, $n \geq 2$ *then the transcendence degree of the set* $\alpha^\beta, \alpha^{\beta^2}, \ldots, \alpha^{\beta^N}$ *with* $N = 2^{n+1} - 4$ *is at least* n.

Many of these results are special cases of a very general conjecture of Schanuel: if $\alpha_1, \ldots, \alpha_n$ are lineraly independent over $\mathbb{A}$, then the transcendence degree of the set $\alpha_1, \ldots, \alpha_n, e^{\alpha_1}, \ldots, e^{\alpha_n}$ is at least n.

12. *Elliptic functions.*

This section can be considered as a continuation of the previous one and the general remarks made there are also valid here. Suppose the Weierstrass functions $\wp(z)$ and $\zeta(z)$ have algebraic invariants g_2 and g_3. Let ω_1 and ω_2 be a fundamental pair of periods. Put $\eta_1 = \zeta(\omega_1/2)$, $\eta_2 = \zeta(\omega_2/2)$. Schneider (1934 II, 1937) used Gel'fond's method to prove the transcendence of $\wp(\alpha)$ at algebraic points α and also to show that $\wp(z)$ and $\zeta(z)$ cannot be algebraic at the same point z. In particular, $\omega_1, \omega_2, \eta_1, \eta_2$ are transcendental. Schneider further proved that ω_1/ω_2 is either transcendental or imaginary quadratic irrational. In the latter case we say that $\wp(z)$ has complex multiplications. Other results of Schneider imply the transcendence of ω_i/η_i and ω_i/π for $i = 1, 2$. The results of Schneider also include the transcendence of the circumference of an ellipse with algebraic axis lengths, of the values of the modular function $J(\tau)$ at all algebraic points τ which are not imaginary quadratic irrationals and of the values of the Beta function $B(p,q) = \Gamma(p)\Gamma(q)/\Gamma(p+q)$ for all rational, but not integral, p and q. For more results see Schneider (1957, pp. 58-64) and Fel'dman and Shidlovskii (1967, pp. 34-35). Gel'fond's method was also used by Fel'dman (1951, 1958, 1968 b) and Bundschuh (1971) to obtain inequalities for the transcendence measures of some numbers related to $\wp(z)$. Ramachandra (1968) applied

Schneider's method of proving Theorem 2 A to derive some analogues for $\wp(z)$ of the results mentioned in the previous section. For example, he proved that if $\wp(z)$ has complex multiplications and b is any transcendental number, then at least one of the numbers $\wp(\omega_1 b), \wp(\omega_1 b^2), \wp(\omega_1 b^3)$ is transcendental.

Baker (1969 b, 1970) made his method applicable to Weierstrass $\wp$-functions. His results were generalized by Coates (1970 b, 1971 a, 1971 b). As an example we mention the following theorem, which includes some of Schneider's results given above:

THEOREM 12 A. *Every non-vanishing linear form in* $1, \omega_1, \omega_2, \eta_1, \eta_2, 2\pi i$ *with algebraic coefficients is transcendental.*

Masser (1973) has given a quantitive version of this result. It follows from Legendre's relation $\omega_1\eta_1 - \omega_2\eta_2 = 2\pi i$ that the numbers are not algebraically independent.

Very recently Masser (197?) has proved a generalization of a result of Schneider which is an analogue for $\wp(z)$ of Baker's results on linear forms in logarithms. We define an algebraic point of $\wp$ to be a complex number u such that either u is a pole of $\wp$ or $\wp(u) \in \mathbb{A}$.
(Similarly, the algebraic points of the exponential function are just the logarithms of algebraic numbers). Masser proves:

THEOREM 12 B. *Let* $\wp(z)$ *have complex multiplications and put* $K = \mathbb{Q}(\omega_1/\omega_2)$. *A finite set of algebraic points of* $\wp(z)$ *is linearly independent over* $\mathbb{A}$ *if and only if it is linearly independent over* K.

Masser also gives an effective, quantititative form of this result. Weaker, but more general, ineffective results of this type were proved by Coates (1971 c). The effective version of Theorem 12 B with $K = \mathbb{Q}$ in case $\wp(z)$ has no complex multiplications would be very desirable.

13. *Algebraic generalizations and applications.*

Schneider (1949) gave a general criterion under which a meromorphic function satisfying a certain type of algebraic differential equation with algebraic coefficients can take on values in a number field at only a finite number of points. His criterion includes above mentioned results on exponential functions and the Weierstrass functions. Schneider's result was simplified and generalized by Lang (1966). Lang also formulated an analogous theorem for functions of severable variables. An unnatural condition in this theorem was removed by Bombieri (1970) by using deep techniques from the theory of several complex variables. These and related results can be found in survey articles by Lang (1971) and Waldschmidt (1973 c).

Siegel (1929) proved that there exist only a finite number of integer points on any curve of genus ≥ 1. An effective upper bound for the number of integer points on any curve of genus equal to 1 was given by Baker and Coates (1970). The general case remains open. For related applications see Sirovich (1969) and Baker and Schinzel (1971).

On using a $\mathfrak{p}$-adic analogue of the main theorem on logarithms, Brumer (1967) succeeded in solving a problem of Leopoldt on the non-vanishing of the $\mathfrak{p}$-adic regulator of an Abelian number field.

Osgood (1973) applied Baker's method to derive an effective lower bound for a certain type of approximation of algebraic functions by rational functions.

Generalizations of Schneider's method to fields of characteristic $p \neq 0$ were given by Wade (1946) and Geijsel (1971, 1973).

14. *Literature.*

Monographs on transcendental number theory have been written by Siegel (1949), Gel'fond (1952), Schneider (1957), and Lang (1966). These books fully describe the Gel'fond-Baker method. A complete account of results obtained by this method, up to 1966, can be found in the survey paper of Fel'dman and Shidlovskii. One can obtain a useful insight into the method from the Gel'fond obituary by Levin, Fel'dman and Šidlovski (1971). There is also a recent account by Gel'fond himself, see Gel'fond (1969).

There will shortly be a book on transcendental number theory written by Baker. This will probably be the first monograph which contains a description of the full Gel'fond-Baker method. Survey papers on the method and its applications have been written by Baker (1971 c, 1971 d, 1973 c) and Lang (1971).

ADAMS, W.W. (1967). Transcendental numbers in the P-adic domain. *Amer.J.Math.* 88, 279 - 308.

BAKER,A. (1964).Rational approximations to $\sqrt[3]{2}$ and other algebraic numbers. *Quart.J.Math.Oxford Ser.* (2) 15, 375 - 383.

——— (1966). Linear forms in the logarithms of algebraic numbers. *Mathematika* 13, 204 - 216.

——— (1967). Linear forms in the logarithms of algebraic numbers II,III. *Mathematika* 14, 102 - 107 , 220 - 228.

——— (1968 a). Linear forms in the logarithms of algebraic numbers IV. *Mathematika* 15, 204 - 216.

——— (1968 b). Contributions to the theory of Diophantine equations. I: On the representation of integers by binary forms. II: The Diophantine equation $y^2 = x^3 + k$. *Phil.Trans.Royal Soc. London* A 263, 173 - 208.

——— (1968 c). The Diophantine equation $y^2 = ax^3 + bx^2 + cx + d$. *J.London Math.Soc.* 43, 1-9.

——— (1969 a). Bounds for the solutions of the hyperelliptic equation. *Proc.Cambridge Philos.Soc.* 65, 439 - 444.

——— (1969 b). On the quasi-periods of the Weierstrass ζ-function. *Nachr.Akad.Wiss.Göttingen, II.Math.-Phys.Kl.* (1969)145-157.

——— (1970). On the periods of the Weierstrass $\wp$-function. *Sympos.Math. IV*, Rome. pp. 155 - 174.

——— (1971 a). Imaginary quadratic fields with class number 2. *Ann.of Math. II Ser.* 94, 139 - 152.

——— (1971 b). On the class number of imaginary quadratic fields. *Bull. Amer.Math.Soc.* 77, 678 - 684.

——— (1971 c). Effective methods in Diophantine problems. *1969 Number Theory Institute,Proc.Sympos.Pure Math.* 20,Amer.Math.Soc., pp. 195 - 205.

——— (1971 d). Effective methods in the theory of numbers. *Actes,Congrès intern.math. 1970*.Gauthier-Villars, Paris, Tome 1, pp. 19-26.

——— (1972). A sharpening of the bounds for linear forms in logarithms. *Acta Arith.* 21, 117 - 129.

——— (1973 a). A sharpening of the bounds for linear forms in logarithms II. *Acta Arith.* 24, 33-36.

——— (1973 b). A central theorem in transcendence theory. *Diophantine approximation and its applications. Proc.Conf.Washington D.C. 1972*, Acad.Press, New York, pp. 1 - 23.

——— (1973 c). Effective methods in Diophantine problems II. *Analytic Number Theory,Proc.Sympos.Pure Math.* 24, Amer.Math.Soc.,pp. 1-7.

——— (197?). A sharpening of the bounds for linear forms in logarithms. *Acta Arith.*, to appear.

BAKER,A., B.J. BIRCH and E.A. WIRSING (1973). On a problem of Chowla. *J. Number Theory* 5, 224 - 236.

BAKER,A. and J. COATES (1970). Integer points on curves of genus 1.*Proc. Camb.Philos. Soc.* 67, 595 - 602.

BAKER, A. and H. DAVENPORT (1969). The equations $3x^2 - 2 = y^2$ and $8x^2 - 7 = z^2$. *Quart. J. Math., Oxford, Ser. (2) 20*, 129 - 137.

BAKER, A. and A. SCHINZEL (1971). On the least integers represented by the genera of binary quadratic forms. *Acta Arith. 18*, 137 - 144.

BOMBIERI, E. (1970). Algebraic values of meromorphic maps. *Invent. Math. 10*, 267 - 287.

BROWNAWELL, W.D. (1974). The algebraic independence of certain numbers related by the exponential function. *J. Number Theory 6*, 22 - 31.

——— (197?). Gel'fond's method for algebraic independence. *Trans. Amer. Math. Soc.*, to appear.

BRUMER, A. (1967). On the units of algebraic number fields. *Mathematika 14*, 121 - 124.

BUNDSCHUH, P. (1971). Ein Approximationsmass für transzendente Lösungen gewisser transzendenter Gleichungen. *J. reine angew. Math. 251*, 32 - 53.

——— (1973). Zum Franklin-Schneiderschen Satz. *J. reine angew. Math. 260*, 103 - 118.

CASSELS, J.W.S. (1953). On the equation $a^x - b^y = 1$. *Amer. J. Math. 75*, 159 - 162.

——— (1963). On a class of exponential equations. *Arkiv f. Mat. 4*, 231 - 233.

CATALAN, E. (1844). Note extraite d'une lettre adressée à l'éditeur. *J. reine angew. Math. 27*, 192.

CHOODNOVSKI, G.V. (197?). *Mat. Zametki*, to appear.

CIJSOUW, P.L. (1972). *Transcendence measures. Thesis*, Univ. of Amsterdam, Amsterdam, Netherlands.

——— (1974 a). Transcendence measures of exponentials and logarithms of algebraic numbers. *Compositio Math. 28*, 163 - 178.

——— (1974 b). Transcendence measures of certain numbers whose transcendency was proved by A. Baker. *Compositio Math. 28*, 179 - 194.

COATES, J. (1969). An effective p-adic analogue of a theorem of Thue. *Acta Arith. 15*, 279 - 305.

——— (1970 a). An effective p-adic analogue of a theorem of Thue. II: The greatest prime factor of a binary form. III: The diophantine equation $y^2 = x^3 + k$. *Acta Arith. 16*, 399 - 412, 425 - 435.

——— (1970 b). An application of division theory of elliptic functions to diophantine approximation. *Inventiones Math. 11*, 167 - 182.

——— (1971 a). The transcendence of linear forms in $\omega_1, \omega_2, \eta_1, \eta_2, 2\pi i$. *Amer. J. Math. 93*, 385 - 397.

——— (1971 b). Linear forms in the periods of the exponential and elliptic functions. *Invent. Math. 12*, 290 - 299.

——— (1971 c). An application of the Thue-Siegel-Roth theorem to elliptic functions. *Proc. Cambridge Philos. Soc. 69*, 157 - 161.

ELLISON, W.J., F. ELLISON, J. PESEK, C.E. STAHL and D.S. STALL (1972). The diophantine equation $y^2 + k = x^3$. *J. Number Theory 4*, 107 - 117.

ERDŐS, P. (1955). On consecutive integers. *Nieuw Archief voor Wiskunde III 3*, 124 - 128.

ERDŐS, P. and J.L. SELFRIDGE (1971). Some problems on the prime factors of consecutive integers II. *Proc.Washington State Univ. Conf. Number Theory*, Washington State Univ., Pullman, pp. 13 - 21.

EULER, L. (1748). *Introductio in analysin infinitorum*, Lausanne, 1748, Chap. VI. *Opera Omnia, VIII and IX.*

FEL'DMAN, N.I. (1951). The approximation of some transcendental numbers II. *Izv.Akad.Nauk SSSR, Ser. Mat. 15*, 153 - 176. *Amer.Math.Soc.Transl.* (2) 59, 246 - 270 (1966).

——— (1958). On the simultaneous approximation of the periods of elliptic functions by algebraic numbers. *Izv.Akad.Nauk SSSR, Ser.Mat. 22*, 563 - 576. *Amer.Math.Soc.Transl.* (2) 59, 271 - 284 (1966).

——— (1968 a). Improved estimate for a linear form of the logarithms of algebraic numbers. *Mat.Sb. 77(119)*, 423 - 436. *Math.USSR Sb. 6*, 393 - 406.

——— (1968 b). The elliptic analogue of an inequality of Gel'fond's. *Trudy Moskov.Mat.Obskch. 18*, 65 - 76. *Trans. Moscow Math. Soc. 18*, 71 - 84 (1969).

——— (1971). An effective refinement of the exponent in Liouville's theorem. *Izv.Akad.Nauk SSSR, Ser.mat. 35*, 973 - 990. *Math.USSR Izv. 5*, 985 - 1002 (1972).

——— (1972). An inequality connected with linear forms in logarithms of algebraic numbers. *Dokl.Akad.Nauk SSSR 207*, 41-43. *Soviet Math.Dokl. 13*, 1464 - 1467.

FEL'DMAN, N.I. and A.B. SHIDLOVSKII (1967). The development and present state of the theory of transcendental numbers. *Uspekhi Mat.Nauk 22-3*, 3 - 81. *Russian Math.Notes 22-3*, 1 - 79.

GAUSS, C.F. (1802). *Disquisitiones arithmeticae.* English ed.: Yale Univ. Press, New Haven, 1966.

GEIJSEL, J.M. (1971). *Transcendence properties of the Carlitz-Bessel-functions. M.C. - report ZW 2/71*, Math.Centrum, Amsterdam, Netherlands.

——— *Schneider's method in fields of characteristic* $p \neq 0$. *M.C. report ZW 17/73*, Math.Centrum, Amsterdam, Netherlands.

GEL'FOND, A.O. (1929). Sur les nombres transcendants. *C.r.Acad.Sci. (Paris) 189*, 1224 - 1226.

——— (1934). On Hilbert's seventh problem. *Dokl.Akad.Nauk SSSR 2*, 1 - 6. *Izv.Akad.Nauk SSSR 7*, 623 - 630.

——— (1935). On the approximation of transcendental numbers by algebraic numbers. *Dokl.Akad.Nauk SSSR 2*, 177 - 182.

——— (1939). On the approximation by algebraic numbers of the ratio of the logarithms of two algebraic numbers. *Izv.Akad. Nauk SSSR, Ser.Mat. 5 - 6*, 509 - 518.

GEL'FOND, A.O.(1940). On the divisibility of the difference of the powers of two integers by a power of a prime ideal. *Mat.Sb.* 7, 7 - 25.

——— (1949). On the algebraic independence of algebraic numbers in certain classes. *Uspekhi Mat.Nauk 4:5*, 14 - 48. *Amer. Math.Soc.Transl. (1) 2*, 125 - 169 (1962).

——— (1952). *Transcendental and algebraic numbers*. English ed.: Dover, New York, 1960.

——— (1969). The seventh problem. *The Hilbert problems, Proc.Conf. Moscow*. German ed.: *Die Hilbertschen Probleme*, Ostwalds Klassiker der exakten Wissenschaften 252, Geest & Portig, Leipzig.

GEL'FOND, A.O. and Yu.V. LINNIK (1948). On Thue's method and the problem of effectiveness in quadratic fields. *Dokl.Akad.Nauk SSSR 61*, 773 - 776.

HERMITE, Ch. (1873). Sur la fonction exponentielle. *C.r.Acad.Sci.(Paris) 77*, 18 - 24, 74 - 79, 226 - 233, 285 - 293. *Oeuvres III*, 150 - 181.

HILBERT, D. (1893). Ueber die Transzendenz der Zahlen e und π. *Nachr.Ges. Wiss.Göttingen (1893)*, 113 - 116. *Math.Ann. 43*, 216 - 220.

——— (1900). Mathematische Probleme (*Nachr.Ges.Wiss.Göttingen (1900)*, 253 - 297. *Bull.Amer.Math.Soc. 8*, 437 - 479 (1902). *Ges. Werke III*, 290 - 329.

KAUFMAN, R.M. (1971. The evaluation of the linear form of logarithms of algebraic numbers in the p-adic metric. *Vestnik Moskov.Univ., Ser. I 26*, 3 - 10.

KEATES, M. (1969). On the greatest prime factor of a polynomial. *Proc. Edinburgh Math.Soc., II. Ser. 16*, 301 - 303.

KUŽMIN, R.O. (1930). On a new class of transcendental numbers. *Izv.Akad. Nauk SSSR, Ser.mat. 3*, 585 - 597.

LAMBERT, J.H. (1768). Mémoire sur quelques propriétés remarquables des quantités transcendantes circulaires et logarithmiques. *Histoire Acad.roy.sci. et belles lettr., Berlin, Année 1761*, 265 - 322. *Opera Math. II*, 112 - 159.

LANG, S. (1966). *Introduction to transcendental numbers*. Addison Wesley, Reading, Mass.

——— (1971). Transcendental numbers and diophantine approximations. *Bull.Amer.Math.Soc. 77*, 635 - 677.

LEVIN, B.V., N.I. FEL'DMAN and A.B. ŠIDLOVSKI (1971). Alexander O. Gelfond. *Acta Arith. 17*, 315 - 336.

LINDEMANN, F.(1882). Ueber die Zahl π. *Math.Ann. 20*, 213 - 225.

LINNIK, Yu.V. and N.G. CHUDAKOV (1950). On a class of completely multiplicative functions. *Dokl.Akad.Nauk SSSR 74*, 193 - 196.

LIOUVILLE, J. (1844). Sur les classes très étendus de quantités dont la valeur n'est ni algébriques, ni même reductible à des irrationelles algébriques. *C.r.Acad.Sci. (Paris) 18*, 883 - 885, 910 - 911. See also *J.Math. pures appl. (1) 16*, 133 - 142 (1851).

MASSER, D.W. (1973). On the periods of the exponential and elliptic functions. *Proc.Camb.Philos.Soc. 73*, 339 - 350.

MASSER, D.W. (197?). Algebraic points of an elliptic function. To appear.

OSGOOD, C.F. (1973). An effective lower bound on the "Diophantine approximation" of algebraic functions by rational functions. *Mathematika 20*, 4 - 15.

RAMACHANDRA, K. (1968). Contributions to the theory of transcendental numbers. *Acta Arith. 14*, 65 - 88.

——— (1971). A note on numbers with a large prime factor III. *Acta Arith. 19*, 77 - 90.

——— (197?). Application of Baker's theory to two problems considered by Erdős and Selfridge. To appear.

RAMACHANDRA, K. and T.N. SHOREY (1973). On gaps between numbers with a large prime factor. *Acta Arith. 24*, 99 - 111.

REID, C. (1970). *Hilbert*. Springer Verlag, Berlin.

ROTH, K.F. (1955). Rational approximation to algebraic numbers. *Mathematika 2*, 1 - 20, 168.

SCHINZEL, A. (1967). On two theorems of Gelfond and some of their applications. *Acta Arith. 13*, 177 - 236.

SCHMIDT, W.M.(1971). Approximation to algebraic numbers. *L'Enseignement Math. 17*, 188 - 253.

SCHNEIDER, Th. (1934). Transzendenzuntersuchungen periodischer Funktionen. I: Transzendenz von Potenzen. II: Transzendenzeigenschaften elliptischer Funktionen. *J. reine angew.Math. 172*, 65 - 69, 70 - 74.

——— (1937). Arithmetische Untersuchungen elliptischer Integrale. *Math.Ann. 113*, 1 - 13.

——— (1949). Ein Satz über ganzwertige Funktionen als Prinzip für Transzendenzbeweise. *Math.Ann. 121*,131 - 140.

——— (1957). *Einführung in die transzendenten Zahlen*. Springer Verlag, Berlin.

SHOREY, T.N. (1972 a). Algebraic independence of certain numbers in the p-adic domain. *Nederl.Akad.Wetensch.Proc. 75 = Indag.Math. 34*, 423-435.

——— (1972 b). P-adic analogue of a theorem of Tijdeman and its application. *Nederl.Akad.Wetensch.Proc.75= Indag.Math. 34*, 436 - 442.

——— (1974). On gaps between numbers with a large prime factor II. *Acta Arith. 25*, 365 - 373.

SIEGEL, C.L. (1929). Ueber einige Anwendungen Diophantischer Approximationen. *Abh.Preuss.Akad.Wiss., Phys-Math.Kl. (1929)*,1 - 70.

——— (1949). *Transcendental numbers*. Annals of Math.Studies no. 16, Princeton Univ.Press,Princeton, N.J.

SIROVICH, C. (1969). On the distribution of elements belonging to certain subgroups of algebraic numbers. *Trans.Amer.Math.Soc. 141*, 93 - 98.

ŠMELEV, A.A. (1968). On the algebraic independence of certain numbers. *Mat.Zametki 4*, 525 - 532. *Math. Notes 4*, 805 - 809.

——— (1971). A.O. Gel'fond's method in the theory of transcendental numbers.*Mat.Zametki 10*, 415 - 426.

ŠMELEV, A.A. (1972 a). On the arithmetical properties of the roots of some transcendental equations. *Vestnik Moskov.Univ.Ser. I 27*, 5 - 14.

——— (1972 b). On the question of the algebraic independence of algebraic powers of algebraic numbers. *Mat.Zametki 11*, 635 - 644. *Math. Notes 11*, 387 - 392.

SPRINDŽUK, V.G. (1971 a). On the greatest prime divisor of a binary form. *Dokl.Akad.Nauk BSSR 15*, 389 - 391.

——— (1971 b). On rational approximations of algebraic numbers. *Izv.Akad.Nauk SSSR 35*, 991 - 1007. *Math. USSR Izv.* 35, 1003 - 1009 (1972).

——— (1972). On an estimate for solutions of Thue's equation. *Izv.Akad.Nauk SSSR 36*, 712 - 741. *Math.USSR Izv. 36*, 705 - 734.

——— (1973). Squarefree divisors of polynomials and class numbers of ideals of algebraic number fields. *Acta Arith. 24*, 143 - 149.

STARK, H.M. (1967). A complete determination of the complex quadratic fields of class number one. *Michigan Math.J. 14*, 1 - 27.

——— (1969). On the problem of unique factorization in complex quadratic fields. *Number Theory, Proc.Sympos.Pure Math. 12*, Amer.Math.Soc., pp. 41 - 56.

——— (1971 a). A transcendence theorem for class number problems. *Ann. of Math. II. Ser. 94*, 153 - 173.

——— (1971 b). Recent advances in determiningall complex quadratic fields of a given class number. *1969 Number Theory Institute, Proc.Sympos.Pure Math. 20*, Amer.Math.Soc., 401 - 414.

——— (1973 a). Further advances in the theory of linear forms in logarithms. *Diophantine approximations and its applications, Proc.Conf.Washington D.C. 1972*, Academic Press, New York, pp. 255 - 293.

——— (1973 b). Effective estimates of solutions of some diophantine equations. *Acta Arith. 24*, 251 - 259.

THUE, A. (1909). Ueber Annäherungswerte algebraischer Zahlen. *J. reine angew.Math. 135*, 284 - 305.

TIJDEMAN, R. (1971 a). On the numbers of zeros of general exponential polynomials. *Nederl.Akad.Wetensch.Proc. 74 = Indag.Math. 33*, 1 - 7.

——— (1971 b). On the algebraic independence of certain numbers. *Nederl.Akad.Wetensch.Proc. 74 = Indag. Math. 33*, 146 - 162.

——— (1972). On the maximal distance of numbers with a large prime factor. *J. London Math.Soc. (2) 5*, 313 - 320.

——— (1973 a). An auxiliary result in the theory of transcendental numbers. *J. Number Th. 5*, 80 - 94.

——— (1973 b). On integers with many small prime factors. *Compositio Math. 26*, 319 - 330.

——— (1974). On the maximal distance between integers composed of small primes. *Compositio Math. 28*, 159 - 162.

——— (197?). On the equation of Catalan. To appear.

WADE, L.I. (1946). Transcendence properties of the Carlitz ψ -functions. *Duke Math.J.* *13*, 79 - 85.

WALDSCHMIDT, M. (1971). Indépendance algébrique des valeurs de la fonction exponentielle. *Bull.Soc.Math.France* *99*, 285 - 304.

——— (1973 a). Propriétés arithmétiques des valeurs de fonctions méromorphes algébriquement indépendantes. *Acta Arith.* *23*, 19 - 88.

——— (1973 b). Solution du huitième problème de Schneider. *J. Number Th.* *5*, 191 - 202.

——— (1973 c). A propos de la méthode de Baker. *Journées arithmétiques françaises* (*1973*),Grenoble.

WALLISSER, R. (1973). Ueber Produkte transzendenter Zahlen. *J. reine angew. Math.* *258*, 62 - 78.

WEIERSTRASS, K. (1885). Zu Lindemann's Abhandlung "Ueber die Ludolph'sche Zahl", *S.-B. Preuss.Akad.Wiss.* (*1885*), 1067 - 1085. *Math.Werke* *II*, 341 - 362.

R. Tijdeman
Mathematisch Instituut
R.U. Leiden
Leiden
The Netherlands

Proceedings of Symposia in Pure Mathematics
Volume 28, 1976

HILBERT'S 8th PROBLEM: AN ANALOGUE

E. Bombieri

The 8th problem of Hilbert is concerned with the distribution of primes and the Riemann Hypothesis. Here I will consider an important analogue of it, the Riemann Hypothesis for curves over finite fields.

A fundamental property of the rational field $\mathbb{Q}$ is that every completion $\mathbb{Q}_v$ of $\mathbb{Q}$ with respect to a valuation v is locally compact commutative field, of course with a non-discrete topology. New locally compact commutative fields can be only of two types:

(a) a finite extension of $\mathbb{Q}_v$,

(b) a field K of characteristic $p > 0$, with a maximal compact subring R which is a regular local ring of dimension 1, with maximal ideal $\mathfrak{m}$ and $R/\mathfrak{m}$ is a finite field $\mathbb{F}_q$ with q elements.

So if we look at commutative fields K such that

(i) for every valuation v of K, K_v is a locally compact field;

(ii) for every $x \in K$ with $x \neq 0$, the product formula $\prod_v |x|_v = 1$ holds,

then K can be only

(A) an algebraic number field of finite degree over $\mathbb{Q}$;

(B) a function field of dimension 1, with constant field a finite field $\mathbb{F}_q$.

The analogy between (A) and (B) is perhaps best explained with an example. Let $k = \mathbb{F}_q$ and let t be an indeterminate.

We have

$\mathbb{Q}$	$k(t)$
$\mathbb{Z}$	$k[t]$
primes, $\pm p$	irreducible polynomials, $\mathfrak{p}$
$\zeta(s) = \sum n^{-s}$	$\zeta(s) = \sum_{(\alpha)} (q^{\deg \alpha})^{-s}$
$= \prod_p (1 - p^{-s})^{-1}$	$= \prod_{\mathrm{irr.}(\mathfrak{p})} (1 - (q^{\deg \mathfrak{p}})^{-s})^{-1}$

The factors appearing in the Euler product can be viewed as local zeta functions of the complete field associated to the valuations determined by the prime ideals of p and $\mathfrak{p}$. It is convenient in the number field case to take $\pi^{-s/2}\Gamma(s/2)$ and $(2\pi)^{-s}\Gamma(s)$ as the local zeta functions for $\mathbb{R}$ and $\mathbb{C}$; then we define

$$\zeta(s, K) = \prod \zeta(s, K_v)$$

we have that $\zeta(s,K)$ is meromorphic in the whole complex plane, of order 1, and satisfies a functional equation

$$(N\mathfrak{d})^{-s/2}\zeta(s,K) = (N\mathfrak{d})^{-(1-s)/2}\zeta(1-s,K)$$

where N denotes the norm, and $\mathfrak{d}$ is the different of K in the number field case, and a canonical divisor in the function field case.

From now on, I will consider only the function field case. It is convenient to change our notation a bit and look at things more geometrically. Let K be a function field as before, let $k = \mathbb{F}_q$ be its field of constants and let $\bar{k}$ be the algebraic closure of k. The field K is the field of functions defined over k on a complete nonsingular projective curve C, defined over k (so C is a nonsingular model of K). Since C is defined over k, the mapping $x \to x^q$, raising the coordinates of a point of C to the qth power, is an automorphism of C, called the Frobenius automorphism. A divisor over k is a linear combination

$$\mathfrak{a} = \sum n_x \cdot x$$

with integral n_x, almost all n_x being 0, of points x on C, and it will be over k if it is invariant by the Frobenius map. A divisor is positive if all n_x are positive, not all being 0; a positive divisor $\mathfrak{a}$ over k is called a prime divisor (over k) if it is not the sum of two positive divisors over k. Divisors in function fields correspond to fractional ideals in number fields, and the degree of a divisor

$$\deg \mathfrak{a} = \sum n_x$$

is the analogue of the logarithm of the norm in number fields. The norm of a divisor over k is given by

$$N\mathfrak{a} = q^{\deg \mathfrak{a}}.$$

Next, since the zeta function is a function of q^{-s}, it is convenient to make the change of variable $q^{-s} = t$ and consider

$$Z(t, C/k) = \Pi_{\mathfrak{p}}(1 - t^{\deg \mathfrak{p}})^{-1}$$

as the zeta function of K/k, the product being over all prime divisors over k. We have

$$\begin{aligned}\frac{Z'}{Z}(t, C/k) &= \sum_m \sum_{\mathfrak{p}} (\deg \mathfrak{p})\, t^{m \deg \mathfrak{p} - 1}\\ &= \sum \Big(\sum_{\deg \mathfrak{p} \mid m} \deg \mathfrak{p}\Big)\, t^{m-1}\\ &= \sum_m N_m\, t^{m-1}\end{aligned}$$

where now the coefficient N_m can be interpreted as the number of points of C, defined over the field $k_m = \mathbb{F}_{q^m}$, the extension of k of degree m.

The Riemann Hypothesis for C/k asserts that all zeros of $\zeta(s,K)$ have real part

$\frac{1}{2}$, or in other words all zeros of $Z(t, C/k)$ have absolute value $q^{-\frac{1}{2}}$. Since the only poles of $Z(t, C/k)$ are at $t = q^{-1}$ and $t = 1$, and these are simple poles, taking the logarithmic derivative we see that if the Riemann Hypothesis is true then

$$\frac{Z'}{Z}(t, C/k) - q(1 - qt)^{-1} - (1 - t)^{-1}$$

is regular in the disk $|t| < q^{-\frac{1}{2}}$; conversely, if this is the case, all zeros of $\zeta(s, K)$ must have real part $\leq \frac{1}{2}$, and hence $= \frac{1}{2}$ by the functional equation. It follows that the Riemann Hypothesis is equivalent to an inequality

$$|N_m - (q^m + 1)| \leq c(\epsilon)\,(q^{\frac{1}{2}+\epsilon})^m$$

for every fixed $\epsilon > 0$, some constant $c(\epsilon)$ and all $m \geq 1$.

Let g be the genus of C. The inequality

$$|N_m - (q^m + 1)| \leq 2gq^{m/2}$$

was proved for the first time by Weil, using the theory of correspondences on C; the special case $g = 1$ had been obtained a few years before by Hasse. Variations on Weil's theme were proposed by several authors, but only recently two essentially new proofs came. One is Deligne's, and this proof has the fundamental merit of extending to higher dimensional analogues; you will hear about it in one of Katz' talks. The other one in Stepanov's, and it will form the heart of my talk.

First of all, we note that the Riemann Hypothesis for the curve C/k is equivalent to the Riemann Hypothesis for the curve C/k_m, so that we may suppose as well that $m = 1$, $q > (g + 1)^4$ and $q = p^a$ where a is even. Now we prove about one half of the result:

THEOREM. If q is as before then $N_1 \leq q + 1 + (2g + 1)q^{\frac{1}{2}}$.

PROOF. We may assume that C has a point P over k, since otherwise $N_1 = 0$. Let R_m denote the space of functions on C over $\bar{k}$, which are regular outside P and which have at P a pole of order $\leq m$, that is $\mathrm{ord}_P f \geq -m$. The following facts are either trivial or well known, and in any case easy to prove:

(i) $\dim R_{m+1} \leq 1 + \dim R_m$

(in old-fashioned language, a zero of f at a given point imposes not more than one linear condition);

(ii) if $m > 2g - 2$, then $\dim R_m = m + 1 - g$

(this is Riemann's part of the theorem of Riemann-Roch, and may be taken as a proposition-definition for the genus g);

(iii) if $f(x) \in R_m$ then $f(x^q) \in R_{mq}$.

Next we note that

(iv) there is a basis $f_1, \ldots, f_r$ of R_m such that $\mathrm{ord}_P f_i < \mathrm{ord}_P f_{i+1}$ for $i = 1, \ldots, r - 1$.

This is due to Clebsch; the proof is easy; we have a filtration $(0) \subset \bar{k} = R_0 \subset R_1 \subset$

$\ldots \subset R_m$ so that

$$R_m = \bigoplus_{i=0}^{m} R_i/R_{i-1}.$$

By property (i), we have $\dim R_i/R_{i-1} \leqq 1$, and the result follows, since a basis can be obtained by picking up for each i, when possible, one element of R_i not in R_{i-1}.

Let n, b be nonnegative integers and let $s_1, \ldots, s_r$ be elements of R_n. Consider the auxiliary function

$$F(x) = s_1^{p^b}(x)\, f_1(x^q) + \ldots + s_r^{p^b}(x)\, f_r(x^q).$$

We have

(v) if $np^b < q$, we have that $F(x)$ is identically 0 only if all the $s_i(x)$ are identically 0.

In fact, suppose that $F(x)$ is identically 0 and that $s_h(x)$ is the first s_i which is not identically 0. Taking the order at P of both sides of the identity

$$s_h^{p^b}(x)\, f_h(x^q) = - s_{h+1}^{p^b}(x)\, f_{h+1}(x^q) - \ldots - s_r^{p^b}(x)\, f_r(x^q)$$

we obtain, using property (iv),

$$p^b \operatorname{ord}_P s_h + q \operatorname{ord}_P f_h \geqq \min_{i>h} (p^b \operatorname{ord}_P s_i + q \operatorname{ord}_P(f_i))$$

$$\geqq -p^b n + q \operatorname{ord}_P f_{h+1}$$

therefore

$$p^b \operatorname{ord}_P s_h \geqq -p^b n + q(\operatorname{ord}_P f_{h+1} - \operatorname{ord}_P f_h)$$

$$\geqq -p^b n + q > 0.$$

This means that s_h vanishes at P, and thus is a function with no poles and at least one zero, hence s_h is identically 0, contradicting our initial assumption.

(vi) if m, $n > 2g - 2$ and if $(n + 1 - g)(m + 1 - g) > p^b n + m + 1 - g$ then we can choose the s_i not all identically 0, such that

$$s_1^{p^b}(x)\, f_1(x) + \ldots + s_r^{p^b}(x)\, f_r(x)$$

is identically 0.

In fact, this function is regular outside P and at P has at most a pole of order $p^b n + m$, whence by (ii) these functions form a space of dimension at most $p^b n + m + 1 - g$. Since each s_i can vary in a vector space of dimension $n + 1 - g$, and since, again by (ii), we have $r = m + 1 - g$, we get statement (vi).

New recall that if x is a point of C rational over k then $x^q = x$; in view of (v) and (vi) we obtain that if $m, n > 2g - 2$; $np^b < q$; $(n + 1 - g)(m + 1 - g) > np^b + m + 1 - g$; then we can construct the auxiliary function $F(x)$ so that it is not identically 0, while it vanishes at every rational point of C over k, except at P. Also, since $F(x)$ by construction is a p^b-th power, we see that it must vanish there with multiplicity at least p^b. Hence F has at least $p^b(N_1 - 1)$ zeros.

On the other hand, F is regular outside P and the order of pole there cannot exceed $np^b + mg$. Thus we have proved that if m, n, b satisfy $m, n > 2g - 2$; $np^b < q$; $(n + 1 - g)(m + 1 - g) > np^b + m + 1 - g$ then we have the inequality

$$p^b(N_1 - 1) \leqq np^b + mq.$$

We choose $p^b = q^{\frac{1}{2}}$, $n = q^{\frac{1}{2}} - 1$, $m = q^{\frac{1}{2}} + 2g$; the conditions are satisfied if $q > (g + 1)^4$, which we may suppose, and we obtain at once the conclusion of the theorem.

In order to complete the proof of the Riemann Hypothesis, we have to prove that $N_1 = q + O(q^{\frac{1}{2}})$, in other words we have to get a good lower bound for N_1.

This will be obtained as a consequence of the Theorem we have already proved and of a little Galois theory.

The function field $\bar{k}(C)$ of C over $\bar{k}$ contains a purely transcendental subfield $\bar{k}(t)$ such that $\bar{k}(C)$ is a separable extension of $\bar{k}(t)$. Hence there is a normal extension of $\bar{k}(t)$ which is also normal over $\bar{k}(C)$; geometrically, we have a situation

$$C' \to C \to \mathbb{P}^1$$

where $C' \to C$ and $C' \to \mathbb{P}^1$ are Galois coverings, with Galois groups H and G respectively, where H is a subgroup of G. Though this situation needs not be realized over the field k, it will always be realized over a finite extension of it and therefore for our purposes we may as well assume that it is in fact realized over k. If x is a point of $\mathbb{P}^1$ over k and unramified in $C' \to \mathbb{P}^1$, and if y is a point of C' lying over x, we have

$$y^q = g(y)$$

for some $g \in G$, called the Frobenius substitution of G at the point y. Let $N_1(C', g)$ be the number of such points of C' with Frobenius substitution g. The same argument used in the proof of the theorem gives

$$N_1(C', g) \leqq q + (2g' + 1)\, q^{\frac{1}{2}} + 1,$$

where $g' =$ genus of C'; alternatively, one may note that $N_1(C', g) = N_1(C'_g)$, where C'_g is a curve over k isomorphic to C' over an extension k_m, where $m =$ order of g (C'_g is thus a certain twisting of C' by means of $g \in G$), and the previous theorem can be applied directly. We have

$$\sum_{g \in G} N_1(C', g) = |G|\, N_1(\mathbb{P}^1) + O(1)$$

(the $O(1)$ takes care of the branch points of the covering) and since $N_1(\mathbb{P}^1) = q + 1$ the upper bound for $N_1(C', g)$ implies

$$N_1(C', g) = q + O(q^{\frac{1}{2}}).$$

We have also

$$\sum_{h \in H} N_1(C', h) = |H| \, N_1(C) + O(1),$$

whence

$$N_1(C) = q + O(q^{\frac{1}{2}}),$$

and this completes the proof of the Riemann Hypothesis.

We conclude with the remark that this proof can be made even more elementary, in the sense that it is sufficient to have the Riemann-Roch theorem in the very weak form

$$\dim R_m = m + O(1);$$

the upper bound $m + 1$ is obvious by comparison with $\mathbb{P}^1$, and the lower bound can be easily obtained by considering a plane model (with singularities) of C.

The proof I have given is my own version of Stepanov's approach (Séminaire Bourbaki 1973, exposé 430); it uses freely finite extensions of the constant field k but does not use derivations in the construction of the auxiliary function $F(x)$. Another proof of the Riemann Hypothesis, more in line with Stepanov's original approach, has been obtained by W. Schmidt (Acta Arithmetica 24(1973), 347-368) and Stepanov himself had previously treated cases of considerable generality (Izv. Akad. Nauk SSSR Ser. Mat. 36(1972), 683-711). The case of hyperelliptic function fields has been the object of a sharp analysis by H. Stark (Proc. Sympos. Pure Math., vol. 24, Amer. Math. Soc., Providence, R.I., 1973, pp. 285-302).

Proceedings of Symposia in Pure Mathematics
Volume 28, 1976

AN OVERVIEW OF DELIGNE's PROOF OF THE RIEMANN HYPOTHESIS FOR VARIETIES OVER FINITE FIELDS

Nicholas M. Katz

Table of Contents

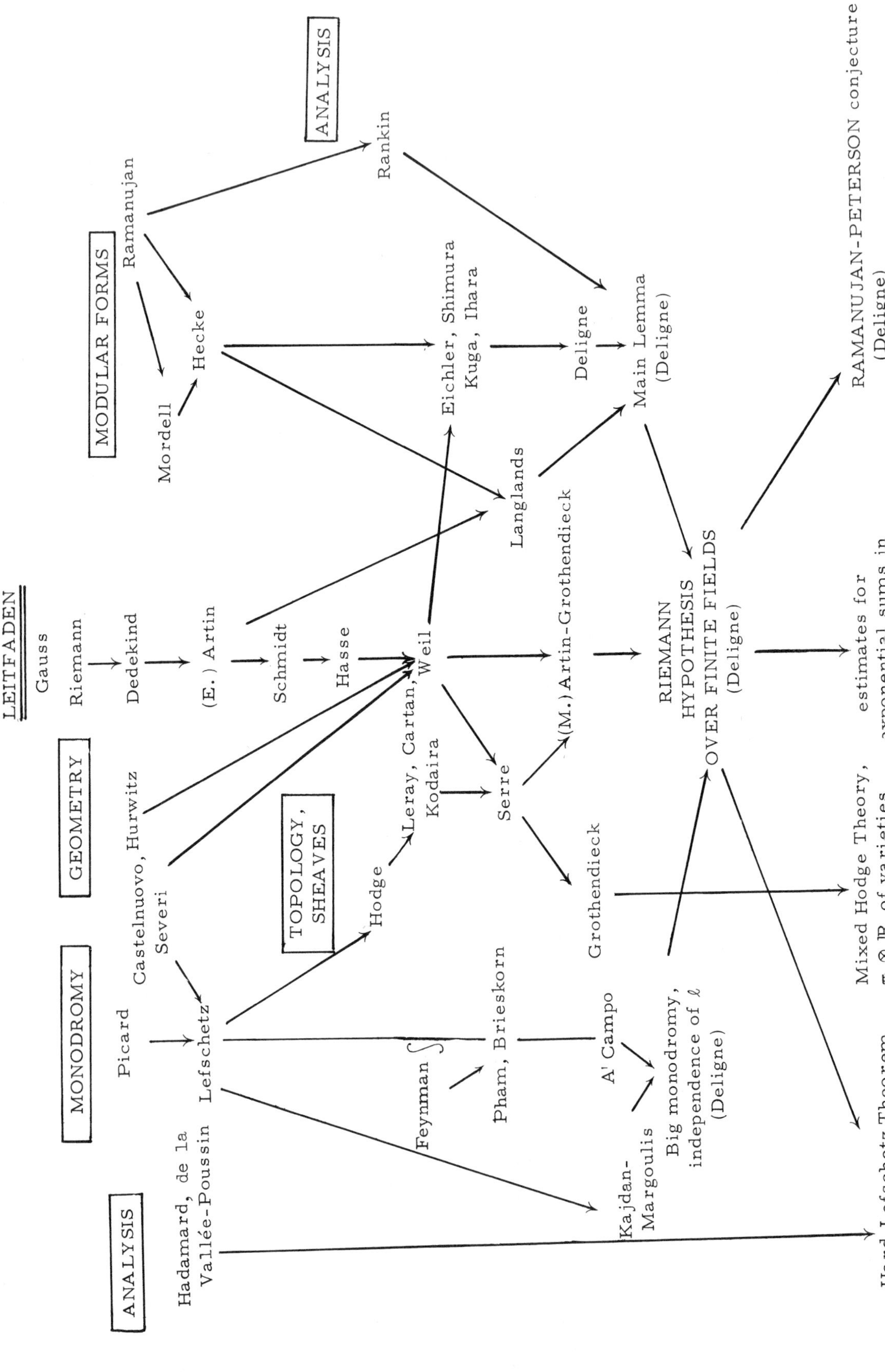
LEITFADEN
Gauss
Riemann
Dedekind
(E.) Artin
Schmidt
Hasse
Weil
GEOMETRY
Castelnuovo, Hurwitz
Severi
MONODROMY
Picard
Lefschetz
ANALYSIS
Hadamard, de la Vallée-Poussin
MODULAR FORMS
Ramanujan
Mordell
Hecke
ANALYSIS
Rankin
TOPOLOGY, SHEAVES
Hodge
Leray, Cartan, Kodaira
Eichler, Shimura
Kuga, Ihara
Langlands
Serre
(M.) Artin-Grothendieck
Grothendieck
Deligne
Main Lemma
(Deligne)
Feynman ∫
Pham, Brieskorn
A'Campo
Kajdan-Margoulis
Big monodromy, independence of ℓ
(Deligne)
RIEMANN HYPOTHESIS OVER FINITE FIELDS
(Deligne)
RAMANUJAN-PETERSON conjecture
estimates for
Mixed Hodge Theory,

Introduction to congruences

The fundamental problem in number theory is surely that of solving equations in integers. Since this problem is still largely inaccessible, we shall content ourselves with the problem of solving polynomial congruences modulo p. The idea of looking at congruences comes up naturally in trying to prove that a given equation has no solutions in integers. For example, the equation

$$y^2 = x^3 - x - 1$$

can have no solutions in integers, because the corresponding congruence modulo three has no solutions in integers modulo three (the left side is 0 or 1 mod 3, the right side is -1 mod 3).

Now when we look at a congruence modulo p,

$$f(x_1, \ldots, x_n) \equiv 0 \quad \mod p$$

it is most natural to look for solutions not only in the prime field $\mathbb{F}_p$ but also in all of its finite extension fields $\mathbb{F}_{p^n}$. If we identify solutions which are conjugate over $\mathbb{F}_p$, we arrive at the notion of a "prime divisor" $\mathfrak{y}$ (a maximal ideal of $\mathbb{F}_p[x_1, \ldots, x_n]/(f)$). The norm of such a prime divisor, noted $N\mathfrak{y}$, is the cardinality of its residue field. Thus $N\mathfrak{y} = p^{\deg \mathfrak{y}}$, where $\deg\mathfrak{y}$ is the number of conjugate solutions which $\mathfrak{y}$ "is."

The Zeta Function

Mindful of the analogy with the Riemann zeta function, we introduce with E. Artin [2] the infinite product

$$\prod_{\mathfrak{y}} (1 - N\mathfrak{y}^{-s})^{-1} \qquad \text{convergent for } \mathrm{Re}(s) >> 0$$

If we make the change of variable $T = p^{-s}$, we obtain

$$\prod_{\mathfrak{y}} (1 - T^{\deg \mathfrak{y}})^{-1}$$

whose logarithm is easily computed to be

$$\sum_{n \geq 1} \frac{T^n}{n} N_n,$$

where

$$N_n = \text{the number of solutions with coordinates in } \mathbb{F}_{p^n}.$$

So for any algebraic variety X over a finite field $\mathbb{F}_q$ (q some power of p), we introduce its zeta function

$$Z(X/\mathbb{F}_q, T) \overset{\text{dfn}}{=\!=} \exp\left(\sum \frac{T^n}{n} N_n\right); \; N_n = \# X(\mathbb{F}_{q^n})$$

$$= \prod_{\mathfrak{y}} (1 - T^{\deg \mathfrak{y}})^{-1}$$

$$= \prod_{\mathfrak{y}} (1 - N\mathfrak{y}^{-s})^{-1} \qquad T = q^{-s}$$

as a formal series in T with $\mathbb{Z}$-coefficients. It contains all of the diophantine information that X has to offer.

Let's compute an easy example. Let $X = \mathbb{A}^r$, the r-dimensional affine space over $\mathbb{F}_q$, whose points with values in $\mathbb{F}_{q^n}$ are simply the r-tuples of elements of $\mathbb{F}_{q^n}$. There are q^{rn} such r-tuples, whence

$$N_n = q^{rn}$$

$$Z(\mathbb{A}^r/\mathbb{F}_q, T) = \exp\left(\sum \frac{T^n}{n} q^{rn}\right) = \frac{1}{1 - q^r T}$$

In fact, E. Artin (1924) had introduced the zeta function only for the function fields of curves over finite fields, as an analogue of the Dedekind zeta function of an arbitrary algebraic number field, in its s-form

$$\zeta(s) = \prod_{y} (1 - Ny^{-s})^{-1} = \sum_{\mathfrak{A}} N\mathfrak{A}^{-s}$$

the last sum extended over all "integral divisors" of the function field. It was only seven years later (1931) that F. K. Schmidt [40] showed that the

Riemann-Roch theorem on the curve itself could be used to establish that for a curve $X/\mathbb{F}_q$ of genus g, its zeta function is a rational function of $T = q^{-s}$ which has the precise form

$$\zeta(s) = \frac{P(q^{-s})}{(1-q^{-s})(1-q^{1-s})} = \frac{P(T)}{(1-T)(1-qT)}$$

where $P(T) = \prod_{i=1}^{2g}(1-\alpha_i T)$ is a polynomial of degree $2g$ with $\mathbb{Z}$-coefficients, whose roots are permuted by $\alpha \longmapsto q/\alpha$. In terms of the complex variable s, this is a functional equation under $s \longmapsto 1-s$. The "Riemann Hypothesis", first formulated by E. Artin [2], asserts that the zeroes of this zeta function $\zeta(s)$ all lie on the line $\mathrm{Re}(s) = 1/2$, or equivalently that

$$|\alpha_i| = \sqrt{q}$$

If we take the logarithms of both sides of the equality

$$\exp\left(\sum \frac{N_n}{n} T^n\right) = \frac{\prod_{i=1}^{2g}(1-\alpha_i T)}{(1-T)(1-qT)}$$

we obtain

$$N_n = 1 + q^n - \sum_{i=1}^{2g} \alpha_i^n$$

the expression of the distribution of primes in terms of the zeroes of the zeta function. The Riemann Hypothesis becomes equivalent to the diophantine statement

$$|N_n - 1 - q^n| \leq 2g\sqrt{q}^{\,n},$$

an equivalence first pointed out by Hasse [23].

The first special case of the Riemann Hypothesis had been done by Gauss, for the lemniscatic elliptic curve $y^2 = x^4 - 1$ over any $\mathbb{F}_p$. E. Artin ([2]) verified some more special cases, and in 1933 Hasse was able to prove it for arbitrary elliptic (genus $g=1$) curves.

Hasse (1933 and 1934) gave two quite different proofs. The first [22] was based on the theory of complex multiplication, and consisted in lifting the elliptic curve with its Frobenius endomorphism to characteristic zero. His second proof ([24],[25]) was explicitly geometric, based on a direct study of the endomorphism ring of the elliptic curve. Hasse and Deuring pointed out ([14 bis],[26]) the relevance of the theory of correspondences to doing the case of curves of higher genus.

Weil (1940 and 1941) then sketched two different proofs of the Riemann Hypothesis for a curve of arbitrary genus g over a finite field. The first ([51]) attacked the problem by using the points of finite order prime to p on the Jacobian of the curve as a sort of first homology group of the curve. Following Hurwitz, he interpreted a correspondence of the curve with itself as giving rise to an endomorphism of the Jacobian, which allowed him to attach an ℓ-adic $2g \times 2g$ matrix to the correspondence, and to interpret its trace in terms of the number of fixed points of the correspondence. He then deduced the Riemann Hypothesis from the positivity of the "Rosatti involution" (cf., [38],[61]). The second proof [52], which dispensed with Hurwitz's "transcendental" theory (i.e., with ℓ-adic matrices and the Jacobian), was based entirely upon Severi's theory of correspondences of the curve with itself. The zeta function was easily expressed in terms of intersection-numbers of correspondences. In terms of these intersection numbers, Weil defined a "trace function" on the ring of all (suitable equivalence classes of) correspondences. The Riemann Hypothesis then followed from a positivity property of this trace (Castelnuovo's inequality, cf., [61],[37]). In the case of a curve of genus <u>one</u>, this proof reduces essentially to Hasse's geometric proof.

Although the correspondence formalism and the positivity statements upon which Weil based his proofs were "well known" in Italian algebraic geometry, and their complex analogues rigorously proven by transcendental methods (cf., [61], pp. 552-5), the lack of adequate foundations for abstract (in the sense of [61]) algebraic geometry led Weil to write his Foundations of Algebraic Geometry [53]. This done, he gave complete accounts of his two proofs, the second in Sur les courbes algébriques et les variétes qui s'en déduisent [54] and the first, generalized to arbitrary abelian varieties, in Variétés abéliennes et courbes algébriques [55].

The Weil conjectures

Then in 1949, Weil conjectured what should be true for higher dimensional varieties [57]. Let X be an n-dimensional projective non-singular variety over $\mathbb{F}_q$. Then

(1) $Z(X/\mathbb{F}_q, T)$ is a rational function of T

(2) Moreover, $Z(X\mathbb{F}_q, T) = \dfrac{P_1(T)\,P_3(T)\ldots P_{2n-1}(T)}{P_0(T)\,P_2(T)\ldots P_{2n}(T)}$

where $P_i(T) = \prod_{j=1}^{b_i}(1 - \alpha_{ij}T)$, $\quad |\alpha_{ij}| = \sqrt{q}^{\,i}$,

the last equality being the "Riemann Hypothesis" in this setting.

(3) Under $\alpha \longmapsto q^n/\alpha$, the $\alpha_{i,j}$ are carried bijectively to the $\alpha_{2n-i,j}$. In terms of the complex variable s, this is a functional equation for $s \longmapsto n - s$.

(4) In case X is the "reduction modulo p" of a non-singular projective variety $\mathbb{X}$ in characteristic zero, then b_i is the i'th topological Betti number of $\mathbb{X}$ as complex manifold.

The moral is that the topology of the complex points of $\mathbb{X}$, expressed through the classical cohomology groups $H^i(\mathbb{X}, \mathbb{C})$, determines the form of the zeta function of X, i.e., determines the diophantine shape of X. There is a heuristic argument for this, as follows (cf., [61]). Among all elements of the algebraic closure of $\mathbb{F}_p$, the elements of $\mathbb{F}_q$ are singled out as the fixed points of the Frobenius morphism $x \longrightarrow x^q$. More generally, if

$\underline{x} = (,\ldots,x_i,\ldots)$ is a solution of some equations which are defined over $\mathbb{F}_q$, the $F(\underline{x}) \overset{\text{dfn}}{=\!=} (,\ldots,x_i^q,\ldots)$ will also be a solution of the same equations, and the point $\underline{x}$ will have its coordinates in $\mathbb{F}_q$ precisely if $F(\underline{x}) = \underline{x}$. Thus F is an endomorphism of our variety X over $\mathbb{F}_q$, and

$$N_n = \#\ \mathrm{Fix}\ (F^n)\ ;$$

$$Z(X/\mathbb{F}_q, T) = \exp\ (\Sigma\ \frac{T^n}{n}\ \#\ \mathrm{Fix}\ (F^n))$$

Suppose we consider instead a compact complex manifold $\mathbb{X}$, and an endomorphism $\mathbb{F}$ of $\mathbb{X}$ with reasonable fixed points. Then the Lefschetz fixed point formula would give us

$$\#\ \mathrm{Fix}\ (\mathbb{F}^n) = \Sigma(-1)^i\ \mathrm{trace}\ (\mathbb{F}^n|H^i(\mathbb{X},\mathbb{C})),$$

which is formally equivalent to the identity

$$\exp\left(\sum_{n\geq 1} \frac{T^n}{n}\#\,\mathrm{Fix}\ (\mathbb{F}^n)\right) = \prod_{i=0}^{2n} \det\left(1 - T\mathbb{F}|H^i(\mathbb{X},\mathbb{C})\right)^{(-1)^{i+1}}$$

The search for a "cohomology theory for varieties over finite fields" which could justify this heuristic argument has been responsible, directly and indirectly, for much of the tremendous progress made in algebraic geometry these past twenty-five years. Weil's proofs of the Riemann Hypothesis for curves over finite fields had already necessitated his Foundations. Around the same time, Zariski had also begun emphasizing the need for an abstract algebraic geometry; his disenchantment with the lack of rigor in the Italian school had come after writing his famous monograph Algebraic Surfaces [63], which gave the "state of the art" as of 1934 (cf., [64). The possibility of transposing to abstract algebraic varieties with their "Zariski topology" the far-reaching topological and sheaf-theoretic methods which had been developed by Picard, Lefschetz, Hodge, Kodaira, Leray, Cartan, ... in dealing with complex varieties was implicit in Weil's 1949 Chicago lecture notes "Fibre Spaces in Algebraic Geometry" [58]. The transposition was carried out by Serre in his famous article "Faisceaux Algébriques Cohéhents" [41]. From the point of view of the Weil conjectures, however, this theory was still inadequate, for when applied to varieties in characteristic p it gave cohomology groups which were vector spaces in characteristic p, so could only give "mod p" traces formulas, i.e., could only give "mod p" congruences for numbers of rational points.

ℓ-adic cohomology

After some false starts (e.g. Serre's Witt vector cohomology [42],[43]) and Dwork's "unscheduled" (because apparently non-cohomological) proof [16] of the rationality conjecture (1) for any variety over $\mathbb{F}_q$, M. Artin and A. Grothendieck developed a "good" cohomology theory [3], based on the notion of étale covering space, and generalizing Weil's ℓ-adic matrices. In fact, they developed a whole slew of theories, one for each prime number $\ell \neq p$, whose coefficient field was the field $\mathbb{Q}_\ell$ of ℓ-adic numbers. Each theory gave a factorization of zeta

$$Z(T) = \prod_{i=0}^{2n} P_{i,\ell}(T)^{(-1)^{i+1}}$$

into an alternating product of $\mathbb{Q}_\ell$-adic polynomials, satisfying conjecture (3). In the case when X could be lifted to $\mathbb{X}$ in characteristic zero, they proved that $P_{i,\ell}$ was a polynomial of degree $b_i(\mathbb{X})$. They did not prove that the $P_{i,\ell}$ in fact had $\mathbb{Q}$-coefficients, nor à fortiori that the $P_{i,\ell}$ were independent of ℓ. This meant that in the factorization of an individual $P_{i,\ell}$

$$P_{i,\ell}(T) = \prod_{j=1}^{b_i} (1 - \alpha_{i,j,\ell} T)$$

the roots $\alpha_{i,j,\ell}$ were only algebraic over $\mathbb{Q}_\ell$, but possibly not algebraic over $\mathbb{Q}$, and so they might not even have archimedean absolute values. (Of course, by a theorem of Fatou, the actual reciprocal zeroes and poles of the rational function $Z(T)$ are algebraic integers, the problem is that there might be cancellation between the various $P_{i,\ell}$ in the ℓ-adic factorization of zeta.)

So the question became how to introduce archimedean considerations into the ℓ-adic theory. Even before the ℓ-adic theory had been developed, Serre (1960), following a suggestion of Weil ([61], p. 556), had formulated and proved a Kahlerian analogue of the Weil conjectures, by making essential use of the Hodge Index theorem. In part inspired by this, in part inspired by his own earlier (1958) realization that the Castelnuovo inequality used by Weil was a consequence of the Hodge Index theorem on a surface ([20], [37]), Grothendieck in the early sixties formulated some very difficult positivity and existence conjectures about algebraic cycles, the so-called "standard conjectures" (cf., [15], [31]) whose truth would imply both the independence of ℓ and the Riemann Hypothesis.

Much to everyone's surprise at the time of Deligne's proof, Deligne managed to avoid these conjectures altogether, except to deduce one of them, the "hard" Lefschetz theorem giving the existence of the "primitive decomposition" of cohomology of a projective non-singular variety, a result previously known only over $\mathbb{C}$, and there by Hodge's theory of harmonic integrals. The rest of the "standard conjectures" remain open. In fact, the generally accepted dogma that the Riemann Hypothesis could not be proven before these conjectures had been proven (cf., [15], I, p 224 for example) probably had the effect of delaying for a few years the proof of the Riemann Hypothesis.

It is quite striking to note that in Deligne's deduction of the hard Lefschetz theorem from the Riemann Hypothesis for varieties over finite fields, he makes essential use of a famous piece of classical analysis, the Hadamard-de Vallé Poussin method of proving that the usual Riemann zeta function has no zeroes on the line $\mathrm{Re}(s) = 1$. He was led to the method in studying Yoshida's proof [62] of the function field analogue of the Sato-Tate conjecture about the distribution of the angles of the eigenvalues of Frobenius in families of elliptic curves. Yoshida needed to show that a certain L-function had no zero on the line $\mathrm{Re}(s) = 1$, and did so using some powerful estimates of Selberg. Deligne realized that Selberg's results gave much more than was needed for the equidistribution question, and checked that the original classical argument of Hadamard-de Vallée Poussin could be used instead. He went on to notice that the method could be used to slightly improve the Lang-Weil inequality for the absolute values of the eigenvalues of Frobenius on H^2 of a projective smooth surface from $|\alpha| \leq q^{3/2}$ to $|\alpha| < q^{3/2}$. (The Riemann Hypothesis in this case is $|\alpha| = q$.)

The new ingredients

So what was it that finally allowed the Riemann Hypothesis for varieties over finite fields to be proven? There were two principal ingredients.

(1) Monodromy of Lefschetz pencils. In the great work of Lefschetz [35] on the topology of algebraic varieties, he introduced the technique of systematically "fibering" a projective variety by its hyperplane sections, and then expressing the cohomology of that variety in terms of the cohomology of those fibres. The general Lefschetz theory was successfully transposed into ℓ-adic cohomology, but it didn't really bear diophantine fruit until Kajdan-Margoulis [30] proved that the "monodromy group" of a Lefschetz pencil of odd fibre dimension was as "large as possible." Deligne realized that if the same result were true in even fibre dimension as well, then it would be possible to inductively prove the independence of ℓ and the rationality of the $P_{i,\ell}$ of X, by recovering them as generalized "greatest common divisors" of the $P_{i,\ell}$ of the hyperplane sections. But the Kajdan-Margoulis proof was Lie-algebra theoretic in nature, via the logarithms of the various Picard-Lefschetz transformations in the monodromy group. The restriction to odd fibre dimension

was necessary because in that case the Picard-Lefschetz transformations were unipotent, so had interesting logarithms, while in even fibre-dimension they were of finite order. Soon after, A'Campo [1] found a counterexample to a conjecture of Brieskorn that the local monodromy of isolated singularities should always be of finite order. Turning sorrow to joy, Deligne realized that A'Campo's example could be used to construct (non-Lefschetz) pencils which would have unipotent local monodromy. These he used to make the Kajdan-Margoulis proof work in even fibre-dimension as well, and so to establish the "independence of ℓ" and rationality of the $P_{i,\ell}$ (cf., [50]).

With this result, the importance of monodromy considerations for diophantine questions was firmly established.

(2) Modular forms, Rankin's method, and the cohomological theory of L-series.

In the years after the Weil conjectures, experts in the theory of modular forms began to suspect a strong relation between the Weil conjectures and the Ramanujan conjecture on the order of magnitude of $\tau(n)$. Recall that the $\tau(n)$ are the q-expansion coefficients of the unique cusp form Δ of weight twelve on $SL_2(\mathbb{Z})$:

$$\Delta(q) = q\left(\prod_{n\geq 1}(1-q^n)\right)^{24} = \Sigma\tau(n)\cdot q^n$$

As an arithmetic function, $\tau(n)$ occurs essentially as the error term in the formula for the number of representations of n as a sum of 24 squares. The Ramanujan conjecture is that

$$|\tau(n)| \leq n^{11/2}\cdot d(n),\ d(n) = \#\text{ divisors of } n.$$

According to the Hecke theory (which had been "prediscovered" by Mordell for Δ) the Dirichlet series corresponding to Δ admits an Euler product:

$$\sum_{n\geq 1}\tau(n)\cdot n^{-s} = \prod_p\left(\frac{1}{1-\tau(p)\cdot p^{-s}+p^{11-2s}}\right)$$

The truth of the Ramanujan conjecture for all $\tau(n)$ is then a formal consequence of its truth for all $\tau(p)$ with p prime:

$$|\tau(p)| \leq 2p^{11/2}$$

This last inequality may be interpreted as follows. Consider the polynomial $1 - \tau(p)T + p^{11}T^2$ and factor it:

$$1 - \tau(p)T + p^{11}T^2 = (1 - \alpha(p)T)(1 - \beta(p) T).$$

Then the Ramanujan conjecture for $\tau(p)$ is equivalent to the equality

$$|\alpha(p)| = |\beta(p)| = p^{11/2}$$

If there were a projective smooth variety X over $\mathbb{F}_p$ such that the polynomial $1 - \tau(p)\cdot T + p^{11}T^2$ divided $P_{11}(X/\mathbb{F}_p, T)$, then the Riemann Hypothesis for X would imply the Ramanujan conjecture for $\tau(p)$ The search for this X was carried out by Eichles, Shimura, Kuga, and Ihara (cf., [29],[32]). They constructed an X which "should have worked," but because their X was not compact and had no obvious smooth compactification, its polynomial P_{11} did not necessarily have all its roots of the correct absolute value. Deligne then showed how to compactify their X and how to see that the Hecke polynomial $1 - \tau(p) T + p^{11}T^2$ divided a certain factor of P_{11}, the roots of which factor would have the "correct" absolute values if the Weil conjectures were true. Thus the truth of the Ramanujan conjecture became a consequence of the universal truth of the Riemann Hypothesis for varieties over finite fields.

In 1939, Rankin [39] had obtained the then-best estimate for $\tau(n)$ (namely $\tau(n) = O(n^{29/5})$) by studying the poles of the Dirichlet series

$$\sum(\tau(n))^2 \cdot n^{-s}$$

Langlands [34] pointed out that the idea of Rankin's proof could easily be used to prove the Ramanujan conjecture, provided one knew enough about the location of the poles of an infinite collection of Dirichlet series formed from Δ by forming even tensor powers: for each even integer 2n one needed to know the poles of the function represented by the Euler product

$$\prod_p \prod_{i=0}^{2n} \left(\frac{1}{1 - \alpha(p)^i \beta(p)^{2n-i} p^{-s}} \right)^{\binom{2n}{i}}$$

Deligne studied Rankin's original paper in an effort to understand the remarks of Langlands. He realized that for L-series over curves over finite fields (instead of over Spec. ($\mathbb{Z}$)), Grothendieck's cohomological theory [19] of such L-series together with the Kajdan-Margoulis monodromy result gave an a priori hold on the poles: Rankin's methods could therefore be combined with Lefschetz pencil-monodromy techniques to yield the Riemann Hypothesis for varieties over finite fields, and with it the Ramanujan-Peterson conjecture as a corollary.

Deligne's proof in a special case

Formulation of the problem

I would now like to explain the idea of Deligne's proof by treating the special case of odd-dimensional non-singular hypersurfaces (so including the case of non-singular plane curves!). This special case, which was in fact the first case that Deligne treated, illustrates the main ideas without overwhelming the cohomological novice. In the general case, the ideas explained here occur as the "Main Lemma."

Let's consider a non-singular hypersurface $X_0 \subset \mathbb{P}^{2n}$ of degree d, over the field $\mathbb{F}_q$. Its zeta function is of the form

$$Z(X_0/\mathbb{F}_q, T) = \frac{P(X_0/\mathbb{F}_q, T)}{\prod_{i=0}^{2n-1} (1 - q^i T)}$$

where $P(X_0/\mathbb{F}_q, T)$ is a polynomial with integral coefficients and constant term one, whose degree $b = \deg P$ is the middle Betti number of any smooth degree d complex hypersurface of dimension $2n-1$. Explicitly,

$$b = \frac{((d-1)^{2n} - 1)(d-1)}{d}$$

Over $\mathbb{C}$, we may factor it

$$P(X_0/\mathbb{F}_q, T) = \prod_{i=1}^{b} (1 - \alpha_i T)$$

According to the functional equation, $\alpha \longrightarrow q^{2n-1}/\alpha$ permutes the α_i. The Riemann Hypothesis for X_0 is the assertion that the complex absolute values of the α_i are all given by

$$|\alpha_i| = \sqrt{q}^{\,2n-1} \qquad i=1,\ldots,b$$

In view of the fact that $\alpha \longmapsto q^{2n-1}/\alpha$ permutes the α_i, these equalities are equivalent to the inequalities

$$|\alpha_i| \leq \sqrt{q}^{\,2n-1} \qquad i=1,\ldots,b.$$

If we equate the cohomological and diophantine expressions for the zeta function, we get

$$Z(X_0/\mathbb{F}_q, T) = \exp\left(\sum_{r\geq 1} \frac{T^r}{r} N_r\right) = \frac{\prod_{i=1}^{b} (1-\alpha_i T)}{\prod_{i=0}^{2n-1} (1-q^i T)}$$

Equating coefficients of the logarithms, we get

$$N_r - \sum_{i=0}^{2n-1} q^{ri} = - \sum_{i=1}^{b} \alpha_i^r ,$$

so that the Riemann Hypothesis for X_0 is equivalent to the diophantine estimates

$$\left|N_r - \sum_{i=0}^{2n-1} q^{ri}\right| \leq b \sqrt{q^r}^{\,2n-1} \quad , \ r=1,2,\ldots$$

However, the most fruitful equivalent formulation of the Riemann Hypothesis for X_0 turns out to be the following form * of the inequality $|\alpha_i| \leq \sqrt{q}^{\,2n-1}$:

* The power series $\dfrac{1}{P(X_0/\mathbb{F}_q, T)} \in \mathbb{Q}[[T]]$ converges (in the archimedean sense) for $|T| < 1/\sqrt{q}^{\,2n-1}$.

A geometric construction

The first step in proving the Riemann Hypothesis for X_0 is to consider not only X_0 itself, but an entire one parameter family (in fact, a Lefschetz pencil) X_t of hypersurfaces in the same ambient projective space. The idea of simultaneously proving the Riemann Hypothesis for all varieties in a suitable family containing the one of initial interest was suggested to Deligne by Bombieri's relating to him that Swinnerton-Dyer had obtained weak estimates in the case of elliptic curves by considering certain L-series attached to "the" universal family of elliptic curves, and relating them to modular forms (!).

Suppose that X_0 is defined by the vanishing of a homogeneous form F of degree d. Choose any other form G of the same degree also defined over $\mathbb{F}_q$, and consider the one parameter family of forms $F + tG$. Denote by X_t the corresponding hypersurface.

It is not difficult to see that, possibly after replacing $\mathbb{F}_q$ by a finite extension, we can choose G in such a way that

a) the hypersurface of equation G is smooth, and intersects X_0 transversely.

b) for all but finitely many values of t in the algebraic closure of $\mathbb{F}_q$, the hypersurface X_t is smooth, while for the remaining values it has one and only one singular point, which is an "ordinary double point."

Let us denote by $\mathbb{A}^1$ the affine t-line over $\mathbb{F}_q$, and by $S \subset \mathbb{A}^1$ the finite set of exceptional parameter values. We will simultaneously prove the Weil conjectures for all the X_t, $t \in \mathbb{A}^1 - S$, by making use of the ℓ-adic "glue" which holds them together. This glue is a certain ℓ-adic representation of a certain arithmetic fundamental group.

The role of monodromy

The classical setting. Suppose first that the ground field is $\mathbb{C}$ rather than $\mathbb{F}_q$. Then the various X_t, $t \in \mathbb{A}^1 - S$, fit together to form a fibration over

$\mathbb{A}^1 - S$ which is locally (on $\mathbb{A}^1 - S$) trivial in the C^∞ sense. The middle-dimensional cohomology groups $H^{2n-1}(X_t, \mathbb{Q})$ therefore form a <u>local coefficient system</u> on $\mathbb{A}^1 - S$, or, what is the same once we pick a base-point $t_0 \in \mathbb{A}^1 - S$, they give a <u>representation</u> of $\pi_1(\mathbb{A}^1 - S)$ on $H^{2n-1}(X_{t_0}, \mathbb{Q})$. This representation <u>respects</u> the alternating intersection-form $\langle , \rangle$ on $H^{2n-1}(X_{t_0}, \mathbb{Q})$, so gives a homomorphism of $\pi_1(\mathbb{A}^1 - S)$ to the symplectic group $Sp = \mathrm{Aut}(H^{2n-1}(X_{t_0}, \mathbb{Q}), \langle , \rangle)$. The Kajdan-Margoulis theorem asserts that the image of π_1 in Sp is <u>Zariski-dense</u>: any polynomial function on Sp which vanishes on the image of π_1 is identically zero.

<u>The ℓ-adic setting</u>. Over the ground field $\mathbb{F}_q$, if we fix a prime number ℓ prime to q, Grothendieck's theory of ℓ-adic cohomology provides us with a similar but even richer structure.

Recall that the arithmetic fundamental group π_1^{arith} of $\mathbb{A}^1 - S$ is a compact totally disconnected group, defined as the quotient of the galois group of the algebraic closure $\overline{\mathbb{F}_q(t)}$ over $\mathbb{F}_q(t)$ by the closed subgroup generated by the inertial subgroups attached to all places of $\overline{\mathbb{F}_q(t)}$ lying over points of $\mathbb{A}^1 - S$. The subgroup $\pi_1^{\text{geom}} \subset \pi_1^{\text{arith}}$ is the corresponding quotient of the galois group of $\overline{\mathbb{F}_q(t)}$ over $\overline{\mathbb{F}}_q(t)$. It is this "geometric fundamental group" which is the analogue of the fundamental group in the classical case. It sits in π_1^{arith} as a closed normal subgroup, with quotient group $\mathrm{gal}(\overline{\mathbb{F}}_q/\mathbb{F}_q) \simeq \hat{\mathbb{Z}}$, the canonical generator being the Frobenius automorphism $a \longrightarrow a^q$ of $\overline{\mathbb{F}}_q$.

$$0 \longrightarrow \pi_1^{\text{geom}} \longrightarrow \pi_1^{\text{arith}} \xrightarrow{\text{"degree"}} \hat{\mathbb{Z}} \longrightarrow 0$$

Just as in algebraic number theory, there is associated to each closed point $x \in \mathbb{A}^1 - S$ and to each place $\overline{x}$ of $\overline{\mathbb{F}_q(t)}$ lying over it, a well-defined Frobenius element $\mathcal{F}_{\overline{x}} \in \pi_1^{\text{arith}}$ whose degree in $\hat{\mathbb{Z}}$ is the integer $\deg(x) \overset{\text{dfn}}{=\!=}$ degree $(\mathbb{F}_q(x)/\mathbb{F}_q)$. If we change $\overline{x}$ but keep x fixed, the element

$\mathcal{F}_{\bar{x}}$ changes by conjugation. Unfortunately, we will need to consider not the $\mathcal{F}_{\bar{x}}$ but their inverses, which we christen $F_{\bar{x}} \overset{\text{dfn}}{=\!=} (\mathcal{F}_{\bar{x}})^{-1}$. We denote by F_x the conjugacy class in π_1^{arith} of all $F_{\bar{x}}$ for points $\bar{x}$ lying over a fixed x; the degree of F_x in $\hat{\mathbb{Z}}$ is the integer $-\deg(x)$.

The ℓ-adic theory provides us with the following data:

1) a b-dimensional $\mathbb{Q}_\ell$-space V with a continuous representation of π_1^{arith} in $\mathrm{Aut}(V)$, such that for every closed point $x \in A^1 - S$, we recover the numerator of the zeta function of $X_x/\mathbb{F}_q(x)$ by the formula

$$\det(1 - TF_x | V) = P(X_x/\mathbb{F}_q(x), T) \in \mathbb{Q}[T]$$

2) an alternating auto-duality intersection form $\langle , \rangle$ on V with values in $\mathbb{Q}_\ell(1-2n)$ which is respected by $\pi_1^{\text{arithmetic}}$. This means that $\langle , \rangle$ is a $\mathbb{Q}_\ell$-valued autoduality such that for $g \in \pi_1^{\text{arith}}$, $v, w \in V$, we have

$$\langle gv, gw \rangle = q^{(1-2n)\deg(g)} \langle v, w \rangle.$$

Thus

$$\begin{cases} \langle \gamma v, \gamma w \rangle = \langle v, w \rangle & \text{for } \gamma \in \pi_1^{\text{geom}} \\ \langle F_{\bar{x}} v, F_{\bar{x}} w \rangle = q^{(2n-1)\deg x} \langle v, w \rangle & \text{for each } F_{\bar{x}} \end{cases}$$

The ℓ-adic version of the Kajdan-Margoulis theorem is that the image of π_1^{geom} in $\mathrm{Aut}(V)$ is Zariski dense (in fact, ℓ-adically open) in the symplectic group $\mathrm{Sp}(V, \langle , \rangle)$.

The proof: a heuristic

We have set out to prove the Riemann Hypothesis for all of the $X_x/\mathbb{F}_q(x)$, $x \in \mathbb{A}^1 - S$. As we have already noted, this is equivalent to proving that for each closed point $x \in \mathbb{A}^1 - S$, the series

$$\frac{1}{P(X_x/\mathbb{F}_q(x), T)} \in \mathbb{Q}[[T]]$$

is convergent archimedeanly for $|T| < 1/\sqrt{q^{\deg x}}^{\,2n-1}$, or equivalently, (if we replace T by $T^{\deg x}$) that the series

$$\frac{1}{\det(1 - T^{\deg x} F_x | V)} = \frac{1}{P(X_x/\mathbb{F}_q(x), T^{\deg x})} \in \mathbb{Q}[[T]]$$

is convergent archimedeanly for $|T| < 1/\sqrt{q}^{\,2n-1}$.

To clarify the basic idea, let us begin by explaining how we could directly deduce this last estimate if two apparently false suppositions were simultaneously true (cf., the Remark at the end of the proof). The first supposition is that each of the series $1/\det(1 - T^{\deg x} F_x | V) \in \mathbb{Q}[[T]]$ has <u>positive</u> coefficients. The second is that the infinite product

$$L(V, T) \overset{\text{dfn}}{=\!=} \prod_x \frac{1}{\det(1 - T^{\deg x} F_x | V)} \in \mathbb{Q}[[T]]$$

is archimedeanly convergent for $|T| < 1/\sqrt{q}^{\,2n-1}$ when viewed as a power series. Granting these suppositions, we would simply remark that as each of the factors $1/\det(1 - T^{\deg x} F_x | V)$ is a series with constant term <u>one</u>, the supposition that its coefficients are <u>positive</u> means that the power series for $L(V, T)$ also has positive coefficients which are term-by-term greater than or equal to the coefficients of any of the factors. Therefore, the supposed archimedean convergence of $L(V, T)$ for $|T| < 1/\sqrt{q}^{\,2n-1}$ would imply that each of the factors $1/\det(1 - T^{\deg x} F_x | V)$ is itself archimedeanly convergent for $|T| < 1/\sqrt{q}^{\,2n-1}$.

<u>The actual proof: "squaring"</u>

We must now explain how to get around the fact that our suppositions are <u>not</u> simultaneously true. The non-positivity of the coefficients of the individual factor $1/\det(1 - T^{\deg x} F_x | V)$ is eliminated by <u>replacing</u> V by any of its <u>even</u>

tensor powers $V^{\otimes 2k}$. (Replacing V by $V^{\otimes 2}$ is analogous to Rankin's replacing $\sum \tau(n)\cdot n^{-s}$ by $\sum (\tau(n)^2 \cdot n^{-s}$.) To see that $1/\det(1-T^{\deg x}F_x|V^{\otimes 2k})$ has positive coefficients, we argue as follows. For any integer $m \geq 1$, we have the formula

$$1/\det(1-T^{\deg x}F_x|V^{\otimes m}) = \exp\left(\sum_{n\geq 1} \frac{T^n}{n} \operatorname{trace}(F_x^n|V^{\otimes m})\right)$$

$$= \exp\left(\sum_{n\geq 1} \frac{T^n}{n} (\operatorname{trace}(F_x^n|V))^m\right)$$

For $m = 1$, this formula, together with the fact that the series $1/\det(1-T^{\deg x}F_x|V)$ has rational coefficients shows that all of the numbers $\operatorname{trace}(F_x^n|V)$ are rational. This rationality, together with the above formula for $m = 2k$, shows that $1/\det(1-T^{\deg x}F_x|V^{\otimes 2k})$ is the exponential of a series with positive coefficients, and therefore has positive coefficients itself.

Review of L-series

Thus in order to apply the argument, we need information on the radius of convergence of the power series

$$L(V^{\otimes 2k}, T) \overset{\text{dfn}}{=\!=} \prod_x \frac{1}{\det(1-T^{\deg x}F_x|V^{\otimes 2k})}.$$

Happily, this information is provided by Grothendieck's cohomological expression for the L-function $L(M,T)$ associated to _any_ continuous finite-dimensional $\mathbb{Q}_\ell$-adic representation of π_1^{arith}. Grothendieck gives a _formula_ for $L(M,T)$ which shows it to be a rational function of T and which, more important for us, gives an à priori hold on its poles, as follows.

Let $M_{\pi_1^{\text{geom}}}$ denote the largest quotient space of M on which π_1^{geom} acts trivially:

$$M_{\pi_1^{\text{geom}}} = M \Big/ \sum_{\gamma\in\pi_1^{\text{geom}}} (1-\gamma)M$$

This factor space is a representation of $\pi_1^{arith}/\pi_1^{geom} \simeq \hat{\mathbb{Z}}$, so the unique element $F \in \hat{\mathbb{Z}}$ of degree -1 (the <u>inverse</u> of the automorphism $a \longrightarrow a^q$ of $\overline{\mathbb{F}}_q$) acts on it. Grothendieck's formula for L-series asserts that the product

$$\det(1-qTF|M_{\pi_1^{geom}})\cdot L(M,T)$$

is a <u>polynomial</u>.

<u>The end of the proof</u>

We now apply this to $M = V^{\otimes 2k}$. The key point is to compute $(V^{\otimes 2k})_{\pi_1^{geom}}$ and the action of F upon it, for then we will know the poles of $L(V^{\otimes 2k}, T)$. Because the image of π_1^{geom} in $Sp = \mathrm{Aut}(V, \langle,\rangle)$ is Zariski-dense (by the Kajdan-Margoulis theorem), their covariants are the same:

$$(V^{\otimes 2k})_{\pi_1^{geom}} \simeq (V^{\otimes 2k})_{Sp}$$

The tensor covariants of the symplectic group, or rather their dual, are well known from classical invariant theory. By the definition of covariants, linear forms on $(V^{\otimes 2k})_{Sp}$ are the same thing as Sp-invariant 2k-linear forms on V, and these are all sums of "complete contractions": for each partition of the set $\{1, \ .., 2k\}$ into two ordered sets $\{a_1, \ldots, a_k, b_1, \ldots, b_k\}$, the corresponding complete contraction is the 2k-linear form on V

$$(v_1, \ldots, v_{2k}) \longrightarrow \prod_{i=1}^{k} \langle v_{a_i}, v_{b_i}\rangle$$

If we remember the action of π_1^{arith}, then the intersection form $\langle,\rangle$ on V takes values in $\mathbb{Q}_\ell(1-2n)$, and we see that the product $\prod_{i=1}^{k} \langle v_{a_i}, v_{b_i}\rangle$ lies in $\mathbb{Q}_\ell(k(1-2n))$. So if we pick a maximal linearly independent set of complete contractions, we get an isomorphism

$$(V^{\otimes 2k})_{Sp} \simeq \oplus\ \mathbb{Q}_\ell(k(1-2n));$$

a space on which F acts as multiplication by $q^{k\cdot(2n-1)}$.

Referring back to Grothendieck's theorem, the denominator of $L(V^{\otimes 2k}, T)$ is at worst given by

$$\det(1 - qTF | \oplus \mathbb{Q}_\ell(k(1-2n))) = \text{a power of } (1 - q^{1+k(2n-1)}T).$$

Therefore the series $L(V^{\otimes 2k}, T)$ converges archimedeanly for $|T| < 1/q^{1+k(2n-1)}$.

The positivity argument then shows that each factor $\dfrac{1}{\det(1 - T^{\deg x} F_x | V^{\otimes 2k})}$ converges archimedeanly for $|T| < 1/q^{1+k(2n-1)}$.

Now suppose that $\alpha(x)$ is an eigenvalue of F_x on V. We must prove that $|\alpha(x)| \leq 1/\sqrt{q^{\deg x}}^{\,2n-1}$. But $\alpha(x)^{2k}$ will be an eigenvalue of F_x on $V^{\otimes 2k}$, and therefore $1/\det(1 - T^{\deg x} F_x | V^{\otimes 2k})$ will have a pole at $T = 1/\alpha(x)^{2k/\deg x}$. But as this series converges for $|T| < 1/q^{1+k(2n-1)}$, we must have the inequality

$$|1/\alpha(x)^{2k/\deg x}| \geq 1/q^{1+k(2n-1)}$$

or equivalently

$$|\alpha(x)|^{2k/\deg x} \leq q^{1+k(2n-1)}$$

or equivalently

$$|\alpha(x)| \leq \sqrt{q^{\deg x}}^{\,(2n-1+1/k)}.$$

Letting k tend to $+\infty$, we obtain

$$|\alpha(x)| \leq \sqrt{q^{\deg x}}^{\,2n-1}. \qquad \text{Q.E.D.}$$

Remark The cohomological expression for $L(V^{\otimes 2k+1}, T)$ together with the fact that the symplectic group has no covariants in any odd tensor power of its standard representation shows that in fact each of the L series $L(V^{\otimes 2k+1}, T)$ is a polynomial, so has infinite radius of convergence. But it is only for the even tensor powers $V^{\otimes 2k}$ that we can be sure that each local factor $1/\det(1 - T^{\deg x} F_x | V^{\otimes 2k})$ has positive coefficients.

Applications

The most striking arithmetic consequence is the generalized Ramanujan-Peterson conjecture on the order of magnitude of the coefficients of cusp forms of weight two or more on congruence subgroups of $SL_2(\mathbb{Z})$. (Deligne and Serre have also proven the conjecture for forms of weight one (unpublished), but the proof is logically independent of the Riemann Hypothesis.)

Another arithmetic application is the estimation of exponential sums in several variables. Though technically difficult, the idea goes back to Weil [56], who showed how the Riemann Hypothesis for curves over finite fields gave the "good" estimate for exponential sums in one variable.

As for geometric applications, we have already mentioned the hard Lefschetz theorem. There is also a whole chain of ideas built around the "yoga of weights," Grothendieck's catch-phrase for deducing results on the cohomology of arbitrary varieties by assuming the Riemann Hypothesis for projective non-singular varieties over finite fields (cf., [7]). The whole of Deligne's "mixed Hodge theory" for complex varieties ([8], [9]), developed before his proof of the Riemann Hypothesis, is intended to prove results about the cohomology of these varieties which follow from the Riemann Hypothesis and from the systematic application of Hironaka's resolution of singularities. The recent work of Deligne, Griffiths, Morgan and Sullivan on the rational homotopy type of complex varieties is also considerably clarified by the use of the Riemann Hypothesis.

Some open questions

1. Independence of ℓ

Let X be an arbitrary variety over an algebraically closed field k. For each prime number ℓ distinct from the characteristic of k, the ℓ-adic cohomology groups $H^i(X, \mathbb{Q}_\ell)$ and the ℓ-adic groups "with compact support"

$H^i_{comp}(X, \mathbb{Q}_\ell)$ are known ([10]) to be finite-dimensional $\mathbb{Q}_\ell$-vector spaces. Let us denote by $b^{comp}_{i,\ell}(X)$ their dimensions. It is unknown whether these numbers are independent of ℓ, except in some special cases, as follows.

When the field k is of characteristic zero, the "comparison theorem" [2] asserts that if we "choose" an embedding of k into the complex number field $\mathbb{C}$, and denote by $X(\mathbb{C})$ the corresponding analytic space over $\mathbb{C}$ with its usual topology, then we have isomorphisms

$$H^i(X, \mathbb{Q}_\ell) \simeq H^i_{sing}(X(\mathbb{C}), \mathbb{Q}) \otimes \mathbb{Q}_\ell$$

$$H^i_{comp}(X, \mathbb{Q}_\ell) \simeq H^i_{comp, sing}(X(\mathbb{C}), \mathbb{Q}) \otimes \mathbb{Q}_\ell$$

where H^i_{sing} (resp. $H^i_{comp, sing}$) denotes classical singular cohomology (resp. with compact support) of usual topological spaces.

When the field k is of characteristic p, an easy specialization argument reduces us to considering only the case when k is the algebraic closure of a finite field $\mathbb{F}_q$, and X/k comes by extension of scalars from a variety $X_0/\mathbb{F}_q$.

In case X/k is proper and smooth, then $b_{i,\ell}(X) = b^{comp}_{i,\ell}(X)$, just because X is already "compact," and we can use the Riemann Hypothesis for $X_0/\mathbb{F}_q$ (extended by Deligne to the proper (not necessarily projective) and smooth case in [6]) to interpret $b_{i,\ell}(X)$ as the number of complex zeroes (if i is odd) or poles (if i is even) of the zeta function $Z(X_0/\mathbb{F}_q, T)$ which lie on the circle $|T| = q^{-i/2}$. Because the zeta function itself is independent of ℓ, this shows that the integer $b_{i,\ell}(X)$ is independent of ℓ, and that the (a priori ℓ-adic) polynomial $\det(1-TF|H^i(X, \mathbb{Q}_\ell))$ has $\mathbb{Q}$-coefficients which are independent of ℓ.

However, when X/k fails to be proper and smooth, this argument breaks down. For arbitrary X/k, Deligne has proven (cf., [6]) that for each $\ell \neq p$, the polynomial $\det(1-TF|H^i_{comp}(X, \mathbb{Q}_\ell))$, whose degree is $b^{comp}_{i,\ell}(X)$, has algebraic numbers as coefficients, and that each of its reciprocal zeroes (the eigenvalues of F on $H^i_{comp}(X, \mathbb{Q}_\ell))$ is an algebraic number α for which there exists an integer $j \leq i$ (j depending upon α) such that α and all of its conjugates over $\mathbb{Q}$ have $|\alpha| = q^{j/2}$. But it is unknown if $\det(1-TF|H^i_{comp}(X, \mathbb{Q}_\ell))$ has $\mathbb{Q}$-coefficients, and a fortiori if it is independent of ℓ. It is the mixing of eigenvalues between the various H^i_{comp} that prevents us from expressing the polynomials $\det(1-TF|H^i_{comp}(X, \mathbb{Q}_\ell))$ in terms of $Z(X_0/\mathbb{F}_q, T)$, as we could do in the proper and smooth case. [A sufficiently strong form of Hironaka's resolution of singularities [27] (at present established only in characteristic zero) would allow the recovery of these characteristic polynomials intrinsically in terms of the zeta functions of the various proper and smooth varieties which would enter into a compactification and resolution of X. But perhaps one can get by without resolution.]

2. An elementary proof (cf., [4])

Now that we know the Riemann Hypothesis for varieties over finite fields, can we give an elementary proof by directly counting rational points? For curves, this has been done recently by Bombieri-Stepanov. An added difficulty in the higher dimensional case is that for the "typical" variety of dimension $d > 1$, the Riemann Hypothesis does not seem to be equivalent to any diophantine statement: the highest cohomology group H^{2d} gives the dominant contribution to the number of rational points, and all the rest of the cohomology is an error term:

$$N_n = q^{dn} + O(q^{n(d-\frac{1}{2})}).$$

This estimate, however, was established in 1953 by Lang-Weil [33] as a consequence of the Riemann Hypothesis for curves. There are, of course, many special classes of varieties (e.g. curves, complete intersections, simply connected surfaces) for which the Riemann Hypothesis is equivalent to a diophantine estimate, and a direct proof valid for these would certainly be of great interest.

3. The Hasse-Weil zeta function (cf., [59], [61])

With the proof of all the Weil conjectures, we may regard the question of number of solutions of equations in finite fields as being fairly well understood. What is not at all understood is the question of solutions of equations in rational numbers. It is expected that the Hasse-Weil zeta function will play an important role in this question. To fix ideas, suppose that X is a projective smooth scheme over $\mathbb{Z}[1/N]$ (i.e., X is a projective non-singular variety over $\mathbb{Q}$ which has "good reduction" at all primes p which are prime to some "conductor" N). Then for each prime p which is prime to N, the "reduction mod p" of X, noted $X(p)$, is a projective smooth variety over $\mathbb{F}_p$. For each integer $0 \leq i \leq \dim(X/\mathbb{Z}[1/N])$, we consider the i'th polynomial $P_i(X(p)/\mathbb{F}_p, T)$ occurring in the zeta function of $X(p)/\mathbb{F}_p$. The i'th Hasse-Weil L-function is defined to be the Dirichlet series with Euler product (over primes not dividing N)

$$L(i; X, s) = \prod_p \frac{1}{P_i(X(p)/\mathbb{F}_p, p^{-s})} .$$

It is convergent for $\mathrm{Re}(s) > 1 + i/2$ by the Riemann Hypothesis for the $X(p)$'s. The Hasse-Weil zeta function is by definition the alternating product of these L-functions.

It is generally conjectured that each of the L functions $L(i, X, s)$ admits a meromorphic extension to the entire s-plane, satisfies a functional equation under $s \longrightarrow i+1-s$, and has all of its zeroes in the half-plane $\mathrm{Re}(s) \leq \frac{i+1}{2}$, with all the zeroes which are not introduced by Γ-factors in the functional equation lying <u>on</u> the line $\mathrm{Re}(s) = \frac{i+1}{2}$. The last conjecture is the "generalized Riemann Hypothesis."

From the point of view of arithmetic algebraic geometry, a variety X over $\mathbb{Z}[1/N]$ is the analogue of a <u>family</u> of varieties parameterized by a curve over a finite field. For such families, the analogues of the Hasse-Weil L-functions are the L-functions associated to the various ℓ-adic representations of π_1^{arith} which the ℓ-adic cohomology provides; the meromorphy (in fact, rationality as a function of p^{-s}) of these latter L functions, and their functional equations, are provided by Grothendieck's theory of such L-functions. The location of their zeroes is a generalized form of the Riemann Hypothesis for varieties over finite fields (which has also been proven by Deligne, but is still unpublished).

What has been proven about the Hasse-Weil L-functions? The meromorphic continuation and functional equation have been established only for very special X (e.g., elliptic curves with complex multiplication [13], [14], diagonal hypersurfaces [60], curves uniformized by modular functions [48], and X of dimension <u>zero</u>). There is not a single known case of the Riemann Hypothesis. In the simplest case, when the variety X over $\mathbb{Z}$ is "a point" (i.e., $X = \mathrm{Spec}\,(\mathbb{Z})$), the Hasse-Weil zeta function $\zeta(X, s)$ becomes the Riemann zeta function!

Bibliography

1. N. A'Campo. Sur la monodromie des singularités isolées d'hypersurfaces complexes. Inv. Math. 20, Fasc. 2 (1973), 147-170.

2. E. Artin. Quadratische Körper in Gebiete der höheren Kongruenzen I and II, Math. Zeit. 19 (1924), 153-246.

3. M. Artin, A. Grothendieck and J.-L. Verdier. SGA 4 Théorie des topos et cohomologie étale des schémas. Lecture Notes in Mathematics 269, 270, 305. Berlin-Heidelberg-New York. Springer-Verlag 1972.

4. E. Bombieri. Counting points on curves over finite fields (d'après A. Stepanov) Séminaire N. Bourbaki 1972/73, June, 1973, Exposé 430. To appear in Springer Lecture Notes in Mathematics.

5. P. Deligne. La conjecture de Weil I. Publ Math. IHES 43 (1974), 273-307.

6. __________. La conjecture de Weil II, to appear; this material was presented by Deligne in his 1973/74 Seminar at IHES.

7. __________. Théorie de Hodge I. Proceedings of the 1970 International Congress of Mathematicians, Nice. Gauthier-Villars, Paris, 1971.

8. __________. Théorie de Hodge II. Pub. Math. IHES 41 (1972).

9. __________. Théorie de Hodge III, to appear in Pub. Math. IHES.

10. __________. Inputs of étale cohomology. Lectures at the 1974 AMS Summer Institute on Algebraic Geometry, Arcata, California.

11. __________. Formes modulaires et représentations ℓ-adiques. Séminaire Bourbaki 1968/69, Exposé 355. Lecture Notes in Mathematics 179, Springer-Verlag, Berlin-Heidelberg-New York, 1971.

12. P. Deligne and N. Katz. SGA 7, part II. Groupes de monodromie en géométrie algébrique. Lecture Notes in Mathematics 340, Springer-Verlag, Berlin-Heidelberg-New York, 1973.

13. M. Deuring. Die Zetafunktion einer algebraischen Kurve vom Geschlechte Eins, Drei Mitteilungen. Nach. Akad. Wiss. Göttingen, 1953, pp. 85-94, 1955 pp. 13-43 and 1956, pp. 37-76.

14. __________. On the zeta function of an elliptic function field with complex multiplication. Proceedings of the International Symposium on Algebraic Number Theory, Tokyo-Nikko, 1956, pp. 47-50.

14 bis. __________. Arithmetische Theorie der Korrespondenzen algebraischen Funktionenkörpes I. J. reine angew. Math. 177 (1937), 161-191.

15. J. Dieudonne. <u>Cours de Géometrie Algébrique</u> I, II. Presses Universitaires de France, Paris, 1974.

16. B. Dwork. On the rationality of the zeta function of an algebraic variety. Amer. J. Math. 82 (1960), 631-648.

17. M. Eichler. Eine Verallgemeinerung der Abelschen Integrale. Math. Zeit 67 (1957), 267-298.

18. A. Grothendieck. SGA 7 part I. Groupes de monodromie en geometrie algébrique. Lecture Notes in Mathematics 288. Berlin-Heidelberg-New York, Springer-Verlag, 1973.

19. __________. Formule de Lefschetz et rationalité des fonctions L.

Séminaire Bourbaki 1964/65, Exposé 279. W. A. Benjamin, New York, 1966.

20. __________. Sur une note de Mattuck-Tate. J. Reine. Angew. Math. 200 Heft 3/4 (1958), 208-215.

21. G. H. Hardy. Ramanujan. Twelve lectures on subjects suggested by his life and work. Chelsea, New York: originally published by Cambridge University Press, 1940.

22. H. Hasse. Beweis des Analogons der Riemannschen Vermutung für die Artinschen und F. K. Schmidschen Kongruenz-zetafunktionen in gewissen elliptischen Fällen. Ges. d. Wiss. Nachrichten. Math-Phys. Klasse, 1933, Heft 3, 253-262.

23. __________. Uber die Kongruenz-zetafunktionen, Sitz-ber. d. Preuss. Akad. d. Wiss, 1934, 250.

24. __________. Abstrakte Begründung der komplexen Multiplikation und Riemannsche Vermutung in Funktionen körpern. Abh. Math. Sem. Univ. Hamburg, 10 (1934), 325-348.

25. __________. Zur Theorie der abstrakten elliptischen Funktionenkörper, I, II and III. J. Reine Angew. Math 175 (1936).

26. __________. Über die Riemannsche Vermutung in Funktionenkörpern, Comptes Rendus du congrès international des mathématiciens, Oslo (1930), pp. 189-206.

27. H. Hironaka. Resolution of singularities of an algebraic variety over a field of characteristic zero, I, II. Annals of Math. 79 (1964), 109-326.

28. W. Hodge. The theory and applications of harmonic integrals, Cambridge University Press, Cambridge, 1941.

29. Y. Ihara. Hecke polynomials as congruence zeta functions in elliptic modular case. Annals of Math. Ser.2 85 (1967), 267-295.

30. Kajdan and Margoulis. Personal communication to P. Deligne, October, 1971.

31. S. Kleiman. Algebraic cycles and the Weil conjectures, in Dix exposés sur la cohomologie des schémas, North Holland, Amsterdam and Paris, 1968.

32. M. Kuga and G. Shimura. On the zeta function of a fibre variety whose fibres are abelian varieties. Annals of Math. Ser.2 82 (1965), 478-539.

33. S. Lang and A. Weil. Number of points of varieties in finite fields. Amer. J. Math., 76 (1954), 819-827.

34. R. Langlands. Problems in the theory of automorphic forms in Lectures in modern analysis and applications III. Lecture notes in Mathematics,

170, pp.18-61, Springer-Verlag, Berlin-Heidelberg-New York, 1970.

35. S. Lefschetz. L'analysis situs et la géométrie algébrique. Gauthier-Villars, Paris, 1924.

36. D. Lehner. Note on the distribution of Ramanujan's tau function. Math. of Comp., vol. 24, no. 111, July 1970, 741-743.

37. A. Mattuck and J. Tate. On the inequality of Castelnuovo-Severi. Abh. Math. Sem., Hamburg, 22 (1958), 295.

38. D. Mumford. Abelian Varieties. Oxford University Press, Bombay, 1970.

39. R. A. Rankin. Contributions to the theory of Ramanujan's function $\tau(n)$ and similar arithmetical functions II. Proc. Comb. Phil. Soc. 35 (1939).

40. F. K. Schmidt. Analytische Zahlenthéorie in Körpern der Characteristik p. Math. Zeit. 33 (1931), 1-32.

41. J.-P. Serre. Faisceaux algébriques cohérents, Annals of Math. 61, no. 2, March 1955, 197-278.

42. ________. Sur la topologie des variétés algébriques en caractéristique p. Contrés de topologie, Mexico, 1956.

43. ________. Quelques propriétédes variétés abéliennes en caractéristique p. Amer. J. Math. LXXX, no. 3 (1958), 715-739.

44. ________. Analogues Kahlériens de certaines conjectures de Weil, Annals of Math. 71, ser. 2 (1960).

45. ________. Rationalité des fonctions ζ des variétés algébriques (d'aprés Bernard Dwork). Séminare Bourbaki 1959/60, Exposé 198. W. A. Benjamin, New York, 1966.

46. ________. Zeta and L functions in Arithmetic Algebraic Geometry, Harper and Row, New York, 1965.

47. ________. Valeurs propres des endomorphismes de Frobenius (d'après P. Deligne). Séminaire Bourbaki 973/74, February 1974, Exposé 446. To appear in Springer Lecture Notes in Mathematics.

48. G. Shimura. Correspondances modulaires et les fonctions ζ de courbes algébriques. M. Math. Soc. Japan 10 (1958), 1-28.

48bis. ________. Sur les intégrales attachés aux former automorphes. J. Math. Soc. Japan 11 (1959), 291-311.

49. G. Shimura and Y. Taniyama. Complex multiplication of abelian varieties and its applications to number theory. Pub. Math. Soc. Japan, no. 6, 1961.

50. J.-L. Verdier. Indépendance par rapport à ℓ des polynomes caractéristiques des endomorphismes de Frobenius de la cohomologie ℓ-adique

(d'après P. Deligne). Séminaire Bombaki 1972/73. Exposé 423. To appear in Springer Lecture Notes in Mathematics.

51. A. Weil. Sur les fonctions algébriques à corps de constantes fine. C. R. Acad. Sci. Paris 210 (1940), 592-594.

52. __________. On the Riemann Hypothesis in function-fields. Proc. Nat. Acad. Sci. U.S.A. 27 (1941), 345-347.

53. __________. <u>Foundations of algebraic geometry</u>, Amer. Math. Soc. Colloq. Pub., Vol. XXIX, New York, 1946.

54. __________. <u>Sur les courbes algébriques et les variétés qui s'en déduisent</u>, Hermann, Paris, 1948.

55. __________. <u>Variétés abeliennes et courbes algébriques</u>, Hermann, Paris, 1948.

56. __________. On some exponential sums. Proc. Nat. Acad. Sci. U.S.A., 34 (1948), 204-207.

57. __________. Number of solutions of equations in finite fields, Bull. A.M.S. 55 (1949), pp. 497-508.

58. __________. Fibre Spaces in algebraic geometry. (Notes by A. Wallace) University of Chicago, 1952.

59. __________. Number theory and algebraic geometry. Proceedings of the 1950 International Congress of Mathematicians, Cambridge, vol. 2, pp. 90-100, A.M.S., Providence, 1952.

50. __________. Jacobi sums as Grossencharaktere, Trans. A.M.S. 73 (1952), 487-495.

61. __________. Abstract versus classical algebraic geometry. Proceedings of the 1954 International Congress of Mathematicians, Vol. III, 550-558, North-Holland, Amsterdam, 1956.

62. H. Yoshida. On an analogue of the Sato conjecture. Inv. Math. 19, 4 (1973), 261-277.

63. O. Zariski. Algebraic Surfaces - Band III, Heft 5, Ergébnisse der Math. und Grenz., 1935 - second supplemented edition, Band 61, Ergébnisse, Springer-Verlag, Berlin-Heidelberg-New York, 1971.

64. __________. <u>Collected Works</u>, vol. II, M.I.T. Press, Cambridge, 1973, especially the preface.

Proceedings of Symposia in Pure Mathematics
Volume 28, 1976

PROBLEMS CONCERNING PRIME NUMBERS

Hugh L. Montgomery

Writings in analytic number theory are generally restricted to presenting what the author is capable of proving, while more ambitious aims are rarely mentioned. On this occasion we take the opposite tack by listing the main conjectures and problems which motivate most of analytic number theory. We need mention only a few questions, because answers to these would provide solutions to a large number of familiar problems. One may feel that the list is not useful in that it does not provide readily accessible research problems, but I have often found it useful to have a larger picture of the suspected truth, against which I can test ideas.

1. The Riemann zeta function. We expect that the Riemann Hypothesis is true, namely that if ρ is a non-trivial zero of $\zeta(s)$ then $\operatorname{Re}\rho = \frac{1}{2}$. Hence we write $\rho = \frac{1}{2} + i\gamma$. We expect that all zeros are simple (so we let $0 < \gamma_1 < \gamma_2 < \dots$ denote the ordinates of the zeros in the upper half-plane), and we ask for a precise lower bound for the differences $\gamma_{n+1} - \gamma_n$. In this direction we put

$$\alpha = \liminf_{n\to\infty} \frac{\log(\gamma_{n+1} - \gamma_n)}{\log n}$$

and suggest that $-1 < \alpha < 0$; perhaps $\alpha = -\frac{1}{3}$. More generally, we suppose that the numbers γ are linearly independent over $\mathbb{Q}$, and we ask for lower bounds for linear forms in the γ's.

AMS (MOS) subject classifications (1970). 10 Hxx.

Let $N(T)$ denote the number of zeros ρ of $\zeta(s)$ with $0 < \gamma \le T$. Then

$$N(T) = \frac{T}{2\pi}\log\frac{T}{2\pi} - \frac{T}{2\pi} + S(T) + \frac{7}{8} + O\left(\frac{1}{T}\right),$$

where $S(t) = \frac{1}{\pi}\arg\zeta(\frac{1}{2}+it)$. The statistical distribution of $S(t)$ was determined by A. Selberg, but the maximum order of $S(t)$ remains unknown. Probably

$$S(t) = O\left(\left(\frac{\log t}{\log\log t}\right)^{\frac{1}{2}}\right).$$

Knowledge of the local distribution of the zeros would have implications concerning prime numbers. For example, it would be useful to know that for fixed $\alpha < \beta$

$$\sum_{\substack{\gamma\le T\\ \gamma'\le T\\ \frac{2\pi\alpha}{\log T}\le\gamma-\gamma'\le\frac{2\pi\beta}{\log T}}} 1 \sim \left(\int_\alpha^\beta 1 - \left(\frac{\sin\pi u}{\pi u}\right)^2 du + \delta(\alpha,\beta)\right)\frac{T}{2\pi}\log T,$$

where $\delta(\alpha,\beta) = 1$ if $\alpha \le 0 \le \beta$, $\delta(\alpha,\beta) = 0$ otherwise. We also expect that

$$\limsup_{n\to\infty}\,(\gamma_{n+1} - \gamma_n)\log\gamma_n = +\infty.$$

In applications it would also be very useful to understand the behavior of sums of the sort $\sum_{|\gamma|\le T} x^{i\gamma}$, for x and T in various ranges.

For Dirichlet L-functions and Dedekind zeta functions the problems are much the same, but some new problems arise, such as that of demonstrating the integrality of, and finding useful formulae for Artin's L-functions.

2. <u>Arithmetic sequences</u>. We know (assuming the Riemann Hypothesis) that $\psi(x) - x$ is never larger than $C\,x^{\frac{1}{2}}(\log x)^2$, and also that it is infinitely often as large as $c\,x^{\frac{1}{2}}\log\log\log x$. It would be interesting to know more precisely how large $\psi(x) - x$ becomes. Similarly we may inquire about the maximum order of $\sum_{n\le x}\mu(n)$, of $Q(x) - \frac{6}{\pi^2}$, where $Q(x)$ is the number of square-free integers not exceeding x, and of $\sum_{n\le x} d(n) - (x\log x + (2C-1)x)$.

Concerning prime numbers in arithmetic progressions we may conjecture that

$$\psi(x;q,a) = \frac{x}{\phi(q)} + O\left(\left(\frac{x}{q}\right)^{\frac{1}{2}+\varepsilon}\right) ,$$

provided $(a,q) = 1$.

We are also interested in the distribution of primes in short intervals. Presumably

$$\psi(x+h) - \psi(x) = h + O(h^{\frac{1}{2}} x^{\varepsilon})$$

for $1 \le h \le x$. This would imply that there is a prime in the interval $(x,x+h)$. This can be made more precise by conjecturing that

$$\lim_{x\to\infty} \sup_{p_n \le x} \frac{p_{n+1} - p_n}{(\log p_n)^2} = 1 ;$$

here p_n denotes the n-th prime. This is slightly stronger than Cramér's conjecture that $\limsup \frac{p_{n+1} - p_n}{(\log p_n)^2} = 1$. Concerning the frequency with which $p_{n+1} - p_n$ is large, it is thought that for fixed $\alpha > 0$

$$\sum_{\substack{p_n \le x \\ p_{n+1} - p_n > \alpha \log p_n}} 1 \sim e^{-\alpha}\pi(x) .$$

In the opposite direction, we may ask how may primes can lie in an interval $(x,x+h)$. Put $\rho(h) = \limsup_{x\to\infty} (\pi(x+h) - \pi(x))$. it seems that for large h , $\pi(h) < \rho(h) < \frac{2h}{\log h}$; it would be instructive to know which of these inequalities is sharp, if either.

3. Additive prime number theory. Given a number of polynomials with integral coefficients, we expect that for infinitely many choices of the integral variables the polynomials will simultaneously take on prime values, unless some local condition holds which makes this obviously impossible. A very general hypothesis of this sort might very well be undecidable, but in the well-known special cases of Goldbach's conjecture, prime k-tuples, primes of the form $n^2 + 1$, etc., we not only expect positive results, but we suppose that the number of solutions is very accurately predicted by the major arc contribution in the Hardy-Littlewood circle method.

4. Diophantine questions. Are the numbers $\zeta(3)$, $\zeta(3)\pi^{-3}$, and C transcendental? Here C denotes Euler's constant. One may inquire similarly about the imaginary parts of the zeros of the zeta function. In a different direction, it would be useful to know that one can approximate to real numbers by rational numbers whose numerators and denominators are both prime. Specifically, is it true that for every irrational number θ and every $\varepsilon > 0$ there are infinitely many prime numbers p , q such that

$$\left|\theta - \frac{p}{q}\right| < q^{-2+\varepsilon} .$$

University of Michigan,
Ann Arbor, MI 48104

Proceedings of Symposia in Pure Mathematics
Volume 28, 1976

PROBLEM 9: THE GENERAL RECIPROCITY LAW

J. Tate

The ninth problem of Hilbert concerns the "most general reciprocity law in an arbitrary algebraic number field". It seems to me that he underestimated its potential when he referred to it as a "more special" problem in number theory and devoted only a short paragraph to it.

Let R be the ring of integers in an algebraic number field. If P is an odd prime ideal in R, i.e., one whose residue class field is of odd characteristic, then the Legendre symbol $(\frac{a}{P})$ is defined, for $a \in R$, to be one less than the number of incongruent solutions, x, of the congruence $x^2 \equiv a (\mathrm{mod}\ P)$. In other words

$$(\frac{a}{P}) = \begin{cases} 1 & \text{if } x^2 \equiv a \pmod{P} \text{ is solvable and } a \not\equiv 0 \pmod{P}, \\ 0 & \text{if } a \equiv 0 \pmod{P}, \\ -1 & \text{if } x^2 \equiv a \pmod{P} \text{ is not solvable.} \end{cases}$$

If we view these congruences (mod P) as equations in the residue field R/P and recall that the multiplicative group $(R/P)^*$ of that field is cyclic of even order, then it is clear that

$$(\frac{a}{P}) = f(a \bmod P),$$

where f is the function on R/P which vanishes at 0 and on $(R/P)^*$ is the unique character of order 2. In particular, the symbol $(\frac{a}{P})$ is multiplicative in a:

$$(\frac{ab}{P}) = (\frac{a}{P})(\frac{b}{P}) .$$

Thus, as function of a, for fixed P, the symbol $(\frac{a}{P})$ is easy to describe. The dual problem - how does $(\frac{a}{P})$ vary with P for fixed a - is the problem of quadratic reciprocity.

Consider the case $R = \mathbb{Z}$, the ring of rational integers. Then $P = p\mathbb{Z}$, where p is a positive odd prime number, and we write $(\frac{a}{p})$ instead of $(\frac{a}{P})$. For fixed p the symbol $(\frac{a}{p})$ depends only on the remainder when the integer a is divided by p and its value is easily tabulated for $0 \leq a < p$. But suppose we fix a and let p vary. What sort of function of p is $(\frac{a}{p})$? If we write

$$a = \pm 2^{m_o} q_1^{m_1} \cdots q_r^{m_r}$$

where the q_i are odd positive primes, then

$$(\frac{a}{p}) = (\frac{\pm 1}{p})(\frac{2}{p})^{m_o}(\frac{q_1}{p})^{m_1} \cdots (\frac{q_r}{p})^{m_r} .$$

Thus the problem for an arbitrary integer a reduces immediately to its special cases $a = -1$, $a = 2$, and $a = q$, a positive odd prime $\neq p$. In these cases $(\frac{a}{p})$ as function of p is given by the classical quadratic reciprocity law conjectured by Euler and proved by Gauss, as follows:

$$(\frac{-1}{p}) = (-1)^{\frac{p-1}{2}}, \qquad (\frac{2}{p}) = (-1)^{\frac{p^2-1}{8}}, \qquad \text{and}$$

$$(\frac{q}{p}) = (\frac{p}{q})(-1)^{\frac{p-1}{2}\cdot\frac{q-1}{2}} = (\frac{p^*}{q}), \qquad \text{where } p^* = (-1)^{\frac{p-1}{2}} p.$$

An easy consequence is that $(\frac{a}{p})$ depends only on the residue class of p (modulo 4a), a fact which is not at all obvious from the definition of $(\frac{a}{p})$.

Hilbert reinterpreted the quadratic reciprocity law and generalized it to an arbitrary algebraic number field K. To do this he introduced a new symbol

$$\left(\frac{a,b}{P}\right) = \begin{cases} 1, & \text{if } ax^2 + by^2 = 1 \text{ has a solution } x,y \in K_P, \\ -1, & \text{if not.} \end{cases}$$

Here a and b are non-zero elements of K, and K_P denotes the completion of K at P. Contrary to the Legendre symbol, this Hilbert symbol is defined not only for odd primes, but for <u>all</u> primes P of K, odd or even, and also for the "infinite primes", for which K_P is the real or complex field. Hilbert proved:

(i) For fixed P, the symbol $\left(\frac{a,b}{P}\right)$ is symmetric and bimultiplicative in a and b.

(ii) If P is an odd prime and a an integer in K not divisible by P, then

$$\left(\frac{a,b}{P}\right) = \left(\frac{a}{P}\right)^{v_P(b)}$$

where $v_P(b)$ is the exponent to which P appears in the prime factorization of b.

(iii) For fixed a and b

$$\prod_P \left(\frac{a,b}{P}\right) = 1.$$

This last product is really finite because by (ii) the P-factor in it is 1 if P is odd and prime to a and to b.

It is easy to see that Hilbert's product formula (iii) generalizes the classical quadratic reciprocity law. If $K = \mathbb{Q}$, the rational field, and we take for a and b two distinct odd positive prime numbers p and q, the product formula (iii) becomes

$$\left(\frac{p,q}{\infty}\right)\left(\frac{p,q}{2}\right)\left(\frac{p,q}{p}\right)\left(\frac{p,q}{q}\right) = 1,$$

for by (ii) we have $\left(\frac{p,q}{r}\right) = 1$ if r is an odd prime different from p and q. The first, "∞", factor is 1 because p and q are positive. Using (ii) and symmetry to evaluate the last two factors we find therefore

$$\left(\frac{q}{p}\right)\left(\frac{p}{q}\right) = \left(\frac{p,q}{2}\right).$$

By (i) the symbol $\left(\frac{p,q}{2}\right)$ depends only on p and q (mod 8), because integers $\equiv 1 \pmod 8$ are 2-adic squares. Taking special values of p and q we can therefore simply verify that

$$\left(\frac{p,q}{2}\right) = (-1)^{\frac{p-1}{2}\cdot\frac{q-1}{2}}.$$

The laws for $\left(\frac{-1}{p}\right)$ and $\left(\frac{2}{p}\right)$ follow similarly now, taking $a = -1$ or 2 and $b = p$.

During the 19th century some special ℓ^{th}-power reciprocity laws were discovered for $\ell > 2$. Suppose K contains a primitive ℓ^{th} root of unity. For a prime ideal P not dividing ℓ the group U_ℓ of ℓth roots of unity injects into the group $(R/P)^*$, which is cyclic of order NP-1. Hence we can define an ℓth power-residue symbol at such primes P by

$$\left(\frac{a}{P}\right)_\ell \in U_\ell \cup \{0\}, \qquad \text{and} \qquad \left(\frac{a}{P}\right)_\ell \equiv a^{\frac{NP-1}{\ell}} \pmod{P}.$$

The symbol $(\frac{a}{P})_\ell$ is then multiplicative in a, and is 1 if and only if the congruence $x^\ell \equiv a \pmod{P}$ is solvable and $a \not\equiv 0 \pmod{P}$. The classical cubic and quartic reciprocity laws for K the field of third or fourth roots of unity were stated in terms of this symbol for $\ell = 3$ or 4 and are of the same general nature as the quadratic law for the rational field.

Hilbert's 9th problem was to prove the reciprocity law for ℓ^{th} power residues for an arbitrary number field K and for arbitrary $\ell > 2$. It's not clear to me whether he meant implicitly that K should contain the ℓth roots of unity or whether he really anticipated the freeing of the problem from that assumption. But at any rate it's clear from his writings at the time that he expected the solution to come from a generalization to arbitrary abelian extensions L/K of his own theory of quadratic extensions. (An abelian extension is a normal field extension with abelian Galois group.)

Building on work of Hilbert and Weber, Takagi created in the period 1900-1920 a general theory of abelian extensions of number fields. To state Takagi's fundamental results I will use the modern notion of the idele class group, instead of the generalized ideal class groups of Weber with which Takagi worked. With every number field K there is associated its idele class group C_K, a locally compact abelian group. This group C_K is defined as the quotient of the restricted product of the multiplicative groups of the completions of K by the multiplicative group of K. Each finite prime P of K determines a subset [P] of C_K which we will call the class of P. It is the set of images in C_K of the local parameters in K_P. If L is a finite extension of K, then there is a norm homomorphism $N_{L/K}\colon C_L \longrightarrow C_K$.

Takagi's main results can be summed up as follows:

1. The correspondence $L \longmapsto N_{L/K}C_L$ is a bijection from the set of finite abelian extensions L of K (in a given algebraic closure $\bar{K}$) to the set of open subgroups of finite index in C_K. For two abelian extensions L and L',

$$L \subset L' \iff N_{L/K}C_L \supset N_{L'/K}C_{L'} .$$

2. For each L the abelian groups

$$\mathrm{Gal}(L/K) \qquad \text{and} \qquad C_K/N_{L/K}C_L$$

are abstractly isomorphic, and for each finite prime P of K, the way in which P decomposes in L is entirely determined by the image $[P]_{L/K}$ in $C_K/N_{L/K}C_L$ of the class [P] of P.

With this work of Takagi the theory of abelian extensions - "class field theory" - seemed in some sense complete, yet there was still no general reciprocity law. It remained for Artin to crown the edifice with such a theorem. He conjectured in 1923 and proved in 1927 that for each abelian extension L/K there is a natural isomorphism

$$C_K/N_{L/K}C_L \xrightarrow{\sim} \mathrm{Gal}(L/K)$$

which is characterized by the fact that for each finite prime P of K it carries $[P]_{L/K}$ onto the set $\mathrm{Frob}_{L/K}(P)$ of "Frobenius substitutions" attached to P. (In the abelian case here under consideration, $\mathrm{Frob}_{L/K}(P)$ is the set of automorphisms $s \in \mathrm{Gal}(L/K)$ such that $sx \equiv x^{NP} \pmod{P}$ for each x in the ring of integers of L; it consists of a single element if P is unramified in L.) When the ℓth roots of unity lie in K and $L = K(a^{1/\ell})$, it follows immediately from the definitions that $\mathrm{Frob}_{L/K}(P)$ is essentially the ℓ-power residue symbol $(\frac{a}{P})_\ell$, and Artin's theorem implies then, that for fixed a the value of $(\frac{a}{P})_\ell$ depends only on the class $[P]_{L/K}$ of P. Making this dependence explicit yields all known reciprocity laws. For this reason Artin said when he first conjectured his theorem that it had to be viewed as the general reciprocity law, for arbitrary fields, with or without roots of unity, even though it sounds a bit strange at first.

How did Artin guess his reciprocity law? He was not looking for it, not trying to solve a Hilbert problem. Neither was he, as would seem so natural to us today, seeking a canonical isomorphism, to make Takagi's theory more functorial. He was led to the law in trying to show that a new kind of L-series which he introduced really was a generalization of the usual L-series. The usual L-series,

$$L(s,X) = \prod_P (1 - X([P])NP^{-s})^{-1},$$

were associated with idele class characters $X\colon C_K \longrightarrow \mathbb{C}^*$. Artin's L-series

$$L(s,R) = \prod_P \det(1 - R(\mathrm{Frob}_{L/K}(P))NP^{-s})^{-1}$$

were associated to representations $R\colon \mathrm{Gal}(L/K) \longrightarrow GL(n,\mathbb{C})$ of Galois groups over K, abelian or not. (In these Euler products, the factors corresponding to ramified primes need special definitions.) Clearly, in order to identify his L-series with the usual L-series in case of an abelian extension L/K, Artin needed an identification of the abelian Galois group Gal(L/K) with a quotient of C_K which would make Frob(P) correspond to [P]; then each representation R of degree 1 would correspond to a character X such that $X([P]) = R(\mathrm{Frob}(P))$. Not only was the idea of Artin's reciprocity law inspired by analysis, but also its proof. Artin thanks Tschebotarov for one of the basic ideas of his proof, the use of cyclotomic fields in the way Tschebotarov had used them in proving his famous density theorem.

With Artin's general reciprocity law, the basic structure of class field theory was complete. There soon followed, by Hasse, the generalization of Hilbert's norm-residue symbol $(\frac{a,b}{P})_\ell$ to arbitrary ℓ, when the ℓth roots of 1 are in K. Also the structure of the Brauer Group and central simple algebras over number fields was determined by Hasse with R. Brauer and E. Noether in the early 1930's. From then until the end of World War II, the main progress in these matters was Chevalley's introduction of ideles and idele classes which enabled smoother formulations and clarified the local-global aspect of class field theory.

In the rest of this talk I want to discuss three post-war developments, each of which represented an extension or generalization of an important aspect of the reciprocity law.

A generalization of the group isomorphism aspect of reciprocity came about as a result of the introduction of the methods of group cohomology into class field theory, by Hochschild, Nakayama and Weil. Artin and I polished these methods in his seminars and, using a trick of Nakayama's, I showed that the cup-product with the fundamental class gives isomorphisms

$$H^{r-2}(G,\mathbb{Z}) \xrightarrow{\ \sim\ } H^r(G,C_L), \quad \text{for all } r \in \mathbb{Z},$$

where G is the Galois group of an arbitrary finite Galois extension L/K. Here the cohomology groups are those for a finite G, in which homology is interpreted as negative dimensional cohomology: $H_r(G,M) = H^{-1-r}(G,M)$ for

$r \geq 1$. Moreover, $H^0(G,M) = M^G/N_G M$ is the "reduced" H^0, consisting of fixed elements modulo norms. For G abelian, we have $H^{-2}(G,\mathbb{Z}) = H_1(G,\mathbb{Z}) = G$, and for $r = 0$, the isomorphism displayed above is just the reciprocity law

$$G \xrightarrow{\sim} C_K/N_{L/K}C_L.$$

Further developments by Nakayama and me led to a general duality theorem for the cohomology of infinite Galois groups like those of the maximal algebraic extension unramified outside a certain set of primes, and these Galois-cohomological results have been reinterpreted and generalized in terms of etale cohomology by M. Artin and B. Mazur.

When the ground field K contains the ℓth roots of unity, the explicit reciprocity law for ℓ^{th} power residues involves the norm residue symbols $(\frac{a,b}{P})_\ell$ for primes P dividing ℓ, just as the original quadratic reciprocity law involves the symbol $(\frac{p,q}{2})_2 = (-1)^{\frac{p-1}{2}\frac{q-1}{2}}$ for odd p and q, and also $(\frac{p,2}{2}) = (-1)^{\frac{p^2-1}{8}}$. In the 1960's a completely new light was shed on the norm residue symbol as a result of the development of algebraic K-theory. The norm residue symbol satisfies several formal identities, the most important of which are, in abbreviated notation, the following:

$$(a_1a_2,b) = (a_1,b)(a_2,b), \qquad (b,a) = (a,b)^{-1}, \qquad \text{and} \qquad (a,1-a) = 1.$$

Work of Steinberg on central extensions of algebraic groups led him to consider all functions of two non-zero variables in an arbitrary field with values in an abelian group satisfying the above identities, and such a function is called a Steinberg cocycle. Milnor defined for every associative ring A with 1 an abelian group K_2A, and interpreted Steinberg's results as showing that, when A is a commutative field the group K_2A is generated by elements (a,b), for $a,b \in A^*$, which do satisfy the above relations. Matsumoto proved that these relations were enough to define K_2A for a field A; in other words K_2A is the target group for the universal Steinberg cocycle. Suppose F_P is the completion of a number field F at a prime P, and let U_P denote the group of roots of unity in F_P. Excluding the case $F_P \approx \mathbb{C}$, the group U_P is finite, say of order m_P. C. Moore showed that the most general continuous Steinberg cocycle on F_P is the norm residue symbol $(\frac{a,b}{P})_{m_P}$, and that, viewing these as cocycles on F for varying P, the only relation among them was that coming from the mth power reciprocity law, where m is the order of the group of roots of unity U_F in F. This result can be expressed by means of an exact sequence,

$$0 \longrightarrow \mathrm{Ker}\ h \longrightarrow K_2F \xrightarrow{\ h\ } \bigoplus_{P \text{ non-complex}} U_P \longrightarrow U_F \longrightarrow 0,$$

where h is the map induced by the symbols $(\frac{a,b}{P})_{m_P}$. The new thing in this is that the order m_P of the local symbols considered varies with P; classically, one never considered symbols of different order simultaneously.

After Moore's determination of Coker h, it was natural to consider the structure of Ker h. Bass told me about the problem in case $F = \mathbb{Q}$, the rational field, and asked me if I could see any relation to classical number theory. I thought not, but a few days later found a proof that $\mathrm{Ker}\ h = 0$ for $F = \mathbb{Q}$, and soon realized that the method, an induction over the primes, using the euclidean algorithm, was just the formal part of Gauss' first proof of quadratic reciprocity; Gauss had determined the structure of $K_2\mathbb{Q}/2K_2\mathbb{Q}$ without realizing it! Using a result of Bass, Garland proved Ker h finite by showing K_2A finite, by analytic methods, where A is the ring of integers in F. Later, Quillen gave a definition of higher K-groups K_r which enabled him to prove a general finiteness theorem and to prove the exactness of a localization sequence which showed in particular that Ker h was the kernel of the map of K_2A onto its easily computable "tame part" (and not only the image of that kernel, as Bass had shown earlier).

These connections between the reciprocity law and cohomology theory and algebraic K-theory which we have just discussed are interesting and important, but I think the biggest problem after Artin's reciprocity law was to extend it in some way to non-abelian extensions, or, from an analytic point of view, to identify his non-abelian L-series. Artin conjectured that these functions are holomorphic in the whole s-plane except for a pole of known order at s = 1. He proved that they satisfied a functional equation and that a power of each was meromorphic. R. Brauer proved in the early 1940's that they were meromorphic. Only in recent years has there been further progress.

In 1967, Langlands was studying the analytic theory of automorphic forms on general reductive algebraic groups and saw a formal relation between Artin's L-series and some Euler products arising in the theory of Eisenstein series. This led him to some general conjectures, of which the following is a superspecial case:

Let $R: \mathrm{Gal}(L/\mathbb{Q}) \longrightarrow GL(2,\mathbb{C})$ be an irreducible representation of degree 2 of the Galois group of a finite Galois extension L of $\mathbb{Q}$. Let N be the conductor of R, as defined by Artin, and let $X: \mathbb{Z} \longrightarrow \mathbb{C}$ be the Dirichlet character such that for each prime number p unramified in L we have $X(p) = \det R(\mathrm{Frob}_{L/Q}(p))$. (The existence of X follows from the famous theorem of Kronecker-Weber which states that every abelian extension of $\mathbb{Q}$ is contained in a cyclotomic extension.) Suppose $X(-1) = -1$. Let the Artin L-series attached to R be

$$L(s,R) = \sum_{n\geq 1} a_n n^{-s},$$

and put

$$f_R(z) = \sum_{n\geq 1} a_n e^{2\pi inz} .$$

Then f_R should be a holomorphic "new-form" of weight 1, character X, and level N. In particular, f_R is holomorphic in the upper half-plane $\mathrm{Im}\, z > 0$, and should satisfy

$$f_R\left(\frac{az+b}{cz+d}\right) = (cz+d)X(d)f_R(z), \qquad \text{for } \begin{pmatrix} a & b \\ c & d \end{pmatrix} \in SL(2,\mathbb{Z}) \text{ and } c \equiv 0(N).$$

Moreover, every holomorphic new-form f of weight 1 should be of the form f_R for a Galois representation R as above.

In general, Langlands sees a pattern in which there should be such a statement for representations of any degree n of Galois groups over any global field K. For $n = 1$, and any K, the relationship envisaged by Langlands is just Artin's reciprocity law. For $n = 2$ and $K = \mathbb{Q}$, it is the conjecture above, if the representation R is "odd" in the sense that $X(-1) = -1$. (For "even" representations the picture involves non-holomorphic modular forms.) In fact, Langlands' vision goes farther, embracing arbitrary reductive algebraic groups, not only $GL(n)$, and the ℓ-adic representations of Galois groups arising from the cohomology of algebraic varieties as well as Artin representations. A. Weil had the idea in special cases, especially concerning the L-series attached to elliptic curves.

What is the situation regarding proofs of these conjectures? Almost nothing is known beyond $GL(2)$, and I will limit the discussion here to the case of $GL(2)$ and Artin representations. In 1969 Langlands proved that the function f_R above is a new form if and only if the L-series $L(s,RR_1)$ is entire (i.e., satisfies Artin's conjecture), for every representation

R_1 of degree 1 of Galois groups over $\mathbb{Q}$; in fact, he proved the analog of that statement for arbitrary Artin-representations of degree 2 over any global field. To do this he had "only" to show that the constant in the functional equation of $L(s,RR_1)$ behaved in a certain way as function of R_1, and then apply the generalization of Hecke's fundamental theorems relating Dirichlet series with functional equations to automorphic forms which had been developed by Jaquet-Langlands and, in a different form, by Weil. In 1973, Deligne and Serre proved that every holomorphic new form f of weight 1 was of the form f_R for an "odd" Galois representation R over $\mathbb{Q}$ of degree 2 as above. This result, which can be viewed as a degree 2 analog of the Kronecker-Weber theorem, is proved by methods which work only over $\mathbb{Q}$, not over an arbitrary K. Deligne had associated ℓ-adic representations to modular forms of weight ≥ 2. Starting from a form of weight 1, Serre considered for each prime ℓ the Deligne representation R_ℓ corresponding to a form f_ℓ of higher weight which was congruent to f (mod ℓ). Assuming the Petersson conjecture for f, Serre could then show that an infinity of these R_ℓ's were congruent (mod ℓ) to one fixed Artin representation R, whose L-series corresponded to f. Deligne then realized that Serre's construction of R would go through with a weak form of the Petersson conjecture which had been proved essentially by Rankin. This gave the theorem; and as a by-product, a tortuous proof of the full Petersson conjecture in weight 1, where it is <u>not</u> a consequence of the conjectures of Weil which were proved by Deligne.

Putting together the results of Langlands and Deligne-Serre we see that there is a one-to-one correspondence between "odd" Artin representations R over $\mathbb{Q}$ of degree 2 <u>such that</u> RR_1 <u>satisfies the Artin Conjecture for all</u> R_1 <u>of degree 1</u>, and holomorphic new forms <u>of weight 1</u>. If we try to find explicit examples of this correspondence, there is a catch on each side, indicated by the underlined phrase. It is easy to construct representations R, but in general we don't know they satisfy Artin's Conjecture. On the other hand, while it is relatively easy to construct modular forms of weight $k > 1$, and the Riemann-Roch theorem tells us exactly how many of them there are at each level, it is not so easy to exhibit forms of weight 1, and the Riemann-Roch formula fails to predict how many of them there are at a given level. Of course if the representation R is monomial, that is, is induced from a degree 1 representation R' over a quadratic extension K of $\mathbb{Q}$, then the Artin Conjecture is satisfied (because $L(s,R)$ is equal to $L(s,R')$, an ordinary abelian L-series for K) and consequently, by Langlands' result, f_R is a modular form. But this "dihedral" case was known to Hecke in 1926!

He constructed the f_R's by theta-functions belonging to binary quadratic forms - forms associated with ideals in the quadratic field K. At the time he wrote quite explicitly that the way to get modular forms is from arithmetic, and in particular, from L-series with a functional equation in which the Γ-factor, which in general is of the form

$$\Gamma(\tfrac{s}{2})^a \, \Gamma(\tfrac{s+1}{2})^b \, \Gamma(s)^c \ ,$$

is of the very simple form $\Gamma(s)$. He then listed all the cases of abelian L-series with such a Γ-factor, thus getting just the dihedral "odd" R's (besides the reducible "odd" R's which give Eisenstein series). Hecke wondered whether in this way he got all modular forms of weight 1, though he doubted this was so. Now, at the same time, in the same University, Hamburg, Artin was defining his non-abelian L-series $L(s,R)$ and proving they satisfied a functional equation, in which the Γ-factor was sometimes $\Gamma(s)$, namely, when R is "odd" of degree 2. Thus, if either had really understood what the other was doing, the above special conjecture might very well have been found by Artin and Hecke 40 years before Langlands hit on it. Perhaps mathematics wasn't ready. As Hilbert wrote in the introduction to his Mathematical Problems:

> "Wenn uns die Beantwortung eines mathematischen Problems nicht gelingen will, so liegt häufig der Grund darin, dass wir noch nicht den allgemeineren Gesichtspunkt erkannt haben, von dem aus das vorgelegte Problem nur als einzelnes Glied einer Kette verwandter Probleme erscheint."

The way Langlands found the conjecture seems a good example of what Hilbert had in mind.

Another reason for the relationship's eluding Artin and Hecke may be the fact that explicit non-dihedral numerical examples are hard to find. Indeed at the time of the DeKalb conference, none was known! I concluded the oral presentation of this paper there by explaining that, in the hope of finding one, I had looked for non-dihedral R's of low conductor, N, and had found an R with $N = 133 = 7 \cdot 19$ which I hoped might be amenable to computation. After the talk, Atkin suggested that the labor involved might be considerably reduced by systematic use of the involutions w_7 and w_{19}. Armed with his theory of the w's, four Harvard students, D. Flath, R. Kottwitz, J. Tunnell, J. Weisinger and I succeeded in the next month in proving (by relatively easy hand computation) the existence of the corresponding new form f_R of weight 1 and level 133 predicted by

Langlands. By the theorem of Deligne-Serre, this produced the first example of an Artin L-series which is known to be holomorphic in spite of the fact that no power of it is a product of abelian L-series.

Proceedings of Symposia in Pure Mathematics
Volume 28, 1976

HILBERT'S TENTH PROBLEM. DIOPHANTINE EQUATIONS: POSITIVE ASPECTS OF A NEGATIVE SOLUTION

Martin Davis[1], Yuri Matijasevič, and Julia Robinson

ABSTRACT

Applications (including the negative solution of Hilbert's tenth problem) and extensions are surveyed of the Main Theorem on Diophantine sets: Every listable (recursively enumerable) set is Diophantine. Key steps in the proof of the Main Theorem are outlined and applied to obtain prime representing polynomials, a universal Diophantine equation, and a sharp form of Gödel's incompleteness theorem. Many famous problems are reduced to the solvability of Diophantine equations. The number, size and effectiveness of solutions are discussed. Relationships are explored with the theory of algorithms (recursion theory), model theory, and algebraic number theory.

INTRODUCTION

Hilbert's tenth problem asked for an algorithm to test polynomial Diophantine equations for solvability in integers (or equivalently in the natural numbers $N = \{0,1,2,3,\ldots\}$). The development of the theory of algorithms (also called recursion theory or computability theory) since the 1930's suggested that the tenth problem might be solved negatively, i.e. that it might be shown that there is no algorithm of the kind sought. Work in this direction had its fruition in 1970 when Matijasevič [34] gave just such a negative solution. Full accounts of this work are available in the expository articles in various languages: Azra [4], Davis [13], Manin [32], and Matijasevič [36]. Other expository articles on the subject are: Fenstad [16], Havel [21], Hermes [22], Ruohnen [52].

The methods used to achieve the negative solution of Hilbert's tenth problem have their positive side as well, and it is this positive side that we propose to stress in this article. (The key negative results are summarized in this introductory section.) From the perspective suggested by these methods, a Diophantine equation is viewed primarily as defining a set or relation on the natural numbers, in a manner which is explained below. This leads to considering the class of Diophantine sets, consisting of all

AMS (MOS) subject classifications (1970). Primary 02F25, 02F50, 10B99, 10N05; Secondary 10B05, 10B10, 12L05.

[1]Supported by the National Science Foundation under Grant DCR71-02039A 03.

sets which can be so defined. The Main Theorem is that the Diophantine sets are precisely those sets which can be *listed* (again this is explained more precisely below) by means of some definite algorithms. The equivalence of these two notions, one arising from the theory of numbers, the other in the theory of algorithms (also known as recursive function theory and as computability theory) permits a quite surprising interplay between the two fields, and leads to an entirely novel collection of problems and results in the theory of Diophantine equations.

In this paper we survey some of the problems and results which arise from this viewpoint. Although familiarity with one of the expository papers on Hilbert's tenth problem cited above will be helpful, we have not in fact assumed such knowledge on the reader's part. Nor does this paper presuppose any knowledge of the theory of algorithms.

If $P(a_1,\ldots,a_\mu,z_1,\ldots,z_\nu)$ is a polynomial with integer coefficients, in the variables $a_1,\ldots,a_\mu,z_1,\ldots,z_\nu$ we may think of the Diophantine equation

$$P(a_1,\ldots,a_\mu,z_1,\ldots,z_\nu) = 0 \qquad (1)$$

as *defining* the set of μ-tuples $\langle a_1,\ldots,a_\mu\rangle$ of natural numbers for which (1) has a solution in natural numbers in the unknowns $z_1,\ldots,z_\nu$. (Here $a_1,\ldots a_\mu$ are the *parameters* of the equation.) Any set which can be defined in this way is called *Diophantine*.

We shall also have occasion to deal with *exponential Diophantine equations*, that is equations of the form (1) in which P is the sum of terms $c\alpha_1^{\beta_1}\alpha_2^{\beta_2}\cdots\alpha_k^{\beta_k}$ where c is an integer and where each α_i and each β_i is either a positive integer or one of the variables $a_1,\ldots,a_\mu,z_1,\ldots,z_\nu$. In such a case we speak of an *exponential Diophantine definition* of an *exponential Diophantine set*.

Now if S is the set of μ-tuples of natural numbers defined by (1), we can set up the following procedure for making a *list* of the members of S: Fix some definite enumeration of all $(\mu+\nu)$-tuples of natural numbers. (One such enumeration is defined later in this introduction.) Proceeding in order through the $(\mu+\nu)$-tuples, compute the value of P for each $(\mu+\nu)$-tuple; and whenever $P = 0$ for $\langle a_1,\ldots,a_\mu,z_1,\ldots,z_\nu\rangle$, place the μ-tuple $\langle a_1,\ldots,a_\mu\rangle$ on the list. Thus the members of S and only the members of S will be placed on the list (perhaps more than once).

A set S of μ-tuples of natural numbers is called *listable* (or *recursively enumerable)* if there is a well defined algorithm for making a list of precisely the members of S. Of course only finitely many members of S can be placed on the list in any finite number of steps. However what is crucial is that each member will *eventually* be placed on the list and that no nonmember will ever be placed on the list. So we have at once:

Every Diophantine set is listable.

Now while we will all agree that the method given above of listing a Diophantine set is an algorithm, we need a definition of algorithm to make our definition of listable set precise. Suitable definitions can be found in books on the theory of algorithms (e.g. Davis [9], Malcev [31], Markov [33], Rogers [51]). However readers of this paper who wish to learn the nature and scope of the results presented will manage perfectly well if they rely on the heuristic account just given.

The main result in the theory of Diophantine sets (hereafter called the Main Theorem) is (cf. Matijasevič [34]):

THEOREM. *Every listable set is Diophantine.*

Thus every set of μ-tuples of natural numbers which can be listed by means of any algorithm whatever can, in fact, be listed in the manner described above where P is a particular polynomial with μ parameters.

Actually the proof of this theorem is constructive and we will use this fact in what follows. Hence we restate the theorem indicating this.

MAIN THEOREM. *There is a procedure which can be applied to an arbitrary algorithm for lising a set* S *of* μ*-tuples of natural numbers to obtain a polynomial* P *with integer coefficients such that* $P(a_1,\ldots,a_\mu,x_1,\ldots,x_\nu) = 0$ *has a solution in natural numbers if and only if* $\langle a_1,\ldots,a_\mu\rangle \in S$.

A set $S \subseteq N$ is *computable* if there is an algorithm to determine in a finite number of steps whether an arbitrary natural number belongs to S. It is easy to see that if S is computable then both S and N-S are listable. Conversely, if both S and N-S are listable, then each number will occur on one or the other list eventually. So we can combine the algorithms for listing S and N-S into an algorithm for computing S, i.e. for telling whether an arbitrary natural number belongs to S.

A basic result in the theory of algorithms (see any of the books referred to above) is:

THEOREM. *There is a listable set* $K \subset N$ *which is not computable.*

It is this theorem which has made it possible to prove that various mathematical problems are algorithmically unsolvable. In §5, we will sketch the construction of such a K. Since K is listable, K is Diophantine. Let

$$P(a,z_1,\ldots,z_\nu) = 0 \qquad (2)$$

be a *Diophantine definition* of K. That is, (2) has a solution in natural numbers $z_1,\ldots,z_\nu$ for a given $a \in N$ if and only if $a \in K$. Then we have

COROLLARY. *There is no algorithm for telling whether* (2) *has a solution in* $z_1,\ldots,z_\nu$ *for given values of the parameter* a.

This corollary gives a negative solution of Hilbert's tenth problem in a strong form: Not only is there no algorithm for testing Diophantine equations in general for solvability, but there is not even such an algorithm for a particular one parameter family of Diophantine equations.

Notice that this result has little practical effect on the problem of finding which Diophantine equations have solutions. This is because we can only hope to find the answer for sufficiently simple equations in general. For example, a is composite if and only if

$$(x + 2)(y + 2) = a \tag{3}$$

has a solution. Now there is a trivial algorithm to determine whether (3) has a solution, since it is sufficient to try all values of x less than $\sqrt{a}$. But it is clear that this method cannot in fact be used to show that a is prime if a has say 50 digits. However the largest known prime $2^{19937} - 1$ is much larger than that and was found using a more sophisticated algorithm which also has its limitations [63]. Thus a given algorithm can be carried out only for a finite set of values of the parameters. Also we may be able to devise ad hoc methods for particular equations without a general algorithm. Thus the negative solution to Hilbert's tenth problem does not preclude our eventually being able to determine exactly which of the equations that can be written with one line of type have solutions.

By using a simple device due to Hilary Putnam [40], Diophantine sets of natural numbers can be defined in a particularly striking manner. Thus let S be the set of natural numbers a for which the equation $P(a,z_1,\ldots,z_\nu) = 0$, has a solution. Let

$$Q(x,z_1,\ldots,z_\nu) = (x+1)\{1 - P^2(x,z_1,\ldots,z_\nu)\} - 1 .$$

Then, S *consists precisely of the nonnegative values assumed by* Q *as* $x,z_1,\ldots,z_\nu$ *assume all values in* N. (For, $P(a,z_1,\ldots,z_\nu) = 0$ implies $Q(a,z_1,\ldots,z_\nu) = a$; moreover $Q(x,z_1,\ldots,z_\nu) = a \geq 0$ implies $1-P^2(x,z_1,\ldots z_\nu) > 0$ which implies $P(x,z_1,\ldots,z_\nu) = 0$ and $a = x$.) Since all the familiar sets of natural numbers which occur in the theory of numbers (such as the primes, the square-free numbers, or the perfect numbers) are listable and hence Diophantine, they can each be represented as the set of nonnegative values assumed by a polynomial.

It is important to remark that a Diophantine set can be defined by a simultaneous system of Diophantine equations just as well as by a single equation. This is because the system $P_1 = 0$, $P_2 = 0$, ..., $P_n = 0$ is equivalent to the single equation $P_1^2 + P_2^2 + \ldots + P_n^2 = 0$. (And, of course, the same holds for exponential Diophantine equations.) From the point of view of logic, this remark means that Diophantine definitions can be combined using the operation "&". The same is true for "v". Namely, the condition

$$P_1 = 0 \vee P_2 = 0 \vee \ldots \vee P_n = 0$$

is equivalent to the single equation

$$P_1 P_2 \cdots P_n = 0 .$$

We will have occasion to make use of Cantor's function:

$$J(x,y) = (1 + 2 + \ldots + (x+y)) + x$$
$$= ((x+y)^2 + 3x + y)/2 .$$

The equation $J(x,y) = a$ obviously has a unique solution $\langle x,y\rangle$ for each natural number a: x is determined as the excess of a over the largest "triangular" number which is $\leq$ a. Thus, the function $J(x,y)$ provides an enumeration of all ordered pairs ($J(x,y)$ is the number of the pair $\langle x,y\rangle$ in the enumeration). The unique solution of $J(x,y) = a$ is written $x = K(a)$,

$y = L(a)$, and the functions J, K, L are called *Cantor's pairing functions*.

Proceeding inductively, we define:

$$J_1(z_1) = z_1$$

$$J_{n+1}(z_1,\ldots,z_{n+1}) = J(J_n(z_1,\ldots,z_n),z_{n+1}) \ .$$

Then $J_2(z_1,z_2) = J(z_1,z_2)$, and generally the function J_n provides a one-one correspondence between the natural numbers and the n-tuples of natural numbers. It is important to notice that for each n, $J_n(z_1,\ldots,z_n)$ is a *polynomial with rational coefficients*. Hence if $J_n(z_1,\ldots,z_n)$ is substituted for a variable in a Diophantine equation, the result will also be a Diophantine equation.

The notions of computability and listability apply not only to μ-tuples of natural numbers, but also to any collection of objects to which algorithmic procedures can apply. What this turns out to mean is that we can speak of computability and listability in connection with objects which have natural representations as *expressions on an alphabet*, i.e. as finite sequences of a finite or countably infinite set of objects, called *symbols*. For example we can speak of a computable or listable set of strings on the symbols *a*, *b*, *c* (e.g. the set of such strings in which no *b* is immediately followed by *c* is computable and hence listable).

An instance (which is in some sense prototypical) is provided by *formalized mathematical theories*. Here the symbols are those needed for logical deductions ($\sim$ & $\vee$ $\rightarrow$ $\Leftrightarrow$ $\forall$ $\exists$ =, parentheses, and a supply of variables) and those needed to express the basic concepts of the part of mathematics being "formalized" [in the case of number theory, these symbols might be e.g. 0, S (for the successor function), +, $\cdot$, <]. Among all of the expressions which can be written using these symbols a certain computable subset is singled out as the *formulas* (or w.f.f.'s) of the theory [e.g. $(\forall x)(x \cdot x = x \rightarrow (x = 0 \vee x = S(0)))$]. And once again a particular computable set of these formulas is singled out as the *mathematical axioms* of the theory (e.g. Peano's axioms). It is clear of course that in order to use a set of axioms we must be able to check that an alleged axiom really is one -- hence the requirement that the axioms form a computable set is essential. Finally the *theorems* of the theory are the formulas obtained by beginning with the axioms and applying rules of logical deduction. Now what is important is that *the set of theorems of a formalized mathematical theory is listable* (though in general it may not be computable).

To see this it is necessary to consider one of the various (equivalent) formulations of the rules of logic. For our purposes it is sufficient to remark that there exist formulations of logic (cf. Rosser [51a]) which begin by specifying a computable set of formulas called the *logical axioms*. One then employs the rule of *detachment* (also called modus ponens) which proceeds from A and (A $\rightarrow$ B) to obtain B, where A and B are each formulas of the theory. Thus in this formulation the *theorems* of the theory are those formulas obtained from the (logical and mathematical) axioms by iteratively employing the rule of detachment.

Now it is easy to describe an algorithm for making a list of the theorems of a theory: We begin with an algorithm for listing the axioms (thus we only use the listability of the axioms although we have assumed that they are computable). The algorithm we are constructing operates alternatively in two phases. In Phase I it employs the axiom listing algorithm to place one more axiom on our list of theorems. In Phase II it searches through the list of theorems so far obtained for pairs to which detachment can be applied; finding such a pair it applies detachment and adjoins the result to the list. It is clear that all theorems and only theorems will eventually be placed on the list; hence the theorems are listable.

1. DIOPHANTINE REPRESENTATION OF THE SET OF PRIMES.

As we mentioned in the introduction, one consequence of the Main Theorem is that there exists a polynomial P with integer coefficients such that the nonnegative range of P over the natural numbers is the set of primes. This corollary was deduced by Putnam in 1960 from the then conjectured Main Theorem and it was considered by some to be an argument against its plausibility. The set of primes is probably the most popular set in number theory and the quest for prime representing functions is an old one. It is remarkable that until 1970 it was not known that there is a representation of such a simple form. To construct a Diophantine definition of the set of primes, we need only a part of the machinery necessary for an arbitrary listable set. (In particular we do not need the Bounded Quantifier Theorem of §4.) In fact the whole construction is purely number-theoretical, which suggests that it may be more informative than Diophantine definitions of other sets. However to avoid disappointment, we should warn the reader that no magic is involved. The definition is really a direct translation of a familiar fact! However now that we know there is a Diophantine equation, which furnishes a Diophantine definition of the primes, we can look forward to successively simpler equations being found.

Our first aim is to find an exponential Diophantine definition of primes. We give here in an improved form the original definition by Robinson [46]. Clearly, p is prime if and only if

$$p = s + 1 = r + 2 \tag{1}$$

$$q = s! \tag{2}$$

$$ap - bq = 1 \tag{3}$$

has a solution for a, b, s, r, q in $\mathbb{N}$ (we could use Wilson's theorem and replace bq by q). Hence we need only transform (2) into an exponential Diophantine equation.

Now

$$s! = \frac{t(t-1)\dots(t-s+1)}{\binom{t}{s}} \quad \text{for} \quad t \geq s .$$

Replacing the numerator by the exponential function t^s, we have

$$s! \leq s!(1 + \frac{1}{t-1}) \ \dots \ (1 + \frac{s-1}{t-s+1}) = \frac{t^s}{\binom{t}{s}} \rightarrow s!$$

as $t \rightarrow \infty$. It is easy to verify that

$$s! = \left[\frac{t^s}{\binom{t}{s}} \right]$$

provided $t \geq 2\, s^{s+2}$. (As usual $[\alpha]$ is the integer n such that $n \leq \alpha < n+1$.) Hence we can replace (2) by

$$t = 2s^{s+2} \tag{4}$$

$$t^s = qu + w \tag{5}$$

$$u = \binom{t}{s} \tag{6}$$

$$u = w + x + 1 \tag{7}$$

since (5) and (7) have a solution in x, w if and only if $q = [t^s/u]$. It remains to transform (6) into an exponential Diophantine equation.

The equation

$$(y + 1)^t = \sum_{i=0}^{t} \binom{t}{i} y^i \tag{8}$$

may be thought of as defining the binomial coefficients when it is considered an identity in y. Also if y is sufficiently large (greater than 2^t) then the binomial coefficients are just the digits in the y-ary expansion of $(y + 1)^t$. Thus the numbers $\binom{t}{i}$ are uniquely defined by saying that they are nonnegative integers satisfying (8) for at least one value of $y > 2^t$. So we can replace (6) by

$$y = 2^t + 1 \tag{9}$$

$$z = y + 1 \tag{10}$$

$$z^t = \ell y^{s+1} + uy^s + m \tag{11}$$

$$u + v = 2^t \tag{12}$$

$$m + n + 1 = y^s\ . \tag{13}$$

Thus, the system (1), (3)-(5), (7), (9)-(13) is an exponential Diophantine definition of the set of primes. (We can combine the system of equations into one equation as indicated in the Introduction.)

Notice that we have incidentally given an exponential Diophantine definition of the binomial coefficient. Indeed, for t and s not both 0,

$u = \binom{t}{s}$ if and only if (9)-(13) can be satisfied for $y, \ell, m, v, z, n \in N$.

Also we have an exponential Diophantine definition of $q = s!$. Namely,

$q = s!$ if and only if (4),(5),(7); (9)-(13) can be satisfied for $u, t, w, x, y, \ell, m, v, z, n \in N$.

To obtain a Diophantine definition of the set of primes and the relations $u = \binom{t}{s}$ and $q = s!$, it suffices now to obtain a Diophantine definition of the exponential function $a = b^c$. That is we need to find a polynomial A in three parameters and say ν unknowns such that

$$A(a,b,c,x_1,\ldots,x_\nu) = 0$$

has a solution if and only if $a = b^c$. It was shown in Robinson [46] that such an A exists provided that there is some Diophantine relation of exponential growth. The proof involved used elementary properties of the equation (with one parameter a):

$$x^2 + (a^2-1)y^2 = 1 \quad \text{for} \quad a > 0 \qquad (*)$$

As a specially simple Pell equation, much is known about the infinite set of its solutions. We will number the solutions in order of size $x = \chi_a(0)$, $y = \psi_a(0)$; $x = \chi_a(1)$, $y = \psi_a(1)$; ... . Matijasevič [34], [74], showed that the sequence of even Fibonacci numbers (which obviously grows exponentially) formed a Diophantine relation. However shortly afterwards several mathematicians [7], [11], [26] showed that the relation given by $y = \psi_a(k)$ is Diophantine. From this, the Diophantine definability of exponentiation follows directly. Here we will give a definition of $y = \psi_a(k)$ from Matijasevič and Robinson [39] which can be expressed with just 3 unknowns. Here and below we write e.g. $x = \square$, to mean: x is a perfect square.

THEOREM. $c = \psi_a(b)$ *if and only if the system*

$$DFI = \square \quad , \quad F \mid H-c \ , \quad b \leq c \qquad (a > 1)$$

has a solution in the unknowns i *and* j *where* D, F, H, I *are defined as follows:*

$$\begin{aligned} D &= (a^2-1)c^2 + 1 \\ E &= 2(i+1)Dc^2 \\ F &= (a^2-1)E^2 + 1 \\ G &= a + F(F-a) \\ H &= b + 2jG \\ I &= (G^2-1)H^2 + 1 \ . \end{aligned}$$

Finally, the easily obtained estimate for the nth solution of (*),

$$(2a-1)^n \leq \psi_a(n+1) \leq (2a)^n$$

enables us to define the exponential function quite simply in terms of ψ. Clearly, for $x > 0$,

$$\frac{\psi_{Mx}(n+1)}{\psi_M(n+1)} \to x^n \quad \text{as} \quad M \to \infty \ ,$$

and hence for sufficiently large M the equation $y = x^n$ is equivalent to

$$\left(y - \frac{\psi_{Mx}(n+1)}{\psi_M(n+1)}\right)^2 < \frac{1}{4} \ . \qquad (**)$$

The explicit construction of a Diophantine definition of $y = x^n$ with just 5 unknowns is described at length in [39].

In fact, if we are interested only in a Diophantine definition of primes then all we have to do is to find a good approximation to $t^s/\binom{t}{s}$. This can be done directly in terms of ψ and leads to a polynomial in 12 variables whose positive range is the set of primes. In [35], the first such polynomial published had 24 unknowns and was of the 37th degree; this was improved in the English translation to 21 variables and the 21st degree. H. Wada independently found a polynomial with 12 variables. J. P. Jones (cf. [25]) has been trying to minimize the number of symbols necessary to write down such a polynomial. His best result so far is 325 symbols in the following

THEOREM (Jones). *The set of primes is identical with the set of positive values assumed by the polynomial*

$$\begin{aligned}&(k+2)\{1-([wz+h+j-q]^2+[(gk+2g+k+1)\cdot(h+j)+h-z]^2+[16(k+1)^3\cdot(k+2)(n+1)^2\\&+1-f^2]^2+[2n+p+q+z-e]^2+[e^3\cdot(e+2)\cdot(a+1)^2+1-o^2]^2+[(a^2-1)y^2+1-x^2]^2\\&+[16r^2y^4\cdot(a^2-1)+1-u^2]^2+[((a+u^2\cdot(u^2-a))^2-1)\cdot(n+4dy)^2+1-(x+cu)^2]^2\\&+[(a^2-1)\ell^2+1-m^2]^2+[ai+k+1-\ell-i]^2+[n+\ell+v-y]^2+[p+\ell(a-n-1)\\&+b(2an+2a-n^2-2n-2)-m]^2+[q+y(a-p-1)+s\cdot(2ap+2a-p^2-2p-2)-x]^2\\&+[z+p\ell(a-p)+t(2ap-p^2-1)-pm]^2)\}\end{aligned}$$

as the variables range over natural numbers.

Can we learn anything about primes from the fact that they are Diophantine? Definitely yes see §6.

2. FAMOUS PROBLEMS

In this section we shall see that the Main Theorem not only implies that there is no algorithm for solving Hilbert's tenth problem, but also shows that the tenth problem is far more all-embracing than one would have suspected. That is, many famous problems, even some explicitly mentioned in Hilbert's address, can be transformed, by using the Main Theorem, into assertions to the effect that some particular Diophantine equation has no solution.

A simple case is Fermat's conjecture that the equation

$$x^n + y^n = z^n \tag{1}$$

has no positive integer solution with $n > 2$. Of course a positive solution of the tenth problem would have enabled us to settle each instance of (1) with $n = 3,4,\dots$. But (cf. §1) we can give explicitly a Diophantine definition of the exponential function, that is we can exhibit a polynomial $A(a,b,c,w_1,\dots,w_k)$ such that the Diophantine equation

$$A(a,b,c,w_1,\dots,w_k) = 0$$

is solvable in unknowns $w_1,\dots,w_k$ if and only if the parameters a, b, c satisfy the condition $a = b^c$. Thus equation (1) is solvable for the same values of n for which there are solutions of equation

$$A^2(p,x,n,w_1,\dots,w_k) + A^2(q,y,n,v_1,\dots,v_k) + A^2(p+q,z,n,u_1,\dots,u_k)=0 \tag{2}$$

Hence, an equivalent formulation of Fermat's conjecture is the assertion that the Diophantine equation

$$A^2(p,x+1,n+3,w_1,\ldots,w_k) + A^2(q,y+1,n+3,v_1,\ldots,v_k) + A^2(p+q,z,n+3,u_1,\ldots u_k) = 0$$

has no solution in nonnegative integers n, p, q, $u_1,\ldots,u_k$, $v_1,\ldots,v_k$, $w_1,\ldots,w_k$, x, y, z. Thus the validity not only of each special case, but of the entire conjecture could be established by a supposed algorithm for testing an arbitrary Diophantine equation.

We would easily write out (2) explicitly. In fact, Keijo Ruohnen [52] did this, based on the construction of A described in Matijasevič [34]. Ruohnen's equation (more precisely, system of three equations which could be easily combined into one equation) has 72 unknowns and occupies less than a page. Today implementing the ideas of Matijasevič and Robinson [39] where a simpler construction of A is given, one can construct (2) with fewer than a dozen unknowns.

More interesting and less evident is the situation with Goldbach's conjecture mentioned in the eighth problem. The conjecture states that every even integer greater than 2 is the sum of two prime numbers. Now we can construct a Diophantine equation

$$B(p,w_1,\ldots,w_k) = 0 \tag{3}$$

which is solvable in unknowns $w_1,\ldots,w_k$ if and only if p is prime. It is easy to check that the equation

$$(u+1)(1-B^2(p_1,w_1,\ldots,w_k) - B^2(p_2,w_1',\ldots,w_k') - (2u+4-p_1-p_2)^2 - t) = a \tag{4}$$

is solvable for all nonpositive a (with $u = 0$, $t = -a$, $p_1 = p_2 = 2$, values of $w_1,\ldots,w_k$, $w_1',\ldots,w_k'$ determined by the values of a solution of (3) with $p = 2$) and for those and only those positive a for which $2a+2$ is the sum of two primes (with $u = a-1$, $t = 0$, $p_1+p_2 = 2a+2$, values of $w_1,\ldots,w_k$, $w_1',\ldots,w_k'$ determined by the solutions of (3) with $p = p_1$ and $p = p_2$ correspondingly). Thus we can reformulate Goldbach's conjecture as the assertion that the polynomial on the left side of (4) represents every integer when the variables range over the nonnegative integers. Replacing each variable by the sum of squares of four new variables we obtain a new polynomial which represents every integer when the variables range over the integers if and only if the Goldbach conjecture is valid.

So far we have reduced Goldbach's conjecture to Diophantine equations, but the conjecture is still outside of the scope of the tenth problem, since Hilbert asked for treatment of single equations, not infinite systems of them. Nevertheless we can reduce Goldbach's conjecture to the case of a single Diophantine equation. Before dealing with the conjecture itself, we consider this possibility in a general setting.

If we have an algorithm to tell in a finite number of steps whether an arbitrary natural number has a particular property P (written P(n)), then we can reduce the assertion that every natural number has the property P to the unsolvability of a particular Diophantine equation and give instructions for writing it down. Namely, from the given algorithms, the Main Theorem, and

Putnam's device mentioned in the Introduction, we can find polynomials R and S such that

$$P(n) \leftrightarrow (\exists w_1 \ldots w_k)[R(w_1,\ldots,w_k) = n]; \sim P(n) \leftrightarrow (\exists z_1 \ldots z_m)[S(z_1,\ldots z_m)=n].$$

Then the problem under discussion can be reformulated as the assertion that the Diophantine equation

$$R(w_1,\ldots,w_k) = n$$

is solvable for each natural number n or equivalently as the assertion that the Diophantine equation

$$S(z_1,\ldots,z_m) = z_0$$

has no solution at all.

Now returning to the Goldbach conjecture, we can easily give an algorithm for testing whether $2n+4$ is the sum of two primes and hence Goldbach's problem can be formulated not only in the form (4) but also as the assertion that some single Diophantine equation has no solution. Thus a positive solution of the tenth problem would have enabled us (in principle) to settle the Goldbach conjecture.

Many other famous problems even some remote from number theory can be formulated in the form $(\forall n)P(n)$ with a suitable decidable P. For example, the classical four color conjecture in the theory of graphs states that every map on a plane or sphere can be colored in 4 colors in such a manner that no two adjacent countries would be colored the same. (For exact definitions and numerous equivalent formulations, cf. Saaty [53].) It is well known that the four color conjecture is equivalent to the statement that every positive integer n has the property: "every map with fewer than n countries can be properly colored in four colors." Since this property is obviously decidable, we see that there is some particular Diophantine equation whose unsolvability is equivalent to the four color conjecture. In §4 we show how such an equation can be explicitly obtained.

Another example is the problem of the consistency of a given axiomatic system mentioned in general form in Hilbert's second problem. A formalized mathematical theory (cf. the Introduction) is called *consistent* if there is no formula R such that R and $\sim$R are both theorems. We have already seen that the set of theorems of a formalized mathematical theory is a listable set. Now for a given formalized mathematical theory consider the following property P of a natural number n: no contradiction arises when n steps of some definite theorem-listing algorithm (e.g. that outlined in the Introduction) for the theory are carried out. Clearly, we have an algorithm for P: we need only scan a finite list for a pair of formulas, R, $\sim$R. Since the assertion that the theory is consistent is just the statement that every positive integer has property P, the consistency of the given formalized mathematical theory is also equivalent to the assertion that some definite Diophantine equation has no solutions.

Hilbert's first problem is Cantor's continuum hypothesis. The difficulties of settling this problem led to consideration of the questions: is the continuum hypothesis consistent with the traditional Zermelo-Fraenkel axioms of set theory? is its negation? By our remarks, each of these questions (in fact both were answered affirmatively by Gödel [19] and Cohen [6] respectively) is equivalent to the unsolvability of a specific Diophantine equation.

We mention in passing that various so called large cardinal axioms studied by contemporary set theorists imply the consistency of certain sets of axioms. Thus from the existence of a *strongly inaccessible cardinal* it is known that the consistency of Zermelo-Fraenkel set theory follows. Hence there is a Diophantine equation such that the existence of a strongly inaccessible cardinal implies that the equation has no solution, whereas it cannot be proved from the Zermelo-Fraenkel axioms alone that the equation has no solution.

Now we proceed to consider in some detail another famous problem, namely Riemann's conjecture about the ζ function mentioned in Hilbert's eighth problem. It is not as evident as before that the conjecture can be formulated in the desired form $\forall n\ P(n)$. Nevertheless it has long been realized that this is the case. (Kreisel outlined a proof in [28] based on his earlier work on computable estimates for analytic functions.)

Riemann's function $\zeta(s)$ is defined by

$$\zeta(s) = \sum_{n=1}^{\infty} n^{-s} .$$

The series converges for all s with $\mathrm{Re}(s) > 1$ to an analytic function which can be extended to the entire complex plane except for $s = 1$ where ζ has its unique pole. Points $-2,-4,-6,\ldots$ are known to be zeros of ζ. All other zeros are called *nontrivial*. They are known to lie inside the so-called critical strip $0 < \mathrm{Re}(s) < 1$ symmetrically about the line $\mathrm{Re}(s) = 1/2$. The Riemann hypothesis states that all the nontrivial zeros in fact lie on the line $\mathrm{Re}(s) = 1/2$. By symmetry, the Riemann hypothesis is equivalent to the statement that $\zeta(s)$ has no zeros for $\frac{1}{2} < \mathrm{Re}(s) < 1$.

To see heuristically that the Riemann hypothesis has the desired form, note that we can express the region $\frac{1}{2} < \mathrm{Re}(s) < 1$ as the union of countably many bounded regions each with rectilinear boundary. The number of zeros of $\zeta(s)$ in each of the regions is given by the integral of $\zeta'(s)/2\pi i\zeta(s)$ taken about its boundary (if there is a zero on one of the boundaries, we can replace the region by a neighboring one). So the Riemann hypothesis is equivalent to the statement that for each such bounded region the corresponding integral is less than (say) $\frac{1}{2}$. To see whether the n-th region has this property is basically an elementary exercise in numerical analysis, finding appropriate rational approximations for $\zeta'(s)/2\pi i\zeta(s)$ and for its integral. However the details are cumbersome and we propose to use a more indirect method which relies on the classical theory of the distribution of primes.

We introduce the computable function, defined for positive integers:

$$\delta(x) = \prod_{n<x} \prod_{j\le n} \eta(j)$$

where $\eta(j) = 1$ unless j is a prime power, and $\eta(p^k) = p$. Then, we have:

THEOREM. *The Riemann hypothesis is equivalent to the assertion that*

$$\left(\sum_{k\le\delta(n)} \frac{1}{k} - \frac{n^2}{2} \right)^2 < 36\, n^3 \quad \textit{for} \quad n = 1,2,3,\ldots \,. \tag{5}$$

(As the reader will readily verify, the constant 36 can easily be improved.)

Obviously the condition (5) is a decidable property of n. And, it would be quite a simple matter, using the methods of proof of the Main Theorem, to obtain explicitly a Diophantine equation which has no solutions, just in case (5) always holds.

In proving the theorem,[2] we make use of the following facts.

(a) $\frac{\Gamma'(3/2)}{\Gamma(3/2)} = 2 - \gamma - \log 4$, where $\gamma = .577\ldots$ is Euler's constant. (Cf. Magnus, Oberhettingher, & Soni [30], p. 15.)

(b) Letting R be the set of nontrivial zeros of $\zeta(s)$,

$$\frac{\zeta'(s)}{\zeta(s)} = \log 2\pi - 1 - \frac{\gamma}{2} - \frac{1}{s-1} - \frac{1}{2}\, \frac{\Gamma'(\frac{s}{2}+1)}{\Gamma(\frac{s}{2}+1)} + \sum_{\rho\in R} \left(\frac{1}{s-\rho} + \frac{1}{\rho}\right) .$$

$$\lim_{s\to1} \left(\frac{\zeta'(s)}{\zeta(s)} + \frac{1}{s-1}\right) = \gamma.$$ (Cf. Landau [29], pp.165, 316-317.)

(c) For positive real x, we write

$$\psi(x) = \sum_{p^k\le x} \log p = \sum_{j\le x} \log \eta(j)$$

$$\psi_1(x) = \int_1^x \psi(u)\, du ,$$

so that for x integral, $\psi(x) \le \sum_{j\le x} \log x = x \log x \le x^{3/2}$ and $\psi_1(x) = \log \delta(x)$. Then for $x\ge1$,

$$\psi_1(x) = \frac{x^2}{2} - \sum_{\rho\in R} \frac{x^{\rho+1}}{\rho(\rho+1)} - x\, \frac{\zeta'(0)}{\zeta(0)} + \frac{\zeta'(-1)}{\zeta(-1)} - \sum_{r=1}^{\infty} \frac{x^{1-2r}}{2r(r-1)} .$$

(Cf. Ingham [24], p. 73.)

(d) $\zeta'(0)/\zeta(0) = \log 2\pi$, $\zeta'(-1)/\zeta(-1) = 1.985\ldots$. (Cf. Walther [65], p. 400.)

(e) For n a positive integer,

$$\left(\sum_{k\le n} \frac{1}{k} \right) - 1 < \log n < \sum_{k\le n} \frac{1}{k} ,$$

as is readily seen on estimating $\log n = \int_1^n du/u$. We have:

[2] We are grateful to Harold N. Shapiro who suggested this approach and provided most of the technical details.

LEMMA.

$$\sum_{\rho\in R} \frac{1}{\rho(1-\rho)} = \gamma + 2 - \log 4\pi .$$

Proof: Using (a) and (b),

$$\sum_{\rho\in R} \frac{1}{\rho(1-\rho)} = \sum_{\rho\in R} \left(\frac{1}{1-\rho} + \frac{1}{\rho}\right)$$

$$= \lim_{s\to 1} \left(\frac{\zeta'(s)}{\zeta(s)} + \frac{1}{s-1}\right) - \log 2\pi + 1 + \frac{\gamma}{2} + \frac{1}{2}\frac{\Gamma'(3/2)}{\Gamma(3/2)}$$

$$= \gamma + 2 - \log 4\pi .$$

Now, *assume the Riemann hypothesis*; i.e. $\rho \in R$ implies $\mathrm{Re}(\rho) = 1/2$. Then for $\rho \in R$, $\bar{\rho} = 1-\rho$. Hence, using the lemma,

$$\sum_{\rho\in R} \frac{1}{|\rho|\cdot|\rho+1|} \le \sum_{\rho\in R} \frac{1}{|\rho|^2} = \sum_{\rho\in R} \frac{1}{\rho(1-\rho)} = \gamma + 2 - \log 4\pi .$$

We can use this estimate in the expression for $\psi_1(x)$ given in (c), noting that the Riemann hypothesis implies that for $x \ge 1$, $\rho \in R$, we have $|x^{\rho+1}| = x^{3/2}$, so that

$$|\psi_1(x) - \frac{x^2}{2}| \le x^{3/2}\left[\sum_{\rho\in R} \frac{1}{|\rho|\cdot|\rho+1|} + \left|\frac{\zeta'(0)}{\zeta(0)}\right| + \left|\frac{\zeta'(-1)}{\zeta(-1)}\right| + \sum_{r=1}^{\infty} \frac{1}{2r(2r-1)}\right]$$

Here,

$$\sum_{r=1}^{\infty} \frac{1}{2r(2r-1)} = \sum_{r=1}^{\infty} \left(\frac{1}{2r-1} - \frac{1}{2r}\right) = \sum_{n=1}^{\infty} \frac{(-1)^{n+1}}{n} = \log 2 ,$$

so using (d),

$$|\psi_1(x) - \frac{x^2}{2}| \le x^{3/2}\left[\gamma + 2 - \log 4\pi + \log 2\pi + 2 + \log 2\right]$$

$$= x^{3/2}[4 + \gamma]$$

$$< 5x^{3/2}$$

But by (e)

$$\left|\sum_{k\le\delta(x)} \frac{1}{k} - \psi_1(x)\right| < 1 .$$

Hence

$$\left|\sum_{k\le\delta(x)} \frac{1}{k} - \frac{x^2}{2}\right| < 1 + 5x^{3/2} \le 6x^{3/2} .$$

Squaring now gives (5).

Conversely, suppose that (5) holds for all positive integers. Then by (e),

$$|\psi_1(x) - \frac{x^2}{2}| < 1 + 6x^{3/2}$$

for $x = 1,2,3,\ldots$. Thus, for any *real* $x \geq 1$,

$$|\psi_1(x) - \frac{x^2}{2}| \leq \int_{[x]}^{x} \psi(u)\,du + |\psi_1([x]) - \frac{[x]^2}{2}| + |\frac{[x]^2-x^2}{2}|$$

$$< \psi([x]) + 1 + 6x^{3/2} + x \leq 9x^{3/2} ,$$

using the inequality in (c). Now (cf. Ingham [24], p. 18) for $\mathrm{Re}(s) > 1$,

$$-\frac{\zeta'(s)}{\zeta(s)} = s\int_1^\infty \frac{\psi(x)}{x^{s+1}}\,dx$$

$$= s\int_1^\infty \frac{d\psi_1(x)}{x^{s+1}}$$

$$= s(s+1)\int_1^\infty \frac{\psi_1(x)}{x^{s+2}}\,dx .$$

Hence,

$$-\frac{\zeta'(s)}{\zeta(s)} - \frac{s(s+1)}{2(s-1)} = s(s+1)\int_1^\infty \frac{\psi_1(x) - \frac{x^2}{2}}{x^{s+2}}\,dx \tag{6}$$

So, assuming our inequality,

$$\left|\frac{\psi_1(x) - \frac{x^2}{2}}{x^{s+2}}\right| \leq \frac{9}{\left|x^{s+\frac{1}{2}}\right|}$$

Hence, the integral in (6) converges uniformly on the closed region $\mathrm{Re}(s) \geq \sigma_0$ for any $\sigma_0 > 1/2$, (6) must be valid for $1/2 < \mathrm{Re}(s)$, and the function on the left is analytic in this domain. But this implies that $\zeta(s)$ does not vanish for $1/2 < \mathrm{Re}(s) < 1$, i.e. that the Riemann hypothesis holds.

Of course the Main Theorem does not permit us to reduce every famous problem to the unsolvability of a Diophantine equation. For example, let us consider the twin prime conjecture:

$$(\forall n)\{(\exists p)[p>n \;\&\; p \text{ prime} \;\&\; p+2 \text{ prime}]\} . \tag{7}$$

Note that the set of n's satisfying the condition in braces is certainly a computable set. This set is either the set of all natural numbers or it is finite. Nevertheless we cannot reduce the twin prime conjecture to the unsolvability of a particular Diophantine equation because we don't know which of these alternatives holds and therefore don't know a specific algorithm for deciding whether n satisfies the condition. However if we strengthen the conjecture:

$$(\forall n)\{(\exists p)[n+4<p<2^{n+4} \;\&\; p \text{ prime} \;\&\; p+2 \text{ prime}]\} , \tag{8}$$

then we will have an algorithm to determine whether n satisfies the condition in braces. Therefore this strong twin prime conjecture can be transformed into the unsolvability of a particular Diophantine equation. Number theorists certainly believe that both the twin prime conjecture and this stronger version are true.

The most intriguing question concerning such transformations of problems into the unsolvability of particular Diophantine equations is the one of usefulness: Is there any mathematical advantage to be gained by such a translation or does it represent a mere curiosity? One of the charms of mathematics is the constant discovery of unexpected almost unbelievable connections. Whatever is logically possible may be true! Diophantine sets as a class have been studied for only 25 years. Their richness has surprised experts and perhaps their usefulness will also.

Of course not much is to be expected from the routine uniform translation of a famous problem into its equivalent Diophantine equation. An approach which may be more promising is to look for decidable conditions sufficient to imply that a given Diophantine equation has no solution and then try to verify (possibly with the aid of a computer) the validity of one of these conditions for the equations under consideration. The simplest condition of this kind is the unsolvability of the equation modulo p. However our present method of obtaining a Diophantine definition of a listable set uses as an intermediate step inequalities $P \leq Q$ which are then replaced by $P + x = Q$ where x is a new unknown. But the corresponding congruences $P+x \equiv Q \pmod{p}$ always have a solution for x. Thus passing to the equation modulo p eliminates the condition $P \leq Q$ and hence could not lead to a proof of unsolvability of an equation corresponding to a serious problem. This suggests that we should look for new proofs of the Main Theorem in the hope of obtaining equations which would not automatically have solutions modulo p for every p. See [47].

Another possibility is that eventually Diophantine equations and definitions may be classified and studied so that the reduction of a problem to an equation of a particular type may in fact yield additional information. For example sets defined with polynomial test functions (§9) may have certain combinatorial or density properties which then would follow for any set so defined. Of course this can only be expected to work if the translation is simpler than it would be in the general case.

Finally, the translation of a theorem of the appropriate form in some part of mathematics shows that the corresponding Diophantine equation has no solution. Hence whatever methods went into proving the theorem can in fact be used to show that a particular Diophantine equation has no solution. It is possible that the same methods can be used to show that a class of equations including perhaps an equation of interest in itself are unsolvable. Such an example providing a new tool for solving Diophantine equations would be a considerable breakthrough. In any case, any mathematical method that has been used to prove a theorem of the appropriate form has in fact been used to show that a particular Diophantine equation has no solution. Thus all mathematical methods can be tools in the theory of Diophantine equations and perhaps we should consciously attempt to exploit them.

3. EFFECTIVENESS AND EQUATIONS WITH FINITELY MANY SOLUTIONS

The negative solution of Hilbert's tenth problem has given impetus to the treatment of other arithmetical problems from the algorithmic point of view. In this section, we deal with the problem of effectiveness in solving equations in integers.

Suppose that we have somehow proved that the Diophantine equation

$$P(x_1,\dots,x_n) = 0$$

has at most finitely many solutions. Depending on the nature of our proof, we can have very different knowledge about the actual solutions. In the luckiest case we have the list of all the solutions; but if reductio ad absurdum has been used then often we have nothing more than a deduction of the contradiction.

As a classical example of the latter situation we can point to the famous theorem by Thue according to which a Diophantine equation of the form

$$F(x,y) = a \tag{1}$$

has at most finitely many solutions where F is an irreducible form of degree at least 3. The original proof by Thue gives no way to find all the solutions and it is only recently, exploiting new ideas, that Baker [5] has given another proof enabling us to calculate (in principle) all the solutions of (1) for a given a.

The methods of the theory of algorithms enable us to construct very easily a computable function ϕ such that for every value of the parameter a,

$$\phi(a,x_1,\dots,x_n) = 0 \tag{2}$$

has at most a finite number of solutions, but there exists no computable function τ such that

$$\phi(a,x_1,\dots,x_n) = 0 \Rightarrow x_1 + \dots + x_n \leq \tau(a) . \tag{3}$$

Thus there is no uniform procedure which can be used given a, to calculate the list of all the corresponding solutions of (2).

The straightforward method for constructing such a ϕ leads to a very complicated function. The question arises: How simple can a function ϕ with the above properties be? For example, could equation (2) be a Diophantine equation? Baker's result shows that ϕ cannot be of the form $F(x_1,x_2) - a$ with F as above. On the other hand it is shown in Matijasevič [37] that (2) can actually be taken to be an exponential Diophantine equation.

More precisely, it was shown that we can effectively find a polynomial A with integer coefficients such that

(i) for every natural number value of the parameter a the equation

$$A(a,x_1,\dots,x_n) = y + 4^y \tag{4}$$

has at most one solution in natural numbers $x_1,\ldots,x_n,y$; hence for every a there is an integer t_a such that equation (4) implies:

$$x_1 + \ldots + x_n + y < t_a ,$$

(ii) for each computable function τ there are natural numbers $a,x_1,\ldots,x_n,y$ such that (4) holds but $x_1 + \ldots + x_n + y > \tau(a)$.

In a sense, the negative answer to Hilbert's tenth problem justifies the past treatment of individual Diophantine equations by ad hoc methods (instead of developing a general theory which is impossible). In a similar sense the above result justifies our proving the finiteness of the number of solutions of an exponential Diophantine equation by reductio ad absurdum and other noneffective methods; these methods may well be the best available to us because universal methods for giving all the solutions do not exist.

Let us outline the main ideas in constructing the polynomial A. Suppose (as can readily be arranged) that (2) has at most one solution for every a, but still there is no computable τ satisfying (3). Then there is no procedure for determining given a whether (2) is solvable. In fact, otherwise we could define τ in the following way: apply the procedure to find if there is a solution; if there isn't, then put $\tau(a) = 0$, if there is, then look over in succession all the n-tuples until finding one satisfying (2); then put $\tau(a) = x_1 + \ldots + x_n$.

On the other hand, if (2) has for every a at most one solution and there is no procedure to determine whether it is solvable then there is no computable τ satisfying (3). This is clear since there are only finitely many n-tuples satisfying $x_1 + \ldots + x_n \leq \tau(a)$ and only such an n-tuple could be the solution.

The situation here is rather like the situation in Thue's proof. His method enables us to find a bound for all the solutions of (3) as soon as we find one solution which is sufficiently (in an appropriate sense) large. But until Baker's result, no method was known for determining whether such a solution exists.

Now our purpose is to find an algorithmically unsolvable one-parameter equation which never has more than a single solution. We start from a one-parameter Diophantine equation

$$P(a,z_1,\ldots,z_k) = 0 \tag{5}$$

for which there is no procedure for determining given a whether there is a solution in natural numbers $z_1,\ldots,z_k$ (cf. the Introduction). We shall construct an exponential Diophantine equation among whose unknowns will be $z_1,\ldots,z_k$. While both equations will have solutions for the same values of a, the new equation will have at most one solution for given a, and indeed one in which $z_1,\ldots,z_k$ will form the least solution of (5) while the other unknowns can easily be expressed in terms of $z_1,\ldots,z_k$. By "the least solution" we mean the least one in Cantor's enumeration of all the k-tuples, i.e. the one for which $J_k(z_1,\ldots,z_k)$ is least (cf. the Introduction).

The k-tuple $\langle z_1,\ldots,z_k\rangle$ is the least solution if and only if it actually is a solution and for each value of w less than $J_k(z_1,\ldots,z_k)$ the following equation has a solution.

$$(J_k(v_1,\ldots,v_k)-w)^2 + (P^2(a,v_1,\ldots,v_k)-u-1)^2 = 0 .$$

Note that $v_1,\ldots,v_k,u$ are uniquely determined by w.

A method for reducing the problem of the solvability of a Diophantine equation for all values of one of the variables (in this case w) not exceeding some bound to the problem of solving an exponential Diophantine equation has been known since 1961 (cf. Davis, Putnam, Robinson [14]): several improvements of this original method are known today (cf. §4), but they all lead to exponential Diophantine equations which have either infinitely many solutions or no solutions. Nevertheless, the original method can easily be modified to give an exponential Diophantine equation having at most one solution provided that the original Diophantine equation has at most one solution for each value of the variable w up to the bound imposed on it.

Obtaining the desired equation of the special form (4) is now a matter of easy transformations involving the introduction of new unknowns. For details see Matijasevič [37].

The above result suggests that it is worthwhile to introduce the notion of *singlefold* (Russian: odnokratno) *representation*. Namely, we say that an existential representation

$$(\exists x_1 \ldots x_n)\ R(a_1,\ldots,a_m,x_1,\ldots,x_n)$$

for some predicate $P(a_1,\ldots,a_m)$ is a singlefold representation if not only

$$P(a_1,\ldots,a_m) \leftrightarrow (\exists x_1 \ldots x_n)\ R(a_1,\ldots,a_m,x_1,\ldots,x_n) \qquad (6)$$

but also

$$P(a_1,\ldots,a_m) \leftrightarrow (\exists!\ x_1\ldots x_n)\ R(a_1,\ldots,a_m,x_1,\ldots,x_n) \qquad (7)$$

where $(\exists!\ x_1\ldots x_n)$ means as usual "there is a unique n-tuple $x_1,\ldots,x_n$ such that". Note that (7) does not imply (6) and that $(\exists x_1 x_2)\ R(a,x_1,x_2)$ being singlefold implies $(\exists x_1)[(\exists x_2)\ R(a,x_1,x_2)]$ being singlefold but not conversely.

In this terminology the main result of Matijasevič [37] is that every listable predicate has a singlefold exponential Diophantine representation. It would be very interesting to extend this result to singlefold Diophantine representations and thus to show that for (2) one could take a Diophantine equation. To do this it would suffice to show that the relation $a = b^c$ (even with fixed $b \geq 2$) has a singlefold Diophantine representation and then apply the main result of Matijasevič [37]. Unfortunately, all of the known Diophantine definitions of $a = b^c$ are infinitefold, and there is no obvious way to modify any of them into a singlefold or even into a finitefold representation.

OPEN PROBLEM. *Is there a finitefold (or better a singlefold) Diophantine definition of* $a = b^c$?

An application of such a definition is mentioned in §13.

In this connection we wish to draw attention to a result which preceded Matijasevič's discovery of the existence of Diophantine definitions of $a = b^c$. It was shown in Davis [10] that a Diophantine representation for $a = 2^c$ could be constructed if the equation

$$9(u^2+7v^2)^2 - 7(r^2+7s^2)^2 = 2 \tag{8}$$

has no solution except for the trivial solution $u = r = 1$, $v = s = 0$. The conjecture about the uniqueness of the solution of (8) was refuted by Herrmann in [23]. However this does not spoil the approach entirely: indeed it can be shown that a *singlefold* Diophantine representation of $a = 2^c$ can be constructed if the equation (8) has only a *finite number* of solutions.

We conclude this section by suggesting that a particular arithmetical problem be considered from the algorithmic point of view. Thue's theorem mentioned above was obtained by him as a consequence of the following theorem on Diophantine approximation: for every algebraic number θ of degree n the inequality

$$\left|\theta - \frac{p}{q}\right| < \frac{1}{q^\nu} \,. \tag{9}$$

has at most finitely many solutions in integers p, q provided $\nu > n/2 + 1$. This was a sharpening (for the case $n > 2$) of a similar result by Liouville with $\nu > n$. Thue's result was in turn sharpened by various authors. The latest and the best possible result (due to Roth) is for the case $\nu > 2$. Thue's proof and the proofs of all the improvements suffer from noneffectiveness -- they give no way to find all the solutions. It may actually be the case that there is no algorithm given θ and $\nu > 2$ for finding all the solutions of (9). If so, it seems very difficult to prove the fact.

4. THE BOUNDED QUANTIFIER THEOREM

In 1931 Gödel [18] revolutionized mathematical logic when he showed that no system of axioms is sufficient to decide all statements of number theory correctly (see §5). In the course of the proof, he needed an arithmetically definable way of representing arbitrary finite sequences of natural numbers. Gödel's elementary solution of this problem using the Chinese remainder theorem is a cornerstone of the negative solution of Hilbert's tenth problem. We shall state the result here in an explicit form needed later.

GÖDEL'S LEMMA. *For every* a *and sequence* $a_1, \ldots, a_n$ *with each* $a_i < a$ *there is a uniquely determined* b *such that*

$$b < \prod_{i=1}^{n} (1 + n!ai)$$

and

$$a_i = \text{Rem}(b,\ 1+n!ai) \quad \textit{for} \quad i = 1,\dots,n.$$

(Here Rem(b,c) is the least nonnegative remainder when b is divided by c.)

Now suppose $P(a,z_1,\dots,z_\nu) = 0$ defines a Diophantine set S whose complement N-S is not Diophantine. Then

$$a \in N\text{-}S \leftrightarrow (\forall k)(\exists z_0 \dots z_\nu)[k = J_\nu(z_1,\dots,z_\nu) \ \& \ P(a,z_1,\dots,z_\nu)^2 = 1+z_0].$$

Here J_ν is the ν-tuple enumerating function defined in the Introduction. Thus, there is a polynomial Q such that the set T defined by

$$a \in T \leftrightarrow (\forall k)(\exists z_0 \dots z_\nu)[Q(a,k,z_0,\dots,z_\nu) = 0]$$

is not Diophantine i.e. not listable. However we shall show that if we put a bound x on k, we get a Diophantine relation $a \in T_x$ with

$$a \in T_x \leftrightarrow (\forall k)_{\le x}(\exists z_0 \dots z_\nu)[Q(a,k,z_0,\dots,z_\nu) = 0] \qquad (1)$$

whatever polynomial Q we start with. Note that a bound on k implies a corresponding bound on $z_0,\dots,z_\nu$. Thus,

$$a \in T_x \leftrightarrow (\exists y)(\forall k)_{\le x}(\exists z_0 \dots z_\nu)_{\le y}[Q(a,k,z_0,\dots,z_\nu)=0] \quad ,$$

since if (1) is satisfied for a given x there are only a finite number of values of the z's needed to verify it.

We shall prove a slightly more general theorem in which the polynomial may depend on the bounds (see Davis, Putnam, and Robinson [14] for an earlier and less satisfactory version of this theorem).

BOUNDED QUANTIFIER THEOREM. *Let* $P(x,y,k,z_1,\dots,z_\nu)$ *be a polynomial with* x, y, *and* k *among its parameters and* $z_1,\dots,z_\nu$ *its variables. Then*

$$(\forall k)_{\le x}(\exists z_1 \dots z_\nu)_{\le y} [P(x,y,k,z_1,\dots,z_\nu) = 0]$$

$$\leftrightarrow (\exists b_1 \dots b_\nu)\left[\binom{b_1}{y+1} \equiv \dots \equiv \binom{b_\nu}{y+1} \equiv P(x,y,Q!-1,b_1,\dots,b_\nu) \equiv 0 \left(\text{mod} \binom{Q!-1}{x+1}\right)\right] \qquad (2)$$

where Q *is a polynomial such that*

$$Q = Q(x,y) > |P(x,y,k,z_1,\dots,z_\nu)| + 2x + y + 1 \qquad (3)$$

for all $k \le x$, $z_1 \le y$, ..., $z_\nu \le y$. *Also* $b_1,\dots,b_\nu$ *may be chosen all* $< \binom{Q!-1}{x+1}$.

Note. If P has additional parameters not listed then they will also occur in Q.

Remark. Since $x = y!$ and $x = \binom{y}{z}$ are Diophantine as we saw in §1, the relation defined by (2) is also Diophantine. Hence working from the inside out, we can eliminate the bounded universal quantifiers from an arithmetical definition in which all the universal quantifiers are bounded and obtain an equivalent Diophantine definition. This theorem is one of the steps in the proof of the Main Theorem.

PROOF. Looking first at the consequences of (3), we see that

$$\binom{Q!-1}{x+1} = (Q!-1)\left(\frac{Q!}{2}-1\right)\cdots\left(\frac{Q!}{x+1}-1\right) \tag{4}$$

where the factors on the right are all integers since $Q > x+1$. All primes $\leq Q$ divide $Q!/(k+1)$ for $k \leq x$ since $Q \geq 2x+2$. Hence a prime which divides one of the factors is $> Q$. If a prime divides both $(Q!/(i+1))-1$ and $(Q!/(j+1))-1$ then it divides $|i-j|$ which is $\leq x < Q$. Hence $i = j$; thus the factors in (4) are pairwise relatively prime. Finally, let p_k be a prime which divides $(Q!/(k+1))-1$. Then

$$Q!-1 \equiv k \pmod{p_k} \tag{5}$$

so

$$P(x,y,Q!-1,b_1,\ldots,b_\nu) \equiv P(x,y,k,\mathrm{Rem}(b_1,p_k),\ldots,\mathrm{Rem}(b_\nu,p_k)) \pmod{p_k}, \quad k \leq x. \tag{6}$$

(This part of the proof parallels the proof of Gödel's Lemma.)

Suppose the right side of (2) holds. Then $p_k | b(b-1)\ \ldots\ (b-y)$ so $\mathrm{Rem}(b,p_k) \leq y$ for $b = b_1,\ldots,b_\nu$. Hence by (3) the absolute value of the right side of (6) is less than Q and hence less than p_k. Also the left side of (6) is $\equiv 0 \pmod{p_k}$ by hypothesis. Hence

$$P(x,y,k,\mathrm{Rem}(b_1,p_k),\ldots,\mathrm{Rem}(b_\nu,p_k)) = 0$$

and the first half of the theorem holds.

Conversely, suppose there are $z_{1k} \leq y,\ \ldots,\ z_{\nu k} \leq y$ such that

$$P(x,y,k,z_{1k},\ldots,z_{\nu k}) = 0 \quad \text{for} \quad k \leq x\,. \tag{7}$$

Then we can find $b_i < \binom{Q!-1}{x+1}$ for $i = 1,\ldots,\nu$ satisfying the system of congruences

$$b_i \equiv z_{ik} \left(\mathrm{mod}\ \frac{Q!}{k+1} - 1\right) \quad \text{for} \quad k \leq x\,, \tag{8}$$

by the Chinese remainder theorem since the moduli are pairwise relatively prime. Then

$$\frac{Q!}{k+1} - 1 \;\Big|\; b_i(b_i-1)\ \ldots\ (b_i-y) \quad \text{for} \quad i = 1,\ldots,\nu\,, \tag{9}$$

since $z_{ik} \leq y$. Furthermore since the divisors in (9) are pairwise relatively prime, their product $\binom{Q!-1}{x+1}$ divides the right side. Also since all the primes dividing the left side of (9) are greater than Q which is greater than $y+1$, we obtain

$$\binom{Q!-1}{x+1} \;\Big|\; \binom{b_i}{y+1} \quad \text{for} \quad i = 1,\ldots,\nu. \tag{10}$$

Finally, by (7) and (8),

$$P(x,y,Q!-1,b_1,\ldots,b_\nu) \equiv P(x,y,k,z_{1k},\ldots,z_{\nu k}) \quad (\text{mod } \frac{Q!}{k+1} - 1). \tag{11}$$

Since the right side of (11) is zero and the moduli are pairwise relatively prime, we have

$$\binom{Q!-1}{x+1} \Big| \ P(x,y,Q!-1,b_1,\ldots,b_\nu)$$

and the second half of the theorem is proved.

Example. To show the usefulness of the bounded quantifier theorem, we will construct a Diophantine definition of the set S of superpowers, i.e.

$$\{1,2^2,3^{3^3},\ldots\} .$$

Notice that $m \in S$ if and only if there is a sequence $t_0,t_1,\ldots,t_n$ such that $t_0 = 1$, $t_{k+1} = n^{t_k}$ for $k \leq n$, and $t_n = m$. Hence by Gödel's lemma,

$$m \in S \leftrightarrow (\exists\, n,b,d)\ (\forall k)_{\leq n}\ [\text{Rem}(b,\ 1+d) = 1$$

$$\& \ \text{Rem}(b,\ 1+(k+2)d) = n^{\text{Rem}(b,1+(k+1)d)} \tag{12}$$

$$\& \ \text{Rem}(b,\ 1+(n+1)d) = m] .$$

(Here we could take $d = (1+n^m)(n+1)!$ and $b < \prod_{i=1}^{n+2} (1+id)$ by the explicit form of Gödel's lemma stated above.) Since the relations given by $x^y = z$ (cf. §1) and $\text{Rem}(x,y) = z$ are Diophantine, we can transform (12) into the form

$$m \in S \leftrightarrow (\exists\, n,a,d)\ (\forall k)_{\leq n}\ (\exists z_1 \ldots z_\nu)\ [P(m,n,a,d,k,z_1,\ldots,z_\nu) = 0]. \tag{13}$$

Finally by the bounded quantifier theorem, the right side of (13) can be shown to be Diophantine and the set of superpowers is Diophantine.

As another application of the bounded quantifier theorem and Gödel's lemma, we will show explicitly how to reduce the four color hypothesis (already briefly discussed in §2) to the unsolvability of a particular Diophantine equation. In fact we shall construct a Diophantine equation $F = 0$ with one parameter a which has a solution for given $a > 0$ just in case there exists a plane map which *cannot* be colored in a colors.

To begin with, it is sufficient to consider maps contained in a bounded portion of the plane which has been marked off into a lattice of unit squares and to assume that the regions consist of connected sets of these squares. We can then further reduce the problem by coloring just the centers of the squares which we take to be points with integer coordinates. Note that two of these points in the same region can be joined by a lattice path (with adjacent points one unit apart) lying within the region. The conditions on a coloring then become:

(a) If $\langle x,y\rangle$ and $\langle u,v\rangle$ are in the same region, they must have the same color;

(b) If <x,y> and <x+1,y> are in different regions, they must have different colors;

(c) If <x,y> and <x,y+1> are in different regions, they must have different colors.

In particular, it is enough to consider maps contained in $T_0, T_1, \ldots$ where T_n is the triangle given by $0 \leq x+y \leq n$. Note that T_n consists of all points x,y such that $J(x,y) \leq Q$ where

$$Q = J(n,0) = (n^2+3n)/2 . \tag{14}$$

Now any sequence $t_0, \ldots, t_Q$ of natural numbers can be considered a coloring of T_n by associating different colors with different natural numbers where t_k is the number of the color of the point x,y with $J(x,y) = k$. It is important to note that such a coloring determines the corresponding map since two points are in the same region if and only if there is a lattice-path between them with all of its points the same color. Hence two sequences $t_0, \ldots, t_Q$ and $u_0, \ldots, u_Q$ are *colorings of the same map* exactly if adjacent points in one sequence have the same color if and only if they do in the other. That is, if and only if,

$$\begin{aligned} t_{J(x,y)} = t_{J(x+1,y)} &\leftrightarrow u_{J(x,y)} = u_{J(x+1,y)} \\ t_{J(x,y)} = t_{J(x,y+1)} &\leftrightarrow u_{J(x,y)} = u_{J(x,y+1)} \end{aligned} \tag{*}$$

for all x,y with $x+y < n$.

Let $t_0, \ldots, t_Q$ be a sequence corresponding to a coloring of some map in T_n. Now suppose that for every sequence $u_0, \ldots, u_Q$ with $u_i < a$, the conditions (*) are not all satisfied. This would mean that the map of which $t_0, \ldots, t_Q$ is a coloring cannot be colored with a colors. It is this proposition that we will translate into a Diophantine equation with one parameter a.

By Gödel's lemma, given a coloring of T_n, $t_0, \ldots, t_Q$, there are s and t such that

$$t_i = \mathrm{Rem}(t, 1+s(i+1)) \quad \text{for} \quad i = 0, \ldots, Q . \tag{15}$$

Furthermore every sequence $u_0, \ldots, u_Q$ with each $u_i < a$ can be represented by some $u \leq R$ such that

$$u_i = \mathrm{Rem}(u, 1+(Q+1)!\,a(i+1)) \quad \text{for} \quad i = 0, \ldots, Q . \tag{16}$$

where

$$R = (1+(Q+2)!a)^{Q+1}. \tag{17}$$

Hence there is a map which cannot be colored in a colors if and only if

$$(\exists n,t,s)\ (\forall u \le R)\ (\exists x,y)\ [(x+y \le n\ \&\ u_{J(x,y)} \ge a) \vee$$
$$\{x+y\ n\ \&\ [(t_{J(x,y)} = t_{J(x+1,y)}\ \&\ u_{J(x,y)} \ne u_{J(x+1,y)})$$
$$\vee\ (t_{J(x,y)} \ne t_{J(x+1,y)}\ \&\ u_{J(x,y)} = u_{J(x+1,y)})$$
$$\vee\ (t_{J(x,y)} = t_{J(x,y+1)}\ \&\ u_{J(x,y)} \ne u_{J(x,y+1)})$$
$$\vee\ (t_{J(x,y)} \ne t_{J(x,y+1)}\ \&\ u_{J(x,y)} = u_{J(x,y+1)})]\}].$$

Hence using (14)-(17), Diophantine definitions of exponentiation and factorial, and the bounded quantifier theorem, we can obtain a Diophantine equation

$$F(a,n,t,s,\ldots) = 0 \tag{18}$$

which has a solution only if there is a map in T_n which cannot be colored with a colors.

Thus, the situation for (18) is as follows: for $a \ge 5$ (18) has no solution (since every map can be colored using 5 colors); for $0 < a \le 3$ (18) is solvable (since there are maps which require at least 4 colors); *for* $a = 4$ *the four color hypothesis is equivalent to* (18) *having no solutions*. Note that, e.g. for $a = 5$, the fact that the Diophantine equation (18) has no solution was proved using, in effect, topological methods.

5. THE UNIVERSAL EQUATION

In 1936, Turing [64] proved a fundamental result in the theory of algorithms: there exists an all encompassing algorithm. In Turing's development this meant a universal Turing machine, that is a single Turing machine that could be programmed to carry out any calculation that any Turing machine could do. In our case, we obtain:

UNIVERSAL EQUATION THEOREM. *There is a Diophantine equation* $U(a,n,x_1,\ldots,x_\nu) = 0$ *such that to every Diophantine set* $\mathcal{D}$ *there is an* n *with:*

$$a \in \mathcal{D} \leftrightarrow (\exists x_1 \ldots x_\nu)\ [U(a,n,x_1,\ldots,x_\nu) = 0]\ .$$

(Below we write $P_n(a,x_1,\ldots,x_\nu)$ for $U(a,n,x_1,\ldots,x_\nu)$.)

Although this theorem is an immediate consequence of the Main Theorem in the context of the theory of algorithms, we will give a direct proof of it without relying on the theory of algorithms. In fact we will sketch the construction of the polynomial U in such a way that given a Diophantine definition of $\mathcal{D}$, we can calculate a corresonding n.

We have treated Diophantine equations as equations of the form $P = 0$ where P is a polynomial with integer coefficients. Another way of looking at Diophantine equations is to consder them to be equations between terms built up from $0,1,x_0,x_1,\ldots$ by repeated additions and multiplications. Clearly, every Diophantine equation without parameters is equivalent to such an equation. There is a convenient way of numbering equations in this form. We first number the terms as follows:

$$\begin{aligned}
\tau_0 &= 0 ,\\
\tau_1 &= 1 ,\\
\tau_{3k+2} &= x_k ,\\
\tau_{3k+3} &= \tau_{K(k)} + \tau_{L(k)} ,\\
\tau_{3k+4} &= \tau_{K(k)} \cdot \tau_{L(k)} , \quad \text{for } k = 0,1,\ldots,
\end{aligned} \tag{1}$$

where K and L are the Cantor pairing functions defined in the Introduction. Let the nth Diophantine equation $\mathcal{E}_n$ be $\tau_{K(n)} = \tau_{L(n)}$. In order to number the Diophantine sets, we will consider x_0 to be a parameter and define $\mathcal{D}_n$ to be the set of values of x_0 for which $\mathcal{D}_n$ has a solution. Thus $m \in \mathcal{D}_n$ if and only if there is a sequence $t_0,t_1,\ldots,t_n$ of natural numbers such that

$$\begin{aligned}
&t_0 = 0,\ t_1 = 1,\ t_2 = m,\ t_{3k+3} = t_{K(k)} + t_{L(k)} ,\\
&t_{3k+4} = t_{K(k)} \cdot t_{L(k)} ,\quad t_{K(n)} = t_{L(n)} \quad \text{for } k \leq n .
\end{aligned} \tag{2}$$

In fact these conditions yield

$$t_{K(n)}(m,t_5,t_8,\ldots) = t_{L(n)}(m,t_5,t_8,\ldots)$$

where no restriction is put on $t_5,t_8,\ldots$. Hence any solution of

$$\tau_{K(n)}(m,x_1,\ldots) = \tau_{L(n)}(m,x_1,\ldots)$$

is represented by some sequence satisfying the conditions above.

Now we use Gödel's lemma to represent the sequence $t_0,\ldots,t_n$. That is, there are a and d so that

$$t_\ell = \mathrm{Rem}(a,\ 1+(\ell+1)d) .$$

Hence $m \in \mathcal{D}_n$ if and only if there are a, d, q, r so that for all $k \leq n$ the system

$$\begin{aligned}
&k = J(i,j),\\
&n = J(q,r),\\
&\mathrm{Rem}(a,1+d) = 0,\\
&\mathrm{Rem}(a,1+2d) = 1,\\
&\mathrm{Rem}(a,1+3d) = m,\\
&\mathrm{Rem}(a,1+(3k+4)d) = \mathrm{Rem}(a,1+(i+1)d) + \mathrm{Rem}(a,1+(j+1)d),\\
&\mathrm{Rem}(a,1+(3k+5)d) = \mathrm{Rem}(a,1+(i+1)d) \cdot \mathrm{Rem}(a,1+(j+1)d),\\
&\mathrm{Rem}(a,1+(q+1)d) = \mathrm{Rem}(a,1+(r+1)d),
\end{aligned} \tag{3}$$

can be satisfied. Finally since (3) is Diophantine we can replace it by a single equation, say

$$P(m,n,a,d,q,r,k,i,j,z_1,\ldots,z_\nu) = 0 . \tag{4}$$

Since the unknowns used to define $\mathrm{Rem}(a,b) = c$ are $\leq a$, we can apply the bounded quantifier theorem with $x = n$ and $y = n + a$ together with the definitions of the binomial coefficient and factorial from §1 to obtain a universal equation $U(m,n,x_1,\ldots,x_\mu) = 0$. The proof sketched here uses methods developed to prove the Main Theorem and of course was not known until after the Main Theorem was proved. Given an algorithm for listing a set M, we can calculate a particular natural number n_0 so that

$$m \in M \leftrightarrow U(m,n_0,x_1,\ldots,x_\mu) = 0 .$$

For example, M could be the set of primes, the set of superpowers, the set of Gödel numbers of the theorems of set theory, the set of numbers of solvable Diophantine equations, etc. (This last can also be defined by considering m as an unknown in the universal equation.)

Let K be the Diophantine set $\{n: U(n,n,x_1,\ldots,x_\mu) = 0$ has a solution$\}$. Then N-K is not Diophantine. The proof is by contradiction. Suppose $N - K = \{a: U(a,k,x_1,\ldots,x_\mu) = 0$ has a solution$\}$ for some k. Then

$$k \in N - K \leftrightarrow U(k,k,x_1,\ldots,x_\mu) = 0 \text{ has a solution } \leftrightarrow k \in K ,$$

which is a contradiction. Thus, *there is a Diophantine set whose complement is not Diophantine.* Also, in view of the Main Theorem K is a listable set which is not computable. Another example of a listable set which is not computable is the set of numbers of all solvable Diophantine equations. (If this last set were computable, Hilbert's tenth problem would be decidable.)

We will now derive a variant of Gödel's incompleteness theorem. In the first place, we have already sketched a proof (cf. the Introduction) that the set of theorems of a formalized mathematical theory is listable. Moreover as we have just observed the set of unsolvable Diophantine equations is not. Thus there is no formalized mathematical theory which can determine correctly the solvability of Diophantine equations. Roughly, Gödel started with an axiomatization of number theory and then exhibited a statement of number theory which was true by construction but not provable from the given axioms. We can now take as the unprovable proposition the unsolvability of a Diophantine equation which is constructed in such a way as to be unsolvable in natural numbers. (Cf. also the remarks on notation in formalized mathematical theories in the Introduction.)

GÖDEL'S INCOMPLETENESS THEOREM. *Let* A *be a system of axioms in a language including the mathematical symbols* 0 , S, +, $\cdot$, <, *and satisfying*

(a) A *is consistent (*i.e. *it is not possible to prove both* ϕ *and* $\sim\phi$ *for any sentence* ϕ*);*

(b) A *is listable;*

(c) A *is sufficiently strong to prove every true statement of the forms*

$$\alpha + \beta = \gamma$$
$$\alpha \cdot \beta = \gamma$$
$$\alpha < \beta$$

where α, β, γ *are among* 0, S0, SS0, ... *and* S *is the successor function.*

Then we can construct a Diophantine equation $F(x_1,\ldots,x_\nu) = 0$ *corresponding to* A *such that* $F = 0$ *has no solutions in natural numbers but we cannot derive*

$$\sim (\exists x_1 \ldots x_\nu)\ [F(x_1,\ldots,x_\nu) = 0] \qquad (5)$$

from A.

REMARK. Any system of axioms satisfying (c) is sufficient to prove that a Diophantine equation solvable in natural numbers has a solution. For example, the system of identities

$$\begin{aligned} x + 0 &= x \\ x + S(y) &= S(x + y) \\ x \cdot 0 &= 0 \\ x \cdot S(y) &= x \cdot y + x \end{aligned} \qquad (\mathcal{D})$$

is sufficient to prove the addition and multiplication tables for 0, $S0$, $SS0$, $SSS0$, ...

Note also that if any sentence of the form (5) is even *consistent* with all the statements in (c) (or with $\mathcal{D}$) then it is true.

PROOF: Let $A = \{a: \sim(\exists x_1 \ldots)\ [P_a(a,\ldots) = 0]$ is provable from the axioms $A\}$. Then, since the theorems are listable, A is listable and hence A is $\mathcal{D}_k$ for some k. Hence

$k \in \mathcal{D}_k \leftrightarrow$ The equation $P_k(k,x_1,\ldots) = 0$ has a solution,

$k \in \mathcal{D}_k \leftrightarrow$ The formula which expresses $\sim(\exists x_1 \ldots)[P_k(k,x_1,\ldots)=0]$ is provable from the axioms A.

Now if $k \in \mathcal{D}_k$ we have a contradiction derivable from A since if $P_k(k,x_1,\ldots) = 0$ has a solution in N by (c) we can prove in A that it has a solution. But A is consistent by (a) so we must have $k \notin \mathcal{D}_k$. This means that $P_k(k,x_1,\ldots) = 0$ has no solution in N but we cannot prove from the axioms A that it has no solution.

REMARK. Notice that if the axioms A are true statements of number theory then A is incomplete. However, if A is an axiomatization of the field of real numbers, then A is complete but gives the wrong answer for $P_k(k,x_1,\ldots) = 0$. If we replace (c) by the stronger condition that A is an extension of Peano's axioms, then A would be incomplete with respect to unsolvability of Diophantine equations.

6. REDUCTIONS

We say a Diophantine equation or definition can be *reduced* to another of a simpler form in some sense if there is an equivalent equation in the simpler form, that is an equation which has a solution for exactly the same values of the parameters. If α is the degree of a universal equation $U(m,n,z_1,\ldots,z_\nu) = 0$ in $m,z_1,\ldots,z_\nu$ then every Diophantine set can be defined by an equation of degree α with ν unknowns. It was this surprising consequence of the Main Theorem that led many mathematicians to believe prior to 1970 that not all listable sets were Diophantine. In §5, we have given a proof of the universal equation theorem independent of the Main

Theorem, but using methods of proof used in proving the Main Theorem.

REDUCTION OF DEGREE (Skolem [58]). *Every Diophantine equation can be reduced to an equation of degree* ≤ 4.

PROOF: Skolem observed that a Diophantine equation can be reduced to a system of equations of the form

$$\begin{aligned} \alpha + \beta &= \gamma \\ \alpha \cdot \beta &= \gamma \end{aligned} \tag{1}$$

by first writing the equation in the form $A = B$ where A and B are polynomials with positive integer coefficients and then introducing new unknowns for the partial terms needed to build up A and B. Since the equations (1) are of at most degree 2, they can be combined into one equation of degree ≤ 4 by replacing

$$A_i = B_i , \qquad i = 1,\ldots,k$$

by

$$\sum_{i=1}^{k} (A_i - B_i)^2 = 0 .$$

In particular, there is a universal equation of degree 4.

REDUCTION OF THE NUMBER OF UNKNOWNS. (Matijasevič and Robinson [39]). *Given a Diophantine equation* $P(a_1,\ldots,a_\mu,z_1,\ldots,z_\nu) = 0$ *there is a Diophantine equation* $Q(a_1,\ldots,a_\mu,x_1,\ldots,x_{13}) = 0$ *which has a solution for exactly the same* μ*-tuples* $a_1,\ldots,a_\mu$.

COROLLARY. *There is a universal Diophantine equation* $U(a,n,x_1,\ldots,x_{13}) = 0$.

The corollary follows because we already know there is some universal equation. J. P. Jones (cf. [25]) is currently studying pairs (δ,ν) such that there is a universal equation of degree δ in ν unknowns, and has given the following examples: (4,153); (6,129); (8,108); (10,107); (20,86); (44,83); (1952,80). These results show how with present methods we can reduce the degree at the expense of the number of unknowns and conversely. We can think of the pair (δ,ν) as a measure of the complexity of a given Diophantine equation.

The equation $U(J_\mu(a_1,\ldots,a_\mu),x_1,\ldots,x_{13}) = 0$ (where J_μ is as defined in the Introduction) is clearly universal for Diophantine sets of μ-tuples. Note that its degree in the parameters increases to ∞ as $\mu \to \infty$.

OPEN PROBLEM. *Are there absolute constants* δ,ν *such that for every* μ *there is a universal Diophantine equation for Diophantine sets of* μ*-tuples of total (parameters and variables) degree* δ *and in* ν *unknowns?*

Most likely, the answer is negative.

R. M. Robinson [50] proposed considering systems of Diophantine equations of the following sort. Let $W_{j,k}$ be the problem of telling whether a system of equations in $j+k$ variables $u_1,\ldots,u_j,v_1,\ldots,v_k$ is solvable in natural numbers, with the agreement that $u_1,\ldots,u_j$ must have the same values in all the equations, but $v_1, \ldots, v_k$ may have different values in different equations. Viewed as an ordinary system of Diophantine equations, $W_{j,k}$ would have $j+\lambda k$ unknowns where λ is the number of equations in the system. He pointed out that the decision problem for both $W_{n,0}$ and $W_{0,n}$ reduced to that of a single Diophantine equation in n unknowns. This is true for $W_{n,0}$ since a system of equations can be reduced to a single equation. In the case of $W_{0,n}$, a decision method for Diophantine equations could be applied separately to each equation of the system. He proved that there was no decision method for $W_{2,3}$ or $W_{1,4}$. Later Matijasevič [38] improved this to $W_{2,2}$ and $W_{1,3}$. These results were all obtained by showing that an arbitrary Diophantine equation could be reduced to the appropriate $W_{i,j}$.

Another measure of the complexity of a Diophantine equation is the total number of additions and multiplications necessary to verify a solution. We may get a smaller number if we write the equation as an equivalent system of equations so we will allow this. Now what we are really interested in is the complexity of a Diophantine definition of a particular set so we can search for the simplest (or at least simpler) definitions. Here there are many options. For example, we may treat constants, parameters, and unknowns alike. Thus, the operations $x+1$ and $x \cdot y$ would each contribute 1 to the total. Another possibility is to count only operations on two parameters or unknowns since these will generally be large compared to the coefficients of the equation or system. For expediency we might consider that multiplication is so time consuming compared to addition that we count only multiplications of parameters and unknowns. The choice of measure and the resulting minimal definitions is part of the theory of the complexity of computation.

A particularly interesting set in the theory of numbers is the set of prime numbers. If a number c is composite there exists a one-step proof of the fact consisting of multiplying two factors together and obtaining c. The work of searching for the factors is considered scratch work and is thrown away. It is not part of the proof. Now the number of steps (divisions) in the trivial algorithm for proving a number p is prime is $[\sqrt{p}]$. Until the set of primes was shown to be Diophantine, it was not known that *there is a proof that* p *is prime consisting of a bounded number of additions and multiplications*. Given some Diophantine definition of the set of primes, the proof that a prime p is a prime consists in substituting suitable values of the unknowns in the defining equation and verifying that it is a solution. Finding the solution is again considered scratch work to be discarded.

By using a universal equation to define an arbitrary listable set, we obtain a proof that a given number belongs to the set in which the number of steps is independent of the size of the number and of the particular set given.

For example, suppose we number the statements in the language of set theory and let T be the set of numbers of the theorems of some axiom system.

Then as we have seen T is a listable set. Thus, we can find a Diophantine definition of T and calculate its number t_0. Then a statement ϕ with number f is a theorem of set theory if and only if $U(f,t_0,x_1,\ldots,x_{13}) = 0$ has a solution in N. Hence if ϕ is in fact a theorem, then there is a metamathematical proof of ϕ whose length is a constant plus the length of the calculation of f. Hence the length of this proof depends only on the length of ϕ. [The usual way of numbering statements is to number each symbol in the language. Then the number of the statement is some code for the finite sequence of numbers corresponding to the sequence of symbols in the statement.]

Another example is obtained by letting $\mathcal{D}_{p_0}$ be the Diophantine set of numbers of theorems in the theory of Peano's axioms. Then if f is the number of a statement of number theory ϕ we see that ϕ is a consequence of Peano's axioms if and only if

$$(\exists x_1 \ldots x_{13}) \; [U(f,p_0,x_1,\ldots,x_{13}) = 0]$$

is a consequence of the axioms $\mathcal{D}$ in §5. So that even though it is a theorem of Ryll-Nardzewski that Peano arithmetic cannot be finitely axiomatized, there is a complete image of it contained in the fragment derivable from the usual recursive definitions of + and $\cdot$. There are finite axiomatizations of stronger theories than Peano's arithmetic (see for example Robinson [48]).

Another kind of reduction is that carried out by Adler [3]:

REDUCTION OF HOMOGENEOUS EQUATIONS. *Every homogeneous Diophantine equation* $P(x_1,\ldots,x_\nu) = 0$ *can be reduced effectively to a system of homogeneous quadratic equations* $Q_1(y_1,\ldots,y_\lambda) = 0, \ldots, Q_k(y_1,\ldots,y_\lambda) = 0.$

Hence every homogeneous Diophantine equation is equivalent to a single homogeneous quartic equation.

These are especially interesting because an outstanding:

OPEN PROBLEM. *Is there an algorithm to tell whether an arbitrary Diophantine equation has a solution in rational numbers?*

It can be shown that this problem is equivalent to the decision problem for nontrivial solutions in integers of homogeneous Diophantine equations.

Another interesting result of Adler is that an arbitrary Diophantine equation can be reduced to an equation

$$F = 1$$

where F is a homogeneous polynomial of degree 4.

7. THE BOUNDARY BETWEEN THE DECIDABLE AND THE UNDECIDABLE

From what we have stated in the previous section, we see that no algorithm can exist for testing even Diophantine equations in 13 unknowns for solvability in natural numbers. On the other hand it is obvious that an algorithm exists for 1 unknown. For 2 unknowns, a decision method has been given by Baker and his collaborators for a wide class of equations but not yet for the general case and for 3 unknowns, almost nothing is known. Thus the problem of locating the precise boundary between the decidable and the undecidable, with respect to the number of unknowns, remains very much an open question.

From the point of view of the total degree of the Diophantine equation, we are somewhat better off. An algorithm is known (cf. Siegel [57]) for Diophantine equations of degree 2 in arbitrarily many unknowns.

There are two directions in which Siegel's result could be generalized. First, to single equations of a fixed degree. The unsolvability of Hilbert's tenth problem together with Skolem's reduction (see §6) shows that there is no algorithm for testing the solvability of equations of degree 4.

OPEN QUESTION. *Is Hilbert's tenth problem decidable for equations of degree 3?*

Second, the generalization could extend Siegel's result to systems of equations of degree 2. But since every Diophantine equation can be put in the form of a system of equations of degree 2 (cf. §6), we see that there is an N such that there is no decision method for systems of N equations. Today the gap between Siegel's result and the least known value of N is very great.

For yet another form of the problem, we can obtain a sharp answer. It is shown in Matijasevič and Robinson [39] that every Diophantine equation is equivalent to a system of equations of the form

$$[\sigma_1(x,y)] = [\sigma_2(x,y)]$$

where the σ's are rational functions with rational coefficients of x and y. Hence there is no algorithm for telling whether such a system of equations has a solution in N. On the other hand, Straus [61] has given a method for telling whether an arbitrary system of equations of the form

$$[\tau_1(x)] = [\tau_2(x)]$$

where the τ's are rational functions with rational coefficients is solvable in N. Namely, consider first one such equation. Then,

$$\tau_i(x) = P_i(x) + \gamma_i + \delta_i(x) , \qquad i = 1,2$$

where P_i is a polynomial with rational coefficients and without a constant term, γ_i is a rational number, and $\delta_i(x)$ is a rational function of negative degree. Then

$$|P_1(x)-P_2(x)| \to \infty \text{ as } x \to \infty$$

$$a + b = c \leftrightarrow (\exists w)((w \dot{-} a) \dot{-} b = w \dot{-} c \ \& \ 1 \dot{-} (w \dot{-} c) = 0)$$

$$a \cdot b = c \leftrightarrow c + c + a^2 + b^2 = (a + b)^2 .$$

Also a system of equations $\alpha_i = \beta_i$ $(i = 1,\ldots,\nu)$ is equivalent to a single equation

$$(\ldots((1 \dot{-} (\alpha_1 \dot{-} \beta_1)) \dot{-} (\beta_1 \dot{-} \alpha_1)) \dot{-} \ldots) \dot{-} (\alpha_\nu \dot{-} \beta_\nu)) \dot{-} (\beta_\nu \dot{-} \alpha_\nu) = 1$$

Hence we obtain:

There is no algorithm for listing identities (over N*) of the form*

$$F(x_1,\ldots,x_n) = 0$$

where F *is obtained by composition from* $\dot{-}$ *and the operation of squaring applied iteratively to particular natural numbers and* $x_1,\ldots,x_n$.

Also, *there is no algorithm for listing identities in one variable of the form*

$$F(x) = 0$$

where F *is obtained by composition from the functions* $\dot{-}$, $[\sqrt{x}]$, Rem *(the remainder function).* For these results see Matijasevič [38].

Another example comes from the fact that + can be existentially defined in terms of S (successor and $\cdot$ since

$$z = x + y \leftrightarrow S(Sx\cdot Sz)\cdot S(y\cdot Sz) = S(Sz\cdot Sz\cdot S(Sx\cdot y))$$

(see Robinson [45]). Hence

Every listable set is existentially definable in terms of S *and* $\cdot$.

Thus a Diophantine equation can be reduced to a system of equations of the forms $\alpha = S\beta$ and $\alpha\cdot\beta = \gamma$ where α, β, γ are natural numbers, parameters, or unknowns. Furthermore, such a system is equivalent to a single equation since $a_i = b_i$ $(i = 0,\ldots,\nu)$ if and only if

$$\prod_i S(a_i M^{2^i}) = \prod_i S(b_i M^{2^i}) \tag{1}$$

where $M = \prod_i (Sa_i \cdot Sb_i)$. To see this, note that M is so large that (1) can hold only if the corresponding coefficients of the powers of M are equal and that the coefficients of M^{2^i} are a_i and b_i respectively.

It is easy to see that S and $\cdot$ are existentially definable in terms of $z = [x/y]$ where $[x/0] = 0$. In fact, we have

$$y = 2x \ \& \ z = 2x+1 \leftrightarrow (x = 0 \ \& \ y = 0 \ \& \ z = 1) \vee ([\tfrac{y}{2}] = [\tfrac{z}{2}] = x \ \& \ [\tfrac{y}{z}] = 0)$$

$$y = x+1 \leftrightarrow (x = 0 \ \& \ y = 1) \vee ([\tfrac{x}{x}] = 1 \ \& \ (\exists z)\{[\tfrac{z}{2x+1}] = 2x+1 \ \& \ [\tfrac{z}{2(2x)}] = y\})$$

$$z = xy \leftrightarrow (z=0 \ \& \ x=0) \vee (z=0 \ \& \ y=0) \vee (\exists t)\{[\tfrac{z}{y}] = x \ \& \ z=t+1 \ \& \ [\tfrac{t}{y}]+1=x\}.$$

Thus, any Diophantine equation can be reduced to a disjunction of systems of equations of the form $[\alpha/\beta] = \gamma$ where α,β,γ are natural numbers, parameters or unknowns. In particular, *there is no algorithm for testing whether a system of such equations has a solution.*

Kosovskii [27] showed that $\cdot$ is existentially definable in terms of $+$, $|$, and any relation S such that

(a) $$S(x,x^2)$$

and

(b) $$S(x,y) \rightarrow y \leq x^2 \ .$$

It is an immediate consequence that $\cdot$ is existentially definable in terms of $+$, $|$, and $\square$. Indeed, we can define the required relation S satisfying (1) and (2) by

$$S(x,y) \leftrightarrow y = \square \ \& \ y + 2x + 1 = \square \ .$$

This result suggests the:

OPEN PROBLEM. *Is there an algorithm to test whether a system of conditions of the forms*

$$\alpha < \beta, \quad \alpha | \beta, \quad \alpha = \square$$

(where α, β *stand for natural numbers or unknowns) has a solution?*

Recently Edward Schwartz [54] has found various systems of functions and relations in terms of which $+$ and $\cdot$ are existentially definable. Thus he showed that $+$ is existentially definable in terms of $\cdot$, x^2+y^2, and $<$. For,

$$a = p + q \leftrightarrow (\exists b,x,y)\{[p = q \ \& \ a = 2\cdot p] \vee [p=0 \ \& \ a=q] \vee [q=0 \ \& \ a=p]$$
$$\vee \ [(p^2+q^2)^2 + 0^2 = x^2 + (2\cdot(p\cdot q))^2 \ \& \ x = a\cdot b$$
$$\& \ (a^2+b^2)^2 + 0^2 = y^2 + (2\cdot(a\cdot b))^2 \ \& \ y = 4\cdot(p\cdot q) \ \& \ b < a]\} \ .$$

Similar methods were used by Schwartz to show that $+$ and $\cdot$ are existentially definable in each of the following systems of functions and relations:

(1) $+$ and x^2+y^2

(2) x^2+y^2, $2x$, and $<$

(3) $x\cdot y$ and x^2-y^2

(4) $x+y$ and x^2-y^2

(5) x^2-y^2 and x^2+y^2

(6) $x\cdot y$, $<$, and x^2-dy^2, d fixed

(7) $x\cdot y$, $<$, and x^2+dy^2, d fixed.

9. SIZE OF SOLUTIONS

A function ϕ is a *test function* for a Diophantine equation $P(a_1,\ldots,a_\mu,x_1,\ldots,x_\nu) = 0$ if whenever $P(a_1,\ldots,a_\mu,x_1,\ldots,x_\nu) = 0$ has a solution for given values of the parameters $a_1,\ldots,a_\mu$ it has a solution in which each unknown is less than $\phi(a_1,\ldots,a_\mu)$. Clearly, a Diophantine equation defines a computable relation if and only if it has a computable test function. Since there are Diophantine sets which are not computable,

ERRATA

Proceedings of Symposia in Pure Mathematics

Volume 28

"Hilbert's tenth problem. Diophantine equations: Positive aspects of a negative solution"

By Martin Davis, Yuri Matijasevič and Julia Robinson

Pages numbered 357–358 should precede pages numbered 355–356.

unless $P_1 = P_2$ identically. Also $\delta_i(x) \to 0$ as $x \to \infty$. Hence either there is an easily calculated bound on possible solutions or $P_1 = P_2$. Throwing out the trivial cases, we can without loss of generality take

$$\tau_1(x) = \frac{P(x)}{n} + \delta_1(x)$$
$$\tau_2(x) = \frac{P(x)}{n} + \frac{m}{n} + \delta_2(x) \qquad (1)$$

where P is a polynomial with integer coefficients and m,n are particular natural numbers. There is an easily computed x_0 so that for $x > x_0$, $\delta_1(x)$ and $\delta_2(x)$ have a constant sign and have absolute value less than 1/n. Then any solution of (1) is either less than x_0 or lies in one of an easily determined set of residue classes (mod n). Also all numbers greater than x_0 in these residue classes will be solutions of (1). Since it is easy to determine whether a finite set of such conditions can be satisfied simultaneously, we can determine whether a system of such equations has a solution. Thus for this problem, we have a sharp result:

For systems of equations in 2 or more unknowns there is no algorithm while for systems of such equations in one variable there is an algorithm.

8. EXISTENTIAL DEFINABILITY AND UNIVERSAL STATEMENTS

We consider formulas which can be written in a language in which there are variables ranging over the natural numbers, symbols for =, &, ∨, and $(\exists x)$, and for each fixed natural number, and symbols for certain given functions and/or relations on N. Such a formula defines a relation on N in its free variables (parameters). And any relation which is so definable is called *existentially definable* in terms of the given function and relations. Obviously a Diophantine relation is simply one which is existentially definable in terms of + and ·. Now it is clear that if a relation R is existentially definable in terms of given functions and relations and if each of the given functions and relations is in turn existentially definable in terms of a second set of functions and relations then R must be existentially definable in terms of this second set. In particular, any class C of functions and relations in terms of which + and · are existentially definable has the property that any Diophantine relation (and hence any listable relation) is existentially definable in terms of C. Also there can be no algorithm to tell whether such a system of equations and relations can be satisfied in N.

Before turning to examples, we want to remark that the negative solution of Hilbert's tenth problem can also be used to show that there is no algorithm for listing certain types of systems of conditions which hold for all natural number values. For example:

There is no algorithm for listing all polynomials P *and* Q *with positive integer coefficients such that*

$$P(x_1,\ldots,x_n) < Q(x_1,\ldots,x_n) \qquad (1)$$

for all natural number values of $x_1,\ldots,x_n$.

In fact, by the reduction theorem we can take $n = 13$. This is clear since there is no algorithm for listing unsolvable Diophantine equations

$$F(x_1,\ldots,x_{13}) = G(x_1,\ldots,x_{13}) .$$

But $F = G$ has no solution if and only if

$$(F - G)^2 > 0 \quad \text{for all} \quad x_1,\ldots,x_{13} .$$

REMARKS. This result contrasts with two earlier related results. First, the corresponding problem for real values of any number of variables $x_1,\ldots,x_n$ where (1) is equivalent to a Boolean combination of inequalities among polynomials in the coefficients of P and Q by Tarski's elimination of quantifiers theorem. Hence:

There is an algorithm to determine whether

$$P(x_1,\ldots,x_n) < Q(x_1,\ldots,x_n)$$

for all real $x_1,\ldots,x_n$.

Second, consider the problem of recognizing identities (over N) of terms built up from natural numbers and x by additions, multiplications, and exponentiations. Richardson [43] obtained an algorithm to solve this problem. See also Goodstein and Lee [20].

For undecidability results in real analysis see Richardson [42] and Adler [2].

In general if there is no algorithm for telling which of a class of existential formulas can be satisfied then there is no algorithm for listing those of the *dual formulas* (& replaced by ∨, ∨ by &, = by ≠, each relation by its complement) which hold for all natural numbers.

For example,

$$p \neq q$$

is equivalent to the equation

$$(1 \dot{-} (p \dot{-} q)) \dot{-} (q \dot{-} p) = 0$$

where $\dot{-}$ is defined as usual by

$$x \dot{-} y = \begin{cases} x - y & \text{if} \quad x \geq y \\ 0 & \text{otherwise} \end{cases} .$$

Hence

There is no algorithm for listing identities (over N*) of the form*

$$F(x_1,\ldots,x_n) = 0$$

where F *is obtained by composition from* $+$, $\cdot$, $\dot{-}$ *applied to natural numbers and* $x_1,\ldots,x_n$.

This brings up the problem of finding minimal subclasses of identities which cannot be listed. For example we can replace $+$, $\cdot$, $\dot{-}$ in the result above by $\dot{-}$, x^2. To see this note that $+$, $\cdot$ can be existentially defined in terms of $\dot{-}$ and x^2 by there are Diophantine equations with one parameter which do not have computable test functions.

Given a Diophantine equation, we know we cannot in general find a computable test function for it. However there are two natural types of questions about test functions which we can hope to answer.

I. *Given some familiar set* S *of natural numbers such as the set of primes, can we find a Diophantine definition of* S *with a polynomial test function? an iterated exponential test function? etc.*

Here a function is called *iterated exponential* if it can be built up from its arguments and particular natural numbers by addition, multiplication, and exponentiation. Thus a^{a^a} is an iterated exponential function but

$$\phi(a) = a^{a^{\cdot^{\cdot^{\cdot^{a}}}}} \Big\} \text{ a levels}$$

is not.

II. *What can be said about sets which can (or cannot) be defined by Diophantine equations with polynomial test functions? with iterated exponential test functions? etc.*

By examination of the proofs of the Diophantine definitions of $z = x^y$, $z = \binom{x}{y}$, $z = x!$, and the set of primes sketched in §1 and given in detail in the expository articles referred to in the Introduction, we see that all of these relations have definitions with iterated exponential test functions. Also if we look at the definition of the set of superpowers in §4, we see that $t_1,\dots,t_{n+1} \le n^m$, $d \le a$, $n \le m$, and a is less than an exponential function of m by Gödel's lemma. Hence by the bounded quantifier theorem, the set of superpowers can also be defined by an equation with an iterated exponential test function.

An example of a set definable by an equation with a polynomial test function is the set of composite numbers. More generally, the same can be said for a set defined by an equation in which one side is a polynomial with positive coefficients not containing the parameter and the other side is a polynomial in the parameter not containing unknowns, or for a set defined by an equation with one unknown.

Another interesting case of sets with polynomial test functions can be obtained using the Cantor pairing function $J(x,y) = ((x+y)^2+3x+y)/2$ (cf. Introduction). We first note that the condition

$$2J(x,y) + 1 = \square$$

is equivalent to $x = y$. (Obviously $2J(x,x)+1= (2x+1)^2$; conversely if $2J(x,y)+1 = (2a+1)^2$, then $x = y = a$ is a solution, and by the one-one property of the pairing function, the unique solution.) Now we consider an arbitrary Diophantine equation $A(x_1,\dots,x_\nu) = B(x_1,\dots,x_\nu)$ where A and B have positive coefficients and no parameters. Then the equation

$$2J(A(x_1,\ldots,x_\nu),B(x_1,\ldots,x_\nu)) + 1 = a^2 \tag{1}$$

is equivalent to A = B if a is regarded as an unknown. Hence we have:

THEOREM. *There is no algorithm for telling whether an arbitrary polynomial with positive integer coefficients assumes a square value.*

If however a is regarded as a parameter, then (1) has a polynomial test function. In fact, all the x's must be less than a. Furthermore the set defined by (1) is infinite if and only if A = B has infinitely many solutions. This proves:

THEOREM. *There is no algorithm for telling whether an arbitrary polynomial with positive integer coefficients represents infinitely many squares.*

In these examples of equations with polynomial test functions, the test functions are in fact bounds for all of the solutions of the equation. This of course is not the case in general.

If $P(a,x_1,\ldots,x_\nu) = 0$ has a polynomial test function say a^n then we can introduce new unknowns and obtain an equivalent equation which has the identity function for a test function. Namely, let $x_i = u_0^{(i)}+\ldots\ u_{n-1}^{(i)}a^{n-1}$ then it will be sufficient to consider u's less than a. Also if a set S can be defined by an equation with a polynomial test function, then it has a definition of the form

$$a \in S \leftrightarrow (\exists x_1\ldots x_\nu\ y_1\ldots y_\nu)[P(a,x_1,\ldots,x_\nu)=0\ \&\ x_1+y_1=a\ \&\ldots\&\ x_\nu+y_\nu=a]. \tag{2}$$

The conditions in brackets can be combined in the usual way into a single equation which for each a has a finite number of solutions. Thus,

THEOREM. *If a set* S *can be defined by an equation with a polynomial test function, then it has a finitefold Diophantine definition.*

REMARK. It is possible that the converse of this theorem is true. This would eliminate the possibility of singlefold definitions for all Diophantine sets raised in §3. It would also mean that the iterated exponential bounds that Baker obtained for a class of equations in 2 unknowns could actually be replaced by polynomial bounds.

We can list all definitions of the form of (2) and hence all sets which can be defined with polynomial test functions. Let (cf. p. 25) $C_n = \{a: (\exists x_1\ldots)_{\leq a}[P_n(a,x_1,\ldots) = 0]\}$. Then the complement of the diagonal set i.e. $\{n: n \notin C_n\}$ cannot be defined by an equation with a polynomial test function although it is clearly computable.

Similarly, we can list the sets definable by equations with iterated exponential test functions. First number the iterated exponential functions of one variable: $E_1, E_2, \ldots$. Then consider the sets

$$S_{ij} = \{m: (\exists x_1\ldots)_{\leq E_j(m)}\ [P_i(m,x_1,\ldots) = 0]\}.$$

All sets which can be defined by equations with iterated exponential test functions are among the S_{ij}. Also since exponentiation can be defined by an equation with an iterated exponential test function, so also can S_{ij}. Again the set $\{J(i,j): J(i,j) \notin S_{ij}\}$ cannot be defined by an equation with an iterated exponential test function although it is clearly computable.

Obviously the union and intersection of two Diophantine sets can each be defined by an equation with a test function equal to the maximum of given test functions for the two sets. However, we have:

OPEN QUESTION. *Is the complement of a set defined by an equation with a polynomial test function always definable by an equation with a polynomial test function?*

Although this question remains open, we can see that the complement of a set defined by an equation with an iterated exponential test function can be defined by an equation with an iterated exponential test function. Indeed, suppose S is defined by $P(a,x_1,\ldots,x_\nu) = 0$ and $E(a)$ is a test function for $P = 0$. Then $a \notin S$ if and only if there is a sequence $t_0,\ldots,t_n$ such that

$$n = J_\nu(E(a),\ldots,E(a)) \;\&\; (\forall k)_{\leq n}\ (\exists x_1\ldots x_\nu y)$$

$$[k = J_\nu(x_1,\ldots,x\ ) \;\&\; t_k = P^2(a,x_1,\ldots,x_\nu) \;\&\; t_k = 1+y]\ . \qquad (3)$$

Here we can put bounds on the x's, y's, and n which are polynomial in $E(a)$ and hence iterated exponential functions of a. Finally using Gödel's lemma and the bounded quantifier theorem, we can transform (3) into a Diophantine definition with an iterated exponential test function.

10. APPLICATIONS OF THE THEORY OF ALGORITHMS

Since the listable sets are identical with the Diophantine sets, all of the theorems about listable sets in the theory of algorithms can be translated into theorems about Diophantine equations. Of course in most cases, the resulting statements seem quite unnatural, and answer questions about Diophantine equations that no one would have been much inclined to ask. In this section we discuss a few examples that we believe are of some interest.

The most spectacular example has already been discussed in §5: the existence of a universal Diophantine equation.

We begin with an application of the universal equation.

THEOREM. *There is a Diophantine equation*

$$W(a,t,x_1,\ldots,x_\nu) = 0 \qquad (1)$$

such that if C_t *is the set of* a *for which* (1) *has a solution, then* $C_0, C_1, \ldots$ *is an enumeration of all computable sets.*

PROOF. Let $\mathcal{D}_0, \mathcal{D}_1, \ldots$ be the sets listed by the universal Diophantine equation $U(a,n,z_1,\ldots,z_\nu) = 0$ as $n = 0,1,\ldots$. We will construct $C_{J(m,n)}$ in such a way that if $\mathcal{D}_m$ and $\mathcal{D}_n$ are complementary sets then $C_{J(m,n)} = \mathcal{D}_m$ and otherwise $C_{J(m,n)}$ is at least computable.

Let $b \in C_{J(m,n)}$ if the following conditions hold:

I. $b \in \mathcal{D}_m$ is verified with the unknowns in the universal equation equal to $y_1, \ldots, y_\nu$, where no "earlier" values $z_1, \ldots, z_\nu$ of the unknowns (earlier in the sense that $J_\nu(z_1, \ldots, z_\nu) < J_\nu(y_1, \ldots, y_\nu)$) serve to verify that $b \in \mathcal{D}_n$.

II. For each $a < b$, either $a \in \mathcal{D}_m$ or $a \in \mathcal{D}_n$.

Now if $\mathcal{D}_m = N - \mathcal{D}_n$ then $C_{J(m,n)} = \mathcal{D}_m$. Hence every computable set is some C_t. Also if there is a number k which does not belong to either $\mathcal{D}_m$ or $\mathcal{D}_n$ then no number $\geq k$ will be in $C_{J(m,n)}$. Hence it is finite and certainly computable. Finally, if $C_{J(m,n)}$ is infinite we can check whether $b \in \mathcal{D}_n$ is verified before $b \in \mathcal{D}_m$ is and hence tell whether b belongs to $C_{J(m,n)}$. The conditions I and II can easily be written out using bounded universal quantifiers. Namely,

$$\begin{aligned} b \in C_t \leftrightarrow (\exists m,n)\{t = J(m,n) \ \& \ (\exists y y_1 \ldots y_\nu)[y = J_\nu(y_1, \ldots, y_\nu) \& U(b,m,y_1, \ldots, y_\nu) = 0 \\ \& \ (\forall z)_{<y} \ (\exists z_1 \ldots z_\nu)[z = J_\nu(z_1, \ldots, z_\nu) \ \& \ U(b,n,z_1, \ldots, z_\nu) \neq 0]] \\ \& \ (\forall a)_{<b} \ (\exists z_1 \ldots z_\nu)[U(a,m,z_1 \ldots z_\nu) \cdot U(a,n,z_1, \ldots, z_\nu) = 0]\}. \end{aligned}$$

We can now use the bounded quantifier theorem to obtain the required polynomial W.

This theorem is essentially that of Suzuki [62], and could have been obtained immediately by combining Suzuki's result with the Main Theorem.

Notice that we escape from a diagonal counterexample because there is no method of listing the set $\{n: n \notin C_n\}$. On the other hand, there can be no Diophantine equation which lists all and only the graphs of computable functions of one variable because otherwise we could compute the diagonal function, add 1 to it, and have a computable function not on the list.

Now we do the best we can. Namely, we start with Peano's axioms and make a list of all Diophantine equations $P(a,b,x_1,\ldots) = 0$ such that we can prove

$$(\forall a)(\exists! b)(\exists x_1, \ldots)[P(a,b,x_1,\ldots) = 0] .$$

For such a P, we call the function F defined by $P(a,F(a),x_1,\ldots) = 0$ for some $x_1,\ldots$ a *provably computable function*. Using our list of equations, we can define a computable function G which grows faster than any provably computable function. Notice that to construct G, we have to step outside Peano's axioms; otherwise G would be a provably computable function which is impossible.

It has been suggested by Kreisel [28] that possibly the reason that a proof of the infinitude of certain sets has not been found is that the function H which enumerates the set in question in order of size grows faster than any provably computable function, just as our diagonal function G does.

There are proof methods in the theory of algorithms which would be very difficult to translate into intuitive number theoretical arguments. For example, Friedberg using his "priority method" showed that there is a listable set M with an infinite complement such that every listable set which includes M either differs from M or from N by a finite number of elements. Cf. [51], [55]. Hence the corresponding theorem about Diophantine sets is obtained.

THEOREM. *There is a Diophantine set* M *with an infinite complement such that every Diophantine set which includes* M *either differs from* M *or from* N *by a finite set.*

COROLLARY. *Every infinite Diophantine set contains infinitely many elements of* M.

11. RINGS OF ALGEBRAIC INTEGERS

In this section we discuss Diophantine definability in rings of algebraic integers and the related question of the existence of an algorithm to tell whether an arbitrary Diophantine equation has a solution in a given ring. There are two natural generalizations of the notion of Diophantine set: one in which the coefficients of the defining polynomial equation remain rational integers while the parameters and unknowns can be assigned values in the given ring, and one in which the coefficients are chosen from the ring as well. We shall see that the two generalizations are very closely related.

Let I be the ring of integers in a subfield F of the field A of all algebraic numbers. Let G be the group of automorphisms of F (or equivalently, I). The degree of F may be finite or infinite over the rational field Q. A relation $T \subseteq I^n$ is called *Diophantine over* I if there is a polynomial equation with coefficients in I

$$P(a_1,\ldots,a_n,y_1,\ldots,y_m) = 0 \qquad (1)$$

such that $T(a_1,\ldots,a_n)$ holds if and only if (1) has a solution in I for the unknowns $y_1,\ldots,y_m$. The equation $P = 0$ is then called a *Diophantine definition* of the relation T. Furthermore, T is called a *pure Diophantine* relation if it has a Diophantine definition with coefficients in Z (the rational integers). (Note that we are writing, e.g. $T(a_1,\ldots,a_n)$ to indicate that $\langle a_1,\ldots,a_n\rangle \in T$.)

Except possibly for the ring of all algebraic integers, it is immaterial whether we consider sets defined by single equations or by Boolean combinations of equations. In particular, let

$$A(x) = \sum_{i=0}^{n} a_i x^i = 0, \quad a_n \neq 0, \quad a_i \in I.$$

have no solution in F. Let

$$B(x,y) = \sum_{i=0}^{n} a_i x^i y^{n-i}.$$

Then in I, $u = 0$ & $v = 0$ if and only if $B(u,v) = 0$. Clearly, $B(0,0) = 0$. Conversely, suppose $v \neq 0$. Then $B(u,v) = 0$ implies $A(u/v) = 0$ where $u/v \in F$ which contradicts the hypothesis on A. So $v = 0$ and hence $u = 0$. Since an arbitrary finite algebraic extension of $K \subset A$ can always be made by adjoining a root of a single equation to K, the only field F for which there is no equation without a solution is the field A of all algebraic numbers. Finally, for any I, we have:

$$u=0 \vee v=0 \leftrightarrow uv=0$$

and

$$u\neq 0 \leftrightarrow (\exists xyz)\,[ux = (2y+1)(3z+1)]\ .$$

This last equivalence holds because when $u \neq 0$, x can be found so that ux is a positive rational integer which is the product of a power of 4 and an odd number.

The general question which we wish to raise here for each ring I of algebraic integers is:

What sets and relations of I *are Diophantine over* I?

The related question corresponding to Hilbert's tenth problem is:

Is there an algorithm to tell whether a Diophantine equation has a solution in I ?

(Both these questions can also be asked with "pure Diophantine" in place of "Diophantine".)

Now the Main Theorem yields the answer to both questions for I = Z. In the first real advance since 1970, Denef [15] has completely answered both questions for the case where I is the ring of integers in a quadratic extension of Q. Before discussing his results, we need to define the notion of a *listable relation on* I. This can be done in any number of equivalent ways. For example, if I is of finite dimension k as a vector space over Z, an n-ary relation on I can be regarded simply as a kn-ary relation on Z. In general, if the elements of I can be numbered systematically, then a relation $T \subseteq I^n$ is *listable* if the corresponding relation among the numbers of the elements of I is a listable relation of natural numbers. Clearly, as in the case of N, any Diophantine relation over I is listable provided I itself is. We will also say that a relation T is *selfconjugate* if $T(x_1,\ldots,x_n)$ implies that $T(x_1^\sigma,\ldots,x_n^\sigma)$ for all automorphisms σ of I. Note that if a relation is pure Diophantine it must be selfconjugate.

THEOREM (Denef). *A relation is Diophantine over the ring* I *of integers of a quadratic extension of* Q *if and only if it is listable. A relation is pure Diophantine if and only if it is listable and selfconjugate over* I.

COROLLARY (Denef). *There is no algorithm to tell whether an arbitrary pure Diophantine equation has a solution in* I *(the ring of integers of a quadratic extension of* Q*).*

R. M. Robinson [49] settled for all I the question:

Which elements of I *have pure Diophantine definitions?*

THEOREM (R. M. Robinson). *Let* I *be the ring of integers in a field of algebraic numbers and suppose an element* α *of* I *is fixed under all automorphisms of* I *(i.e.* α *is selfconjugate). Then there are polynomials* P *and* Q *with rational integer coefficients such that*

$$x = \alpha \leftrightarrow (\exists y)[P(y) = 0 \;\&\; x = Q(y)] \;. \qquad (2)$$

Note that we have changed Robinson's theorem to conform to our statement of the problem -- the original version applied to arithmetical definitions of elements of an algebraic field. The two equations in (2) can be combined into one except in the case of the ring of all algebraic integers. In that case the rational integers are the only selfconjugate numbers and they can be defined directly.

Some negative results follow from considerations of selfconjugacy. For example,

THEOREM. *The set of real algebraic integers is not Diophantine over the ring of all algebraic integers.*

PROOF: Since there are automorphisms of A which take a real number into a non-real number, the set of real algebraic integers is not pure Diophantine. Suppose $P(a,y_1,\dots,y_m) = 0$ is a Diophantine definition over the ring of all algebraic integers of the ring of real algebraic integers and the coefficients of P lie in $I = Z(\alpha_1,\dots,\alpha_q)$. We can find an automorphism σ of A which leaves I fixed but still takes some real algebraic integer b into a nonreal integer. Hence $P(b,x_1,\dots,x_m) = 0$ implies $P(b^\sigma,x_1^\sigma,\dots,x_m^\sigma) = 0$. But this is impossible since b^σ is not real yet lies in the set defined by $P = 0$.

The next theorem gives the exact connection between pure Diophantine and Diophantine relations.

THEOREM. *Let* I *be the ring of integers of a field* F *of algebraic numbers. A relation* $A \subseteq I^n$ *is pure Diophantine if and only if* A *is Diophantine and selfconjugate.*

PROOF: Suppose A has the Diophantine definition

$$P(a_1,\dots,a_n,y_1,\dots,y_m) = 0$$

with coefficients in Z. Then A is of course Diophantine. If $A(x_1,\dots,x_n)$ holds, so that for suitable $y_1,\dots,y_m \in I$,

$$P(x_1,\dots,x_n,y_1,\dots,y_m) = 0$$

then (since σ is an automorphism leaving Z fixed)

$$P(x_1^\sigma,\dots,x_n^\sigma,y_1^\sigma,\dots,y_m^\sigma) = 0$$

so that $A(x_1^\sigma,\dots,x_n^\sigma)$. Hence A is selfconjugate.

Conversely, let A be Diophantine and selfconjugate, and let

$$P(a_1,\dots,a_n,y_1,\dots,y_m) = 0$$

be a Diophantine definition of A with coefficients in I. Let $\sigma \in G$ and let P^σ be obtained from P by replacing each coefficient by its image under σ. There are only a finite number k of different polynomials P^σ since each coefficient goes into some root of its irreducible equation. Let $\sigma_1,\dots,\sigma_k$ be the corresponding automorphisms which are the identity function on numbers not in the ring generated by the coefficients of P.

Let

$$Q(x_1,\dots,x_n,y_1,\dots,y_m) = \prod_{j=1}^{k} P^{\sigma_j}(x_1,\dots,x_n,y_1,\dots,y_m) .$$

since the different σ's enter symmetrically into the calculation of the coefficients of Q, these coefficients are fixed under G. Hence each coefficient has a pure Diophantine definition of the form (2). Thus, we can obtain a pure Diophantine definition of the same relation that is defined by $Q = 0$. We shall show that this relation is in fact A, thus completing the proof.

Suppose $A(x_1,\dots,x_n)$. Then $P(x_1,\dots,x_n,y_1,\dots,y_m) = 0$ for some $y_1,\dots,y_m \in I$. Hence (since the identity automorphism belongs to G), $Q(x_1,\dots,x_n,y_1,\dots,y_m) = 0$.

Finally, let $Q(x_1,\dots,x_n,y_1,\dots,y_m) = 0$. Then for some j

$$P^{\sigma_j}(x_1,\dots,x_n,y_1,\dots,y_m) = 0$$

Let τ be the inverse of σ_j. Then,

$$P(x_1^\tau,\dots,x_n^\tau,y_1^\tau,\dots,y_m^\tau) = 0 ,$$

so that $A(x_1^\tau,\dots,x_n^\tau)$ holds. Since A is selfconjugate and $x_i = (x_i^\tau)^{\sigma_j}$, we have $A(x_1,\dots,x_n)$.

THEOREM. *Let* F *be a finite extension of* Q *and* I *the ring of integers of* F. *Every listable relation on* I *is Diophantine over* I *if and only if the set* Z *is Diophantine over* I.

PROOF: Since Z is a listable subset of I, we need only show that if Z is Diophantine over I then every listable relation is Diophantine over I.

Let $A \subseteq I^n$ be listable and $\{\eta_1,\dots,\eta_k\}$ be an integral basis for I over Z. Let $R \subseteq Z^{nk}$ be defined by letting $R(u_1^{(1)},\dots,u_k^{(1)},\dots,u_1^{(n)},\dots,u_k^{(n)})$ hold just in case $A(u_1^{(1)}\eta_1+\dots+u_k^{(1)}\eta_k,\dots,u_1^{(n)}\eta_1+\dots+u_k^{(n)}\eta_k)$ holds. Thus R is a listable relation over Z and hence is Diophantine. Let $P(a_1^{(1)},\dots,a_k^{(1)},\dots,a_1^{(n)},\dots,a_k^{(n)},y_1,\dots,y_m) = 0$ be a Diophantine definition of R over Z. Then, we have (*with all variables ranging over* I):

$$A(x_1,\dots,x_n) \leftrightarrow (\exists u_1^{(1)},\dots,u_k^{(1)},\dots,u_1^{(n)},\dots,u_k^{(n)},y_1,\dots,y_m)$$

$$[u_1^{(1)} \in Z \ \& \ \dots \ \& \ u_k^{(1)} \in Z \ \& \ \dots \ \& \ u_1^{(n)} \in Z \ \& \ \dots \ \& \ u_k^{(n)} \in Z \ \&$$
$$y_1 \in Z \ \& \ \dots \ \& \ y_m \in Z \ \&$$
$$x_1 = u_1^{(1)}\eta_1 + \dots + u_k^{(1)}\eta_k \ \&$$
$$\vdots$$
$$x_n = u_1^{(n)}\eta_1 + \dots + u_k^{(n)}\eta_k \ \&$$
$$P(u_1^{(1)},\dots,u_k^{(1)},\dots,u_1^{(n)},\dots,u_k^{(n)},y_1,\dots,y_m) = 0] \ .$$

By hypothesis, there is a Diophantine definition for a condition of the form $\alpha \in Z$. Hence since Diophantine relations over I are closed under '&', we are done.

REMARK. In order to obtain the above theorem for an infinite extension F of Q, we would need some way of representing an arbitrary algebraic integer α of F as a listable relation over Z such that we can identify α as the number represented by means of a Diophantine relation over I. For example in Z, we can represent any algebraic number as a pair of Dedekind cuts. But in the ring of all algebraic integers, as we have seen, there is no way to define the set of real algebraic integers and the same argument also shows that it could not be defined in terms of Z.

Returning to the question of an algorithm to tell whether a Diophantine equation has a solution in I, we have already noted that if Z is definable there can be no such algorithm. It seems likely that for rings of integers of finite extensions of Q it will be possible to define Z and hence that there will be no algorithm in these cases. Thus, the most promising place to look for an algorithm is for the ring of all algebraic integers.

12. THE NUMBER OF SOLUTIONS OF A DIOPHANTINE EQUATION

In a Diophantine equation D with coefficients in Z, the following are all natural questions:

(1) Does D have a solution?

(2) Does D have infinitely many solutions?

(3) Does D have fewer than 100 solutions?

(4) Does D have an even (or a prime, etc.) number of solutions?

Here we will assume that solutions are being sought in some fixed domain R which may be the natural numbers N, the positive integers P, the rational integers Z, or the ring I of a proper subfield of all algebraic numbers. And we will assume that the analogue of Hilbert's tenth problem for R (i.e. question (1) in the above list) is known to be unsolvable. (So on the basis of current knowledge, R could be N, P, Z, or (by Denef's work [15], mentioned in §11) the ring I of integers of a *quadratic* extension of the rationals.) We shall show that all the questions (1)-(4) (and any others like them) are unsolvable. This was originally shown for P in Davis [12]; the present discussion follows the somewhat simplified treatment of Smorynski [60].

Let $a_1,\ldots,a_n$ be n given rational integers. Then we can make up a Diophantine equation with rational integer coefficients and unknowns $x_1,\ldots,x_n$ which has the unique solution in R: $x_1 = a_1, \ldots, x_n = a_n$. To see this we just combine these n equations into one equation using repeatedly the method of combining two equations into one given in §11.

Now, let $P(x_1,\ldots,x_n) = 0$ be an arbitrary Diophantine equation, and let $\langle a_1,\ldots,a_n\rangle$ be the first (in some suitable ordering e.g. using J_n) n-tuple of positive integers for which $P(a_1,\ldots,a_n) \neq 0$. And let $Q(x_1,\ldots,x_n) = 0$ be the equation constructed above whose unique solution in R is $x_1 = a_1, \ldots, x_n = a_n$. Then, we set:

$$T^+(P) = P\cdot Q .$$

We will write $\#(F)$ for the cardinal number of the set of solutions in R of the equation $F = 0$. (Thus for any F, $0 \le \#(F) \le \aleph_0$.) Then T^+ has been constructed so that

$$\#(T^+(P)) = \#(P) + 1 .$$

We define:

$$T^0(P) = P ,$$

$$T^{m+1}(P) = T^+(T^m(P)) .$$

Thus,

$$\#(T^m(P)) = \#(P) + m .$$

Then, we define

$$T^\infty(P) = (2u+1)\cdot P$$

where u is a variable that does not occur in P, and we note that

$$\#(T^\infty(P)) = \begin{cases} 0 & \text{if } \#(P) = 0 \\ \aleph_0 & \text{if } \#(P) \neq 0 \end{cases}$$

since $2u+1 \neq 0$ for all $u \in R$.

THEOREM. *Let* $C = \{0,1,\ldots,\aleph_0\}$. *Let* $A \subseteq C$ *but* $A \neq \emptyset, C$. *Then there is no algorithm for testing a given Diophantine equation* $P = 0$ *to determine whether* $\#(P) \in A$.

PROOF: It is sufficient to consider the following cases.

Case 1. $A = \{0\}$. This is the analogue for R of Hilbert's tenth problem and our assumption is that there is no algorithm to tell whether an arbitrary Diophantine equation has a solution in R. We will show that the other cases can be reduced to this one.

Case 2. $0,\aleph_0 \in A$. Let $m \notin A$. Then

$$\#(P) = 0 \leftrightarrow \#(T^m(T^\infty(P))) \notin A.$$

Case 3. $0 \in A$, $\aleph_0 \notin A$. Then

$$\#(P) = 0 \leftrightarrow \#(T^\infty(P)) \in A.$$

Hence in either Case 2 or Case 3 an algorithm to tell whether $\#(P) \in A$ would also solve Hilbert's tenth problem.

Case 4. $0 \notin A$. Then $C-A$ is either in Case 2 or Case 3, and so has already been treated.

13. DEGREES OF DIOPHANTINE PROBLEMS

In this section we will once again be dealing exclusively with the domain N. We also let $P_0 = 0$, $P_1 = 0$, $P_2 = 0, \ldots$ be an explicit enumeration of all Diophantine equations (e.g., the particular enumeration defined in §5). First, we want to generalize the notion of an algorithm to allow the use of some outside information. We say that A is *computable from* B if there is an algorithm for testing membership in A which however may require answers to (a finite number of) questions of the form: "*Does* $n \in B$?" Notice that this is a broader notion than A is *Diophantine in* B i.e. existentially definable in terms of $+$, $\cdot$, and B. (See §8.) For example, if $A = \{x^2 : x \in B\}$ then A is Diophantine in B as well as computable from B. However if

$$A = \{x: (\exists y)\ [x = \sum_{\substack{b \in B \\ b < y}} b]\},$$

then A is computable from B but not in general Diophantine in B. Note that A is trivially computable from its complement $\bar{A} = N-A$. We also say that A is *listable from* B provided there is an algorithm for listing members of A which may require answers to a finite number of questions of the form "*Does* $n \in B$?". For example, if

$$A = \{n: (b \in B)(P_n(b, x_1, \ldots) = 0 \text{ has a solution})\},$$

then A is listable from B. And of course A is computable from B just in case A and $\bar{A}$ are both listable from B. The customary picturesque way of describing the set A being computable from B is to say that A is computable by means of an oracle which can answer correctly questions of the sort "*Does* $n \in B$?".

In discussing these matters, we need to code a finite amount of information about membership (and nonmembership) in a set as a single integer. This may be done in a variety of ways -- all more or less arbitrary. To make matters definite, we call a natural number n the *code* for S where

$$n = \sum_{k \in S} 2^k .$$

For any set B, let $\tilde{B}$ be the set of codes for *modified initial segments* of B, i.e. sets of the form

$$\{b: (b \in B \ \& \ b < k) \vee b = k\} .$$

Thus,

$$b \in B \leftrightarrow (\exists k, x)[\text{the coefficient of } 2^b \text{ in the binary expansion of } x \text{ is } 1 \ \& \ x \geq 2^{b+1} \ \& \ x \in \tilde{B}] . \quad (1)$$

Clearly, $\tilde{B}$ is computable from B and conversely.

Now instead of asking the original oracle for B a finite number of questions of the form 'Does $n \in B$?' as we proceed in our calculation of A, we can obtain the same information if the oracle for $\tilde{B}$ will disclose a sufficiently large $x \in \tilde{B}$. In order to see how this works, suppose we are given an algorithm for listing A from B. We modify the algorithm to calculate a set R of pairs $\langle a,x\rangle$ so that whenever the original algorithm asks 'Does $n \in B$?', we answer 'yes' if $x \geq 2^{n+1}$ and the coefficient of 2^n in the binary expansion of x is 1; we answer 'no' if $x \geq 2^{n+1}$ and the coefficient of 2^n is 0; if $x < 2^{n+1}$, we go back to the beginning and start over with x+1 in place of x. As the calculation proceeds, whenever the modified algorithm yields that a is in the set being calculated, we put $\langle a,x\rangle$ in R. Thus, R is a listable and hence a Diophantine relation. On the other hand if $\langle a,x\rangle \in R$ and $x \in \tilde{B}$ then $a \in A$. Thus, we have proved:

THEOREM. *A is listable from B if and only if A is Diophantine in $\tilde{B}$ where*

$$\tilde{B} = \{x\colon (\exists n)\ [x = 2^n + \sum_{\substack{k\in B\\ k<n}} 2^k]\}\ . \tag{2}$$

In fact,

$$x \in A \leftrightarrow (\exists x \in \tilde{B})(\exists y_1 \ldots y_n)[P(a,x,y_1,\ldots,y_n) = 0]$$

where $P(a,x,y_1,\ldots,y_n) = 0$ *is a Diophantine definition of the relation* R *above.*

The sets A and B are called *Turing equivalent* if each is computable from the other. Now it is easy to see that this is an equivalence relation. The equivalence classes are called *degrees*. For a degree $\underset{\sim}{a}$, we write $\deg(A) = \underset{\sim}{a}$ for $A \in \underset{\sim}{a}$. The degrees are partially ordered (i.e. $\underset{\sim}{a} \leq \underset{\sim}{b}$ if $\underset{\sim}{a} = \deg(A)$ and $\underset{\sim}{b} = \deg(B)$ where A is computable from B); each pair of degrees $\underset{\sim}{a}$, $\underset{\sim}{b}$ has a supremum written $\underset{\sim}{a} \cup \underset{\sim}{b}$, in this ordering; there is a least degree $\underset{\sim}{0}$; and there is a jump operator which maps each degree $\underset{\sim}{a}$ onto the degree $\underset{\sim}{a}'$ which is the maximum degree of a set listable from a set A with $\deg(A) = \underset{\sim}{a}$, so that for each $\underset{\sim}{a}$ we have $\underset{\sim}{a} < \underset{\sim}{a}'$. In particular $\underset{\sim}{0}'$ is the maximum degree of a listable set. (For an introduction to the extensive theory of degrees see Rogers [51] or Shoenfield [56].)

Let

$$S = \{n \mid P_n(x_0,x_1,\ldots) = 0 \text{ has a solution}\}\ ,$$

where we are using the enumeration of Diophantine equations mentioned above. We will show that

$$\deg(S) = \underset{\sim}{0}'. \tag{3}$$

In the first place, S is listable so $\deg(S) \leq \underset{\sim}{0}'$. Let L be any listable set and let $P_n(k,x_1,\ldots,x\ ) = 0$ be a Diophantine definition of L. Then $k \in L$ if and only if the number (in the enumeration) of the equation $P_n(k,x_1,\ldots,x\ ) = 0$ (or any equivalent equation) belongs to S. Since we can devise an algorithm for computing a number of $P_n(k,x_1,\dots,x_\nu) = 0$ as a function of k (using the numbering of §5, it is enough to calculate the number of $P_n^2 + (x_0-k)^2 = 0$), and can then find out from the oracle for S whether the equation has a solution, we see that

$$\deg(L) \leq \deg(S)$$

for all listable sets L. Hence $\deg(S) = \underset{\sim}{0}'$.

As another example, consider the set F of numbers of equations which have a finite number of solutions. We will show that

$$\deg(F) \leq \underset{\sim}{0}'' \quad (= \deg(S)') . \tag{4}$$

We need to show that F is listable from S. Now all solutions of $P(x_0,\dots,x_\nu) = 0$ satisfy $x_0 < b, \dots, x_\nu < b$ if and only if $P(x_0+b,\dots,x_\nu+b) = 0$ has no solution. Also there is a computable function h(n,b) such that $P_{h(n,b)}(x_0,\dots,x_\nu) = P_n(x_0+b,\dots,x_\nu+b)$. Hence $n \in F$ if and only if $h(n,b) \notin S$ for some b. Therefore F is listable from S and (4) holds.

THEOREM. *Let* $C = \{0,1,\dots,\aleph_0\}$ *and let* $A \subseteq C$, $A \neq \emptyset, C$. *Let* W *be the set of* n *such that* $\#(P_n) \in A$. *Then*

$$\underset{\sim}{0}' \leq \deg(W) \leq \deg(A) \cup \underset{\sim}{0}''. \tag{5}$$

PROOF. The left inequality is a strengthened form of the theorem in the last section, and its proof is easily obtainable by analyzing the proof of the latter. Namely, in each of the cases 2 through 4, we have a computable function g such that either

$$x \in S \leftrightarrow g(x) \in W$$

or

$$x \in S \leftrightarrow g(x) \notin W .$$

In Case 2, for example, g(n) is the number of the equation $T^m(T^\infty(P_n)) = 0$ and the corresponding equivalence can be expressed as

$$n \in S \leftrightarrow g(n) \in W .$$

Hence S is computable from W and hence by (3),

$$\underset{\sim}{0}' \leq \deg(W) .$$

Now for the right inequality in (5), it is sufficient to show that W is computable from A and F. To check whether $n \in W$, we first use F to see whether $P_n = 0$ has a finite or infinite number of solutions. If $n \in F$, then we can find all solutions -- first searching for and finding a bound on the unknowns using S which is of course in $\underset{\sim}{0}''$ and then trying all possible values of the unknowns below this bound -- finally we use A to check whether $\#(P_n) \in A$. If $n \notin F$, then $\#(P_n) \in A$ if and only if $\aleph_0 \in A$. Thus, W is computable from A and F so the theorem is proved.

COROLLARY. *If* A *is a finite set of finite cardinal numbers then* $\deg(W) = \underset{\sim}{0}'$.

PROOF. By the theorem, it is sufficient to show that $\deg(W) \leq \underset{\sim}{0}'$. It is enough to consider the case where A consists of a single number, say k, since a uniform upper bound is preserved under a finite union. Now we can easily give an algorithm for listing the set $W_{\geq k}$ of numbers of equations with at least k solutions so

$$\deg(W_{\geq R}) \leq \underset{\sim}{0}'.$$

But $W = W_{\geq k} \cap \bar{W}_{\geq k+1}$; so $\deg(W) \leq \underset{\sim}{0}'$.

We have presented here all the known information about $\deg(W)$ for $A \subseteq C$. In particular, it is not known whether there exists a computable A for which $\deg(W) > \underset{\sim}{0}'$. There is the natural

CONJECTURE. *If* $A = \{\aleph_0\}$, *then* $\deg(W) = \underset{\sim}{0}''$.

Although this conjecture remains open, it is interesting to note that it would follow easily if every Diophantine set has a singlefold or even finitefold Diophantine definition in the sense of §3.

It is well known that there is no algorithm for listing the infinite listable sets. We can weaken the above conjecture slightly and ask

Show there is no algorithm for listing the Diophantine equations with infinitely many solutions.

14. MODELS OF ARITHMETIC

Here we will be concerned with models of the true sentences of the arithmetic of natural numbers. These models are mathematical structures consisting of a domain together with operations + and •, a relation <, and two individuals 0 and 1 defined on the domain so that specified sentences in the language $L(+,\cdot,<,0,1)$ are true. Skolem [59] first showed that there are models whose domains contain *infinite* numbers but for which all sentences true in the arithmetic of natural numbers are true in these models. An element of the domain is an *infinite* number if it is greater than 1, 1+1, 1+1+1, ... If a set S of (true) sentences of arithmetic is sufficiently strong to imply every true sentence (in N) of the form

$$\alpha + \beta = \gamma$$
$$\alpha \cdot \beta = \gamma$$
$$\alpha < \beta$$
$$x \leq \alpha \rightarrow x = 0 \vee x = 1 \vee \ldots \vee x = \alpha$$
$$x < \alpha \vee x = \alpha \vee x > \alpha$$

where α, β, γ are 0 or sums of 1's, then models of S containing infinite numbers are called *nonstandard* models. The only other models are isomorphic to N itself.

Infinite numbers correspond to sets of natural numbers. In particular, the sentence,

$$(\forall b)(\exists ad)(\forall x)_{\leq b}[d(x+1)+1 \mid a \leftrightarrow \phi(x)]$$

where $\phi(x)$ is any formula of number theory with one free variable x, is a true statement of number theory (namely, a form of Gödel's lemma) and hence true in a model of true arithmetic. (This idea was first used by Stanley Tenenbaum.) Let M be a nonstandard model of true arithmetic and b some infinite number in M. Let $\bar{n} = \sum_{i=1}^{n} 1$ for $n \in N$. Then there are a and d in M so that

$$(d(\bar{n}+1)+1)y = a \tag{1}$$

has a solution for y (in M) if and only if $\phi(\bar{n})$ holds in N. Thus, any arithmetically definable set of natural numbers is existentially definable in a nonstandard model of true arithmetic. Scott [55] exploited this remark to sharpen a theorem of Rabin:

The first order theory of countable models of the system of all predicates of natural numbers is categorical.

Note that (1) is a *generalized Diophantine equation over* M, i.e. an equation of the form

$$R_1(x_1,\ldots,x_\nu) = R_2(x_1,\ldots,x_\nu)$$

where R_1 and R_2 are polynomials with coefficients in M (compare with the discussion at the beginning of §11).

We shall consider models of these classes of true sentences of the arithmetic of natural numbers:

(1) models of *true arithmetic*, i.e. all sentences of $L(+,\cdot,<,0,1)$ true in N;

(2) models of Peano's axioms including an axiom identifying $<$ as the ordering induced by addition (called models of *Peano arithmetic*);

(3) models of all $\forall\exists$-sentences true in N, i.e. sentences of the form

$$(\forall x\ldots x_\mu)(\exists y_1\ldots y_\nu)[\phi(x_1,\ldots,x_\mu,y_1,\ldots,y_\nu)]$$

where ϕ is a quantifier-free formula (called models of $\forall\exists$-*arithmetic*).

One aspect of model theory is the study of these structures. The fact that every listable set is Diophantine can be used both to simplify old proofs and to obtain new results. Here we shall give some samples of these results which can be readily understood rather than go into the technicalities of the sharpest theorems.

I. Suppose M and M' are models of true arithmetic and M is a submodel of M'. Then there is a model M^* obtained by *interpolation* from M, i.e. by adjoining elements which lie between elements of M, such that M' is obtained from M^* by *extrapolation*, i.e. by adjoining elements which are greater than every element of M^*. See Adler [1], Čudnovskii [8], and Gaifman [17]. Gaifman generalized the theorem to models of Peano arithmetic.

II. Let M be a nonstandard model of true arithmetic. Then there is a generalized Diophantine equation over M which has no solution in M but which has a solution in an extension of M which is also a model of true arithmetic. (See Rabin [41] and Gaifman [17].)

A model M of $\forall\exists$-arithmetic is called *existentially complete* if every generalized Diophantine equation over M which has no solution in M also has no solution in every extension of M which is also a model of $\forall\exists$-arithmetic.

III. There is a formula $\phi(a)$ in $L(+,\cdot,<,0,1)$ which holds in an existentially complete model of $\forall\exists$-arithmetic if and only if a is infinite. Hence

$$(\forall a)(\sim\phi(a))$$

holds in N but in no other existentially complete model of $\forall\exists$-arithmetic. (See A. Robinson [44].)

REFERENCES

1. Andrew Adler, *Extensions of nonstandard models of number theory*, Zeitschr. f. Math. Logik und Grundlagen d. Math. 15 (1969), 289-290.

2. __________, *Some recursively unsolvable problems in analysis*, Proc. Amer. Math. Soc. 22 (1969), 523-526.

3. __________, *A reduction of homogeneous diophantine problems*, J. Lond. Math. Soc. (2), 3 (1971), 446-448.

4. Jean-Pierre, Azra, *Relations diophantiennes et la solution négative du 10e problème de Hilbert*, Séminaire Bourbaki (1970/71), exposé 383, 11-28.

5. Alan Baker, *Contributions to the theory of Diophantine equations: I. On the representation of integers by binary forms, II. The Diophantine equation* $y^2 = x^3 + k$, Philos. Trans. Roy. Soc. London Ser. A. 263 (1968), 173-208.

6. Paul Cohen, *The independence of the continuum hypothesis*, Proc. Nat. Acad. Sci. USA, 50 (1963), 1143-1148, 51 (1964), 105-110.

7. G. V. Čudnovskii, *Diophantine predicates* (Russian), Uspehi Mat. Nauk, 25 (1970) No. 4, (154), pp. 185-186.

8. __________, *Certain arithmetical problems* (Russian), Akad. Nauk Ukrain. SSR, Preprint IM-71-3.

9. Martin Davis, *Computability and Unsolvability*, McGraw-Hill, New York, 1958.

10. __________, *One equation to rule them all*, Trans. New York Acad. Sci., Series II, 30 (1968), 766-773.

11. __________, *An explicit Diophantine definition of the exponential function*, Comm. Pure Appl. Math. 24 (1971), 137-145.

12. __________, *On the number of solutions of Diophantine equations*, Proc. Amer. Math. Soc., 35 (1972), 552-554.

13. __________, *Hilbert's tenth problem is unsolvable*, Amer. Math. Monthly, 80 (1973), 233-269.

14. __________, Hilary Putnam, and Julia Robinson, *The decision problem for exponential Diophantine equations*, Ann. Math. 74 (1961), 425-436.

15. J. Denef, *Hilbert's tenth problem for quadratic rings*, Proc. Amer. Math. Soc., to appear.

16. Jens Erik Fensted, *Hilbert's 10th problem* (Norwegian), Nordisk Matema-

tisk Tidskift, 19 (1971), 5-14.

17. H. Gaifman, *A note on models and submodels of arithmetic*, Conf. in Math. Logic (W. Hodges, ed.) London 1970, Springer Verlag, 1972, 128-144.
18. Kurt Gödel, *Über formal unentscheidbare Sätze der Principia Mathematica und verwandter Systeme I*, Monatsh. Math. und Physik, 38 (1931) 173-198. English translations: (1) Kurt Gödel, *On Formally Undecidable Propositions of Principia Mathematica and Related Systems*, Basic Books, 1962. (2) Martin Davis (ed.), *The Undecidable*, Raven Press, 1965, pp. 5-38. (3) Jean Van Heijenoort (ed.), *From Frege to Gödel*, Harvard University Press, 1967, pp. 596-616.
19. ________, *Consistency-proof for the generalized continuum hypothesis*, Proc. Nat. Acad. Sci. U.S.A., 25 (1939), 220-224.
20. R. L. Goodstein, and R. D. Lee, *A class of equations in recursive arithmetic*, Zeitschr. f. Math. Logik und Grundlagen d. Math. 12 (1966), 235-239.
21. Ivan Havel, *On Hilbert's tenth problem* (Czech.), Pokr. mat. fyz. a astron. 18 (1973), 185-192.
22. Hans Hermes, *Die Unlösbarkeit des zehnten Hilbertschen Problems*, L'Enseignement Math. II, 18 (1972), 47-56.
23. O. Herrmann, *A nontrivial solution of the Diophantine equation* $9(u^2+7v^2)^2 - 7(r^2+7s^2)^2 = 2$, Computers in Number Theory (Proc. Atlas Symp., 2), Oxford 1969, 207-217.
24. A. E. Ingham, *The Distribution of Prime Numbers*, Cambridge Tract No. 30, Cambridge University Press, 1932.
25. James P. Jones, Daihachiro Sato, Hideo Wada, and Douglas Wiens, *Diophantine representation of the set of prime numbers*, Amer. Math. Monthly, to appear.
26. N. K. Kosovskii, *Diophantine representations of the sequence of solutions of the Pell equation* (Russian), Zap. Naučn. Sem. Leningrad. Otdel. Mat. Inst. Steklova (LOMI), 20 (1971), 49-59. English translation: J. Sov. Math., 1 (1973), 28-35.
27. ________, *On solutions of systems consisting of both word equations and word length inequalities* (Russian), Ibid., 40 (1974), 24-29.
28. Georg Kreisel, *Mathematical significance of consistency proofs*, J. Symbolic Logic 23 (1958), 155-182.
29. Edmund Landau, *Handbuch der Lehre von der Verteilung der Primzahlen-Erste Band*, B. G. Teubner, 1909.
30. Wilhelm Magnus, Fritz Ohettinger and Raj Pal Soni, *Formulas and Theorems for the Special Functions of Mathematical Physics*, Springer Verlag, 1966.
31. A. I. Malcev, *Algorithms and Recursive Functions* (Russian), Moscow 1965. English translation: Walters-Noordhoff, 1970.
32. Yu. I. Manin, *Hilbert's tenth problem* (Russian), Sovremennye Prob. Mat., 1973, I, 5-37.
33. A. A. Markov, *Theory of Algorithms* (Russian), Trud. Mat. Inst. Steklova 42, Moscow-Leningrad 1954. English translation: Ots 60-51085, U. S.

Dept. of Commerce, 1961.

34. Yuri Matijasevič, *Enumerable sets are diophantine* (Russian), Dokl. Akad. Nauk SSSR, 191 (1970), 279-282. Improved English translation: Soviet Math. Doklady, 11 (1970), 354-357.

35. ________, *Diophantine representation of the set of prime numbers* (Russian), Dokl. Akad. Nauk SSSR, 196 (1971), 770-773. Improved English translation with Addendum: Soviet Math. Doklady, 12 (1971), 249-254.

36. ________, *Diophantine sets*, Uspehi Mat. Nauk 27 (1972), # 5, 185-222; English translation: Russ. Math. Surveys 27 (1972), #5, 124-164.

37. ________, *The existence of noneffectizable estimates in the theory of exponential Diophantine equations* (Russian), Zap. Naučn. Sem. Leningrad. Otdel. Mat. Inst. Steklova (LOMI) 40 (1974), 77-93.

38. ________, *On recursive unsolvability of Hilbert's tenth problem*, Proc. Fourth Int. Congress Logic, Method. and Phil. of Sci., Bucharest 1971 (North-Holland 1973), 89-110.

39. ________ and Julia Robinson, *Reduction of an arbitrary Diophantine equation to one in 13 unknowns*, Acta Arithmetica 27 (1974), 521-553.

40. Hilary Putnam, *An unsolvable problem in number theory*, J. Symb. Logic, 25 (1960),220-232.

41. Michael O. Rabin, *Diophantine equations and nonstandard models of arithmetic*, Logic, Methodology, Philosophy of Science (ed. Nagel, Suppes, Tarski), Proc. of 1960 Stanford Congress, Stanford University Press, 1962, 151-158.

42. Daniel Richardson, *Some undecidable problems involving elementary functions of a real variable*, J. Symbolic Logic 33 (1968), 514-520.

43. ________, *Solution of the identity problem for integral exponential functions*, Zeitschr. f. Math. Logik und Grundlagen d. Math. 15 (1969), 333-341.

44. Abraham Robinson, *Nonstandard arithmetic and generic arithmetic*, Proc. Fourth Int. Congress Logic, Method. and Phil. of Sci., Bucharest, 1971 (North-Holland, 1973), 137-154.

45. Julia Robinson, *Definability and decision problems in arithmetic*, J. Symbolic Logic, 14 (1949), 98-114.

46. ________, *Existential definability in arithmetic*, Trans. Amer. Math. Soc., 72 (1952), 437-449.

47. ________, *Solving diophantine equations*, Proc. Fourth Int. Congress Logic, Method. and Phil. of Sci., Bucharest 1971 (North-Holland, 1973), 63-67.

48. ________, *Axioms for number theoretic functions*, Izbrannye voprocy algebry i logiki (volume dedicated to the memory of A. I. Malcev), Novosibirsk 1973, 253-263.

49. R. M. Robinson, *Arithmetical definability of field elements*, J. Symbolic Logic 16 (1951), 125-126.

50. ________, *Some representations of Diophantine sets*, J. Symbolic Logic 37 (1972), 572-578.

51. Hartley Rogers, Jr., *Theory of Recursive Functions and Effective Computability*, McGraw-Hill, 1967.

51a. J. B. Rosser, *Logic for Mathematicians*, McGraw-Hill, 1953.

52. Keijo Ruohnen, *Hilbert's tenth problem* (Finnish), Arkhimedes (Helsinkfors) 1972, 71-100.

53. T. L. Saaty, *Thirteen colorful variations on Guthrie's four-color conjecture*, Amer. Math. Monthly, 79 (1972), 2-43.

54. Edward Schwartz, *Existential definability in terms of some quadratic functions*, Doctoral Dissertation, Yeshiva University, 1974.

55. Dana Scott, *On a theorem of Rabin*, Indag. Math. 22 (1960), 481-484.

56. Joseph Shoenfield, *Degrees of Unsolvability*, North-Holland/American Elsevier, 1971.

57. Carl Ludwig Siegel, *Zur Theorie der quadratischen Formen*, Nachr. Akad. Wiss. Göttingen Math.-Phys. Kl. II (1972), 21-46.

58. Th. Skolem, *Diophantische Gleichungen*, Ergebnisse d. Math. u. Ihrer Grenzgebiete, Bd. 5, Julius Springer, 1938.

59. Th. Skolem, *Über die Nicht-charakterisierbarkeit der Zahlenreihe mittels endlich oder abzählbar unendlich vieler Aussagen mit ausschliesslich Zahlenvariablen*, Fund. Math. 23 (1934), 150-161.

60. C. Smorynski, *Diophantine decision problems; a brief survey* (unpublished).

61. Ernst Straus, Oral communication.

62. Yoshindo Suzuki, *Enumeration of recursive sets*, J. Symbolic Logic 24, (1959), 311.

63. Bryant Tuckerman, *The 24th Mersenne prime*, Notices Amer. Math. Soc. 18 (1971), 60.

64. Alan Turing, *On computable numbers, with an application to the Entscheidungsproblem*, Proc. London Math. Soc. (2), 42 (1936-7), 230-265; 43 (1937), 544-546.

65. Alurin Walther, *Anschauliches zur Riemannschen Zetafunktion*, Acta Math. 48 (1926), 393-400.

OTHER RELATED ARTICLES

66. Gregory Chaitin, *Information-theoretic limitations of formal systems*, J. Assoc. Comput. Mach. 21 (1974), 403-424.

67. Martin Davis and Hilary Putnam, *Diophantine sets over polynomial rings*, Illinois J. Math., 7 (1963), 251-256.

68. Georg Kreisel, *What have we learned from Hilbert's second problem?* This Symposium.

69. Charles F. Miller III, *Some connections between Hilbert's tenth problem and the theory of groups*, Word Problems (Boone, Cannonito, Lyndon ed.) North-Holland, 1973, 483-506.

70. S. Yu. Maslov, *The concept of strict representability in the general theory of calculi*, Trud. Mat. Inst. Steklov 93 (1967), 3-42. English translation: Proc. Steklov Inst. 93 (1967), 1-50.

71. Yuri Matijasevič, *The connection between Hilbert's tenth problem and systems of equations between words and lengths* (Russian), Zap. Naučn. Sem. Leningrad. Otdel. Mat. Inst. Steklova (LOMI), 8 (1968), 132-144. English translation: Seminars in Mathematics, V. A. Steklov Math. Inst., Leningrad, 8 (1970), 61-67.

72. __________, *Two reductions of Hilbert's tenth problem* (Russian), Ibid., 145-158. English translation: Ibid., 68-74.

73. __________, *Arithmetic representation of powers* (Russian), Ibid., 159-165. English translation: Ibid., 159-165.

74. __________, *Diophantine representation of enumerable predicates*, Izv. Akad. Nauk SSSR, Ser. Mat. 35 (1971), 3-30; English translation, Math. USSR Izv. 5 (1971), 1-28.

75. __________ and Julia Robinson, *Two universal three-quantifier representation of recursively enumerable sets* (Russian), Teopiya algorifmov u matematiceskaja logika. Vycislitel'nyi Centr. A. N. SSSR, Moscow 1974 (volume dedicated to A. A. Markov), 112-123.

76. Julia Robinson, *Diophantine decision problems*, MAA Studies in Mathematics 6 (1969) [Studies in Number Theory, ed. W. J. LeVeque, pp. 76-116].

77. __________, *Unsolvable Diophantine problems*, Proc. Amer. Math. Soc. 22 (1969), 534-538.

78. Paul Wang, *The undecidability of the existence of zeros of real elementary functions*, J. Assoc. Comput. Mach. 21 (1974), 586-589.

COURANT INSTITUTE OF MATHEMATICAL SCIENCES OF NEW YORK UNIVERSITY,
STEKLOV MATHEMATICAL INSTITUTE OF ACADEMY OF SCIENCES OF USSR, LENINGRAD,
AND
UNIVERSITY OF CALIFORNIA, BERKELEY

Proceedings of Symposia in Pure Mathematics
Volume 28, 1976

HILBERT'S ELEVENTH PROBLEM: THE ARITHMETIC THEORY OF QUADRATIC FORMS

O. T. O'Meara

1.

An m-ary classic integral quadratic form in the variables $X_1,\ldots,X_m$ over the integral domain D with quotient field $F = D \div D$ of characteristic not 2 is a homogeneous polynomial of degree 2 of the form

$$f = \sum_{1\leq i,j\leq m} a_{ij}X_iX_j \qquad (a_{ij} = a_{ji} \in D) .$$

If we make a substitution

$$X_i = \sum_{\lambda=1}^{n} t_{i\lambda}Y_\lambda \qquad (t_{i\lambda} \in D)$$

of new variables $Y_1,\ldots,Y_n$ we obtain an n-ary classic integral quadratic form

$$g = \sum_{1\leq i,j\leq n} b_{ij}Y_iY_j \qquad (b_{ij} = b_{ji} \in D) .$$

Any g which can be obtained from f in this way is said to be represented by f over D, and we then write $g \to f$ (over D). If the transforming matrix $T = (t_{ij})$ is invertible over D, we say that g and f are equivalent over D, or in the same class over D, and write $g \cong f$ (over D). The numbers represented by f over D are the numbers obtained by substituting elements of D for the variables $X_1,\ldots,X_m$. A study of the numbers represented by f is a special case of a study of the forms represented by f. A quadratic form is called isotropic over D if it represents 0 non-trivially over D. A quadratic form f is called even over D if all $a_{ii} \in 2D$, i.e. if all numbers represented by f over D fall in 2D.

The symmetric matrix $(a_{ij}) = A$ is called the matrix of the quadratic form f. Representation $B \to A$ (over D), and equivalence $B \cong A$ (over D), can

AMS(MOS) subject classifications (1970). 10-03, 10B40, 10C05, 10C15, 10C20, 10C30, 10J05, 10M20, 12-03, 12A85, 12A95, 15-03, 15A33, 15A36, 15A63, 20G25, 20G30, 20H20, 20H25.

be defined for symmetric matrices by the equation $B = {}^tTAT$, and the corresponding notions for f and g translate into these notions for symmetric matrices when quadratic forms are replaced by their associated matrices. The discriminant disc f of a quadratic form f, or disc A of the associated symmetric matrix A, is simply det A modulo squares of units of D. The discriminant is clearly an invariant of the class of f or A. We assume throughout that all given quadratic forms or symmetric matrices are non-degenerate, i.e. that their discriminants are non-zero. A diagonal quadratic form is one of the form

$$a_1X_1^2 + \dots + a_mX_m^2,$$

i.e. one whose associated matrix is diagonal. Of course

$$g \cong f \text{ (over } D) \quad \Rightarrow \quad g \cong f \text{ (over } D_1)$$

for any $D_1 \supseteq D$.

Observe that

1) Lagrange's theorem that every positive integer is the sum of four integers squared is a theorem about the numbers represented by the quadratic form $X_1^2 + \dots + X_4^2$ over $\mathbb{Z}$;

2) Hermite's theorem that any positive definite quadratic form f over $\mathbb{Z}$ with discriminant 1 and $m \leqq 7$ is equivalent to the quadratic form

$$X_1^2 + \dots + X_m^2,$$

is a theorem on the classification of quadratic forms over $\mathbb{Z}$ (the theorem fails for $m = 8$);

3) The fact that any quadratic form over a field F is equivalent to a diagonal form is a theorem on the diagonalizability of quadratic forms (which fails over arbitrary integral domains, even over $\mathbb{Z}$);

4) Sylvester's theorem that any quadratic form over $\mathbb{R}$ is equivalent to one of the form

$$X_1^2 + \dots + X_p^2 - X_{p+1}^2 - \dots - X_m^2,$$

with p and m unique, is a classification theorem over $\mathbb{R}$;

5) The fact that any quadratic form over a finite field $\mathbb{F}_q$ is equivalent to one of the form

$$X_1^2 + \dots + X_{m-1}^2 + dX_m^2,$$

is a classification theorem over $\mathbb{F}_q$, a complete set of invariants being provided by the quantities m and disc f.

2. HILBERT'S ELEVENTH PROBLEM

When Hilbert delivered his lecture on Mathematical Problems to the International Congress in Paris in 1900 he had already completed his systematic development of algebraic number theory (in 1897) and he had just established the Hilbert Reciprocity Law for number fields (in 1899). Some years earlier Minkowski had shown in a letter to Hurwitz how to classify quadratic forms over the field of rational numbers $\mathbb{Q}$ using the elaborate theory of quadratic forms over the ring of rational integers $\mathbb{Z}$ that was known at the end of the 19th century. With this as background Hilbert posed his eleventh problem on "Quadratic Forms with any Algebraic Numerical Coefficients." The English translation of the problem which appeared in the Bulletin of the American Mathematical Society in 1902 reads as follows.

"Our present knowledge of the theory of quadratic number fields puts us in a position to attack successfully the theory of quadratic forms with any number of variables and with any algebraic numerical coefficients. This leads in particular to the interesting problem: to solve a given quadratic equation with algebraic numerical coefficients in any number of variables by integral or fractional numbers belonging to the algebraic realm of rationality determined by the coefficients."

D. HILBERT, Die Theorie der algebraischen Zahlkörper, Jber. Deutsch. Math.-Verein., 4(1897), pp. 175-546.

D. HILBERT, Über die Theorie des relativquadratischen Zahlkörpers, Math. Ann. 51(1899), pp. 1-127.

H. MINKOWSKI, Über die Bedingungen, unter welchen zwei quadratische Formen mit rationalen Koeffizienten ineinander rational transformiert werden können, Crelles J., 106(1890), pp. 5-26.

3. THE HASSE PRINCIPLE

In a series of five articles in 1923/1924 Hasse set up a theory of quadratic forms over $\mathbb{Q}$ that was independent of the elaborate theory over $\mathbb{Z}$, made extensive use of Hensel's p-adic numbers, gave his own solution to the representation and classification problem over $\mathbb{Q}$, and then extended his solution to any algebraic number field. In order to prove Hasse's theorems one has to use deep results from the theory of numbers, but the theorems themselves are easy to state and to apply. First consider two quadratic forms f and g over the field of rational numbers $\mathbb{Q}$. Then $\mathbb{Q}$ can be considered as a subfield of each $\mathbb{Q}_p$ $(p = 2,3,5,7,\ldots,\infty)$. Here the $\mathbb{Q}_p$ $(p < \infty)$ are the p-adic number fields obtained by completing $\mathbb{Q}$ under each of its essentially distinct non-archimedean valuations, while $\mathbb{Q}_\infty$ is obtained by completing $\mathbb{Q}$ under its essentially unique archimedean valuation and is therefore isomorphic to $\mathbb{R}$. Clearly

$$g \cong f \text{ over } \mathbb{Q} \quad \Rightarrow \quad g \cong f \text{ over } \mathbb{Q}_p \text{ for } p = 2,3,5,\ldots,\infty.$$

But in fact

$$g \cong f \text{ over } \mathbb{Q} \iff g \cong f \text{ over } \mathbb{Q}_p \text{ for } p = 2,3,5,\ldots,\infty .$$

This is the celebrated Hasse-Minkowski theorem for equivalence over $\mathbb{Q}$. It is an illustration of what is now called the Hasse principle. It reduces the classification of quadratic forms over $\mathbb{Q}$ to the classification over each $\mathbb{Q}_p$. What is the classification over the $\mathbb{Q}_p$? For $p = \infty$ it is Sylvester's theorem. So consider $p < \infty$. Here one introduces the Hasse symbol $S_p f$ which is well-defined by replacing f by an equivalent diagonal form

$$a_1X_1^2 + \ldots + a_mX_m^2,$$

and putting

$$S_p f = \prod_{1 \leq i \leq j \leq m} \left(\frac{a_i, a_j}{p}\right)$$

where $\left(\frac{a,b}{p}\right)$ is the Hilbert symbol defined by

$$\left(\frac{a,b}{p}\right) = \begin{cases} +1 & \text{if } a\xi^2 + b\eta^2 = 1 \text{ over } \mathbb{Q}_p \\ -1 & \text{otherwise} . \end{cases}$$

Then quadratic forms over $\mathbb{Q}_p$ $(p < \infty)$ are classified by

$$m, \ \operatorname{disc}_p f, \ S_p f.$$

This completes the classification over $\mathbb{Q}$. Appearances to the contrary, it is a fact that the classification is readily computable in a finite number of steps.

If F is an algebraic number field, i.e. a finite extension of $\mathbb{Q}$, one again considers all the completions $F_{\mathfrak{p}}$ as $\mathfrak{p}$ runs through the set of inequivalent valuations Ω on F. The Hasse-Minkowski theorem

$$g \cong f \text{ over } F \iff g \cong f \text{ over } F_{\mathfrak{p}} \ \forall\, \mathfrak{p} \in \Omega$$

still holds. A finite number of $\mathfrak{p}$ are archimedean, and each corresponding $F_{\mathfrak{p}}$ is then isomorphic to $\mathbb{R}$ or to $\mathbb{C}$; in the first case classification is given by Sylvester's theorem, in the second it is trivial. Each remaining $F_{\mathfrak{p}}$ is non-archimedean, and here classification is given by

$$m, \ \operatorname{disc}_{\mathfrak{p}} f, \ S_{\mathfrak{p}} f,$$

where the Hasse symbol $S_{\mathfrak{p}} f$ is defined in $F_{\mathfrak{p}}$ just as it was defined for $\mathbb{Q}_p$.

The Hasse principle also holds for representations,

$$g \to f \text{ over } F \iff g \to f \text{ over } F_{\mathfrak{p}} \ \forall\ \mathfrak{p} \in \Omega,$$

and for isotropy,

$$f \text{ isotropic over } F \iff f \text{ isotropic over } F_{\mathfrak{p}} \ \forall\ \mathfrak{p} \in \Omega,$$

over an algebraic number field F.

A similar theory holds for global fields where, by definition, a global field is either an algebraic number field, i.e. a finite extension of $\mathbb{Q}$, or an algebraic function field in one variable over a finite constant field, i.e. a finite extension of k(t) with k finite and t transcendental over k.

The validity of the Hasse principle has been studied in many different arithmetic situations. Confining ourselves to algebraic number fields we note that it holds for quadratic forms of countable dimension (O'Meara, 1959), for central simple algebras (Brauer-Hasse-Noether, 1931), not for simultaneous isotropy of two quadratic forms (Iskovskikh, 1971). It does not hold for the equivalence of quadratic forms over $\mathbb{Z}$, but we will see that it fails there only in a finite way. It fails for $3x^3 + 4y^3 + 5z^3 = 0$ over $\mathbb{Q}$. It has been studied for algebraic groups and for elliptic curves - see the survey articles by Kneser in §8, and by Cassels below.

R. BRAUER, H. HASSE and E. NOETHER, Beweis eines Hauptsatzes in der Theorie der Algebren, Crelles J., 167(1931), pp. 399-404.

J. W. S. CASSELS, Diophantine equations with special reference to elliptic curves, J. London Math. Soc., 41(1966), pp. 193-291.

H. HASSE, Über die Darstellbarkeit von Zahlen durch quadratische Formen im Körper der rationalen Zahlen, Crelles J., 152(1923), pp. 129-148.

H. HASSE, Über die Äquivalenz quadratischer Formen im Körper der rationalen Zahlen, Crelles J., 152(1923), pp. 205-224.

H. HASSE, Symmetrische Matrizen im Körper der rationalen Zahlen, Crelles J., 153(1924), pp. 12-43.

H. HASSE, Darstellbarkeit von Zahlen durch quadratische Formen in einem beliebigen algebraischen Zahlkörper, Crelles J., 153(1924), pp. 113-130.

H. HASSE, Äquivalenz quadratischer Formen in einem beliebigen algebraischen Zahlkörper, Crelles J., 153(1924), pp. 158-162.

V. A. ISKOVSKIKH, A counterexample to the Hasse principle for systems of two quadratic forms in five variables, (Russian), Mat. Zametki, 10(1971), pp. 253-257.

O. T. O'MEARA, Infinite dimensional quadratic forms over algebraic number fields, Proc. Amer. Math. Soc., 10(1959), pp. 55-58.

4. SIEGEL'S THEOREMS

What about a Hasse principle for quadratic forms over $\mathbb{Z}$? If $\mathbb{Z}_p$ is the ring of the p-adic integers, i.e. the closure of $\mathbb{Z}$ in $\mathbb{Q}_p$, then clearly

$$g \cong f \text{ over } \mathbb{Z} \quad \Rightarrow \quad g \cong f \text{ over } \mathbb{Z}_p \text{ for } p = 2,3,5,7,\ldots .$$

Accordingly we should ask if

$$g \cong f \text{ over all } \mathbb{Z}_p \text{ and } \mathbb{Q}_\infty \quad \Rightarrow \quad g \cong f \text{ over } \mathbb{Z}?$$

In general the answer is no, as can be seen by taking the forms

$$X^2 + 55Y^2 \quad \text{and} \quad 5X^2 + 11Y^2.$$

So one says that quadratic forms g and f over $\mathbb{Z}$ are in the same genus if they are equivalent over $\mathbb{Z}_p$ for $p = 2,3,5,7,\ldots$, and also over $\mathbb{Q}_\infty$. Two quadratic forms in the same class are clearly in the same genus, so each genus of forms is partitioned into classes. It is a classical theorem that the number of classes in a genus is finite, and this number is called the class number of the given form.

Let us change from the language of quadratic forms to the language of symmetric matrices. Consider $m \times m$ and $n \times n$ symmetric matrices A and B of non-zero determinant over $\mathbb{Z}$ with $m \geq n$. Make the additional assumption that they are positive definite (when regarded as matrices over $\mathbb{R}$). Let

$$\mathbb{A}(A,B)$$

denote the number of representations of B by A over $\mathbb{Z}$, i.e. the number of distinct solutions X over $\mathbb{Z}$ with ${}^tXAX = B$. The assumption of positive definiteness then implies that $\mathbb{A}(A,B)$ is finite. Put $\mathbb{E}(A) = \mathbb{A}(A,A)$. Similarly let $\mathbb{A}_q(A,B)$ denote the number of distinct representations of B by A modulo q. It turns out that the quantity

$$\mathbb{A}_q(A,B) \div q^{mn-\frac{1}{2}n(n+1)}$$

is constant for all high powers $q = p^a$ of a given prime p. Let $\alpha_p(A,B)$ stand for this constant when $m > n$, let it stand for half this constant when $m = n$. Then it is a fundamental theorem that if $A_1,\ldots,A_h$ are representatives of all the distinct classes in the genus of A, then the ratio

$$\left(\frac{\mathbb{A}(A_1,B)}{\mathbb{E}(A_1)} + \cdots + \frac{\mathbb{A}(A_h,B)}{\mathbb{E}(A_h)}\right) \div \left(\frac{1}{\mathbb{E}(A_1)} + \cdots + \frac{1}{\mathbb{E}(A_h)}\right)$$

is equal to the convergent product

$$\varepsilon\mathbb{A}_\infty(A,B) \prod_p \alpha_p(A,B) \qquad (p = 2,3,5,7,\ldots)$$

where $\varepsilon = 1$ when $m > n+1$ or when $m = n = 1$, and $\varepsilon = \frac{1}{2}$ when $m = n+1$ or $m = n > 1$. The quantity $\mathbb{A}_\infty(A,B)$, defined as a certain limit of ratios of volumes determined by the mapping $Y \to {}^tYAY$ of $\mathbb{R}^{mn}$ ($m \times n$ matrices) into $\mathbb{R}^{\frac{1}{2}n(n+1)}$ ($n \times n$ symmetric matrices) is equal to

$$\gamma_{mn}\ (\det A)^{-\frac{n}{2}}\ (\det B)^{\frac{m-n-1}{2}}$$

where γ_{mn} is the quantity

$$\frac{\pi^{\frac{n(2m-n+1)}{4}}}{\Gamma(\frac{m-n+1}{2})\ \Gamma(\frac{m-n+2}{2})\ \cdots\ \Gamma(\frac{m}{2})}$$

For all but a finite number of p, the $\alpha_p(A,B)$ look like

$$\prod_{k=1}^{n/2} (1-p^{2k-m-1})$$

(here we took m odd and n even for the purpose of the illustration). If A corresponds to a quadratic form of class number 1, then

$$\mathbb{A}(A,B) = \varepsilon\mathbb{A}_\infty(A,B) \prod_p \alpha_p(A,B) \qquad (p = 2,3,5,\ldots)$$

and we have an expression for the number of representations of B by A in terms of local representation numbers. If, on the other hand, we take A = B we obtain (when $m > 1$)

$$\frac{1}{\mathbb{E}(A_1)} + \cdots + \frac{1}{\mathbb{E}(A_h)} = \frac{2\Gamma(\frac{1}{2})\Gamma(\frac{2}{2})\cdots\Gamma(\frac{m}{2})\,(\det A)^{\frac{m+1}{2}}}{\pi^{\frac{m(m+1)}{4}} \prod_p \alpha_p(A,A)}$$

where $p = 2,3,5,\ldots$, a fact that was known to Minkowski but for an incorrect power of 2. The quantity

$$M(A) = \frac{1}{\mathbb{E}(A_1)} + \cdots + \frac{1}{\mathbb{E}(A_h)}$$

is called the mass of the genus of the positive definite A.

Let us mention a few of the many applications of the theory. First, the mass formula provides an obvious lower bound on the class number h, and this fact has been used to ascertain that certain class numbers in specific situations are > 1. Similarly, a knowledge of the mass plus enough information about individual $\mathbb{E}(A_i)$'s can be used to verify that one has a complete set of representatives of all classes in a genus, in particular it can be used in certain situations to actually find h. Again, the quadratic form $X_1^2 + \dots + X_9^2$ over $\mathbb{Z}$ must have class number > 1, for if it were 1 we would have

$$\begin{aligned} 18 &= \mathbb{A}(I_9, I_1) \\ &= \mathbb{A}_\infty(I_9,I_1)\cdot\alpha_2(I_9,I_1)\cdot\prod_p \alpha_p(I_9,I_1) \quad (p = 3,5,\dots) \\ &= \frac{\pi^{\frac{9}{2}}}{\Gamma(\frac{9}{2})}\cdot\frac{137}{128}\cdot\prod_p (1+p^{-4}) \\ &= \frac{16\pi^4}{105}\cdot\frac{137}{128}\cdot\frac{16}{17}\,\frac{\zeta(4)}{\zeta(8)} \\ &= \frac{274}{17} \end{aligned}$$

which is absurd.

The above theory is contained in the first of a series of three long articles by Siegel that appeared during 1935/1937. The second article is concerned with the indefinite case - here the quantities $\mathbb{A}(A,B)$ and $\mathbb{E}(A)$ are infinite and things have to be reformulated. The third paper in the series moves everything from $\mathbb{Q}$ to algebraic number fields. Siegel returned to the indefinite theory in 1944, 1951 and 1952.

B. W. JONES, Related genera of quadratic forms, Duke Math J., 9(1942), pp. 723-756.

B. W. JONES, Representations by quadratic forms, Ann. of Math., 50(1949), pp. 884-899.

C. L. SIEGEL, Über die analytische Theorie der quadratischen Formen I,II,III, Ann. of Math.; 36(1935), pp. 527-606; 37(1936), pp. 230-263; 38(1937), pp. 212-291.

C. L. SIEGEL, On the theory of indefinite quadratic forms, Ann. of Math., 45(1944), pp. 577-622.

C. L. SIEGEL, Indefinite quadratische Formen und Funktionentheorie I, II, Math. Ann.; 124(1951), pp. 17-54; 124(1952), pp. 364-387.

5. THE GEOMETRIC APPROACH

In 1937 Witt, in his paper in which he discovered his cancellation law for quadratic forms, introduced a geometric approach which we now describe. (This approach has subsequently been exploited by Dieudonné for all the classical groups.) Consider an m-dimensional vector space V over the given field F (still of characteristic not 2) and assume that V is provided with a symmetric bilinear form

$$q:\quad V \times V \to F.$$

Call this a quadratic space V over F. Orthogonality in V means $q(x,y) = 0$. Put $Q(x) = q(x,x)$ and call the $Q(x)$'s the numbers represented by the quadratic space V. Call V isotropic if $Q(x) = 0$ for some non-zero vector x in V. Attach a symmetric matrix A to V through a base $x_1,\ldots,x_m$ by putting $A = (q(x_i,x_j))$. The discriminant disc V is then the discriminant of A over F. A representation ($\rightarrow$) of one quadratic space into another is a linear map that preserves the underlying bilinear forms, and an isometry ($\cong$) is a bijective representation. These concepts replace the concepts of representation and equivalence of quadratic forms over F. Assume from now on that V is non-degenerate, i.e. that disc $V \neq 0$. The isometries of V onto V form a subgroup of $GL_m(V)$, called the orthogonal group of V, and written $O_m(V)$. We will need two of its normal subgroups. The first is the so-called group of rotations $O_m^+(V) = O_m(V) \cap SL_m(V)$. The second is the so-called spinorial kernel $O_m'(V)$ which we will not define except to say that it should be viewed as the commutator subgroup of $O_m(V)$, and in fact it is equal to the commutator subgroup whenever V is isotropic.

The geometric approach was extended to the domains of number theory by Eichler in his book on quadratic forms (see § 11). For the moment, however, we continue with our general situation $F = D \div D$. A D-lattice M on V is a D-submodule of V which can be expressed in the form

$$M = D_1x_1 + \ldots + D_mx_m$$

for some base $x_1,\ldots,x_m$ for V and some "invertible" ideals $D_1,\ldots,D_m$ of D. If this can be done with all $D_i = D$, then M is called a free lattice on V. (Obviously all lattices are free when D is a principal ideal domain.) Representation and isometry can be defined for lattices in the obvious way, and these concepts generalize the concepts of representation and equivalence of quadratic forms over D, the connection being made via the symmetric matrix $A = (q(x_i,y_j))$ associated with the base $x_1,\ldots,x_m$ of the free module

$$M = Dx_1 + \ldots + Dx_m.$$

Since equivalence over D implies equivalence over F, the usual philosophy when classifying quadratic forms over D is to assume that they are already equivalent over F. In geometric language this means that the lattices in question can be taken on the same space V. Classification then becomes a question of determining whether or not lattices M and M_1 on V are related by an equation

$$M_1 = \sigma M \quad \text{for some} \quad \sigma \in O_m(V).$$

Accordingly one defines the class

$$\text{cls } M = \{\sigma M \mid \sigma \in O_m(V)\}$$

and the proper class

$$\mathrm{cls}^+ M = \{\sigma M \mid \sigma \in O_m^+(V)\}.$$

One puts

$$O_m(M) = \{\sigma \in O_m(V) \mid \sigma M = M\},$$

and one defines $O_m^+(M)$ in a similar way. If M is free with associated matrix A, one defines the discriminant disc M as the quantity det A modulo squares of units of D; and one calls M unimodular if A is a unimodular matrix, i.e. if A and A^{-1} have entries in D. (The quantity $\mathbb{E}(A)$ corresponds to the order of the group $O_m(M)$ in the special situation of § 4.)

We now become much more specific about our underlying rings. From now on F will denote an algebraic number field and D will denote the domain of all algebraic integers in F. Let Ω stand for all the non-trivial inequivalent valuations on F; let $F_\mathfrak{p}$ denote the completion of F at $\mathfrak{p} \in \Omega$; let $D_\mathfrak{p}$ denote the $\mathfrak{p}$-adic integers in $F_\mathfrak{p}$, i.e. the closure of D in $F_\mathfrak{p}$, when $\mathfrak{p}$ is non-archimedean. Thus

$$D = \bigcap_\mathfrak{p} (F \cap D_\mathfrak{p}) \qquad (\mathfrak{p} \text{ non-archimedean}).$$

Let $V_\mathfrak{p}$ denote the quadratic space obtained by extending the field of scalars F to $F_\mathfrak{p}$, at each $\mathfrak{p} \in \Omega$; let $M_\mathfrak{p}$ denote the $D_\mathfrak{p}$-lattice generated by M in $V_\mathfrak{p}$, when $\mathfrak{p}$ is non-archimedean. Note that in this terminology the Hasse-Minkowski theorem states

$$W \cong V \iff W_\mathfrak{p} \cong V_\mathfrak{p} \ \forall\, \mathfrak{p} \in \Omega.$$

The quadratic space V is said to be indefinite if $V_\mathfrak{p}$ is isotropic for at least one archimedean $\mathfrak{p}$ in Ω, otherwise it is called definite. This is consistent with the usual definitions over $\mathbb{Z}$ and $\mathbb{R}$.

E. WITT, Theorie der quadratischen Formen in beliebigen Körpern, Crelles J., 176(1937), pp.31-44.

6. THE GENUS

Two lattices M and N on V are said to be in the same genus if $M_\mathfrak{p}$ and $N_\mathfrak{p}$ are in the same class for all non-archimedean $\mathfrak{p}$ in Ω. This partitions the set of lattices on V into genera, and each genus gen M can then be partitioned into classes

$$\mathrm{cls}\, M_1, \ldots, \mathrm{cls}\, M_h.$$

The number h is called the class number and is finite (compare § 4). We have already noted that $h = 1$, which would correspond to a Hasse principle for integral quadratic forms, does not always hold. Nevertheless there are several situations in which $h = 1$, and in these situations classification over D is reduced to classification over all $D_\mathfrak{p}$, i.e. to a characterization of the genus, which is known. The solution is as follows. First one introduces the concept of a modular lattice over $D_\mathfrak{p}$, i.e. a lattice whose associated matrix is a non-zero scalar multiple of a unimodular matrix over $D_\mathfrak{p}$. Then one shows that $M_\mathfrak{p}$ can always be expressed as an orthogonal direct sum of modular sublattices. By suitably grouping these modular sublattices we see that $M_\mathfrak{p}$ is an orthogonal direct sum of modular "components". Such a decomposition is called a Jordan splitting of $M_\mathfrak{p}$. To give a specific example in matrix language over the p-adic integers $\mathbb{Z}_p$, the existence of a Jordan splitting corresponds to the fact that every symmetric matrix is equivalent over $\mathbb{Z}_p$ to one of the block form

$$\begin{pmatrix} p^{i_1}A_1 & & & & & \\ & p^{i_2}A_2 & & & & \\ & & \cdot & & & \\ & & & \cdot & & \\ & & & & \cdot & \\ & & & & & p^{i_t}A_t \end{pmatrix}$$

where the A_j are unimodular over $\mathbb{Z}_p$, and $i_1 < i_2 < \ldots < i_t$.

First let us characterize the genus in the case $F = \mathbb{Q}$, i.e. over the ring of rational integers $\mathbb{Z}$. This is the same as describing the class over each $\mathbb{Z}_p$. If p is odd, then the Jordan components of M_p are essentially unrelated, so classification is essentially reduced to the case where M_p is unimodular, and here the invariants are rank M_p and disc M_p. If $p = 2$, unimodular classification is given by considering two unimodular lattices M_2 and N_2 on the same 2-adic space V_2, and proving that M_2 and N_2 are in the same class if and only if each is even or neither is even over D_2 (in the sense of § 1). The non-unimodular 2-adic case is more difficult than the p-adic case with $p > 2$ since this time the Jordan components interact with each other, and a classification theorem is obtained by describing this interaction in terms of Hasse symbols associated with the various components in certain ways. The 2-adic classification is due to Jones in 1944 and Pall in 1945.

Now consider D as the ring of algebraic integers in an algebraic number field F. In order to characterize the genus over D we have to describe the class over $D_\mathfrak{p}$. If 2 is a unit in $D_\mathfrak{p}$, the description is the same as for $\mathbb{Z}_p$ with p odd. So assume that 2 is not a unit is $D_\mathfrak{p}$. Once again the Jordan components interact with each other. But this time the numbers represented by $M_\mathfrak{p}$, i.e. the numbers $Q(M_\mathfrak{p})$, also play an essential role. So does the additive group $\mathfrak{g}M_\mathfrak{p}$ generated by $Q(M_\mathfrak{p})$. This group can be expressed in the form

$$gM_{\mathfrak{p}} = aD_{\mathfrak{p}}^2 + bD_{\mathfrak{p}},$$

thereby providing new invariants a,b for $M_{\mathfrak{p}}$. The unimodular classification is as follows. If $M_{\mathfrak{p}}$ and $N_{\mathfrak{p}}$ are unimodular on $V_{\mathfrak{p}}$, then

$$\text{cls } M_{\mathfrak{p}} = \text{cls } N_{\mathfrak{p}} \iff Q(M_{\mathfrak{p}}) = Q(N_{\mathfrak{p}}),$$

i.e. $M_{\mathfrak{p}}$ and $N_{\mathfrak{p}}$ are in the same class if and only if they represent the same numbers. This is equivalent to $gM_{\mathfrak{p}} = gN_{\mathfrak{p}}$. The general case is obtained by tying the Jordan components together using Hasse symbols plus a,b invariants of the above type associated with these components. The dyadic classification, as well as the invariants a,b, were obtained by O'Meara in 1955 and 1957.

Minkowski's characterization of the genus was quite different and relied on so-called Gauss sums. This was over $\mathbb{Z}$. Over D one needs generalized Gauss sums, plus information derived from the a,b invariants given above. This was done by O'Meara in 1957.

The local integral representations of a quadratic form, i.e. the quadratic forms represented by a given quadratic form over $D_{\mathfrak{p}}$, were determined over $\mathbb{Z}_p$ for any p, in fact over $D_{\mathfrak{p}}$ whenever 2 is not "ramified" in $D_{\mathfrak{p}}$, by O'Meara in 1958. The ramified case, with the representing form modular, is due to Riehm in 1964.

H. BRAUN, Geschlechter quadratischer Formen, Crelles J., 182(1940), pp. 32-49.

B. W. JONES, A canonical quadratic form for the ring of 2-adic integers, Duke Math. J., 11(1944), pp. 715-727.

C. JORDAN, Sur la forme canonique des congruences du second degré et le nombre de leurs solutions, J. Math. Pures Appl. 17(1872), pp. 368-402.

O. KÖRNER, Die Masse der Geschlechter quadratischer Formen vom Range ≦ 3 in quadratischen Zahlkörpern, Math. Ann. 193(1971), pp. 279-314.

O. KÖRNER, Quadratsummen und Kongruenzlösungsdichten in $\mathfrak{p}$-adischen Zahlkörpern, Math. Nachr., 57(1973), pp. 15-38.

O. KÖRNER, Integral representations over local fields and the number of genera of quadratic forms, Acta Arith., 24(1973), pp. 301-311.

J. E. McATEE, Modular invariants of a quadratic form for a prime power modulus, Amer. J. Math., 41(1919), pp. 225-242.

O. T. O'MEARA, Quadratic forms over local fields, Amer. J. Math., 77(1955), pp. 87-116.

O. T. O'MEARA, Integral equivalence of quadratic forms in ramified local fields, Amer. J. Math., 79(1957), pp. 157-186.

O. T. O'MEARA, Local characterization of integral quadratic forms by Gauss sums, Amer. J. Math., 79(1957), pp. 687-709.

O. T. O'MEARA, The integral representations of quadratic forms over local fields, Amer. J. Math., 80(1958), pp. 843-878.

G. PALL, On the order invariants of integral quadratic forms, Quart. J. Math., 6(1935), pp. 30-51.

G. PALL, The arithmetical invariants of quadratic forms, Bull. Amer. Math. Soc., 51(1945), pp. 185-197.

G. PALL, The weight of a genus of positive n-ary quadratic forms, Proc. Sympos. Pure Math., 8(1965), pp. 95-105.

H. PFEUFFER, Bemerkung zur Berechnung dyadischer Darstellungsdichten einer quadratischen Form über algebraischen Zahlkörpern, Crelles J., 236(1969), pp. 219-220.

C. RIEHM, On the integral representations of quadratic forms over local fields, Amer. J. Math., 86(1964), pp. 25-62.

7. GENUS, SPINOR GENUS AND CLASS

Two lattices M and N on V are said to be in the same spinor genus if there are $\sigma \in O_m(V)$ and $\Sigma_{\mathfrak{p}} \in O'_m(V_{\mathfrak{p}})$ such that

$$M_{\mathfrak{p}} = \sigma_{\mathfrak{p}} \Sigma_{\mathfrak{p}} N_{\mathfrak{p}}$$

at all non-archimedean $\mathfrak{p}$. (Here $\sigma_{\mathfrak{p}}$ denotes the natural extension of the given σ in $O_m(V)$ to $\sigma_{\mathfrak{p}}$ in $O_m(V_{\mathfrak{p}})$.) In this way we obtain a partitioning of the lattices on V into spinor genera spn M that is finer than the genus but not as fine as the class, i.e.

$$\text{cls } M \subseteq \text{spn } M \subseteq \text{gen } M.$$

The proper spinor genus $\text{spn}^+ M$ is defined by taking the above σ in $O_m^+(V)$ and we then have

$$\text{cls}^+ M \subseteq \text{spn}^+ M \subseteq \text{gen } M.$$

The fundamental theorem between the class and the spinor genus is the fact that they are equal, i.e.

$$\text{cls } M = \text{spn } M, \qquad \text{cls}^+ M = \text{spn}^+ M,$$

provided V is indefinite and $m \geq 3$. The concept of the spinor genus is due to Eichler. The theorem just quoted was originally proved by Eichler over $\mathbb{Q}$, and in certain cases over algebraic number fields, in 1952. The final version over algebraic number fields was established by Kneser in 1956. Its proof rests on a theorem of "strong approximation" in the orthogonal group. In the same paper Kneser showed how to interpret spinor genera group-theoretically in a certain kind of adele group which we will describe in § 8. He deduced, among other things, that if $m \geq 3$ the number of proper spinor genera in a genus gen M is of the form 2^r with $r \geq 0$, and in fact any value of r can be obtained by varying M on the given V. As a particular example of this theory one obtains good sufficient conditions for the class to be equal to the genus in the indefinite case, thereby clarifying and extending work done by Meyer over $\mathbb{Z}$ in the 1890's.

So a genus of indefinite quadratic forms can contain 2^r classes with $r \geq 0$ arbitrary, for any $m \geq 3$. It is interesting to note that in a subject with as long a history as integral quadratic forms, it was speculated as late as 1944 by Siegel (reference in § 4) and as late as 1950 in Jones' book (see § 11), that an indefinite genus might possible have to contain only one class. The first example to the contrary seems to have been provided by Siegel in a letter to Jones in the early 1950's (see references below).

We know that quadratic forms over F can be diagonalized, i.e. that V splits into a direct sum of mutually orthogonal lines. We have already mentioned that this is not the case over D, not even over $\mathbb{Z}$. However, using the theory of the genus and the spinor genus, Watson showed in his book in 1960 (see § 11) that any M has at least one non-trivial orthogonal splitting $M = M_1 \perp M_2$ if the situation is indefinite and $m \geq 12$, with 12 best possible. And Gerstein showed in 1969 that in an indefinite situation over an algebraic number field F there is a similar splitting number m_0, but m_0 is unbounded as F is allowed to vary. We will see that the opposite phenomenon occurs when V is not indefinite.

K.-S. CHANG, Diskriminanten und Signaturen gerader quadratischer Formen, Arch. Math. (Basel), 21(1970), pp. 59-65.

M. EICHLER, Grundzüge einer Zahlentheorie der quadratischen Formen im rationalen Zahlkörper I,II, Comment. Math. Helv.; 20(1947), pp. 9-60; 21(1948), pp. 1-28.

M. EICHLER, Die Ähnlichkeitsklassen indefiniter Gitter, Math. Z. 55(1952), pp. 216-252.

L. J. GERSTEIN, Splitting quadratic forms over integers of global fields, Amer. J. Math., 91(1969), pp. 106-134.

L. J. GERSTEIN, Orthogonal decomposition of modular quadratic forms, Invent. Math., 17(1972), pp. 21-30.

B. W. JONES, An extension of Meyer's theorem on indefinite ternary quadratic forms, Canadian J. Math., 4(1952), pp. 120-128; correction, 5(1953), pp. 271-272.

B. W. JONES, and E. H. HADLOCK, Properly primitive ternary indefinite quadratic genera of more than one class, Proc. Amer. Math. Soc., 4(1953), pp. 539-543.

B. W. JONES and D. MARSH, A proof of a theorem of Meyer on indefinite ternary quadratic forms, Amer. J. Math., 77(1955), pp. 513-525.

B. W. JONES and G. L. WATSON, On indefinite ternary quadratic forms, Canadian J. Math., 8(1956), pp. 592-608.

M. KNESER, Klassenzahlen indefiniter quadratischer Formen in drei oder mehr Varänderlichen, Arch. Math. (Basel), 7(1956), pp. 323-332.

A. MEYER, Mathematische Mittheilungen, Vierteljschr. Naturforsch. Gesellsch. Zürich, 36(1891), pp. 241-250.

A. MEYER, Zur Theorie der indefiniten ternären quadratischen Formen, Crelles J., 108(1891), pp. 125-139.

A. MEYER, Ueber indefinite ternäre quadratische Formen, Crelles J., 113(1894), pp. 186-206; 114(1895), pp. 233-254; 115(1895), pp. 150-182; 116(1896), pp. 307-325.

G. L. WATSON, Representation of integers by indefinite quadratic forms, Mathematika, 2(1955), pp. 32-38.

G. L. WATSON, The equivalence of quadratic forms, Canadian J. Math., 9(1957), pp. 526-548.

8. ADELES

One can associate an adele group G_A with the quadratic space V by starting with a standard lattice M^0 on V (the final definition is independent of the choice of M^0) and defining G_A as the subgroup of the direct product

$$\prod_{\mathfrak{p} \in \Omega} O_m^+(V_\mathfrak{p})$$

consisting of all elements $(\Sigma_\mathfrak{p})_{\mathfrak{p} \in \Omega}$ for which $\Sigma_\mathfrak{p} \in O_m^+(M_\mathfrak{p}^0)$ for almost all non-archimedean $\mathfrak{p}$. One identifies $O_m^+(V)$ and its subgroup $O_m^+(M)$ with subgroups, written G_F and G_o, of G_A via the diagonal map. The subgroup of elements $(\Sigma_\mathfrak{p})_{\mathfrak{p} \in \Omega}$ of G_A with $\Sigma_\mathfrak{p} \in O_m'(V_\mathfrak{p})$ for all $\mathfrak{p}$ in Ω is written G_A'. And the subgroup of $(\Sigma_\mathfrak{p})_{\mathfrak{p} \in \Omega} \in G_A$ with $\Sigma_\mathfrak{p} \in O_m^+(M_\mathfrak{p})$ for all non-archimedean $\mathfrak{p}$ is written G_A^∞. It follows from the definitions involved that the proper classes in a genus gen M are in correspondence with a double coset decomposition

$$G_A = G_F \Sigma_1 G_A^\infty \cup \ldots \cup G_F \Sigma_j G_A^\infty.$$

The interpretation of the finiteness of class number, then, is essentially the finiteness of j. For proper spinor genera one can do better since $G_A' G_F G_A^\infty$ is a normal subgroup of G_A and, in fact, the proper spinor genera then correspond to the elements of the group $G_A / G_A' G_F G_A^\infty$.

Siegel's theorems can also be interpreted using adeles. Consider, for purposes of illustration, a positive definite situation over $\mathbb{Z}$ and $\mathbb{Q}$. Then the p-adic topology on $\mathbb{Q}_p$ makes $O_m^+(V_p)$ into a locally compact topological group in a natural way. And the topologies on the $O_m^+(V_p)$ can be used in a certain natural way to make G_A into a locally compact topological group with discrete subgroup G_F. Accordingly there are left invariant Haar measures on the $O_m^+(V_p)$ and on G_A. Since Haar measure is not absolutely unique, the different measures just described are initially unrelated. However, it is possible to construct a set of local measures at the different valuations $p = 2,3,5,\ldots,\infty$, with the measure at p related to the measure at q, and then to derive a global measure on G_A by a process involving absolutely convergent infinite products. It turns out that the global measure constructed in this way is unique. The measure of the homogeneous space G_A/G_F is called the Tamagawa number of the orthogonal group in question. The fundamental theorem is that the Tamagawa number is equal to 2. On the other hand, it follows from the constructions involved, that the Tamagawa number is equal to

$$M(A)\,\alpha_\infty(A,A)\prod_p \alpha_p(A,A) \qquad (p = 2,3,5,\ldots).$$

(Here we have switched to the notation of § 4. The quantity $\alpha_p(A,A)$ corresponds, for example, to a local measure of $O_m^+(M)$.) Putting this quantity equal to 2 gives us the formula for the mass of a genus. So the adele interpretation of the mass formula is the fact that the Tamagawa number is equal to 2.

The idea of interpreting Siegel's theorems using adeles is due to Tamagawa and Kneser. For further extensions of these and other results from the orthogonal group to algebraic groups, and for more references in these directions, see the articles listed below.

A. BOREL, Some finiteness properties of adele groups over number fields, Inst. Hautes Etudes Sci. Publ. Math., 16(1963), pp. 5-30.

A. BOREL and HARISH-CHANDRA, Arithmetic subgroups of algebraic groups, Ann. of Math., 75(1962), pp. 485-535.

M. KNESER, Darstellungsmasse indefiniter quadratischer Formen, Math. Z., 77(1961), pp. 188-194.

M. KNESER, Strong Approximation, Proc. Sympos. Pure Math., 9(1966), pp. 187-196.

M. KNESER, Semi-simple algebraic groups, Algebraic Number Theory (Proc. Instructional Conf., Brighton, 1965), Washington, 1967, pp. 250-265.

T. TAMAGAWA, Adeles, Proc. Sympos. Pure Math., 9(1966), pp. 113-121.

A. WEIL, Adeles and algebraic groups, Institute for Advanced Study, Lecture Notes, 1961, 121 +ii pp.

A. WEIL, Sur la théorie des formes quadratiques, Colloq. Théorie des Groupes Algébriques (Bruxelles, 1962), Paris, 1962, pp. 9-22.

9. POSITIVE DEFINITE QUADRATIC FORMS

In this paragraph we will be concerned with definite classic integral quadratic forms over algebraic number fields. So let F be a totally real algebraic number field (there is only trivial definiteness if F is not totally real). This means that all archimedean completions $F_\mathfrak{p}$ are $\mathbb{R}$, i.e. none is $\mathbb{C}$. Assume that V is definite. So $V_\mathfrak{p}$ is anisotropic, hence either positive definite or negative definite, at each archimedean $\mathfrak{p}$. If all these $V_\mathfrak{p}$ are in fact positive definite, then V is called positive definite. As usual, D is the ring of algebraic integers in F and M is a D-lattice on V. Assume that $q(M,M) \subseteq D$. To say that M is decomposable means that M can be expressed as an orthogonal direct sum $M = M_1 \perp M_2$ of non-zero sublattices. It is clear that M can be split

$$M = M_1 \perp \dots \perp M_t$$

into indecomposable components and in 1952 Eichler proved that the indecomposable components are unique but for their order. So a knowledge of all definite lattices is obtained by knowing the indecomposable ones.

First consider positive definite classic integral quadratic forms of discriminant 1 over $\mathbb{Z}$. We have already mentioned Hermite's result that all these forms are in a single class for each $m \leqq 7$. It was known to Korkine and Zolotareff in 1873 and to Minkowski in 1882 that there were at least two classes when $m = 8$, and Mordell proved that there were exactly two classes in 1938. Ko proved the same result for $m = 9, 10, 11$ in 1938 and 1939. In 1941 Witt showed that there were exactly two even forms when $m = 16$, one of them indecomposable. In 1957 Kneser gave an explicit construction for listing all forms of given rank, and used it to reprove results for $m \leqq 8$, to find indecomposables for $m = 8, 12, 16, 20, \dots$, and to find all unimodular indecomposables in the range $1 \leqq m \leqq 16$ -- there are seven of them. Niemeier applied Kneser's methods in 1973 and determined all unimodular even forms in 24-variables (one of them is associated with the Leech lattice which leads to the Conway simple group). Conway has mentioned to me that he has determined all unimodular lattices up to 24-variables (unpublished). These methods were applied to 3 and 4 dimensions over $\mathbb{Z}[\sqrt{3}]$ by Salamon in 1969.

In 1935 Siegel stated that "It is probably true that for every genus with a given class number, there exist only a finite number of fields..." There have been several moves in this general direction, by Magnus, Maass, Watson, Dzewas, Barner, Peters, Körner, Gerstein, Pohst, and others. We mention two of them. In 1945 Siegel proved that there are just two totally real algebraic number fields in which the natural generalization of Lagrange's theorem holds (all totally positive integers are sums of integral squares), namely $\mathbb{Q}$ and $\mathbb{Q}(\sqrt{5})$. While in 1971, Pfeuffer showed, using Siegel's mass formula, O'Meara's description of the genus, and estimates of Magnus, that "totally positive quadratic forms with three or more variables and class number 1 exist only in a finite number of algebraic number fields. Each field allows only a finite number of such forms with bounded scale." At this point we should mention that Pizer, Ponomarev and Kitaoka have recently obtained results on class numbers of definite quadratic forms in 4-variables.

When are there indecomposable positive definite classic integral quadratic forms of given rank m and given discriminant a? Mordell, Erdös and Ko considered these questions over $\mathbb{Z}$ in the 1930's. It was shown, for example, that non-decomposable (a variation of indecomposable) forms exist whenever $m > 13a + 176$, with a specified. And Kneser's constructions in 1957 found all indecomposables with $a = 2$ and $m \leqq 15$ (there are seven), and with $a = 3$ and $m \leqq 14$ (there are fourteen). In a forthcoming paper O'Meara will show that indecomposables with given m and a exist for any m if a is a prime, but for 23 exceptions; for any m if a is not square-free, with 3 exceptions; and for any $m \geqq 10$, with 5 exceptions. So far over $\mathbb{Z}$. In the case of D in a totally real F, O'Meara shows that an arbitrary M on a definite V always contains an indecomposable lattice L with volume $vL = \mathfrak{p}^2 vM$, where volume is an ideal version of discriminant, and $\mathfrak{p}$ can be any prime ideal which is "non-dyadic with $N\mathfrak{p} > 3$." In particular, all definite spaces support indecomposable lattices. And it follows that if the above D is a principal ideal domain, there are indecomposable positive definite quadratic forms over D with discriminant a for any m if a is not square-free, but for the 3 exceptions already noted, all of them over $\mathbb{Z}$.

K. BARNER, Über die quaternäre Einheitsform in total reellen algebraischen Zahlkörpern, Crelles J., 229(1968), pp. 194-208.

H. BRAUN, Über die Zerlegung quadratischer Formen in Quadrate, Crelles J., 178(1938), pp. 34-64.

M. BROUÉ and M. ENGUEHARD, Une famille infinie de formes quadratiques entières; leurs groupes d'automorphismes, Ann. Sci. École Norm. Sup., 6(1973), pp. 17-52.

J. H. CONWAY, A group of order 8,315,553,613,086,720,000, Bull. London Math. Soc., 1(1969), pp. 79-88.

J. DZEWAS, Quadratsummen in reell-quadratischen Zahlkörpern, Math. Nachr. 21(1960), pp. 233-284.

M. EICHLER, Note zur Theorie der Kristallgitter, Math. Ann., 125(1952), pp. 51-55.

P. ERDÖS and C. KO, Some results on definite quadratic forms, J. London Math. Soc., 13(1938), pp. 217-224.

P. ERDÖS and C. KO, On definite quadratic forms, which are not the sum of two definite or semi-definite forms, Acta Arith., 3(1939), pp. 102-122.

D. ESTES and G. PALL, The definite octonary quadratic forms of determinant 1, Illinois J. Math., 14(1970), pp. 159-163.

L. J. GERSTEIN, The growth of class numbers of quadratic forms, Amer. J. Math., 94(1972), pp. 221-236.

Y. KITAOKA, Two theorems on the class number of positive definite quadratic forms, Nagoya Math. J., 51(1973), pp. 79-89.

Y. KITAOKA, Quaternary even positive definite quadratic forms of prime discriminant, Nagoya Math. J., 52(1973), pp. 147-161.

M. KNESER, Zur Theorie der Kristallgitter, Math. Ann., 127(1954), pp. 105-106.

M. KNESER, Klassenzahlen definiter quadratischer Formen, Arch. Math. (Basel), 8(1957), pp. 241-250.

C. KO, Determination of the class number of positive quadratic forms in nine variables with determinant unity, J. London Math. Soc., 13(1938), pp. 102-110.

C. KO, On the positive definite quadratic forms with determinant unity, Acta Arith., 3(1939), pp. 79-85.

A. KORKINE and G. ZOLOTAREFF, Sur les formes quadratiques, Math. Ann., 6(1873), pp. 366-389.

O. KÖRNER, Formeln für die $\mathfrak{p}$-adischen Masse der quadratischen Einheitsformen in algebraischen Zahlkörpern, Math. Ann., 200(1973), pp. 267-279.

O. KÖRNER, Bestimmung einklassiger Geschlechter ternärer quadratischer Formen in reell-quadratischen Zahlkörpern, Math. Ann., 201(1973), pp. 91-95.

H. MAASS, Modulformen und quadratische Formen über dem quadratischen Zahlkörpern $R(\sqrt{5})$, Math Ann., 118(1941), pp. 65-84.

H. MAASS, Über die Darstellung total positiver Zahlen des Körpers $R(\sqrt{5})$ als Summe von drei Quadraten, Abh. Math. Sem. Univ. Hamburg, 14(1941), pp. 185-191.

H. MAASS, Quadratische Formen über quadratischen Körpern, Math. Z., 51(1948), pp. 233-254.

W. MAGNUS, Über die Anzahl der in einem Geschlecht enthaltenen Klassen von positiv-definiten quadratischen Formen, Math. Ann., 114(1937), pp. 465-475; correction, 115(1938), pp. 643-644.

L. J. MORDELL, The representation of a definite quadratic form as a sum of two others, Ann. of Math., 38(1937), pp. 751-757.

L. J. MORDELL, The definite quadratic forms in eight variables with determinant unity, J. Math. Pures Appl., 17(1938), pp. 41-46.

H.-V. NIEMEIER, Definite quadratische Formen der Dimension 24 und Diskriminante 1, J. Number Theory, 5(1963), pp. 142-178.

R. E. O'CONNOR and G. PALL, The construction of integral quadratic forms of determinant 1, Duke Math. J., 11(1944), pp. 319-331.

O. T. O'MEARA, The construction of indecomposable positive definite quadratic forms, Crelles J., to appear.

M. PETERS, Quadratische Formen über Zahlringen, Acta Arith., 24(1973), pp. 157-164.

H. PFEUFFER, Quadratsummen in totalreellen algebraischen Zahlkörpern, Crelles J., 249(1971), pp. 208-216.

H. PFEUFFER, Einklassige Geschlechter totalpositiver quadratischer Formen in totalreellen algebraischen Zahlkörpern, J. Number Theory, 3(1971), pp. 371-411.

A. K. PIZER, Class numbers of positive definite quaternary quadratic forms, preprint.

M. POHST, Mehrklassige Geschlechter von Einheitsformen in total reellen algebraischen Zahlkörpern, Crelles J., 262/3(1973), pp. 420-435.

P. PONOMAREV, Class numbers of definite quaternary forms with square discriminant, J. Number Theory, to appear.

P. PONOMAREV, Arithmetic of quaternary quadratic forms, preprint.

R. SALAMON, Die Klassen im Geschlecht von $x_1^2 + x_2^2 + x_3^2$ und $x_1^2 + x_2^2 + x_3^2 + x_4^2$ über $\mathbb{Z}[\sqrt{3}]$, Arch Math. (Basel), 20(1969), pp. 523,530.

C. L. SIEGEL, Lectures on the analytical theory of quadratic forms, Institute for Advanced Study, Lecture Notes, 1935, 239 pp.

C. L. SIEGEL, Sums of mth powers of algebraic integers, Ann of Math., 46(1945), pp. 313-339.

G. L. WATSON, Transformations of a quadratic form which do not increase the class-number, Proc. London Math. Soc., 12(1962), pp. 577-587.

G. L. WATSON, The class-number of a positive quadratic form, Proc. London Math. Soc., 13(1963), pp. 549-576.

G. L. WATSON, Positive quadratic forms with small class numbers, Proc. London Math. Soc., 13(1963), pp. 577-592.

G. L. WATSON, One-class genera of positive quadratic forms, J. London Math. Soc., 38(1963), pp. 387-392.

G. L. WATSON, One-class genera of positive ternary quadratic forms, Mathematika 19(1972), pp. 96-104.

G. L. WATSON, One-class genera of positive quaternary quadratic forms, Acta Arith., 24(1974), pp. 461-475.

E. WITT, Eine Identität zwischen Modulformen zweiten Grades, Abh. Math. Sem. Univ. Hamburg, 14(1941), pp. 323-337.

10. RELATED AREAS OF INTEREST

There are several areas of research that are one or two steps removed from Hilbert's problem on the arithmetic theory of quadratic forms but which should be mentioned. I will do so by title only.

1. Reduction theory.
2. Composition theory and binary quadratic forms.
3. Orthogonal groups over fields and arithmetic domains; their generators, normal subgroups and isomorphisms.
4. Local integral versions of Witt's cancellation and prolongation theorems.
5. The quadratic form associated with an algebraic number field via the trace.
6. Quadratic forms over various special fields.
7. Quadratic forms over arbitrary fields.
8. Quadratic forms and algebraic K-theory.
9. Hermitian forms along the lines of quadratic forms.

11. BOOKS ON THE ARITHMETIC THEORY OF QUADRATIC FORMS

One can get an idea of the status of the theory of quadratic forms at the end of the 19th century by browsing through Minkowski's collected works

H. MINKOWSKI, Gesammelte Abhandlungen I,II, Teubner Verlag, Leipzig, 1911; 371 + xxxi pp.; 467 + iv pp.

and also

P. BACHMANN, Die Arithmetik der quadratischen Formen, Teubner Verlag, Leipzig; 1898, 668 + xvi pp.; 1923, 538 + xxii pp.

Note, however, that parts of Bachmann's books are unreliable. Dickson puts it this way in the preface to his book

L. E. DICKSON, Studies in the theory of numbers, The University of Chicago Press, 1930, 230 + x pp.

on ternary and quaternary quadratic forms. "...Proofs of most of these new results depend upon known facts which are here developed in detail and with minute care. Previous statements of several of these facts have been erroneous and their proofs quite inadequate. It was no small task to write a satisfactory exposition. Practically no help was to be had from Bachmann's two large volumes on Die Arithmetik der quadratischen Formen (1898 and 1923), which contain the same errors as the original articles, and which impose very restrictive conditions on the topics treated (but without explicit mention in the theorems),..." For a history up to 1921 see

L. E. DICKSON, History of the theory of numbers, Carnegie Institute of Washington, vol. III, 1923, 313 + v pp.

The two main books on quadratic forms over the rational numbers $\mathbb{Q}$ and the rational integers $\mathbb{Z}$ are

B. W. JONES, The arithmetic theory of quadratic forms, Carus Monographs, Buffalo, 1950, 212 + x pp.

and

G. L. WATSON, Integral quadratic forms, Cambridge University Press, 1960, 143 + xii pp.,

their central message being the development from scratch of the local-global classification theory over $\mathbb{Q}$, the genus and local integral classification over $\mathbb{Z}_p$ with p odd, and some reduction theory. Watson also develops the important, and at that time newly discovered, theory of the spinor genus, and he pursues a study of the decomposition of quadratic forms. Jones uses p-adic numbers, Watson does not. Both use the language of forms and matrices. The reader will notice a distinct difference in their styles. In addition to Jones and Watson one should mention two books on number theory,

Z. I. BOREVICH and I. R. SHAFAREVICH, Number Theory, Academic Press, 1966, 435 + x pp.

and

J.-P. SERRE, A course in arithmetic, Springer-Verlag, New York Heidelberg Berlin, 1973, 115 + viii pp.,

which include a study of quadratic forms over $\mathbb{Q}$ and $\mathbb{Q}_p$. Borevich and Shafarevich also discuss binary quadratic forms, while Serre discusses quadratic forms of discriminant ± 1 over $\mathbb{Z}$. On the other hand

J. MILNOR and D. HUSEMOLLER, Symmetric bilinear forms, Springer-Verlag, New York Heidelberg Berlin, 1973, 147 + viii pp.

is devoted to quadratic forms, but not to the arithmetic theory; nevertheless it discusses questions about quadratic forms over $\mathbb{Z}$ (often without proof), including facts about their relationship with the geometry of numbers, Gauss sums, Siegel's theorem, and the Witt rings $W(\mathbb{Z})$ and $W(\mathbb{Q})$. Similarly

T. Y. LAM, The algebraic theory of quadratic forms, W. A. Benjamin, Massachusetts, 1973, 334 + xi pp.,

while concentrating on the algebraic rather than the arithmetic aspects of the theory of quadratic forms, has a chapter on local-global theory and the Witt ring $W(\mathbb{Q})$.

There are two books on the arithmetic theory of quadratic forms over algebraic number fields (in fact over global fields) F and their rings of algebraic integers D, namely

M. EICHLER, Quadratische Formen und orthogonale Gruppen, Springer-Verlag, Berlin Göttingen Heidelberg, 1952, 220 + xii pp.,

and

O. T. O'MEARA, Introduction to quadratic forms, Springer-Verlag, Berlin Heidelberg New York; 1963, 1971, 1973; 342 + xii pp.

Both authors use the geometric language of quadratic spaces and lattices, and the valuation-theoretic approach to the underlying algebraic number field F. Eichler touches on the genus, unfolds one more chapter in the theory of the class and the spinor genus that he was developing at the time, establishes a local-global theory over $F_{\mathfrak{p}}$ and F, and gives his own approach to Siegel theory and the mass. O'Meara develops what is needed from number theory and class field theory, establishes the local-global theory over $F_{\mathfrak{p}}$ and F, gives a complete account of the genus and a comprehensive account of the class and the spinor genus, and touches on positive definite quadratic forms over $\mathbb{Z}$.

Finally let us mention that all but one of our references to papers by Siegel can be found in his collected works

C. L. SIEGEL, Gesammelte Abhandlungen I,II,III, Springer-Verlag, Berlin Heidelberg New York, 1966; 548 pp.; 491 pp.; 484 pp.

UNIVERSITY OF NOTRE DAME
NOTRE DAME, INDIANA

Proceedings of Symposia in Pure Mathematics
Volume 28, 1976

SOME CONTEMPORARY PROBLEMS WITH ORIGINS IN THE JUGENDTRAUM

R. P. Langlands

The twelfth problem of Hilbert reminds us, although the reminder should be unnecessary, of the blood relationship of three subjects which have since undergone often separate developments. The first of these, the theory of class fields or of abelian extensions of number fields, attained what was pretty much its final form early in this century. The second, the algebraic theory of elliptic curves and, more generally, of abelian varieties, has been for fifty years a topic of research whose vigor and quality shows as yet no sign of abatement. The third, the theory of automorphic functions, has been slower to mature and is still inextricably entangled with the study of abelian varieties, especially of their moduli.

Of course at the time of Hilbert these subjects had only begun to set themselves off from the general mathematical landscape as separate theories and at the time of Kronecker existed only as part of the theories of elliptic modular functions and of cyclotomic fields. It is in a letter from Kronecker to Dedekind of 1880,[1] in which he explains his work on the relation between abelian extensions of imaginary quadratic fields and elliptic curves with complex multiplication, that the word Jugendtraum appears. Because these subjects were so interwoven it seems to have been impossible to disentangle the different kinds of mathematics which were involved in the Jugendtraum, especially to separate the algebraic aspects from the analytic or number theoretic. Hilbert in particular may have been led to mistake an accident, or perhaps necessity, of historical development for an "innigste gegenseitige

[1] Gesammelte Werke, Bd V.

Berührung." We may be able to judge this better if we attempt to view the mathematical content of the *Jugendtraum* with the eyes of a sophisticated contemporary mathematician.

An elliptic curve over a field k is a curve A in some projective space $\mathbb{P}^n$, defined say by equations

$$g_i(x_0, \ldots, x_n) = 0$$

together with a rational map

$$z_j = f_j(x_0, \ldots, x_n; y_0, \ldots, y_n)$$

from $A \times A$ to A which turns the set of points on A into a group. Roughly speaking – the adverb is to be taken seriously – an elliptic curve over an arbitrary commutative ring R, which we always take to be noetherian, is defined in the same way except that the coefficients of f_j and g_i are to lie in R. If one has an elliptic curve over B_1 and a homomorphism $\varphi : B_1 \longrightarrow B_2$ then replacing the coefficients of f_j and g_i by their images under φ we obtain an elliptic curve over B_2. In this way the sets $\mathcal{A}(B)$ of isomorphism classes of elliptic curves over a commutative noetherian ring B become a covariant functor on the category of such rings.

In the theory of complex multiplication one introduces a subfunctor. Take E to be an imaginary quadratic field and let O be the ring of integers in E. We are now interested only in rings B together with a homomorphism $\psi : O \longrightarrow B$ and maps $B_1 \longrightarrow B_2$ for which

$$\begin{array}{ccc} & O & \\ {}^{\psi_1}\swarrow & & \searrow^{\psi_2} \\ B_1 & \longrightarrow & B_2 \end{array}$$

is commutative. The tangent space $T(A)$ to an elliptic curve over B at the zero is a B-module. We are interested in abelian varieties A over B together with an action of the elements of O as endomorphisms of A so that the associated action of $x \in O$ on $T(A)$ is just multiplication by $\psi(x) \in B$. This gives us a new functor $B \longrightarrow \mathcal{A}^O(B)$. If n is a positive integer and if we consider only rings B in which n is invertible, we can introduce a refinement. We can let $A_n(B)$ be the points of A with coefficients from B

whose order divides n and introduce as additional datum an isomorphism of O-modules

$$\lambda : O/nO \longrightarrow A_n(B) .$$

This defines a new function $B \longrightarrow \mathcal{A}_n^O(B)$.

The methods of contemporary algebraic geometry, with which the present author is as yet only superficially acquainted, allow one to prove the existence of a universal object for this functor. This is a ring B_n, a homomorphism $O \longrightarrow B_n$, an abelian variety A' over B_n, an action of O on A', and an isomorphism

$$\lambda' : O/nO \longrightarrow A'_n(B_n)$$

so that the conditions imposed above are satisfied and so that for any B any element of $\mathcal{A}_n^O(B)$ is obtained by functoriality from A', λ' and a uniquely determined homomorphism $B_n \longrightarrow B$. This is not quite true for small n but the difficulty can be obviated by some technical considerations and is not worth stressing here.

The methods not only establish the existence but also allow one to read off properties of the ring B_n from properties of the functor $\mathcal{A}_n^O$, that is, of elliptic curves over rings. For example, the notion of smoothness or, in the language of algebraic number theory, lack of ramification, is translated into a notion of deformability. The deformation theory of elliptic curves, and of abelian varieties, is well understood, and one can show that $F_n = B_n \otimes_O E$ is a finite direct sum $\oplus E_i$ of finite algebraic extensions of E unramified away from the primes dividing n and that if O_i is the ring of integers in E_i then B_n, a subring of F_n, is equal to

$$\oplus O_i[\tfrac{1}{n}] .$$

Here $O_i[\frac{1}{n}]$ is the subring of E generated by O_i and $\frac{1}{n}$.

If we imbed E into $\overline{\mathbb{Q}} \subseteq \mathbb{C}$ then the algebra F_n is determined by the action of $\mathcal{G}(\overline{\mathbb{Q}}/E)$ on the set of its E-homomorphisms into $\mathbb{C}$, which is also $\mathcal{A}_n^O(\overline{\mathbb{Q}}) = \mathcal{A}_n^O(\mathbb{C})$. If the action is transitive the algebra is a field. Before investigating it we introduce some automorphisms of F_n. These are defined by automorphisms of the functor restricted for the moment to rings in which every positive integer is invertible. This means that B_n is to be replaced by F_n.

Let I_f be those ideles of E whose component at infinity is 1. We may imbed $E^\times$ in I_f. We are going to define an action of I_f on the functor $\mathcal{A}_n^O$. Let O_f be the ring of adeles which are integral everywhere and have component 1 at infinity. Suppose first that $g \in I_f \cap O_f$. There is a positive integer m and an $h \in I_f \cap O_f$ so that $gh = m$. Suppose A, λ in $\mathcal{A}_n^O(B)$ is given. There is an extension B' of B and an isomorphism (of sheaves!)

$$\lambda' : O/n'O \longrightarrow A_{n'}(B')$$

so that

$$m\lambda'(x) = \lambda(x) \quad .$$

g acts on $O/n'O$ and we define a new elliptic curve A_1 by dividing by

$$\{\lambda'(gx) \mid x \in nO\} \quad .$$

There is then an isogeny $\psi : A \longrightarrow A_1$ with this kernel and we define λ_1 by

$$\lambda_1(x) = \psi(\lambda'(gx)) \quad .$$

A_1, λ_1 actually define an element of $\mathcal{A}_n^O(B)$. The action of g takes A, λ to A_1, λ_1. Since elements of O are easily seen to act trivially we can extend the action to all of I_f by letting that of ℓg, with ℓ a positive integer, be the same as that of g.

The action on $\mathcal{A}_n^O(\mathbb{C})$ can easily be made explicit. If $g \in I_f$ let gO be the ideal $gO_f \cap E$. We have imbedded E in $\mathbb{C}$, and the quotient of $\mathbb{C}$ by the lattice gO is an elliptic curve A^g on which O acts. Moreover

$$A_n^g(\mathbb{C}) = \frac{gO}{n}/gO \quad .$$

If we regard O/nO as O_f/nO_f we may define λ^g as

$$x \longrightarrow \frac{gx}{n} \quad .$$

If

$$K^n = \{k \in I_f \mid k \equiv k^{-1} \equiv 1 (\mathrm{mod}\ n)\}$$

then A^g, λ^g and A^h, λ^h are isomorphic if and only if

$$h \in E^{\times} g K^n$$

so that as a set $\mathcal{A}_n^0(\mathbb{C})$ is just the quotient space $E^{\times}\backslash I_f/K^n$. The action of I_f is the obvious action on the quotient space.

By functoriality the action of $\mathcal{G}(\overline{\mathbb{Q}}/E)$ on $\mathcal{A}_n^0(\overline{\mathbb{Q}}) = \mathcal{A}_n^0(\mathbb{C})$ commutes with that of I_f. Therefore there is a unique homomorphism $\sigma \longrightarrow \varphi(\sigma)$ of $\mathcal{G}(\overline{\mathbb{Q}}/E)$ into $E^{\times}\backslash I_f/K^n$ so that the actions of σ and $\varphi(\sigma)$ are the same. It follows in particular that $\mathcal{G}(\overline{\mathbb{Q}}/E)$ acts through an abelian quotient. To understand the homomorphism $\sigma \longrightarrow \varphi(\sigma)$ we have only to identify $\varphi(\sigma)$ when σ is the Frobenius at a prime $\mathfrak{p}$ of E which does not divide n.

Let $E_{\mathfrak{p}}$ be the completion of E at $\mathfrak{p}$, $\overline{E}_{\mathfrak{p}}$ an algebraic closure of $E_{\mathfrak{p}}$, and $\overline{O}_{\mathfrak{p}}$ the ring of integers in $\overline{E}_{\mathfrak{p}}$. Fix an imbedding $\overline{\mathbb{Q}} \hookrightarrow \overline{E}_{\mathfrak{p}}$. This yields

$$\mathcal{A}_n^0(\overline{\mathbb{Q}}) \xrightarrow{\sim} \mathcal{A}_n^0(\overline{E}_{\mathfrak{p}}) = \mathrm{Hom}_O(B_n, \overline{E}_{\mathfrak{p}}) = \mathrm{Hom}_O(B_n, \overline{O}_{\mathfrak{p}}) .$$

Since B_n is unramified at $\mathfrak{p}$, we may use the map $\overline{O}_{\mathfrak{p}} \longrightarrow \overline{\kappa}_{\mathfrak{p}}$, the algebraic closure of the residue field $\kappa_{\mathfrak{p}}$ of O at $\mathfrak{p}$, to obtain

$$\mathrm{Hom}_O(B_n, \overline{O}_{\mathfrak{p}}) \simeq \mathrm{Hom}_O(B_n, \overline{\kappa}_{\mathfrak{p}}) = \mathcal{A}_n^0(\overline{\kappa}_{\mathfrak{p}}) .$$

All these isomorphisms do not affect the action of the Frobenius. Because p is not invertible in $\overline{\kappa}_{\mathfrak{p}}$, the group I_f no longer operates, at least not quite as before. However I_f^p, consisting of those ideles which are 1 at infinity and at p, continues to operate, because for these ideles we can take the auxiliary integer m prime to p, and the difficulties attendant upon the anomalous behavior of p-division points in characteristic p do not appear. Actually because of the simplicity of the present situation, it is rather easy to define an action of the missing part of I_f, namely I_p, the multiplicative group of $O \otimes \mathbb{Q}_p$. However, we want to avoid all ad hoc techniques. What is needed is an understanding of the finite subgroups, in the scheme-theoretic sense, of an elliptic curve over a field of characteristic p with order a power of p. The general method is the theory of the Dieudonné module. I do not want to give its definition here. It is a module $D(A)$ functorially associated to A. The action of I_p is replaced by the action of the O-automorphisms of $D(A) \otimes \mathbb{Q}$. This group turns out however, because of the special situation with which we are dealing, to be I_p so that I_f does operate once again. Moreover $E^\times$ and K^n still act trivially. Since I_f is generated by $E^\times$, I_f^p, and K^n its action is compatible with the isomorphisms of sets introduced above.

If ϖ is a generator of the maximal ideal of $O_{\mathfrak{p}}$ then $\varpi \in E^\times_{\mathfrak{p}} \subseteq I_p$. The theory of Dieudonné modules acquired, it is immediate that the action of ϖ on $\mathcal{A}_n^O(\bar{\kappa}_{\mathfrak{p}})$ is the same as that of the Frobenius. It follows that F_n is a field and is the abelian extension of E associated to $E^\times I_\infty K^n \subseteq I$ by class field theory. Moreover the homomorphism

$$\mathfrak{G}(\overline{\mathbb{Q}}/E) \longrightarrow \mathfrak{G}(F_n/E) \simeq I/E^\times I_\infty K^n \simeq I_f/E^\times K^n$$

given by class-field theory is just $\sigma \longrightarrow \varphi(\sigma)$. So far we have gotten by without any real arithmetic; only the arithmetic of finite fields has played a role. However, it is an essential part of the Jugendtraum that every abelian extension of E is contained in some F_n. For this we appeal to class-field theory.

But no elliptic modular functions have yet appeared. Let $V(\mathbb{Z})$ be the module of column vectors of length two over $\mathbb{Z}$. We can consider the functor which associates to B the isomorphism classes of elliptic curves over B provided with an isomorphism

$$\lambda : V(\mathbb{Z}/n\mathbb{Z}) \longrightarrow A_n(B) \ .$$

This functor is also represented by a universal object over a ring J_n. The morphism $\mathcal{A}_n^O \longrightarrow \mathcal{A}_n$ obtained by fixing an isomorphism

$$O \simeq V(\mathbb{Z})$$

and then forgetting the action of O yields a homomorphism $\eta : J_n \longrightarrow B_n$. If we imbed $B_n \longrightarrow \mathbb{C}$ over E then of course the image generates a class-field, as described above. Composing the imbedding with η yields a homomorphism of J_n or of $J_n \otimes \mathbb{C}$ into $\mathbb{C}$.

$J_n \otimes \mathbb{C}$ is the ring of rational functions on an algebraic variety S_n over $\mathbb{C}$ whose points give the homomorphisms of J_n into $\mathbb{C}$, that is, the elements of $\mathcal{A}_n(\mathbb{C})$. In particular to obtain ψ we have to evaluate the elements of J_n at some point of $S_n(\mathbb{C})$. There is, at least from the analytic standpoint, a more concrete way of viewing $S_n(\mathbb{C})$ and hence $J_n \otimes \mathbb{C}$. Let G be the group $GL(2)$. Let J_0 be the matrix

$$\begin{pmatrix} 0 & -1 \\ 1 & 0 \end{pmatrix} .$$

If $g = (g_\infty, g_f)$ belongs to $G(\mathbb{A})$ with $g_\infty \in G(\mathbb{R})$, $g_f \in G(\mathbb{A}_f)$ we set

$$g_f V(\mathbb{Z}) = g_f V(\mathbb{Z}_f) \cap V(\mathbb{Q}) .$$

Here $\mathbb{Z}_f$ is the closure of $\mathbb{Z}$ in $\mathbb{A}_f$ and $V(\mathbb{Z}_f) = V(\mathbb{Z}) \otimes \mathbb{Z}_f$. $g_f V(\mathbb{Z})$ is a lattice in $V(\mathbb{R})$. Let $J = g_\infty J_0 g_\infty^{-1}$. We turn $V(\mathbb{R})$ into a one-dimensional space over $\mathbb{C}$ by defining multiplication by $\sqrt{-1}$ to be J. Then

$$V(\mathbb{R})/g_f V(\mathbb{Z})$$

is an elliptic curve A^g over $\mathbb{C}$. Also

$$A_n^g(\mathbb{C}) = \frac{g_f V(\mathbb{Z})}{n} \Big/ g_f V(\mathbb{Z})$$

so we may take λ^g to be

$$x \longrightarrow \frac{g_f x}{n} .$$

The isomorphism class of A^g, λ^g is determined solely by the image of g in the double coset space

$$G(\mathbb{Q}) \backslash G(\mathbb{A}) / K_\infty K^n$$

if K_∞ is the centralizer of J_0 in $G(\mathbb{R})$ and K^n is

$$\{k \in G(\mathbb{Z}_f) \mid k \equiv 1 (\mathrm{mod}\ n)\} .$$

This double coset space has a natural complex structure and may now be identified with $S_n(\mathbb{C}) = \mathcal{A}_n(\mathbb{C})$.

Analyzing the double cosets more carefully one sees that $S_n(\mathbb{C})$ consists of finitely many connected pieces each of which is the quotient of the Poincaré half-plane by a congruence subgroup. The elements of $J_n \otimes \mathbb{C}$, in particular the elements of J_n, are functions on these pieces and are in fact just the elliptic modular functions of level n. The points of $S_n(\mathbb{C})$ corresponding to the homomorphism ψ introduced above are easily found explicitly. Summing up, we conclude that the class field F_n is generated by the values of the elliptic modular functions in J_n at a certain easily found point of

$$G(\mathbb{Q})\backslash G(\mathbb{A})/K_\infty K^n \ .$$

As we said, any connected piece of this space is a quotient of the Poincaré half-plane by a discrete group. If we lift the functions in J_n to the half-plane they become transcendental.

This aspect, the generation of class fields by the values of transcendental functions, has been emphasized by Hilbert who suggests, in the twelfth problem, that it may be possible to find for an arbitrary number field transcendental functions with a similar property. Whether justly or not, the twelfth problem has received very little attention. Any progress made on it has been an incidental result of research with quite different ends, although it too has its origins in the Jugendtraum. The bulk of this research is due to Shimura.

A characteristic of the number theory of the twentieth century has been the dominant role played by zeta-functions and L-functions, especially at a conjectural level. The analytic properties of the L-functions associated to an algebraic variety over a number field have been particularly difficult, usually impossible, to determine. But Shimura has studied very deeply certain varieties, which, like the varieties defined by elliptic modular functions, are closely related to algebraic groups. For various reasons it is to be expected that the L-functions associated to these Shimura varieties can be expressed in terms of the L-functions associated to automorphic forms on the group defining the variety and on certain related groups. This in itself is not enough to establish the analytic properties but it is a first step. Shimura, inspired by earlier work of Eichler, has been able to confirm the expectation for some of his varieties, basically those which are curves.

But many problems remain. I want to discuss one of them, rather casually, in the remainder of the lecture. There are various notions of a reciprocity law, all of them implicit in the laws of class-field theory. For example, one can view a theorem asserting that an L-function defined by diophantine data, that is, by an algebraic variety over a number field, is equal to an L-function defined by analytic data, that is, by an automorphic form, as a reciprocity law. There is good reason for this, for the Artin reciprocity law is such an assertion. The results of Eichler and Shimura are also of this form. There is nonetheless a more concrete notion available.

Suppose one has an algebraic variety S defined over a number field E.

Suppose in fact that equations defining S have been chosen whose coefficients are integral outside of some finite set of primes Q. If $\mathfrak{p} \notin Q$ and $\kappa_{\mathfrak{p}}$ is the residue field of E at $\mathfrak{p}$ we can reduce the equations modulo $\mathfrak{p}$ and then speak of the set $S(\bar{\kappa}_{\mathfrak{p}})$ of points of S with coefficients in $\bar{\kappa}_{\mathfrak{p}}$. $S(\bar{\kappa}_{\mathfrak{p}})$ is given together with an action on it, that of the Frobenius $\Phi_{\mathfrak{p}}$. An explicit description of the sets $S(\bar{\kappa}_{\mathfrak{p}})$ and of the actions of $\Phi_{\mathfrak{p}}$ for all $\mathfrak{p} \notin Q$ could also be viewed as a reciprocity law. For example, if E = Q and S is defined by the equation

$$x^2 + 1 = 0$$

then $S(\bar{\kappa}_p)$ for $p \nmid 2$ is a set with two elements and Φ_p acts trivially or not according as $p \equiv 1$ or $p \equiv 3$ modulo 4. This is the first supplement to the law of quadratic reciprocity.

It is very likely that Shimura varieties admit a reciprocity law in this sense. I want to describe explicitly the form the law will most probably take. The description is speculative, but I have verified its correctness, in so far as my limited command of the necessary techniques allows, for those varieties which arise as solutions of moduli problems for abelian varieties.

To know the zeta-function of a variety, at least in the sense of knowing the factors of its Euler product expansion for almost all $\mathfrak{p}$, one just has to know the number of points in $S(\kappa_{\mathfrak{p}}^n)$ for all positive n, if $\kappa_{\mathfrak{p}}^n$ is the extension of $\kappa_{\mathfrak{p}}$ of degree n. This is just the number of fixed points of $\Phi_{\mathfrak{p}}^n$ in $S(\bar{\kappa}_{\mathfrak{p}})$. One might expect that this could be determined from the explicit description of $S(\bar{\kappa}_{\mathfrak{p}})$ and of the action of $\Phi_{\mathfrak{p}}$; so that from a reciprocity law in the second sense for the Shimura variety S one could obtain one in the first sense, at least for its zeta-function. However, difficult combinatorial problems arise which have not yet been seriously broached. But I have been able to make the transition in a limited number of cases, among which are included varieties of arbitrary large dimension.

The work of Shimura has been exposed in a remarkably clear fashion by Deligne,[2] who also added improvements of his own. One begins with a reductive algebraic group G over $\mathbb{Q}$ and a homomorphism $h_0 : GL(1) \longrightarrow G$ defined over $\mathbb{C}$. The pair (G, h_0) is subject to some simple formal conditions. If R is the torus over $\mathbb{R}$ obtained from GL(1) over $\mathbb{C}$ by

[2] Séminaire Bourbaki, 1970/71.

restriction of scalars so that over $\mathbb{C}$

$$R \simeq GL(1) \times GL(1)$$

then the composition

$$h : R \xrightarrow{\sim} GL(1) \times GL(1) \longrightarrow G$$

where the second map is $(x, y) \longrightarrow h_0(x)^{\rho} h_0(y)$ with ρ the complex conjugation, is to be a homomorphism defined over $\mathbb{R}$. The centralizer of $h(R)$ in $G_{der}(\mathbb{R})$ is to be a maximal compact subgroup of $G_{der}(\mathbb{R})$ and if K_∞ is the centralizer of $h(R)$ in $G_{der}(\mathbb{R})$ then the quotient $G(\mathbb{R})/K_\infty$ is to carry an invariant complex structure, specified by h_0.

It is in fact not h_0 which is significant but the collection of $\mathrm{ad}\, g \circ h_0$, $g \in G(\mathbb{R})$. If T is a Cartan subgroup of G defined over $\mathbb{Q}$ with $T(\mathbb{R}) \cap G_{der}(\mathbb{R})$ compact we may choose $h_0' = \mathrm{ad}\, g \circ h_0$ so that it factors through T. We then denote $h_0' : GL(1) \longrightarrow T$ by $\hat{\mu}$; it is a coweight of T. If E is defined to be the fixed field of the set of all $\sigma \in \mathfrak{G}(\overline{\mathbb{Q}}/\mathbb{Q})$ for which $\sigma\hat{\mu} = \omega\hat{\mu}$ with ω in the Weyl group of T then E, which is a finite extension of $\mathbb{Q}$ in $\mathbb{C}$, plays an important role in the study of Shimura varieties.

If K is an open compact subgroup of $G(\mathbb{A}_f)$ then the complex manifold

$$S_K(\mathbb{C}) = G(\mathbb{Q})\backslash G(\mathbb{A})/K_\infty K$$

is the set of complex points of an algebraic variety over $\mathbb{C}$. It has been conjectured, hesitantly by Shimura, openly by Deligne, that this family of algebraic varieties should have models S_K over E. The precise conjecture actually demands certain further properties of the S_K, which serve to characterize them uniquely. These properties are patterned on the Jugendtraum; so that implicit in any proof of the existence of these canonical models S_K is a partial solution to the twelfth problem. The conjecture, colloquially referred to as the Shimura conjecture, has been solved for many groups but by no means all. My suggestions will only make sense for those groups for which the Shimura conjecture is acquired.

The group $G(\mathbb{A}_f)$ operates on

$$\varprojlim_K S_K(\mathbb{C}) \ .$$

It is demanded that this be reflected in an action of $G(\mathbb{A}_f)$ on

$$\varprojlim_K S_K$$

defined over E.

Fix a prime $\mathfrak{P}$ of E and let p be the prime of $\mathbb{Q}$ it divides. I shall suppose that the group G is quasi-split over $\mathbb{Q}_p$ and split over an unramified extension. Recall that if G_{sc} is the simply-connected form of the derived group G_{der} then Bruhat and Tits have associated a building to $G_{sc}(\mathbb{Q}_p)$ on which $G(\mathbb{Q}_p)$ acts. A special maximal compact subgroup of $G(\mathbb{Q}_p)$ is the intersection of the stabilizer in $G(\mathbb{Q}_p)$ of a special vertex of the Bruhat-Tits building with

$$\{g \in G(\mathbb{Q}_p) \mid |\chi(g)| = 1 \text{ for all rational characters of } G \text{ defined over } \mathbb{Q}_p\} .$$

We shall only be interested in K of the form

$$K = K^p K_p$$

where $K^p \subset G(\mathbb{A}_f^p)$ and K_p is a special maximal compact of $G(\mathbb{Q}_p)$.

The varieties S_K are defined over E and hence over $E_{\mathfrak{P}}$. Suppose $O_{\mathfrak{P}}$ is the ring of integers of $E_{\mathfrak{P}}$. To speak of $S_K(\bar{\kappa}_{\mathfrak{P}})$ we need models over $O_{\mathfrak{P}}$. At the moment I do not know how they should be characterized. Presumably if S_K/E is proper and smooth then $S_K/O_{\mathfrak{P}}$ should also be proper and smooth. But if S_K/E is not proper, some attention will have to be paid to the behavior at infinity. I simply ignore the difficulty for now and go on to describe the anticipated structure of $S_K(\bar{\kappa}_{\mathfrak{P}})$. It is enough to consider that of

$$\varprojlim_{K^p} S_K(\bar{\kappa}_{\mathfrak{P}}) = S_{K_p}(\bar{\kappa}_{\mathfrak{P}})$$

provided that we know how $G(\mathbb{A}_f^p)$ acts on the right-hand side, for

$$S_K(\bar{\kappa}_{\mathfrak{P}}) = S_{K_p}(\bar{\kappa}_{\mathfrak{P}})/K^p .$$

The set $S_{K_p}(\bar{\kappa}_{\mathfrak{P}})$ should be the union of certain subsets invariant under $G(\mathbb{A}_f^p)$ and $\Phi = \Phi_{\mathfrak{P}}$. Each of them is constructed from the following data:

(i) A group H over $\mathbb{Q}$ and an imbedding $H(\mathbb{A}_f^p) \hookrightarrow G(\mathbb{A}_f^p)$;

(ii) A group $\bar{G}$ over $\mathbb{Q}_p$ and an imbedding $H(\mathbb{Q}_p) \hookrightarrow \bar{G}(\mathbb{Q}_p)$;

(iii) A space X on which $\overline{G}(\mathbb{Q}_p)$ and Φ act, the two actions commuting with each other.

The imbeddings $H(\mathbb{A}_f^p) \hookrightarrow G(\mathbb{A}_f^p)$, $H(\mathbb{Q}_p) \hookrightarrow \overline{G}(\mathbb{Q}_p)$ when combined with the diagonal imbedding $H(\mathbb{Q}) \hookrightarrow H(\mathbb{A}_f)$ yield an action of $H(\mathbb{Q})$ on $G(\mathbb{A}_f^p) \times X$. The subsets to which I referred have the form

$$Y = H(\mathbb{Q})\backslash G(\mathbb{A}_f^p) \times X \ .$$

$G(\mathbb{A}_f^p)$ acts in the obvious way to the right and Φ acts through its action on X.

Before venturing a general prescription for H, G, and X we should orient ourselves with a brief glance at $G = GL(2)$ with h given by

$$(a+ib,\ a-ib) \longrightarrow \begin{pmatrix} a & -b \\ b & a \end{pmatrix} \qquad a,\ b \in \mathbb{C},\ a^2+b^2 \neq 0 \ .$$

For mnemonic reasons, I adhere to a slightly different convention than Deligne, so that my h is the inverse of his. For this pair G, h there is one subset for each imaginary quadratic extension F of $\mathbb{Q}$. H is the group F^* over $\mathbb{Q}$ associated in the usual way to F so that $H(\mathbb{Q}) = F^\times$, $\overline{G}$ is also H, and X is the quotient of $H(\mathbb{Q}_p) = (E \otimes \mathbb{Q}_p)^\times \simeq \mathbb{Q}_p^\times \times \mathbb{Q}_p^\times$ by $H(\mathbb{Z}_p)$, the group of units, $\mathbb{Z}_p^\times \times \mathbb{Z}_p^\times$. If $\mathfrak{q}$ is one of the prime divisors of p in E and ϖ the corresponding uniformizing parameter then Φ is multiplication by $\varpi \in (E \otimes \mathbb{Q}_p)^\times$. There is one additional subset. For it, H is the multiplicative group of the quaternion algebra over $\mathbb{Q}$ split everywhere except at infinity and p and $\overline{G}$ is H. X is the quotient $\overline{G}(\mathbb{Q}_p)/\overline{G}(\mathbb{Z}_p)$, if $\overline{G}(\mathbb{Z}_p)$ is the multiplicative group of the maximal order in the completion of the algebra at p. Φ is multiplication by any ϖ in this order which generates the maximal ideal.

There is an alternative description of the X for the final subset which yields more insight into the general situation. Let $\mathfrak{K}$ be the completion of the maximal unramified extension of $\mathbb{Q}_p$ and $\mathcal{O}$ its ring of integers. Denote by $a \longrightarrow {}^\sigma a$ the action of the Frobenius. Let $\mathcal{H}$ be the set of $\mathcal{O}$-lattices in the space of column vectors of length 2 over $\mathfrak{K}$. $\mathcal{H}$ is the set of vertices in the Bruhat-Tits building of $G(\mathfrak{K})$. Set

$$b = \begin{pmatrix} 0 & 1 \\ p & 0 \end{pmatrix} \ .$$

Define an action of Φ on $\mathcal{H}$ by

$$\Phi\mathcal{L} = b^{\sigma}\mathcal{L} .$$

Then

$$\overline{G}(\mathbb{Q}_p) = \{g \in G(\mathcal{K}) \mid b^{\sigma}gb^{-1} = g\}$$

and X is the set of all $\mathcal{L}$ in $\mathcal{H}$ for which

$$p\mathcal{L} \subsetneq \Phi\mathcal{L} \subsetneq \mathcal{L} .$$

Geometrically this means that the images of $\mathcal{L}$ and $\Phi\mathcal{L}$ in the Bruhat-Tits building of $G_{sc}(\mathcal{K}) = SL(2, \mathcal{K})$ are joined by an edge. To verify that the two descriptions of X are not essentially different, one uses the fact that the Bruhat-Tits building is a tree. It is an amusing exercise.

To define H, $\overline{G}$, and X in general we fix $\overline{\mathbb{Q}} \hookrightarrow \mathbb{C}$ and then choose, once and for all, an imbedding $\overline{\mathbb{Q}} \hookrightarrow \overline{\mathbb{Q}}_p$ so that the prime of E it defines is $\mathfrak{P}$. Suppose γ belongs to $G(\mathbb{Q})$ and is semi-simple. Suppose moreover that all the eigenvalues of γ have absolute value 1 away from infinity and p. Let

$$H^{o} = \{g \in G \mid g\gamma^{m} = \gamma^{m}g \text{ for some } m \neq 0 \text{ in } \mathbb{Z}\} .$$

H^{o} is connected and of course defined over $\mathbb{Q}$. Suppose $h^{o} : R \longrightarrow H^{o}$ and the composition $R \xrightarrow{h^{o}} H^{o} \hookrightarrow G$ is conjugate under $G(\mathbb{R})$ to h. If T is a Cartan subgroup of H^{o} defined over $\mathbb{Q}$ with $T(\mathbb{R}) \cap G_{der}(\mathbb{R})$ compact, then, as before, replacing h^{o} by $\operatorname{ad} g \circ h^{o}$, $g \in H^{o}(\mathbb{R})$, if necessary, we may suppose h^{o} factors through T. The associated

$$h_0^{o} : GL(1) \longrightarrow T$$

is a coweight $\mu^{\wedge}$ of T. $\mu^{\wedge}$ is not uniquely determined by h^{o}, but its orbit under the Weyl group of T in H^{o} is; and that suffices for the following.

If $L(T)$ is the $\mathbb{Z}$-module

$$\operatorname{Hom}(T, GL(1))$$

and

$$L^{\wedge}(T) = \operatorname{Hom}(GL(1), T)$$

then $L^{\wedge}(T)$ is also

$$\mathrm{Hom}(L(T), \mathbb{Z}) \ .$$

Define $\lambda^{\wedge}(\gamma) \in L^{\wedge}(T)$ by

$$|\lambda(\gamma)|_p = p^{-\langle \lambda, \lambda^{\wedge}(\gamma)\rangle} \qquad \lambda \in L(T) \ .$$

Let M be the lattice of rational characters of H^o defined over $\mathbb{Q}_p$. We say that the pair (γ, h^o) is of Frobenius type if there is an $r > 0$ in $\mathbb{Q}$ so that $\lambda^{\wedge}(\gamma) - r\mu^{\wedge}$ is orthogonal to M.

Later an equivalence relation will be introduced on pairs of Frobenius type. To each equivalence class will be associated H, $\overline{G}$, and X, as well as

$$Y = H(\mathbb{Q})\backslash G(\mathbb{A}_f^p) \times X \ .$$

For each equivalence class we will also define a multiplicity d. If dY is the disjoint union of d copies of Y then, as a set on which Φ and $G(\mathbb{A}_f^p)$ act, $S_{K_p}(\overline{\kappa}_{\mathfrak{p}})$ should be isomorphic to the disjoint union over equivalence classes of pairs of Frobenius type of the sets dY.

For the moment fix γ and h^o. H will be obtained from H^o by an inner twisting. Since the Hasse principle is valid for the adjoint group H^o_{ad}, it is enough to specify the twisting locally! Of course it has also to be verified that there is a global twisting with the specified local behavior; but this turns out to be a matter of standard techniques. The twisting is trivial except at infinity and p. At infinity it is so arranged that $H_{der}(\mathbb{R})$ is compact. Before describing the twisting at p, we introduce a subgroup $\overline{G}^o$ of G defined over $\mathbb{Q}_p$. It is the connected subgroup whose Lie algebra is spanned by those elements V in the Lie algebra of G satisfying

$$\mathrm{Ad}\, \gamma(V) = \varepsilon V$$

with $\varepsilon \in \overline{\mathbb{Q}}_p$ and $|\varepsilon|_p = 1$. $\overline{G}^o$ contains H^o. $\overline{G}$ will be a twisted form of $\overline{G}^o$.

We shall in fact twist $\overline{G}^o$ and H^o simultaneously. If T is as above, let T_{ad} be its image in H^o_{ad} and $\overline{T}_{ad}$ its image in $\overline{G}^o_{ad}$. We choose T so that T_{ad} is anisotropic over $\mathbb{Q}_p$. Choose a finite Galois extension k of $\mathbb{Q}$ in $\overline{\mathbb{Q}}$ over which T splits. Suppose $a_{\sigma,\tau}$ is a fundamental 2-cocycle for $k_p/\mathbb{Q}_p$. Since

$$T(k_p) = L^\wedge(T) \otimes k_p^\times$$

we may introduce the 1-cochain

$$\sigma \longrightarrow a_\sigma = \Sigma_{\tau \in \mathfrak{G}(k_p/\mathbb{Q}_p)} \sigma\tau\mu^\wedge \otimes a_{\sigma,\tau} .$$

It takes values in $T(k_p)$ but is not a 1-cocycle. However its image in $T_{ad}(k_p)$ or $\overline{T}_{ad}(k_p)$ is. Composing with the maps

$$H^1(\mathfrak{G}(k_p/\mathbb{Q}_p),\ T_{ad}(k_p)) \longrightarrow H^1(\overline{\mathbb{Q}}_p,\ H^0)$$
$$H^1(\mathfrak{G}(k_p/\mathbb{Q}_p),\ \overline{T}_{ad}(k_p)) \longrightarrow H^1(\overline{\mathbb{Q}}_p,\ \overline{G}^0)$$

we obtain the twisting cocycles for H^0 and $\overline{G}^0$ at p. One must of course verify that the twistings are independent of all auxiliary data.

The homomorphism $H^0 \longrightarrow G_{der}\backslash G$ yields $H \longrightarrow G_{der}\backslash G$. The multiplicity d is the number of elements in $H^1(\overline{\mathbb{Q}}, H)$ which are trivial at every place except p, including infinity, and which lie in the kernel of

$$H^1(\overline{\mathbb{Q}}, H) \longrightarrow H^1(\overline{\mathbb{Q}}, G_{der}\backslash G) .$$

It may be, however, a little rash to predict d on the basis of the examples studied, for the groups involved have special cohomological properties.

The set X is the object must complicated to define. Set

$$\nu^\wedge = \Sigma_{\tau \in \mathfrak{G}(k_p/\mathbb{Q}_p)} \tau\mu^\wedge$$

and denote

$$\nu^\wedge \otimes x \in L^\wedge(T) \otimes k_p^\times = T(k_p)$$

by $x^{\nu^\wedge}$. We define the Weil group, $W_{k_p/\mathbb{Q}_p}$, by means of the cocycle $a_{\sigma,\tau}$. If $w = (x, \sigma) \in W_{k_p/\mathbb{Q}_p}$, with $x \in k_p^\times$, $\sigma \in \mathfrak{G}(k_p/\mathbb{Q}_p)$, set

$$b_w = x^{\nu^\wedge} a_\sigma .$$

Then $w \longrightarrow b_w$ is a 1-cocycle. Let D be the maximal torus in the centre of H^0 split over $\mathbb{Q}_p$. Let $\mathfrak{k}$ be the maximal unramified extension of $\mathbb{Q}_p$. It turns out that if we enlarge k_p to some k'_p and inflate b_w to $W_{k'_p/\mathbb{Q}_p}$ then we may represent its class by a cocycle $\{\overline{b}_w\}$ such that

$$\bar{b}_w = \bar{b}'_w \bar{b}''_w$$

where $\bar{b}'_w \in T(\mathfrak{k})$, $\bar{b}''_w \in D(\bar{\mathbb{Q}}_p)$ and

$$|\lambda(\bar{b}''_w)|_p = 1$$

for all rational characters of D. Moreover if W^o is the kernel of $W_{k_p/\mathbb{Q}_p} \longrightarrow \mathfrak{G}(\bar{\mathbb{F}}_p/\mathbb{F}_p)$ then we may take $\bar{b}'_w = 1$ for $w \in W^o$. If w is any element of $W_{k'_p/\mathbb{Q}_p}$ which maps to the Frobenius in $\mathfrak{G}(\bar{\kappa}_{\mathfrak{p}}/\kappa_{\mathfrak{p}})$ set

$$b = \bar{b}'_w \ .$$

We regard b as an element of $G(\mathfrak{k})$. Any other choice of the auxiliary data replaces b by $cb^{\sigma}c^{-1}$ if σ is the Frobenius on $\mathfrak{k}$. Such a change is irrelevant for our purposes. Observe that we may realize $\bar{G}(\mathbb{Q}_p)$ as

$$\{g \in G(\mathfrak{k}) \mid b^{\sigma}gb^{-1} = g\} \ .$$

The group K_p determines a special vertex in the Bruhat-Tits building of $G_{sc}(\mathbb{Q}_p)$ and hence of $G_{sc}(\mathfrak{k})$ which in turn determines a parahoric subgroup $K_p(\mathfrak{k})$ of $G(\mathfrak{k})$. Set

$$\mathcal{H} = G(\mathfrak{k})/K_p(\mathfrak{k}) \ .$$

Let F be the map $\mathcal{H} \longrightarrow \mathcal{H}$ which takes the point represented by g to the point represented by $b^{\sigma}g$.

There is a bijection between conjugacy classes of parabolic subgroups of G and conjugacy classes of parahoric subgroups with a representative in $K_p(\mathfrak{k})$. Let $\mathcal{J}$ be the class determined by the parabolic subgroup generated by T and the family of one-parameter subgroups corresponding to roots α with $\langle \alpha, \mu^\wedge \rangle \geq 0$. Any point $\mathcal{V}$ of $\mathcal{H}$ determines a special spoint $\mathcal{V}_i$ in the Bruhat-Tits building of each simple factor $G_i(\mathfrak{k})$ of $G_{sc}(\mathfrak{k})$. We consider only those $\mathcal{V}$ such that if $\mathcal{W} = F\mathcal{V}$ then, for each i, $\mathcal{V}_i$ and $\mathcal{W}_i$ are either the same or are joined by an edge. Then $\mathcal{V}_i$ and $\mathcal{W}_i$ determine a parahoric subgroup of $G_i(\mathfrak{k})$ and, as a consequence, $\mathcal{V}$ and $\mathcal{W}$ determine a parahoric subgroup of $G(\mathfrak{k})$. X consists of those $\mathcal{V}$ for which this parahoric subgroup lies in the class $\mathcal{J}$. $G(\mathbb{Q}_p)$ acts on X. If $r = [E_{\mathfrak{p}} : \mathbb{Q}_p]$ we define the action of $\Phi_{\mathfrak{p}}$ to be F^r.

The correct conditions defining the equivalence of two pairs (γ_1, h_1^o), (γ_2, h_2^o) seem to be local, one condition at each finite place, but none at the infinite place. There should be positive integers m and n and a δ in the centre of $G(\mathbb{Q})$ with every eigenvalue a unit so that first of all γ_1^m and $\delta\gamma_2^n$ are conjugate in $G(\mathbb{Q}_\ell)$ for each $\ell \neq p$. They should also be conjugate in $G(\mathfrak{K})$. Let

$$\delta\gamma_2^n = g\gamma_1^m g^{-1} \qquad g \in G(\mathfrak{K}) .$$

Suppose b_1 and b_2 in $H_1^o(\mathfrak{K})$ and $H_2^o(\mathfrak{K})$ are associated to (γ_1, h_1^o) and (γ_2, h_2^o) as above. Then $gb_1{}^\sigma g^{-1} \in H_2^o(\mathfrak{K})$. The final condition for equivalence is that there be a c in $H_2^o(\mathfrak{K})$ so that

$$cgb_1{}^\sigma g^{-1}\,{}^\sigma c^{-1} = b_2 .$$

In order to define the Γ-factors which should appear in the functional equation of the zeta-function of a Shimura variety one must also know something about their behavior at the infinite places of E. Two problems arise. If τ is an automorphism of $\overline{\mathbb{Q}}$ over $\mathbb{Q}$ we may apply τ to the family S_K over E to obtain a family of varieties ${}^\tau S_K$ over ${}^\tau E$. This new family should be again just the canonical models for the Shimura varieties defined by some new pair $({}^\tau G, {}^\tau h_0)$. There is an obvious guess. ${}^\tau G$ should be obtained from G by an inner twisting which is trivial at every finite place. If ρ is the complex conjugation and T is chosen as above then the twisting cocycle at infinity should be $\rho \longrightarrow t_\rho$ with

$$\lambda(t_\rho) = (-1)^{\langle \lambda, \tau\mu^\wedge - \mu^\wedge \rangle} \qquad \lambda \in L(T) .$$

Then T may also be regarded as a Cartan subgroup of ${}^\tau G$ over $\mathbb{R}$. The homomorphism ${}^\tau h_0$ should just be the composition

$$GL(1) \xrightarrow{\tau\mu^\wedge} T \hookrightarrow {}^\tau G .$$

If the field E is real then the complex involution acts on $S_K(\mathbb{C})$ which as a complex manifold is isomorphic to

$$G(\mathbb{Q})\backslash G(\mathbb{A})/K_\infty K .$$

It should be possible to define the resulting involution on the double coset space explicitly. E can be real only if $\rho\hat{\mu} = \omega\hat{\mu}$ with ω in the Weyl group of T' in G. If this is so then ω can be realized by an element w in the normalizer of T in $G(\mathbb{R})$. w will normalize K_∞ so that the map $g \longrightarrow gw$ may be transferred to the quotient. This should give the involution.

Shimura and Shih are working on these two problems, which are deeper than they appear at first glance.

Proceedings of Symposia in Pure Mathematics
Volume 28, 1976

THE 13-TH PROBLEM OF HILBERT

G. G. Lorentz[1)]

This paper consists of two parts. In Part I, we review the present-day situation of the 13-th problem. Part II is technical, we prove there a new theorem about linear superpositions.

Part I.

1. The 13-th problem of Hilbert.

Hilbert formulated his 13-th problem in the following way:
(I) "Prove that the equation of the seventh degree $x^7+ax^3+bx^2+cx+1=0$ *is not solvable with the help of any continuous functions of only two variables".*

Examining this sentence, we see that its first part is purely algebraic, its second part purely analytic. No wonder that it is much better to split the problem into two, an algebraic, and an analytic one. Here we will be concerned only with the latter, and reformulate it as follows:
(II) "Prove that there are continuous functions of three variables, not representable by continuous functions of two variables".
Somewhat arbitrarily, we shall call this "Hilbert's conjecture".

We are allowed here to use functions of functions, or *superpositions.* For example,

$$f(x,y,z) = F(g(x,y),h(\phi(x),\psi(z))) \tag{1.1}$$

is a superposition of functions of one and of two variables. Sometimes we consider superpositions where certain functions are fixed, constant, while others are arbitrary, variable. Of importance are sometimes *linear superpositions.* These are superpositions *which are linear in the variable functions.* For example,

$$f(x,y) = \sum_{i=1}^{m} p_i(x,y)g_i(q_i(x,y)),\quad p_i,q_i \text{ fixed} \tag{1.2}$$

is a linear superposition.

Hilberts question (II) asks then: are there at all functions of three variables? Let us take a good Calculus text and examine what genuine functions of two or three variables can we find. The function $x+y$ will be found, but the product xy is nothing new, it reduces to the sum and functions of one variable, $xy = e^{\log x+\log y}$, similarly for $x+y+z = (x+y)+z$. So perhaps there do not exist any other functions?

AMS (MOS) subject classifications (1970). Primary 26A72; Secondary 41A65, 46E15.

1) Research supported by the National Science Foundation under grant GP-23566 and by a fellowship from A.v.Humboldt Stiftung.

The astonishing result of Kolmogorov (1957) confirms this. Kolmogorov proved:

Theorem A. There exist fixed continuous increasing functions $\phi_{pq}(x)$, *on* $I = [0,1]$ *so that each continuous function* f *on* I^n *can be written in the form*

$$(1.3) \qquad f(x_1,\dots,x_n) = \Sigma_{q=1}^{2n+1} g_q \left(\sum_{p=1}^{n} \phi_{pq}(x_p) \right),$$

where g_q *are properly chosen continuous functions of one variable.*

Several authors have improved the representation (1.3). Lorentz [5] showed that one can replace the g_q by one single function g; Sprecher [10] replaced functions ϕ_{pq} by $\lambda_p \phi_q$, where λ_p are constants. Thus, one can replace (1.3) by

$$(1.4) \qquad f(x_1,\dots,x_n) = \Sigma_{q=1}^{2n+1} g\,(\Sigma_{p=1}^{n} \lambda_p \phi_q(x_p)).$$

Here is the geometric meaning of Kolmogorov's theorem. Consider the mapping H of I^n into R^{2n+1} given by

$$(1.5) \qquad y_q = \sum_{p=1}^{n} \lambda_p \phi_q(x_p), \; q=1,\dots,2n+1$$

The properties of Kolmogorov's functions ϕ_q imply that H is a homeomorphism. Theorem A gets the following interpretation:

There exists a homeomorphic embedding H *of* I^n *into* R^{2n+1} *so that on the image* $H(I^n) = B$, *each continuous function* f *takes the special form*

$$f(y_1,\dots,y_{2n+1}) = \sum_{q=1}^{2n+1} g(y_q).$$

The functions ϕ_q in (1.4) can be assumed to possess some smoothness properties. Lorentz [6] noticed that they can be selected to belong to the class Lipα for all $\alpha<1$, Fridman established this even with $\alpha=1$. According to a remark of J.P.Kahane, this fellows in a trivial way from our geometric interpretation of Theorem A. The set B is a direct sum $B=C_1+\dots+C_n$ of n curves, $C_p : y_q = \lambda_p \phi_q(x_p)$, $q=1, 2n+1$. Each of them is a multiple of the single curve $C : y_q = \phi_q(x)$, $q=1, 2n+1$, $0 \le x \le 1$.This curve is rectifiable, since the functions ϕ_q are monotone. If we introduce on C the natural parameter s, $0 \le s \le 1$-a multiple of the arc lenght, we obtain a different representation for C, $C : y_q = \psi_q(s)$, $q=1,\dots,2n+1$, $0 \le s \le 1$ with $\psi_q \in$Lip1. Returning to (1.4), we obtain a representation with fixed functions ψ_q instead of ϕ_q.

There exist now very simple proofs of Kolmogorov's theorem (see Kahane [4]), which do not attempt to construct the functions ϕ_q (as the proof in [6] does) but instead shows by means of Baire's theorem, that most selections of $2n+1$ tuples of increasing functions ϕ_q will do.

Another component of all proofs of the theorem is the following simple geometric fact. Let us call a "quasi-covering" of $I = [0,1]$ any set $U = \{I_i, i=1,\dots,N\}$ of disjoint closed subintervals of I of equal lenght. By U^n we denote the set of all cubes $I_{i_1}x\dots xI_{i_n}, i_k=1,\dots,N$, $k=1,\dots,n$, so that U^n is a quasi-covering of I^n by cubes. Now for each $\delta>0$ there exist $2n+1$ quasi-coverings $U_j, j=1,\dots,2n+1$ of I, consisting of intervals of lenght $<\delta$, and a δ-net of points x in I^n, with the property that each point x is "more often covered than not". More precisely, each point x of the δ-net is contained in a square of at least n+1 of the 2n+1 quasi-coverings $U_j, j=1,\dots,2n+1$. I do not know of any applications of Kolmogorov's theorem, except perhaps that the multi-variable case of the Weierstrass approximation theorem follows from its single variable case. Perhaps Kolmogorov's theorem is of the nature of a pathological example, whose main purpose is to disprove hopes that are too optimistic.

We note that Doss [1] showed that for n=2, at least five terms are necessary in (1.3), if the functions ϕ are assumed to be increasing.

2. Hilbert's conjecture in true if all functions are continuously differentiable.

Theorem A disproves conjecture (II) of Hilbert, but does not solve his problem completely. The situation can be described as follows. Assume that some early Analyst would claim that interval and square are different because there does not exist a one to one mapping of one onto the other. It would not be enough then to refute this statement by means of a counter example. No, our duty would be also to prove that the statement is true, if the mapping is continuous.

Examining Hilbert's conjecture (II), we should check whether it becomes true, if the functions involved are assumed to possess some degree of smoothness. And indeed, this is the case.

Theorem B. There are $r(r=1,2,\dots)$ times continuously differentiable functions of $n\geq 2$ variables, not representable by superpositions of r times continuously differentiable functions of less than n variables; there are r times continuously differentiable functions of two varables which are not representable by sums and continuously differentiable functions of one variable.

This is the "Fundamental Theorem of the Differential Calculus", which asserts the non-emptiness of its basic categories.

The most remarkable protagonist of this trend - to prove results similar to Hilbert's conjecture - has been A.G.Vituškin, a very original mathematician. His ideas are deep, proofs usually very complicated. Sometimes they can be simplified. In the field of non-linear approximation, this happened in the papers of H.Shapiro [7], Lorentz [9]. At other times his concrete methods of prof compare favorably with abstract approach found in other schools. The book of Zalcman [14] gives an absorbing description of this for complex approximation. In the present article, we shall review Vistuškin's work concerning the 13-th problem of Hilbert.

It is clear, that conjecture (II) is based on a sound principle, namely that not all "bad" functions (functions of a large class) can be represented by "good" functions (those of a narrow class). We may try to express the quality, or"the measure of goodness" of a function f by some numerical characteristic, $\chi(f)$. The trouble with conjecture (II) is that the characteristic $\chi(f) = 1/n$, where n is the number of variables, does not work. One can take $\chi(f) = r$, if f is r-times continuously differentiable. This characteristic works, as is well known from Calculus, but does not produce anything new. Are there other characteristics? We shall say that a function f of n variables, defined on a closed bounded region in $\mathbb{R}^n$, belongs to the class $W^{n,q}$, $q>0$, with $q=r+\alpha$, $r=0,1,\ldots$, $0<\alpha\leq 1$, if f is r times continuously differentiable, with all its partial desivatives of order r belonging to the class Lipα. Vituškin puts

$$\chi(f) = \frac{q}{n} . \tag{2.1}$$

In 1954, before Kolmogorov's theorem, he showed:

Theorem C. Not all functions of a given characteristic $\chi_0 = q_0/k_0 > 0$ *can be represented by superpositions of functions of characteristics* $\chi = q/k > \chi_0$, $q \geq 1$.

Theorem B is an immediate censequence of this.

3. Representation of functions by linear superpositions.

Other results of Vituškin and of his associate Henkin from the years 1964-67 ([11],[12],[3],[13]) are intimely connected with the theorem of Kolmogorov. The main conclusion is that this theorem, while valid for Lipschitz functions ϕ_{pq}, ceases to be true for continuously differentiable fuctions ϕ_{pq}.

Let G be a closed domain in R^n, that is, the closure of an open connected bounded set. For a given m, we consider fixed continuous functions p_i, q_i in G, $i=1,\ldots,m$, with q_i continuously differentiable. Let F be the family of all linear superpositions

$$f(x_1,\ldots,x_n) = \sum_{i=1}^{m} p_i(x_1,\ldots,x_n) g_i(q_i(x_1,\ldots,x_n)), \tag{3.1}$$

where g_i are arbitrary continuous functions of one variable. Vituškin and Henkin have the following theorems:

Theorem D. The family F given by (3.1) is nowhere dense in the space C(G).

In particular, there exist polynomials in $x_1,\ldots,x_n$, not representable in form (3.1).

Theorem E. For each family F, there exists a closed subdomain $D\subset G$, *so that the restrictions of functions* $f\in F$ *to* D *form a closed supspace* F_D *of* C(D).

Theorem E is not only a technical theorem needed for the proof of Theorem D, but is also of independent interest.

We shall give a very imperfect description of the proof which Vituškin and Henkin supply for their results. They have carried out the proofs for n=2. The first method consists in a careful study of level curves of functions q and of the dependence of these functions from each other. One can modify the functions q_i, p_i (and the number m) so that on a properly chosen subregion $D \subset G$, the space of functions f_i for which the linear combination $\Sigma p_i f_i(q_i)$ is identically zero on D, is only finitely dimensional. This is still true if we supplement p_i, q_i by means of $p_{m+1}=1$, $q_{m+1}=x_1+ax_2$, where a is some natural number. This would mean that the space of continuous functions f for which $f(x_1+ax_2) \equiv -\Sigma p_i f(q_i)$, is also finitely dimensional. Hence for at least one b, the function $f(x) = -x^b$ does not have the above representation. It follows that there exists even a polynomial $(x_1+ax_2)^b$, not representable in form (1.7). As a consequence, F is not identical with C(G). Theorem E (whose proof also requires the full strength of the first method) is now used to infer that F is nowhere dense in C(G).

The second method of Vituškin and Henkin [13] relies only little on their analysis of the level curves. Instead, they use the notion of functional *dimension* r(X) of a subspace X of C(D) and prove, for properly chosen D, that $r(F_D) \leq 1$, while $r(C(D)) = 2$. This is even a stronger statement than Theorem D. Other tools used in [13] are the ε-entropy with respect to the norm $\|\cdot\|_\delta$ (see (5.5)), and the notion of ε,δ- distinguishable continuous functions. They [13] state explicitly that their methods do not apply to linear superpositions of the form

$$(3.2) \quad f(x_1,\ldots,x_n) = \sum_{i=1}^{m} p_i(x_1,\ldots,x_n) f_i(q_{i1}(x_1,\ldots,x_n),\ldots,q_{ik}(x_1,\ldots,x_n))$$

where m and k<n are fixed integers.

In Part II we prove

Theorem F. A family F given by (3.2) with fixed continuous p_i, *fixed continuously differentiable* q_{ij} *is a set of first category in* C(G).

It should be noted that Fridman [2], for families (3.2), k=2,n=3 proves that F is even nowhere dense , if all $p_i=1$ and the functions q_{ij} are *twice continuously differentiable*. Our proof of Theorem F is a good introduction to the paper of Vituškin and Henkin, since it is essentially simpler than their proof, and yet uses their notions of functional dimension and of norm $\|\cdot\|_\delta$.

4. *Entropy and capacity.*

For most known proofs of Theorems B - F very useful are the notions of *entropy* and *capacity* due to Kolmogorov. They measure the "size" of a compact subset (or a subset with a compact closure) F of a metric space X.

For each ε>0, the set F can be covered by a finite collection of balls of X of radius ε. The number n of balls in the collection depends upon its choice, but the minimum $N(F) = N_\varepsilon(F) = \min n$ of n for all such coverings expresses an important property of F. Then

$$(4.1) \quad H_\varepsilon(F) = \log N(F)$$

is the *entropy* (or the metric entropy) of F. This is a function of ε>0, which normally tends rapidly to +∞ as ε→0. Its rate of growth describes the size of the compact set F.

A dual notion is the capacity. Points $x_1,\dots,x_m$ in F are called ε-distinguishable, if the distance between any two of them satisfies $\rho(x_i,x_j)>\varepsilon, i\neq j$. For each $\varepsilon>0$, there exists a finite maximum $M_\varepsilon(F) = M(F) = \max m$. Its logarithm

$$C_\varepsilon(F) = \log M_\varepsilon(F) \tag{4.2}$$

is the capacity of F. The general theory of entropy and capacity is not rich. We have

$$C_{2\varepsilon}(F) \leq H_\varepsilon(F) \leq C_\varepsilon(F),\ \varepsilon>0. \tag{4.3}$$

If X is a linear normed space, and $F = F_1 \oplus\dots\oplus F_m$ is a direct sum of compact subset of X, then

$$H_\varepsilon(F) \leq \sum_{i=1}^{m} H_{\varepsilon/m}(F_i). \tag{4.4}$$

Most interesting results about entropy involve its computations for different compacts in functional spaces (spaces of analytic functions, spaces $W^{n,q}$ and so on) and conclusions which one may derive from comparision of entropies of different sets. For example, if F_r is a ball, in the natural norm, of the space $W^{n,q}$, one has

$$H_\varepsilon(F_r) \approx \left(\frac{1}{\varepsilon}\right)^{n/q}. \tag{4.5}$$

Theorem C can be easily derived from this fact.

There exists a general method of computation of metric entropy in linear normed spaces, developed by the author (Lorentz [7]). One can roughly say that it is possible to find the entropy of each set whose approximation properties are known.

Part II.

5. *Further useful notions.*

We shall describe some of the tools used by Vituškin and Henkin [13] in the second proof of Theorem D, because they seem to be of general interest. They will be used in the proof of Theorem F. Families F of form (3.2) are linear spaces, and even balls F_r in F are not compact in the uniform norm, consequently they have infinite entropy. This notion has to be modefied, if it in to be useful in the present situation.

Let G be a closed region in R^n, we consider the space C(G) of continuous functions on G. Two functions f_1, f_2 in C(G) are *ε,δ-distinguishable,* if there exists in G a cube Q_δ, with sides parallel to the axes and side length δ, so that either $f_1(x) - f_2(x)>\varepsilon$ for all $x\in Q_\delta$ or $f_1(x) - f_2(x)<-\varepsilon$ for all $x\in Q_\delta$. For a set A of continuous functions we can define its ε,δ-capacity $C_{\varepsilon,\delta}(A)$ in the following way. Let m stand for the cardinal number of any finite subset of A, consisting of mutually ε,δ-distinguishable elements, let $M_{\varepsilon,\delta}(A) = \sup m$. We put

(5.1) $$C_{\varepsilon,\delta}(A) = \log M_{\varepsilon,\delta}(A).$$

As an example, let us compute the capacity of a ball U_r in $C(G)$ of radius r. This does not depend on the position of its center, which we shall assume to be the origin. Then $||f||\le r$ for $f\in U_r$.

Theorem 1. For the ε,δ*-capacity of* U_r *one has*

(5.2) $$\lim_{\delta\to 0} \frac{\log C_{\varepsilon,\delta}(U_r)}{\log(1/\delta)} = n.$$

for all $\varepsilon>0$, $r>0$.

Proof: We shall establish here only the relation

(5.3) $$\underline{\lim}_{\delta\to 0} \frac{\log C_{\varepsilon,\delta}(U_r)}{\log(1/\delta)} \ge n,$$

which will be used later, leaving the rest of the proof to the reader.

There is a constant $\Theta>0$ with the property that for each $\delta>0$, there exists in G a set of $J = [\Theta\delta^{-n}]$ disjoint closed cubes $Q_j, j=1,\dots,J$, of side length δ. For each selection of signs $\varepsilon_j = \pm 1$, $j=1,\dots,J$, we can construct a continuous function f, $||f||\le r$, with values $f(x) = \varepsilon_j r$ on Q_j. This will give us 2^J functions in U_r, which are ε,δ-distinguishable. Hence $M_{\varepsilon,\delta}(U_r) \ge 2^J$, $\log M_{\varepsilon,\delta} \ge [\Theta\delta^{-n}]$ and $\log C_{\varepsilon,\delta}(U_r) \ge n \log(1/\delta)$ + const., which proves (5.3).

By means of the ε,δ-capacity, Vituškin and Henkin define the functional dimension $r(F)$ of a subspace F in $C(G)$. One can put

$$r(F) = \sup_{\varepsilon,r>0} \lim_{\delta\to 0} \frac{\log C_{\varepsilon,\delta}(F_r)}{\log(1/\delta)}.$$

(Actually, their definition is slightly different). It would be interesting to investigate this notion in more detail.

They also define, for $f\in C(G)$ and some $\delta>0$, the norm

(5.5) $$||f||_\delta = \sup_{U_\delta\subset G} \frac{1}{|U_\delta|} \left|\int_{U_\delta} f(x)\,dx\right|,$$

the supremum taken over all balls U_δ contained in G.

We shall need a generalization of this, with regions Q'_δ of proper shape instead of the U_δ.

For each $\delta>0$, let $\{Q_\delta'\}$ be a family of closed regions contained in G with the following properties:

(a) *Each cube* Q_δ *in* G *contains at least one region* Q'_δ;

(b) *One has, for the volume of* Q'_δ, $|Q'_\delta| \ge c_0\delta^n$, *where* c_0 *does not depend on* δ.

We define the semi-norm on $C(G)$,

(5.6) $$||f||_\delta = \sup_{Q'_\delta\subset G} \frac{1}{|Q'_\delta|} \left|\int_{Q'_\delta} f(x)dx\right|.$$

For a subset $F\subset C(G)$, we define its ε-entropy $H_\varepsilon^\delta(F)$ and ε-capacity $C_\varepsilon^\delta(F)$ with respect to the norm $||\cdot||_\delta$ in the usual way.

If two functions $f_1, f_2 \in F$ are ε,δ-distinguishable, then, by virtue of (a), they are also ε-distinguishable in terms of the norm (5.6). Therefore $M_\varepsilon^\delta(F) \geq M_{\varepsilon,\delta}(F)$ and we obtain:

$$C_{\varepsilon,\delta}(F) \leq C_\varepsilon^\delta(F) \leq H_{\varepsilon/2}^\delta(F). \tag{5.7}$$

6. Estimation of $H_\varepsilon^\delta(F_r)$ *from above.*

Let $p, q_1, \ldots, q_k$, where $k<n$ be fixed continuous functions in G, let the q_j be continuously differentiable. We consider the set F of functions given by

$$p(x_1,\ldots,x_1)g(q_1,\ldots,q_k) \tag{6.1}$$

with variable continuous function g, and compute its entropy. The rank of the Jacobian matrix

$$\left\| \begin{matrix} \frac{\partial q_1}{\partial x_1}, & \ldots, & \frac{\partial q_1}{\partial x_n} \\ \ldots & \ldots & \ldots \\ \frac{\partial q_k}{\partial x_1}, & \ldots, & \frac{\partial q_k}{\partial x_n} \end{matrix} \right\| \tag{6.2}$$

is a function $l(x_1,\ldots,x_n)$ of the point $x = (x_1,\ldots,x_n)\in G$; let l be its maximum.

Theorem 2. There is a subregion $D\subset G$ *so that for the norm* $||\cdot||_\delta$ *given by (5.6) with* G *replaced by* D,

$$\overline{\lim_{\delta\to 0}} \frac{H_\varepsilon^\delta(F)}{\log \frac{1}{\delta}} \leq l, \quad F_r = F\cap U_r \tag{6.3}$$

Proof. We select a point $x = (x_1,\ldots,x_n)$ where $l(x_1,\ldots,x_n) = l$ and a neighborhood D of x for which the following is true. One of the minors of (6.2) is not zero on D; let it be the Jacobian of $q_1,\ldots,q_l$ with respect to $x_1,\ldots,x_l$. The map

$$\begin{aligned} q_1(x_1,\ldots,x_n) &= y_1 \\ &\ldots\ldots \\ q_l(x_1,\ldots,x_n) &= y_l \\ x_{l+1} &= y_{l+1} \\ &\ldots\ldots \\ x_n &= y_n \end{aligned} \tag{6.4}$$

and its inverse Φ are one to one and (6.4) has the Jacobian $\Delta\neq 0$. The functions $q_{l+1},\dots,q_k$ can be expressed on D as functions of $q_1,\dots,q_l$. Without loss of generality, we can assume that D is the image under Φ of a cube Q of the y-space and that the function p does not vanish on D. Let $\|\operatorname{grad}\Phi\| = c_1$. If x is the center of a cube $Q_\delta\subset D$, and $y = \Phi^{-1}(x)$, we define the family Q'_δ to be $Q'_\delta = \Phi(\overline{Q}_{\delta 1})$, where $\overline{Q}_{\delta_1}$ is the cube in Q with center y and side length $\delta_1 = c_1\delta$. Clearly $Q'_\delta\subset Q_\delta$. Conditions (a),(b) are satisfied. To functions f of form (6.1) there correspond in a one to one way functions F of the family F' given by

$$P(y_1,\dots,y_n)G(y_1,\dots,y_l),$$

where $P = p/|\Delta|$ and the G are arbitrary continuous functions. For the norm $\|f\|_\delta$ we obtain, after a change of variables in the integral (5.6)

$$\|f\|_\delta \le \frac{1}{c_o}\sup_{\overline{Q}_{\delta_1}\subset Q}\delta^{-n}\left|\int_{\overline{Q}_{\delta_1}}F(y)dy\right| = \frac{1}{c_2}\|F\|_{\delta_1},\ c_2 = c_o c_1^{-n},$$

where $\|F\|_\delta$ is the norm of type (5.6) on Q:

$$\|F\|_\delta = \sup_{Q_\delta\subset Q}\delta^{-n}\left|\int_{Q_\delta}F(y)dy\right|. \tag{6.5}$$

For the corresponding entropy, $H^\delta_\varepsilon(F_r) \le H^{\delta_1}_{c_2\varepsilon}(F'_{r_1})$, where $r_1 = (\min\Delta)^{-1}r$. It follows that we have to prove the relation

$$\overline{\lim_{\delta\to 0}}\,\frac{H^\delta_\varepsilon(F'_r)}{\log(1/\delta)} \le l,\quad \varepsilon,r>0. \tag{6.6}$$

Let $\varepsilon,\delta>0$ be given. Our purpose will be to construct not too many cubes $Q_j, j=1,\dots,J$ of side length δ with the property that for each $Q_\delta\subset Q$ there is at least one Q_j for which

$$\delta^{-n}\left|\int_{Q_\delta}Fdy - \int_{Q_j}Fdy\right| < \frac{\varepsilon}{3},\quad F\in F'_r. \tag{6.7}$$

We may assume that $Q = [0,a]^n$, $a>0$. Let Q'_δ, Q''_δ denote the projections of a cube Q_δ onto the spaces $y_1,\dots,y_l$ and $y_{l+1},\dots,y_n$, respectively. It will be necessary to treat them in different fashions.

First take a natural number $p>12lr\varepsilon^{-1}$, and consider the points $0, \delta/p, (2\delta)/p,\dots,(N\delta)/p$, where $N = [ap/\delta]$. For each interval $I\subset[0,a]$ of length δ, there is at least one $I'_i=[i\delta/p,(i+p)\delta/p]$ for which the symmetric difference $\Delta(I,I'_i)$ has measure $\le 2\delta/p$. There are $\le ap\delta^{-1}$ such intervals I'_i.

Let $Q'_j, j=i,\dots,J_1$ be all products $I'_{i_1}\times\dots\times I'_l$; we have $J_1\le(\frac{ap}{\delta})^l$.

For each Q'_δ in $[0,a]^l$, there exists a Q'_j with $|\Delta(Q'_\delta-Q'_j)| < 2l\delta^l/p$. Therefore we shall have for all $F\in F'_r$ and all $y_{l+1},\dots,y_n$,

$$\delta^{-l}\left|\int_{Q'_\delta}Fdy_1,\dots,dy_l - \int_{Q'_j}Fd_{y_1}\dots d_{y_l}\right| \le \frac{2lr}{p} < \frac{1}{6}\varepsilon. \tag{6.8}$$

Let $\omega(t)$ be the modulus of continuouity of the function P on Q. Since P does not varnish, $|P|\geq c>0$ on Q, and for each $F\in F_r'$, $|G|\leq rc^{-1}=M$. We take $h>0$ so small that $\omega(nh)<\varepsilon/(6M)$. Assuming that δ is smaller than h, we consider all intervals $I_i''=[ih,ih+\delta]$, $i=0,\dots,q-1$, which lie in $[0,a]$; their number q does not exceed $ah^{-1}+1$. Each interval I of length δ in$[0,a]$ is obtainable by a translation of lenght <h from one of the I_i''. Consider the squares $Q_j''=I_{i_1}''x\dots xI_{i_{n-l}}''$, $j=1,\dots,J_2$ in $[0,a]^{n-l}$; each Q_δ'' is obtainable from one of them by a translation of length <nh; there are $J_2\leq(ah^{-1}+1)^n$ squares Q_j''. For each $F\in F_r'$, since G does not depend upon $y_{l+1},\dots,y_n$,

$$(6.9)\qquad |\int_{Q_\delta''}F\,dy_{l+1}\dots dy_n-\int_{Q_j''}Fdy_{l+1}\dots dy_n|\leq M\omega(nh)<\tfrac{1}{6}\,\varepsilon.$$

We define now

$$Q_j=Q_{j_1}'\times Q_{j_2}''\ ,\ j=1,\dots,J,\qquad J=J_1J_2.$$

For any Q_δ in Q, if Q_δ', Q_δ'' are its projections, let Q_{j_1}'', Q_{j_2}'' be defined by (6.8) and (6.9). Then

$$\delta^{-n}|\int_{Q_\delta}-\int_{Q_j}|\leq\delta^{-n}|\int_{Q_\delta'xQ_\delta''}-\int_{Q_\delta'xQ_j''}|$$

$$+\delta^{-n}|\int_{Q_\delta'}\times Q_j''-\int_{Q_j'xQ_j''}|<\tfrac{1}{3}\,\varepsilon.$$

Relation (6.7) has the following consequence. For two functions $F_1,F_2\in F_r'$, let

$$(6.10)\qquad \delta^{-n}|\int_{Q_j}F_1\,dy-\int_{Q_j}F_2\,dy|<\tfrac{1}{3}\,\varepsilon,\ j=1,\dots,J\ ;$$

then $||F_1-F_2||_\delta<\varepsilon$.

After this preparation, we can easily estimate the entropy of F_r'. For $F\in F_r'$, the values of $\delta^{-n}\int_{Q_\delta}Fdy$ lie between $-r$ and r. There are at most $\Theta r/\varepsilon$ intervals I of length $\frac{1}{3}\varepsilon$ which cover $[-r,r]$, where Θ is some constant. We consider the covering of F_r' by sets A defined as follows.

The set A is determined by giving for each Q_j an interval I_j (of the above set); $F\in A$ means that

$$\delta^{-n}\int_{Q_j}Fdy\in I_j\ ,\qquad j=1,\dots,J.$$

For two different F_1,F_2 belonging to the same set A, we have $||F_1-F_2||_\delta<\varepsilon$. It follows that each A is contained in a ball of radius ε. There are at most

$$\left(\frac{\Theta r}{\varepsilon}\right)^J$$

sets A. Thus, $N_\varepsilon^\delta(F_r)\leq(\Theta r/\varepsilon)^J$, and $H_\varepsilon^\delta(F_r)\leq J\log(Cr/\varepsilon)$. If $\varepsilon,r>0$ are fixed and $\delta\to0$, then, since p and h are also constants,

$$\log H_\varepsilon^\delta(F_r)\leq l\,\log\frac{1}{\delta}+\text{const.},$$

and (6.6) follows.

7. Completion of the proof of Theorem F.

We consider again the family F of functions given by (3.2); by $F^{(i)}$, $i=1,\dots,m$ we denote the family (6.1) with $p,q_1,\dots,q_k$ replaced by $p_i,q_{i1},\dots,q_{ik}$. We apply the inequality (4.4) to $G_r = F_r^{(1)}\oplus\dots\oplus F_r^{(m)}$ and obtain

$$H_\varepsilon^\delta(G_r) \le \sum_{i=1}^m H_{\varepsilon/m}^\delta (F_r^{(i)}).$$

Theorem 2 insures that $H_\varepsilon^\delta(G_r) < (l+\sigma)\log(1/\delta)$, for each $\sigma>0$ and all small $\delta>0$.

Let $U\rho$ be an arbitrary ball of radius ρ in C(D). From Theorem 1, (5.3) and inequalities (5.6) we conclude that $H_\varepsilon^\delta(U_\rho)>(n-\sigma)\log(1/\delta)$. This implies the G_r is not dense in each ball U_ρ in C(D). The same is true for the space C(G): the sets G_r are nowhere dense. But $F = U_{r-1}^\infty G_r$, hence F is a set of first category in C(G).

The importance of Theorem E is now clear. If we would have this statement for our family F, it would even follow from the above that F is nowhere dense. A further, still stronger conculsion would be that $r(F) \le k$.

REFERENCES

1. R. Doss, *On the representation of continuous functions of two variables by means of addition and continuous functions of one variable*, Colloqium Math. 10 (1963), 249-259.

2. B.L.Fridman, *The nowhere density of the space of linear superpositions of functions of several variables*, Izv. Akad. Nauk, Ser. Mat. 36 (1972), 814-846.

3. G.M.Henkin, *Linear superpositions of continuously differentiable functions*, Dokl. Akad. Nauk SSSR, 157 (1964), 288-290.

4. J.-P.Kahane, *Sur le théorème de superposition de Kolmogorov*, to appear in J. Approx. Theory.

5. G.G.Lorentz, *Metric entropy, widths, and superpisitions of functions*, Amer. Math. Monthly 69 (1962), 469-485.

6. G.G.Lorentz, *Approximation of Functions*, Holt, Rinehart and Winston, New York, 1966.

7. G.G.Lorentz, *Metric entropy and approximation*, Bull. Amer. Math. Soc. 72 (1966), 903-937.

8. D.A.Ostrand, *Dimension of metric spaces and Hilbert's problem 13*, Bull. Amer. Math. Soc. 71 (1965), 619-622.

9. H.S.Shapiro, *Some negative theorems of approximation theory*, Michigan Math. J. 11 (1964), 211-217.

10. D.A.Sprecher, *On the structure of continuous functions of several variables*, Trans. Amer. Math. Soc., 115 (1964), 340-355.

11. A.G.Vituškin, *Some properties of linear superpositions of smooth functions*, Dokl. Akad. Nauk SSSR, 156 (1964), 1003-1006.

12. A.G.Vituškin, *Proof of the existence of analytic functions of several complex variables which are not representable by linear superpositions of continuously differentiable functions of fewer variables*, Dokl. Akad. Nauk SSSR 156 (1964), 1258-1261.

13. A.G.Vituškin and G.M.Henkin, *Linear superpositions of functions*, Uspehi Mat. Nauk 22 (1967), 77-124. (=Russian Math. Surveys, 22 (1967), 77-125.)

14. L.Zalcman, *Analytic capacity and rational approximation*, Lecture Notes in Math. no. 5o, Springer Verlag, Berlin, 1968.

University of Texas
Austin, Texas 78712

Proceedings of Symposia in Pure Mathematics
Volume 28, 1976

HILBERT'S FOURTEENTH PROBLEM - THE FINITE GENERATION OF SUBRINGS SUCH AS RINGS OF INVARIANTS

David Mumford[1]

1. INTRODUCTION

The precise statement of the problem is this:

Let k be a field

Let K be a subfield of the rational functions in n-variables over k:

$$k \subset K \subset k(x_1,\cdots,x_n).$$

(n.b. all such K are automatically finitely generated over k as fields)

Is the ring:

$$K \cap k[x_1,\cdots,x_n]$$

finitely generated over k?

The motivation for this question came from its affirmative answer by Hilbert and others in certain very interesting cases: e.g., say $\text{char}(k) = 0$, suppose $G = SL(m)$ is acting linearly on k^n, and suppose K is defined as the field of G-invariant rational functions. Then $K \cap k[x_1,\cdots,x_n]$ is just the ring of G-invariant polynomials and Hilbert had proven that this was finitely generated. Unfortunately, it turns out that the answer is, in general, NO: $K \cap k[x_1,\cdots,x_n]$ may require an infinite number of generators. A beautiful counter-example was discovered by M. Nagata [13] in 1959. It would appear that after Hilbert's discovery of the extremely general finiteness principle on which his proof in the $SL(m)$-invariant case was based, namely "Hilbert's basis theorem" on the finite generation of all ideals in $k[x_1,\cdots,x_n]$, Hilbert was overly optimistic about finiteness results in other algebraic contexts. However my belief is that it was not at all a blind alley: that on the one hand its failure reveals some very significant and far-reaching subtleties in the category of varieties; and that the search for cases where it and

---

AMS (MOS) subject classifications (1970) 13B99, 14C20, 14E05.

[1]Research supported by the National Science Foundation under grant GP-36269X2.

related geometric questions are correct is a very important area of research in algebraic geometry. In fact, my guess is that it was Hilbert's idea to take a question that heretofore had been considered only in the narrow context of invariant theory and thrust it out into a much broader context where it invited geometric analysis and where its success or failure had to have far-reaching algebro-geometric significance. We will discuss the problem in 3 sections – first in the case of invariant theory where K is the field of G-invariant functions for some G, second in its geometric form involving linear systems formulated and analyzed first by Zariski [23], and thirdly as a special case of the general problem of forming quotient spaces of varieties by algebraic equivalence relations.

2. INVARIANT THEORY

Hilbert's proof of the finiteness when K is the field of G-invariant functions, $G = SL(m)$, $char(k) = 0$ is so very elegant and simple that it should really be part of every mathematician's bag of tricks. So I would like to begin by running through this marvelous proof: to begin with, it is known that if V is any finite-dimensional polynomial representation of SL(m) in char. 0, then V is completely reducible. In particular, there is a unique decomposition:

$$V = V^G \oplus V_1$$

where V^G is the subspace of invariant vectors and V_1 is a G-stable subspace containing no invariants. Let ρ_V be the projection of V onto V^G with kernel V_1. Next, let $R = k[X_1,\cdots,X_n]$, and let $R^G \subset R$ be the ring of invariants. R and R^G are graded rings, i.e.,

$$R = \oplus R_k, \qquad R_k = \text{vector space of homogeneous degree } k \text{ polynomials}$$

$$\text{and } R^G = \oplus R_k^G, \qquad R_k^G = G\text{-invariants in } R_k.$$

Thus the operators

$$\rho_{R_k}: R_k \longrightarrow R_k^G$$

patch together into a projection

$$\rho_R: \ R \longrightarrow R^G.$$

A simple argument using the uniqueness of ρ shows that ρ_R satisfies the identity:

$$\rho_R(fg) = f\rho_R(g), \quad f \in R^G,\ g \in R.$$

Now we let

$$R_+^G = \bigoplus_{k>0} R_k^G$$

and let $I = R_+^G.R$ be the <u>ideal</u> in R generated by all invariants of positive degree. Hilbert's Basis Theorem asserts that

$$I = \sum_1^N f_i.R$$

for some $f_1,\cdots,f_N \in I$; we can assume if we like that each f_i is in fact in R^G and homogeneous of some degree d_i. Then Hilbert asserts that these f_i generate R^G as ring! He proves this by induction on degree: choose $g \in R_n^G$ and assume all $h \in R_{n'}^G$, for $n' < n$ are polynomials in the f_i's. Then $g \in I$, hence there is an expression:

$$g = \sum_1^N a_i f_i, \quad a_i \in R_{n-d_i}.$$

Apply ρ_R:

$$g = \rho_R g = \sum \rho_R(a_i f_i) = \sum (\rho_R a_i) f_i.$$

Then $\rho_R a_i \in R_{n-d_i}^G$ which is a polynomial in the f_i's by induction, hence so is g!

What was the history of invariant theory after Hilbert? First of all, Hilbert did not give the above abstract description of ρ, but rather an explicit construction of ρ, called "Cayley's Ω-process" in which ρ appears in the Universal enveloping algebra of $s\ell(m)$. As mentioned in Hilbert's problem itself, A. Hurwitz [7] had already observed and H. Weyl was later to use effectively the fact that if $k = \mathbb{C}$, (and we can reduce easily any char. 0 case to the case $k = \mathbb{C}$), then

$$\rho x = \int_{g \in SU(m)} g^*(x)\cdot dg$$

$SU(m)$ = special unitary group

dg = Haar measure

Via the fact that any reductive algebraic group over $\mathbb{C}$ has a Zariski-dense compact subgroup, this gives us an explicit construction for the projection ρ for any such groups, hence a proof of finiteness. The final step - to observe that no explicit formula for ρ is needed but one merely must know the complete reducibility of all finite-dimensional representations to construct ρ abstractly - was taken by M. Schiffer in 1933 (unpublished; it appeared in H. Weyl's "Classical Groups" [22], Supplement C).

In char. p, no semi-simple group has the property that all its representations are completely reducible. For instance, think of SL(2) acting on the 3-dimensional space of quadratic forms

$$V = k\cdot x^2 + k\cdot xy + k\cdot y^2.$$

In char. 2, $k\cdot x^2 + k\cdot y^2$ is an invariant subspace with no complement. Therefore the Schiffer-Hilbert method breaks down. However, very recently, W. Haboush [25] has succeeded in proving the following Theorem which I conjectured in [9]:

THEOREM: If a semi-simple (or even reductive) algebraic group G acts on a vector space V and leaves fixed a vector $v \in V$, there is a polynomial function f on V such that:

i) $f(v) \neq 0$

ii) f is G-invariant.

In char. 0, f exists and may be taken linear by complete reducibility. Seshadri [17] had previously proven that such f's exist when G = SL(2). Nagata [14] has proven that if G has the property of the Theorem (this is sometimes stated as "G is semi-reductive"), then the ring of G-invariants* is finitely generated, i.e., whenever G acts linearly on $kx_1 + \cdots + kx_n$, then $k[x_1,\cdots,x_n]^G$ is finitely generated. Therefore, it follows that the ring of invariants is finitely generated for G reductive.

---

*We have not made precise before whether by G-invariants we meant polynomials $f(x_1,\cdots,x_n)$ which were identically invariant, i.e.,

$$f(g(x)) - f(x) \equiv 0 \quad \text{as function of } g,$$

or f's which were invariant separately under every $g \in G(k)$ (the k-rational points of G). If k is infinite, G(k) is Zariski-dense in G and there is no difference between these 2 concepts. But if k is finite there is a difference: in this case G(k) is finite and I wish G-invariant to mean identically invariant.

For groups G which are not semi-simple or reductive (i.e., which have a "unipotent radical"), very little is known even in char. 0 about finiteness of the ring of invariants. I know of only 2 results --
a) Weitzenbock [21] (cf. Also [16]) proved $k[x_1,\cdots,x_n]^G$ finitely generated if $G = \mathbb{G}_a$ (i.e., $\mathbb{G}_a$ = the additive group of the ground field),
b) Nagata's counter-example [13] is a non-finitely generated ring $k[x_1,\cdots,x_n]^G$ where G is commutative, but G is a product of many groups $\mathbb{G}_a$ and many groups $\mathbb{G}_m$ (here $\mathbb{G}_m$ = the multiplicative group of the ground field*).

3. ZARISKI'S FORMULATION WITH LINEAR SYSTEMS

We recall that if X is a non-singular projective variety (or more generally if X is normal) and D is a positive divisor on X (i.e., $D = \Sigma n_i E_i$, $E_i \subset X$ a subvariety of codimension 1 and $n_i \geq 0$), then we define:

$$\mathcal{L}(D) = \left\{ \begin{array}{l} \text{vector space of rational functions f on X} \\ \text{with poles bounded by D, i.e., } \forall\ E \subset X \text{ of} \\ \text{codimension 1,} \\ \quad \text{ord}_E\, f \geq -(\text{mult. of E in D}). \end{array} \right\}$$

(Either $\mathcal{L}$ (D) or the family of divisors that occurs as the zeroes of the functions $f \in \mathcal{L}(D)$ is called a linear system on X.) Zariski introduced the 2 rings:

$$R(D) = \bigcup_{n=0}^{\infty} \mathcal{L}(nD) = \left\{ \begin{array}{l} \text{ring of rational functions f with} \\ \text{poles of any order but only on D} \end{array} \right\}$$

$$R^*(D) = \bigoplus_{n=0}^{\infty} \mathcal{L}(nD).$$

---

*In concrete terms, a representation of $\mathbb{G}_a^n$ is a commutative group of matrices all of the form

$$\begin{pmatrix} 1 & & * \\ & \ddots & \\ 0 & & 1 \end{pmatrix}$$

in a suitable basis of $kx_1 + \cdots + kx_n$. A representation of $\mathbb{G}_m^n$ is a commutative group of diagonal matrices

$$\begin{pmatrix} * & & 0 \\ & \ddots & \\ 0 & & * \end{pmatrix}$$

in a suitable basis of $kx_1 + \cdots + kx_n$.

The ring $R^*(D)$, though apparently much bigger than $R(D)$, is easily shown to be isomorphic to $R(D_1)$ for a suitable divisor D_1 on a variety X_1 which is a $\mathbb{P}^1$-bundle over the variety X you start with. So the class of rings $R^*(D)$ is really a subset of the class of rings $R(D)$. More generally, for any divisors* $D_1,\cdots,D_k$, we can define a k-times graded ring:

$$R^*(D_1,\cdots,D_k) = \bigoplus_{n_1=0}^{\infty} \cdots \bigoplus_{n_k=0}^{\infty} \mathcal{L}(\Sigma n_i D_i)$$

and this is also isomorphic to $R(D_1)$ for a suitable D_1 on an X_1 (which is now a $\mathbb{P}^k$- bundle over X.) In his penetrating article [23], Zariski showed that Hilbert's rings $K \cap k[x_1,\cdots,x_n]$ were isomorphic to rings of the form $R(D)$ for a suitable X and D; asked more generally whether all the rings $R(D)$ might not be finitely generated; and proved $R(D)$ finitely generated if $\dim X = 1$ or 2. I want to outline the procedure for finding X and D such that:

$$K \cap k[x_1,\cdots,x_n] \cong R(D).$$

First of all, X is to be a suitable projective variety with function field K^{**}. For any such X, the inclusion of fields

$$K \subset k(X_1,\cdots,X_n)$$

defines a "rational map"

$$\pi: \mathbb{P}^n \longrightarrow X$$

i.e., π is a many-valued map whose graph in $\mathbb{P}^n \times X$ is a subvariety and which is single-valued on a Zariski -open subset $U \subset \mathbb{P}^n$. Let $r = \dim X = \text{tr.d.}_k K$. Then if X is chosen "sufficiently blown up", one can make π^{-1} nice in the sense:

$\forall\, x \in X$, the full inverse image $W_x = \pi^{-1}[x]$ has dimension $n-r$.

Roughly speaking, we have a fibration of $\mathbb{P}^n$ by $(n-r)$-dimensional algebraic sets W_x such that K is the field of rational functions constant on each

---

* If D has some negative coefficients, an $f \in \mathcal{L}(D)$ should have corresponding zeroes of order at least that coefficient.

** If resolution is known for this dimension and characteristic one would take X non-singular; if not, one takes X to be normal and $R(D)$ is defined as before.

W_x, i.e., invariant generically under the equivalence relation defined by belonging to the same W_x. Of course, these W_x's may become singular and in general will meet at certain "bad" points of $\mathbb{P}^n$, namely where the map π is not single-valued. Now let $D_1, \cdots, D_k$ be the subvarieties of X of codimension 1 such that $\pi^{-1}[D_i] \subset$ (the hyperplane at ∞, $\mathbb{P}^n - \mathbb{A}^n$). Then for all rational functions f on X, f has poles only on $\bigcup D_i$ if and only if $f \circ \pi$ has poles only at ∞, hence

$$K \cap k[x_1, \cdots, x_n] = R(\sum D_i).$$

Unfortunately, it was precisely by focusing so clearly the divisor-theoretic content of Hilbert's 14th problem that Zariski cleared the path to counter-examples. The history is this -

i) Rees [15] in 1958 found a 3-dimensional X and a D with R(D) infinitely generated. His X was birational to $\mathbb{P}^2 \times E$ (E an elliptic curve).

ii) Nagata [13] in 1959 found that for suitable points $P_1, \cdots, P_r \in \mathbb{P}^2$, if X is the surface obtained by blowing up each P_i into a rational curve E_i, then

$$R^*(\ell, -\sum_1^r E_i) \quad (\ell \text{ a line not through any } P_i)$$

is infinitely generated; and that this ring was a ring of invariants $k[x_1, \cdots, x_{2r}]^G$ as mentioned in §1.

iii) Zariski [24] in 1962 returned to the problem and pursuing some constructions which had been considered in different contexts by Grauert [3] and Nagata [12], found that it was not at all uncommon for $R^*(D)$ to be infinitely generated when dim X = 2 (hence for R(D) to be infinitely generated when dim X = 3).

I would like to describe the situation Zariski looked at because it is a very useful source of counter-examples to several problems and illustrates some basic facts about the category of algebraic varieties. Suppose you have

a) a non-singular surface X,

b) a curve $E \subset X$ of genus $g > 0$ such that

i) $(E^2) < 0$ (i.e., the normal bundle to E in X has negative curvature)

ii) Pic X $\longrightarrow$ Pic E is injective (i.e., if a line bundle L on X is trivial on E, then it is trivial on X).

Such a situation is not hard to obtain: start with any sufficiently general hypersurface section H_o on X_o and blow up enough generic points on H_o to make its normal bundle negative. First of all, here is what Grauert observed about this situation: analytically, E can be blown down, i.e., there is a normal analytic surface X_1 and $\pi: X \longrightarrow X_1$ mapping E to a point x but bijective elsewhere. But X_1 is not a variety: if it were, x would have an affine neighborhood $U \subset X$, hence $C = X_1 - U$ would be a curve not containing x, hence $\pi^{-1}(C)$ would be a curve on X disjoint from E, hence "twisting by $\pi^{-1}(C)$" we get a line bundle $\mathcal{O}_X(\pi^{-1}C)$ trivial on E but not trivial on X: contradiction.

Zariski did this: let $H \subset X$ be a hyperplane section, let $a = (H.E)$, (the intersection number of H and E), let $(E^2) = -b$. Then he showed

$$R^*(bH + aE)$$

is not finitely generated. The reason is this — look for functions f on X with poles $kbH + kaE$, some $k \geq 1$. If at some $P \in E$, $x = 0$ is the local equation of E, expand f:

$$f = \frac{g_o}{x^{ak}} + \frac{g_1}{x^{ak-1}} + \cdots\cdots$$

and consider the function g_o on E. Suitably interpreting what g_o means, g_o comes out as a section of a line bundle on E; in fact the line bundle $\mathcal{O}_X(kbH+kaE)$ on X restricted to E. This has degree 0 but by assumption (b ii) is not trivial. So it has no sections and $g_o \equiv 0$, i.e., f can have at most poles of type $kbH + (ka-1)E$. On the other hand, Zariski showed that there is a fixed k_o such that for all k, there are functions f with poles of type $kbH + \max(0, ka-k_o)E$. To see the implications of this, say for instance that $k_o = 1$: then for all k, let

$$f_k \in \mathcal{L}(kbH + kaE)$$

have a pole $kbH + (ka-1)E$. Then for all k,

$$f_k \notin \begin{pmatrix} \text{subring of } R^*(bH + aE) \text{ generated by} \\ 1, f_1, \cdots, f_{k-1} \end{pmatrix}$$

since every function in the degree k piece of the subring has a pole of at most $kbH + (ka-2)E$! Taking into account that $R^*(bH+aE)$ is <u>graded</u>, it requires at least one generator in each degree, hence is not finitely generated!

Are there any positive results asserting that $\mathbf{R}(D)$ and $R^*(D)$ are finitely generated in some cases? When $\dim X = 2$, Zariski's paper [24] gives a thorough analysis of when $R^*(D)$ is finitely generated. In higher dimensions, at the moment, the best results are numerical criteria on D implying that D is ample, which in turn implies very quickly that both $R(D)$ and $R^*(D)$ are finitely generated. Here "D ample" means that for some $n \geq 1$, nD is a hyperplane section of X in a suitable projective embedding. These criteria use intersection numbers and are as follows:

1.) Nakai's Criterion: if for every subvariety $Y \subset X$, $(Y.D^r) > 0$ where $r = \dim Y$, then D is ample.
2.) Seshadri's Criterion: if there is an $\epsilon > 0$ such that for every curve $C \subset X$, $(C.D) > \epsilon \cdot \left[\max_{P \in C}(\text{mult. of } P \text{ on } C)\right]$, then D is ample.

For proofs, see Hartshorne's book [4], Chapter I.

4. QUOTIENT SPACES BY ALGEBRAIC EQUIVALENCE RELATIONS

Another way of generalizing Hilbert's problem is to ask: given a variety X, and

$$R \subset X \times X, \quad R = \begin{cases} 1) \text{ a finite union of subvarieties of } X \times X \\ 2) \text{ set-theoretically, an equivalence relation on } X \end{cases}$$

when is there another variety Y and a surjective morphism $f\colon X \longrightarrow Y$ such that

$$R = \left\{ (x_1,x_2) \,\middle|\, f(x_1)=f(x_2) \right\}?$$

For short, we speak* of Y as X/R. Two cases of particular interest are i) a group G acts on X and $R = \left\{ (x,gx) \,\middle|\, x \in X, g \in G \right\}$, and ii) E is a subvariety of X to be "blown down" and $R = (\text{diagonal}) \cup (E \times E)$. In Hilbert's case, $X = \mathbb{A}^n$ (affine space) but one is given R only generically by specifying the subfield K (i.e., $R = \left\{ (x_1,x_2) \,\middle|\, x_1,x_2 \text{ belong to some } W_x \right\}$ in the notation of §2); Hilbert's problem can be broken up into 2 steps – first extend this equivalence relation nicely to one on all of $\mathbb{A}^n$, second prove $\mathbb{A}^n/R$ exists and is an affine variety, in which case Hilbert's ring $k[X_1,\cdots,X_n] \cap K$ is just the affine coordinate ring of $\mathbb{A}^n/R$.

Returning to the general case, it is always possible to find a Zariski open subset $U \subset X$ stable under R such that U/R exists (this may be proven for instance using Chow coordinates of the equivalence classes).

---

* The requirements do not determine Y uniquely, but in all cases that arise, there are natural extra conditions one imposes that make Y unique if it exists at all.

Equivalently the field of rational functions K on X/R is easy to construct and then any model Y of K realizes X/R on some sufficiently small Zariski-open U in X. The real problem is a birational one of finding a Y which works everywhere. However, as in Zariski's divisor formulation of the problem, one is confronted straightway by a raft of counter-examples:

1.) Grauert's example [3] described in §3 of an $E \subset X$, where $\dim X = 2$, $(E^2) < 0$* and E can be blown down analytically but not algebraically,

2.) Hironaka [6] found a beautiful example of a complete (though non-projective) variety X on which $\mathbb{Z}/2\mathbb{Z}$ acts freely, but $X/(\mathbb{Z}/2\mathbb{Z})$ is not a variety at all,

3.) Nagata and I found ([9], p. 83) examples of $PGL(n)$ acting freely on quasi-projective varieties X such that the orbit space $X/PGL(n)$ is not a variety.

In rough outline, here is the idea of Hironaka: take a 3-dimensional projective variety X_o with 2 curves C_1, C_2 in it crossing transversely at 2 points P_1, P_2 and with $\mathbb{Z}/2\mathbb{Z}$ acting on X_o interchanging the C's and the P's:

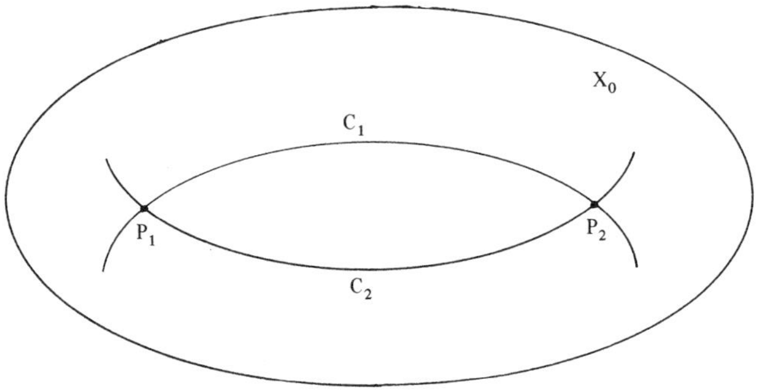

We then blow up C_1 and C_2 in X_o to obtain X. However, where the C's cross, we must specify the order in which the C's are blown up — so at P_1, we blow up C_1 first, then in the resulting variety we blow up C_2; at P_2,

*If $(E^2) \geq 0$, then E cannot be blown down even analytically so of course one cannot construct X/R, $R = (\text{diag}) \cup (E \times E)$, algebraically. For general equivalence relations R one asks first that R have some reasonable properties ensuring that X/R exists in the analytic context.

we blow up C_2 first, then in the result, we blow up C_1. Then $\mathbb{Z}/2\mathbb{Z}$ still acts on X. However, say Y = X mod ($\mathbb{Z}/2\mathbb{Z}$) were a variety. Since X_o is projective, it can be shown (cf. e.g. [10], p. 111) that X_o mod($\mathbb{Z}/2\mathbb{Z}$) is a variety Y_o. In Y_o, C_1 and C_2 have the same image D and P_1 and P_2 have the same image Q. Then Y would be obtained from Y_o by blowing up D; but at Q, the 2 branches of D must be blown up in a definite order. As D is an irreducible curve, these 2 branches cannot be distinguished by rational functions! This turns out to mean that Y in fact does not exist in the category of algebraic varieties.

Confronted with these counter-examples, people have had 2 reactions: a) find criteria for X/R to exist as a variety, or b) instead enlarge the category you are working in. The ploy (b) was most notably successful in Weil's hands in his 2nd proof of the Riemann hypothesis for curves over finite fields [20]. His idea here required the construction of the Jacobian variety of such a curve. At that time, only affine and projective varieties had been considered. Weil invented the category of what he called abstract varieties – now called simply varieties – and constructed the Jacobian as one of these. Subsequently he and Chow independently showed that the Jacobian was actually a projective variety; however, at the time, Weil instead developed the theory of "abstract" varieties far enough to by-pass the question of projectivity and prove the Riemann Hypothesis using these Jacobians. Matsusaka [8] made an initial attempt at enlarging the category even further. However it was M. Artin who found, I believe, the most natural enlargement: he calls these new objects algebraic spaces (cf. [1] and [2]). One way to define these is simply to introduce them as formal quotients X/R, where X is a scheme and R is an étale equivalence relation, i.e., $R \subset X\times X$ is a subscheme such that the projection

$$p: R \longrightarrow X$$

is étale — essentially makes R into an unramified covering over X. Artin then went on to show that the category of algebraic spaces is closed under apparently all "reasonable" further quotient operations $X \longrightarrow X/R$. For details we refer the reader to his papers, which make algebraic spaces into a very effective and powerful tool.

Still you may have a sentimental attachment to familiar old varieties. It would appear especially that projective varieties play such a central technical role in algebraic geometry that it may be virtually impossible to eliminate their use even if you wanted to. In any case, it is very interesting to prove, when possible, that X/R is actually a projective variety. I would like to state one such result concerning orbit spaces:

Suppose:

X = a projective variety over k

G = a semi-simple (or more generally reductive) algebraic group over k, acting on X

$X \subset \mathbb{P}^n$: an embedding such that the action of G on X extends to an action on $\mathbb{P}^n$.

Then there are canonical open subsets

$X_s \subset X$: the set of "stable" points

$X_{ss} \subset X$: the set of "semi-stable" points

such that X_s, X_{ss} are G-invariants, and there is a diagram:

$$\begin{array}{ccccc} X_s & \subset & X_{ss} & \subset & X \\ \downarrow & & \downarrow & & \\ X_s/G & \subset & \overline{X_s/G} & & \end{array}$$

where $\overline{X_s/G}$ is a projective variety, X_s/G is an open subset of $\overline{X_s/G}$, $X_s \longrightarrow X_s/G$ makes X_s/G into an orbit space by G, and $X_{ss} \longrightarrow \overline{X_s/G}$ makes $\overline{X_s/G}$ into the quotient of X_{ss} by a cruder equivalence relation $\sim$ defined by:

$$x \sim y \text{ if } \overline{O^G(x)} \cap \overline{O^G(y)} \cap X_{ss} \neq \emptyset$$

(here $O^G(x)$ = G-orbit of x). This theorem is proven in my book [9] when char. = 0, and the part about X_s is proven by Seshadri [18] when char. = p. This X_{ss}-part in char. p follows from the recent results of Haboush[25] discussed in §1. See [11] for examples and a discussion of this result. This result has proven very useful for proving that various moduli spaces are quasi-projective varieties (and not "just" algebraic spaces).

The above theorem is in fact a natural extension of Hilbert's own ideas about the ring of invariants, especially as developed in his last big paper on the subject, "Über die vollen invariantensystemen" [5]. To indicate this, let me define X_{ss}. Assume for simplicity that there is actually a representation of G on k^{n+1} which induces the action of G on the $\mathbb{P}^n$ ambient to X. We then make the definition:

If $x \in X$, then

$$x \in X_{ss} \Longleftrightarrow \begin{cases} \forall \text{ homomorphisms } \lambda: \mathbb{G}_m \longrightarrow G, \text{ let } x(\lambda) = \lim_{t \to 0} \lambda(t)(x). \\ \text{Let } x(\lambda)^* \in k^{n+1} \text{ be homogeneous coordinates for } x(\lambda), \\ \text{so that } \lambda(t)[x(\lambda)^*] = t^r \cdot [x(\lambda)^*] \text{ for some } r \in \mathbb{Z}. \\ \text{We ask } r < 0 \text{ for all } \lambda. \end{cases}$$

Now let R be the homogeneous coordinate ring of X, and let R_+^G be the invariants with no constant term. Then we have Hilbert's result:

$$X - X_{ss} = V(R_+^G.R).$$

Contrary to the usual credo that Hilbert eliminated the interest in studying special cases in invariant theory, my belief is that some of the most challenging problems still open in invariant theory concern special cases. I would like to raise two rather broad questions:

PROBLEM Let S be the parameter space for a family $\{X_s | s \in S\}$ of non-singular projectively normal subvarieties $X_s \subset \mathbb{P}^n$. Assume PGL(n+1) acts on S so that for all $g \in PGL(n+1)$, $X_{g(s)} = g(X_s)$. Assume this action is proper. Then is the quotient S/PGL(n+1) always a variety?

PROBLEM Now that we have computers, is there a practical way to actually find generators of such classical rings of invariants as those of a binary or ternary n-ic (i.e., SL(2) or SL(3) acting on the space of homogeneous degree n polynomials in 2 or 3 variables)? After an extraordinary effort, Shioda [19] only recently found these for binary octics.

Added in proof: Independent of W. Haboush's work, E. Formanek and C. Procesi have recently in a preprint entitled "Mumford's Conjecture for the general linear group" given another very beautiful proof of the semi-reductivity of GL(n) and SL(n).

REFERENCES

[1] M. Artin, Algebraic Spaces, Yale Math. Monographs, Yale Univ. Press

[2] M. Artin, Algebraization of formal moduli, I, in Global Analysis, Princeton Univ. Press, 1969; II in Annals of Math., 91 (1970), p. 88.

[3] H. Grauert, Über Modifikationen und exzeptionelle analytische Mengen, Math. Annalen, 146 (1962), p. 331.

[4] R. Hartshorne, Ample Subvarieties of Algebraic Varieties, Springer Lecture Notes 156, Springer, 1970.

[5] D. Hilbert, Über die vollen Invariantensystemen, Math. Annalen, 42 (1893), p. 313.

[6] H. Hironaka, An example of a non-kählerian deformation of kählerian complex structures, Annals of Math., 75 (1962), p. 190.

[7] A. Hurwitz, Über die Erzeugung der Invarianten durch Integration, Nachr. Gött. Ges. Wissensch. 1897, p. 71.

[8] T. Matsusaka, Theory of Q-varieties, Publ. Math. Soc. of Japan, 1965.

[9] D. Mumford, Geometric Invariant Theory, Springer-Verlag, 1965.

[10] D. Mumford, Abelian Varieties, Tata Inst. Studies in Math., Oxford Univ. Press, 1970.

[11] D. Mumford and K. Suominen, Introduction to the theory of moduli, in Algebraic Geometry, Oslo 1970, ed. by F. Oort, Noordhoff, 1972.

[12] M. Nagata, Existence theorems for nonprojective complete algebraic varieties, Ill. J. Math., 2 (1958), p. 490.

[13] M. Nagata, On the 14th problem of Hilbert, Am. J. Math., 81 (1959), p. 766.

[14] M. Nagata, Invariants of a group in an affine ring, J. Math. Kyoto Univ. 3 (1964), p. 369.

[15] D. Rees, On a problem of Zariski, Ill. J. Math., 2 (1958), p. 145.

[16] C.S. Seshadri, On a theorem of Weitzenböck in invariant theory, J. Math. Kyoto Univ., 1 (1962), p. 403.

[17] C.S. Seshadri, Mumford's conjecture for GL(2) and applications, in Algebraic Geometry, Tata Inst. Studies in Math., Oxford Univ. Press, 1969.

[18] C.S. Seshadri, Quotient spaces modulo reductive algebraic groups, Annals of Math., 95 (1972), p. 511.

[19] T. Shioda, On the graded ring of invariants of binary octavics, Am. J. Math., 89 (1967), p. 1022.

[20] A. Weil, Variétés Abéliennes et Combes Algébriques, Hermann, 1948.

[21] R. Weitzenbock, Über die Invarianten von Linearen Gruppen, Acta Math., 58 (1932), p. 230.

[22] H. Weyl, The Classical Groups, Princeton Univ. Press, 2nd Ed. 1946.

[23] O. Zariski, Interprétations algébrico-géometriques du 14ième problème de Hilbert, Bull. Sci. Math., 78 (1954), p. 155; collected papers vol. II.

[24] O. Zariski, The theorem of Riemann-Roch for high multiples of an effective divisor on an algebraic surface, Annals of Math., 76 (1962), p. 560; collected papers vol. II.

[25] W.J. Haboush, Reductive groups are semi-reductive, preprint.

Proceedings of Symposia in Pure Mathematics
Volume 28, 1976

PROBLEM 15. RIGOROUS FOUNDATION OF SCHUBERT'S ENUMERATIVE CALCULUS

Steven L. Kleiman[1]

ABSTRACT

Schubert's calculus was first interpreted and rigorously justified by van der Waerden (1929) by means of the calculus of algebraic cohomology classes developed by Lefschetz. Entirely algebraic treatments of the foundations of Schubert's calculus have become possible through the jumbled efforts of a great many mathematicians, who have contributed to the constructions of algebraic intersection rings to replace the topological cohomology ring. However, this work does not constitute a complete solution to Hilbert's fifteenth problem; for, in the statement and explanation of the problem, Hilbert makes clear his interest in the effective computability and actual verification of the geometrical numbers of classical enumerative geometry. Due primarily to Schubert (1886), the classical method of obtaining certain numbers, like the number $(1!\,2!\,\ldots\,d!\,h!)/(n-d)!\,\ldots\,n!$ of d-planes in n-space meeting $h = (d+1)(n-d)$ general $(n-d-1)$-planes, was vindicated topologically by Ehresmann (1934) and algebraically by Hodge (1941, 1942) by means of an explicit determination of the cohomology ring, and respectively, of an equivalent algebraic intersection ring, on the Grassmann manifold. In the offing, there is the exciting hope of the development in algebraic geometry of a general enumerative theory of singularities of mappings, a theory of Thom polynomials, which will, among other things, unify and justify the classical work dealing with prescribed conditions of intersection and contact imposed on linear spaces. Classically, conditions of intersection and contact were imposed on other figures as well. For example, Schubert (1879), in his book, obtains the number 666, 841, 048 of quadric surfaces tangent to 9 given quadric surfaces in space, and the number 5, 819, 539, 783, 680 of twisted cubic space curves tangent to 12 given quadric surfaces. Today, we cannot vouch for the accuracy of these two spectacular numbers, nor do we even know whether Schubert's method is basically sound.

AMS (MOS) subject classifications (1970). Primary 14-03, 14-01, 14N10; Secondary 14C15, 14M15, 14E99, 14M99.

[1]Research partially supported by the National Science Foundation under grant no. P 28936.

PROBLEM 15. RIGOROUS FOUNDATION OF SCHUBERT'S ENUMERATIVE CALCULUS

"The problem consists in this: To establish rigorously and with an exact determination of the limits of their validity those geometrical numbers which Schubert[1, 2] especially has determined on the basis of the so-called principle of special position, or conservation of number, by means of the enumerative calculus developed by him.

"Although the algebra of to-day guarantees, in principle, the possibility of carrying out the processes of elimination, yet for the proof of the theorems of enumerative geometry[3] decidedly more is requisite, namely, the actual carrying out of the process of elimination in the case of equations of special

---

[1]The following biographical sketch of Schubert was kindly offered by Irving Kaplansky, who combined the entries in the Poggendorf volumes for 1898, 1904 and 1926: Hermann Cäsar Hannibal Schubert was born May 22, 1848 in Potsdam. He died July 20, 1911 in Hamburg. [The rest Kaplansky reproduced untranslated.] Dr. Phil, 1870 Halle; ward 1872 gymnasiallehrer zu Hildesheim, seit 1876 Oberlehrer und Prof. am Johanneum zu Hamburg [Or]. Erhielt 1875 d. grosse gold Med. v.d. Dän Akad. f.s. "Chacter. der Raumcurven 3. Ordn." Danach in Ruhestd. Kaplansky added, "The 1939 Poggendorf lists two obituary notices, neither of which is at present accessible to me."

David Hilbert mentions Schubert in a memorial address [24] for Adolf Hurwitz, saying, "Adolf Hurwitz wurde am 26. März 1859 in Hildesheim geboren. Hier besuchte er das städtische Realgymnasium, an welchen damals der Fachkreisen später bekannt gewordene Mathematiker Hannibal Schubert den mathematischen Unterricht erteilte. Schubert führte den jungen Hurwitz schon auf der Sekunda in den "Kalkül der abzählenden Geometrie" ein, eine damals neu emporkommende Disziplin, deren systematische Bearbeitung und Ausbildung Schubert sich zu seiner Lebensaufgabe gemacht hatte. Hurwitz wurde durch diesen persönlichen Verkehr mit Schubert sehr frühzeitig zu sebständigem Forschen angeregt und veröffentlichte bereits als 17-jähriger Schüler mit seinem Lehrer zusammen in den Nachrichten unserer Gesellschaft eine Arbeit über den Chasles'schen Satz $\alpha\mu + \beta\nu$.

"Auf Schuberts Rat begann Hurwitz 1877 sein Studium bei Klein, ..."

Schubert wrote a series of mathematical essays and recreations, designed for the general public, which appeared in an English translation in 1899 and again in 1903 in a second edition [89]. Interesting in and of themselves, they also offer a glimpse at Schubert's philosophical view of mathematics and the impressive extent of his scholarship.

[2]Hilbert refers here to Schubert's book [84].

[3]The encyclopedia article [118] by Zeuthen and Pieri is an excellent general historical introduction and guide to the classical literature of enumerative geometry. Chapter I of Severi's monograph [102] gives a concise "historical-critical" introduction to enumerative geometry up to 1940.

form in such a way that the degree of the final equations and the multiplicity of their solutions may be foreseen."[4]

Consider a typical enumerative problem, the problem of finding the number of lines meeting 4 given lines in 3-space. Schubert solves it in his book [84, p. 13] this way. He alters the position of the 4 given lines so that the first and second lines intersect and third and fourth lines intersect. Then there are clearly 2 lines meeting the 4 lines, namely, the line joining the 2 points of intersection, and the line of intersection of the 2 planes spanned.

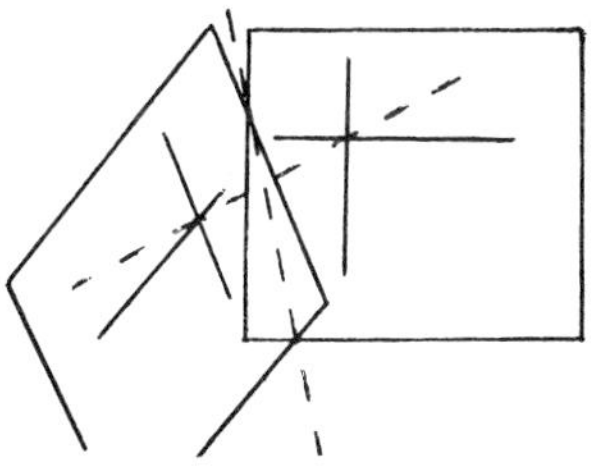

Appealing to the principle of conservation of number, Schubert concludes without further ado that the number of lines meeting 4 given lines in 3-space is 2 (or infinite).

The ambient 3-space is, of course, complex projective 3-space. The 2 planes above, if parallel, intersect at infinity. Imaginary numbers are available for use if necessary, for example, if the 4 given lines have equations with real coefficients while the 2 lines meeting them do not.[5] The use of imaginary numbers and of points at infinity in geometry, although begun to some extent much earlier, was boldly developed by Monge, Carnot, and Poncelet around the turn of the 19th century.[6]

---

[4] This is Hilbert's full statement of the problem, as it appears on p. 464 in Newson's translation [23].

[5] For example, take 3 of the 4 lines to be rulings in one of the 2 families of rulings of a suitable quadric surface, such as $X_0^2 + X_1^2 = X_2^2 + X_3^2$, and take the 4th line to be the line joining two conjugate points on the surface. Conjugation leaves the 4th line invariant, so it can be defined by real equations. The 2 lines meeting these 4 lines are clearly the 2 rulings in the other family, determined by the conjugate points. Conjugation interchanges the 2 lines, so neither can be defined exclusively by real equations.

[6] An excellent concise history of the development of complex projective geometry is found in section 1 of Schoenflies and Tresse's encyclopedia article [83].

The principle of conservation of number[7] was first extensively used, it is generally agreed,[8] by Poncelet [71, 1822], who called it the principle of continuity. It was attacked bitterly by Cauchy even before the publication of Poncelet's treatise [71], and Cauchy's judgement lead to some prejudice. Nevertheless, the principle came into widespread use.

Schubert heightened interest in the principle. To circumvent the prejudice, he renamed it in 1874 the principle of special position. The new name did not adequately capture the spirit of continuity, so in 1876 he began calling it the principle of conservation of number. Although Schubert felt the principle had not yet received the thorough justification it deserved,[9] nevertheless he made sophisticated use of it, greatly enriching enumerative geometry.

Theoretically, an enumerative problem can always be solved by working with the essential equations. Suppose the problem can be described by n homogeneous equations in $n + 1$ homogeneous unknowns. Eliminating[10] the unknowns leads to a single homogeneous equation in 2 homogeneous unknowns, whose roots correspond to the solutions of the original system. So the number of solutions, counted with their multiplicities, if finite, is equal to the degree of the single equation, and that degree is equal to the product of the degrees of the original n equations. Thus the weighted number of solutions remains constant when the parameters of the problem are varied.

---

[7] What follows and more about the history of the principle can be found in Zeuthen and Pieri's encyclopedia article [118], especially in sections 5-11, and in Coolidge's book [10], especially on pages 185-188.

[8] Poncelet himself [71, p. XIV] said that the principle was regarded as an axiom by the most knowledgeable geometers.

[9] Schubert expressed this feeling in his book [85, p. 12]. Afterwards though he continued by saying that, interpreted algebraically, the principle asserts simply that the number of roots of an equation remains constant when the coefficients are varied. Earlier Poncelet [71, XIV-XV] had said that the principle could be demonstrated easily by means of algebra, but that he had refrained from using this resource because the principle ought to be viewed in a purely geometric light. Meanwhile, de Jonquières [13] had tried to base the principle on the fundamental theorem of algebra (every equation of degree n has n roots).

[10] The elimination theory required here is not as naive as it sounds because of the possibility of extraneous roots; it was first developed by Bézout (1764) (see [68]; 57, 129-132]). A glimpse at classical elimination theory and its use in enumerative geometry is given by Brill [6, 1904].

Most enumerative problems cannot be described so simply, and the principle of conservation of number was furiously attacked. Notably, Study [104, 1903; 105, 1904] and Kohn [46, 1903] criticized the way the principle was formulated and applied, and they gave examples[11] where the principle purportedly failed. In response, Sturm [106, 1907] pointed out that Schubert's restrictions on the numbers of parameters simply ruled out these examples. Severi [99, 1912] put the matter definitively[12] in geometric terms; he worked with parameter varieties and considered their dimensions, their intersections, etc.

Let us use geometric terms and deal with the problem of finding the number of lines meeting 4 given lines in 3-space. In general, the d-planes in n-space are parametrized by a certain Grassmann manifold,[13] $G_{d,n}$. It sits in N-space with $N = \binom{n+1}{d+1} - 1$. A d-plane is represented by a point, whose coordinates are called its Plücker[14] coordinates. The Plücker coordinates of a d-plane are not arbitrary but satisfy certain famous quadric relations, discovered by Grassmann in 1844. Conversely, if the quadratic relations are satisfied by the coordinates of a particular point in N-space, the point represents a d-plane. In the case of lines in 3-space, only one quadratic relation is essential, and so $G_{1,3}$ is a quadric hypersurface in 5-space.

We need to express the condition that a line meet a given line. Schubert

---

[11] One of Study's examples is discussed by Coolidge [10, 185-186].

[12] Earlier, Giambelli [18, 1904] had formulated the principle of conservation of number similarly; however, Severi developed the ideas and addressed himself explicitly to the examples of Study and Kohn.

[13] A guide to the classical literature on Grassmann manifolds is found in section 2 of Segre's encyclopedia article [93, 788-794]. A detailed discussion, including proofs, of the facts about Grassmann manifolds and Schubert varieties stated in this report can be found in Hodge and Pedoe's book [28] and in the Monthly article [43]. These sources also give additional information about the geometry of these varieties. After the publication of the two sources, Svanes [107-108] established the rigidity of the cones over the varieties and determined which cones are Gorenstein, Musili [66] characterized the factorial cones in 8 ways and determined the Picard groups of all the varieties, and Doubilet, Rota and Stein [15] developed a new way of deriving properties of the Plücker coordinates.

[14] Plücker (1830) studied configurations of lines in 3-space. However these coordinates were introduced by Grassmann (1844) first and are also called Grassmann coordinates.

did this in his book [84], but there he was concerned only with geometry in 3-space. Subsequently (1886), in a very important series of articles [85; 86; 87], he enlarged his notation and technique to handle arbitrary d-planes. Starting with a flag, or nested sequence of linear spaces,

$$A_0 \subseteq \dots \subseteq A_d ,$$

he introduced the following condition on a d-plane L, now known as a Schubert condition:

$$\dim(A_j \cap L) \geq j \qquad \text{for } j = 0, \dots, d.$$

The condition is satisfied if and only if the Plücker coordinates of the d-plane L satisfy certain linear relations in addition to the quadratic relations. So, the points of the Grassmann manifold $G_{d,n}$ representing the d-planes L satisfying the condition form an algebraic variety. It is known as a Schubert variety. Let us denote it by

$$[A_0, \dots, A_d].$$

For example, the lines in 3-space meeting a given line M are represented by the points of the Schubert variety $[M, \mathbb{P}^3]$. This Schubert variety is defined by one linear equation in addition to the one quadratic equation defining $G_{1,3}$ in 5-space. So the lines meeting 4 given lines, $L_1, \dots, L_4$, are represented by the points of the intersection,

$$[L_1, \mathbb{P}^3] \cap \dots \cap [L_4, \mathbb{P}^3],$$

and this intersection is defined by 4 linear equations and 1 quadratic equation in 6 homogeneous variables. This system of equation has 2 solutions, or 1 double solution, or infinitely many solutions. Therefore, if there are finitely many lines meeting 4 given lines, there are 2 lines, or 1 line counted twice.

The symbolic calculus of conditions, which Schubert developed and enriched with numerous examples, demonstrated its power in practice by its application to enumerative problems where elementary methods fail and in theory by its unifying effect in the first systematic treatment of enumerative geometry, Schubert's book [84]. The calculus is founded on the idea of representing a geometric condition on a figure by an algebraic symbol. If symbols y and z represent independent conditions, then by definition the product yz represents the condition that both be satisfied, and the sum $y + z$ represents the condition that one or the other be satisfied.[15] The operations of addition and multiplication obviously satisfy the commutative, associative and distributive laws. An equation $x = y$ means that the two conditions are the same for enumerative purposes in the sense that both restrict freedom by the same amount and that the same number of figures satisfy xw as satisfy yw for each auxiliary condition w such that these two numbers are finite. (In reality, we are considering equivalence classes of conditions.) In these terms, the principle of conservation of number asserts that continuously varying the parameters of a condition yields an equal condition.

The calculus grew out of Chasles's work (Comptes rendus, 1864) in the enumerative theory of conics. Chasles found[16] that the number of conics in a one parameter family that satisfy a given condition could be expressed in the form,

$$\alpha\mu + \beta\nu,$$

where α and β depend only on the condition- they are called its characteristics- and μ and ν depend only on the family-- in fact, μ is the number of conics in the family passing through a given point (it is independent of the choice of the particular point by the principal of conservation of number) and ν is the number of conics in the family tangent to a given line. Starting with this expression, Chasles developed expressions for the number of conics in an i-parameter family that satisfy i independent conditions. In particular, for the number of all conics satisfying 5 independent conditions with characteristics,

$$\alpha,\ \beta;\ \alpha',\ \beta';\ \alpha'',\ \beta'';\ \alpha''',\ \beta''';\ \alpha^{iv},\ \beta^{iv},$$

---

15 Schubert [84, p. 333, Lit. 3] mentions the similarity of this and the logical calculus.

16 There is some difference of classical opinion on whether Chasles had a proof. Schubert [84, p. 274 and Lit. 51, p. 344] says Chasles found the expression experimentally and presented it without proof; Zeuthen and Pieri [118, p. 287, footnote 103] contend he outlined a proof.

he obtained this expression:

$$\alpha\alpha'\alpha''\alpha'''\alpha^{iv} + 2\,\Sigma\,\alpha\alpha'\alpha''\alpha'''\beta^{iv} + 4\,\Sigma\,\alpha\alpha'\alpha''\beta'\beta^{iv}$$
$$+ 4\,\Sigma\,\alpha\alpha'\beta''\beta'''\beta^{iv} + 2\,\Sigma\,\alpha\beta'\beta''\beta'''\beta^{iv} + \beta\beta'\beta''\beta'''\beta^{iv}.$$

Halphen (Comptes rendus, 1873) remarked that this last expression may be factored into the formal product,

$$(\alpha\mu+\beta\nu)(\alpha'\mu+\beta'\nu)(\alpha''\mu+\beta''\nu)(\alpha'''\mu+\beta'''\nu)(\alpha^{iv}\mu+\beta^{iv}\nu),$$

provided that, when the indicated multiplication is carried out, the term $\mu^r\nu^s$ is replaced by the number of conics passing through r points and tangent to s lines. This remark inspired Schubert tremendously. He published announcements the next year and the year after that, then full-length articles, and then his book- all within 6 years.

The geometric interpretation of the calculus begins with the variety parametrizing the figures of an enumerative problem. A condition on the figures defines a subset of the parameter variety, namely, the subset of points representing the figures satisfying the condition. The subset is algebraic because acceptable conditions are describable by algebraic equations. An i-fold condition, one restricting freedom by i parameters, defines a subset of pure codimension i. Independent conditions yield subsets in general position. The product of independent conditions corresponds to the intersection,[17] or product, of the subsets; the sum of independent conditions corresponds to the union, or sum, of the subsets. Equality of conditions corresponds to what is now called numerical equivalence.

The foundations of the calculus were first secured for all applications by van der Waerden [111, 1930]. Severi [99, 1912; 100, 1916] had already put the matter in geometric terms and shown that many cases are covered by his algebraic intersection theory; however, that theory is at best limited to intersections of hypersurfaces on the parameter variety. Van der Waerden saw that the (simplicial) topological intersection theory, developed by Lefschetz (1924, 1926) from some ideas of Poincaré and Kronecker, has all

[17] The multiplicities with which the components of the intersection appear have to be duly taken into account, of course.

the necessary generality and rigor.

In a topological intersection theory, each algebraic subset of the parameter variety is assigned[18] a cohomology class. Continuously varying the subset yields another subset with the same cohomology class; in other words, the two subsets are homologically equivalent. If two algebraic subsets are in general position, then their intersection[17] is assigned the (cup) product of their cohomology classes and their union is assigned the sum. Therefore, if several algebraic subsets in general position intersect in a finite number of points, the number is conserved when the parameters of the subsets are varied continuously because the number is equal to the degree of the product of the assigned cohomology classes and the classes are invariant; in other words, homological equivalence implies numerical equivalence.[19] If the subsets are defined by the conditions of an enumerative problem, it follows that the number of figures meeting the conditions is conserved when the parameters of the problem are varied continuously. (It always goes without saying that geometric numbers are to be considered valid only if the numbers are finite.) This is a rigorous justification of the principle of conservation of number within the context of an interpretation of the calculus of conditions by means of the calculus of algebraic cohomology classes.

Each cohomology group has a finite basis. Hence also its subgroup of algebraic cohomology classes has a finite basis. Consequently, the i-fold conditions have a finite basis; that is, every i-fold condition is equal to a certain linear combination of finitely many basic i-fold conditions. This conclusion constitutes a general affirmative solution to Schubert's problem of characteristics, and it too was first seen by van der Waerden [111]. Schubert set aside the last chapter of his book to deal with the problem of characteristics, but the problem's spirit pervades the entire volume, indeed, virtually all of Schubert's work in enumerative geometry.

The coefficients $\alpha_1, \ldots, \alpha_p$ in an equation,

---

[18] The difficulty of defining the cohomology class of an algebraic set and of establishing the desired properties varies somewhat with the choice of cohomology theory. At present, the most popular treatment is that of Borel and Haefliger [5], who use "Borel-Moore"-homology. We look forward to having a detailed treatment using étale cohomology soon.

[19] It is generally conjectured that homological equivalence is equal to numerical equivalence. The meaning and form of the conjecture are developed in [39], following some ideas of Grothendieck.

$$C = \alpha_1 C_1 + \dots + \alpha_p C_p,$$

expressing an arbitrary i-fold condition C as a linear combination of basic conditions $C_1, \dots, C_p$ are characters of C. Consider an i-parameter system of the figures. Let $c, c_1, \dots, c_p$ and m denote the cohomology classes of the subsets defined by the conditions C, $C_1, \dots, C_p$ and by the system. Then the number of figures in the system satisfying the condition C is equal to the degree, $\deg(m.c)$. So the number is given by the expression,

$$\alpha_1 \mu_1 + \dots + \alpha_p \mu_p \qquad \text{with } \mu_j = \deg(m.c_j).$$

The integers μ_j obviously depend only on the system, not on the condition C. In fact, μ_j is equal to the number of figures in the system satisfying the basic condition C_j. This is Schubert's definitive way of generalizing Chasles's expression, $\alpha\mu + \beta\nu$.

Schubert solved the problem of characteristics in a number of cases by determining explicit bases.[20] In a number of other cases, he had a good idea of which conditions would form bases, and he worked many examples in which he expressed given conditions in terms of these. Once the basic conditions have been decided on, it is important for later applications to specific enumerative problems to know the numbers of figures satisfying all the various strings of these conditions. With impressive zeal (before the age of the computer), Schubert worked out table after table after table of these fundamental numbers, as he called them, which commonly are in the thousands, tens of thousands, and hundreds of thousands.

Entirely algebraic, rigorous treatments of the foundations of the calculus have become possible with the development of algebraic intersection theories.[21] These theories have evolved nonlinearly[22] through the efforts of a great many mathematicians, including Severi, van der Waerden, Chow,

---

[20] A general method for determining an explicit basis for the algebraic cohomology classes on an algebraic variety is still unknown today and would be worthwhile to have.

[21] Most of these theories extend the range of application of the calculus to problems in geometry over any algebraically closed ground field of any characteristic.

[22] One line of development, that from Severi to Weil, is traced in van der Waerden's interesting historical article [114].

Chevalley, Zariski, Weil, Hodge-Pedoe, Gröbner, Samuel, Matsusaka, Shimura, Grothendieck, Serre, Washnitzer, Monsky, Lubkin, Artin, Kleiman, Jussila, Deligne, Berthelot, Illusie, Gersten, Quillen, Bloch, Ogus, and Fulton, to name a few. At present, the most popular way to assign a multiplicity to a component W of a proper intersection of subvarieties, X and Y, of a smooth variety Z is to use Serre's [95] remarkable tor-formula,[23]

$$i(X.Y, W;Z) = \Sigma\ (-1)^j \text{length}[\text{tor}_j^{\mathcal{O}_{Z,W}}(\mathcal{O}_{Z,X}, \mathcal{O}_{Z,Y})].$$

The formula is simple and easy to use, but it does not appear to embody the old geometric idea of coalescing intersections. At present, the Chevalley seminar [9] is the most popular reference for a detailed treatment of traditional global intersection theory- the theory of intersection rings constructed directly out of the algebraic subsets by means of a local intersection theory and an adequate equivalence relation. The last two decades have witnessed the development of some radically different global intersection theories. Naive algebraic K-theory is especially well-suited for computations. Higher algebraic K-theory promises to be a powerful new tool. The algebraic, or Weil, cohomology theories, which have all the useful formal properties of the topological cohomology theories, provide the basis for the only known, entirely algebraic general existence proof of a solution to Schubert's problem of characteristics.

The development of an intersection theory is unquestionably a very great accomplishment, which is necessary for a rigorous foundation of Schubert's enumerative calculus, yet to claim that the development of an intersection theory provides a complete solution to Hilbert's fifteenth problem is to give the problem a narrow interpretation,[24] for the development of

---

[23] A simple example with a nontrivial higher "tor" is this: let Z be 4-space, X a surface with a point W whose local ring is not Cohen-Macaulay (e.g., X the cone over the twisted cubic space curve and W the vertex), and Y a general 2-plane through W.

[24] Narrow views of this sort have been widespread. In [118, p. 279, footnote 70], [102, German translation, p. 12] and [81, p. 285], the authors say that Hilbert's fifteenth problem asks for a justification of the principle of conservation of number. In [111, p. 337], the author says the problem calls for a rigorous foundation of Schubert's enumerative calculus, and he then leaves

an intersection theory is only the first step on the road toward the validation of the geometrical numbers of classical enumerative geometry, and in both the statement and the explanation of the problem Hilbert makes clear his interest in the efficient production of accurate geometrical numbers.

One theorem, which has considerable practical value in problems in which the figures are linear spaces, details the additive structure and the duality structure of the cohomology ring[25] of the Grassmann manifold. Known as the basis theorem, it asserts that the Schubert cycles, that is, the cohomology classes of the Schubert varieties, form an additive basis of the cohomology ring and that the basis is self-dual under the pairing,

$$x, y \longmapsto \deg(x.y),$$

of Poincaré duality. The basis theorem is essentially Schubert's solution [86, 1886] of the problem of characteristics for linear spaces. The importance of the basis theorem, and of some related results, led to popularity of the false impression[26] that the Schubert calculus amounts to a determination of the explicit structure of the cohomology ring of the Grassmann manifold and of the flag manifold.[27]

The basis theorem was proved rigorously by Ehresmann [17, 1934, 416-422] first. He observed that the Schubert varieties furnish a cellular

---

the impression that this is a call for the development of an intersection theory. In [115, p. viii, bottom], the author says that his development of a local- not global, but local- intersection theory provides a complete solution to the problem.

[25] The cohomology ring and the various other intersection rings are essentially the same on a Grassmann manifold, on a flag manifold, and on the other familiar rational manifolds with cellular decompositions (cf. [20] and [40; p. 69, Corollary (3.13)]).

[26] This belief is expressed in so many words in the middle of page 331 of the second edition of the influential book, Foundations of Algebraic Geometry [115], (no mention is made of the conflict between this belief and the author's interpretation of Hilbert's fifteenth problem, which is given at the bottom of page viii). The same false impression of the Schubert calculus is left by the Monthly article, "Schubert calculus" [43].

[27] Schubert [84; 90; 91] did a certain amount of work on the problem of characteristics, and on some related matters, for some special kinds of flags.

decomposition[28] of the Grassmann manifold, and he derived the additive structure of the cohomology ring by applying some general results about cell complexes, which he established for this purpose. He proved the duality part by a simple direct computation involving explicit local equations of suitably chosen Schubert varieties. Later, Hodge[29] [25, 1941] proved a purely algebraic version of the first and hardest part of the basis theorem. He used a method of degenerating a subvariety of the Grassmann manifold ultimately into a combination of Schubert varieties by means of linear transformations, which is like Schubert's original method. Then he [20, 1942] proved the duality assertion by the same direct computation Ehresmann used. Recently, Laksov [47, 1972] gave a new, more formal proof of the basis theorem.

To formulate the basis theorem in more detail, let $(a_0, \ldots, a_d)$ denote the class of a Schubert variety $[A_0, \ldots, A_d]$ such that A_j has dimension a_j for $j = 0, \ldots, d$.[30] The class is independent of the particular choice of $A_0, \ldots, A_d$ because, given a second choice $B_0, \ldots, B_d$, there exists an invertible linear transformation carrying A_j isomorphically onto B_j for $j = 0, \ldots, d$ and so there exists a (connected) continuous family deforming $[A_0, \ldots, A_d]$ into $[B_0, \ldots, B_d]$. The Schubert variety $[A_0, \ldots, A_d]$ has dimension, $i = \Sigma\, (a_j - j)$; hence, its class $(a_0, \ldots, a_d)$ lies in the $2(h-i)$th cohomology group, where

$$h = (d+1)(n-d)$$

is the dimension of the Grassmann variety $G_{d,n}$. The basis theorem asserts that the Schubert cycles $(a_0, \ldots, a_d)$ with $\Sigma\, (a_j - j) = h - i$ form a (free) basis

---

[28] In fact, each Schubert variety contains an open subset, which is isomorphic to an affine space, whose complement is the union of certain smaller Schubert varieties. The cellular decomposition generalizes, by means of the Bruhat decomposition, to the flag manifold and the other quotients of semi-simple algebraic groups by parabolic subgroups.

[29] Hodge's proof is the basis for the proof in Hodge-Pedoe [27, XIV, 5, 337-352].

[30] Schubert [85, 86, 87, 1886] introduced the symbol $(a_0, \ldots, a_d)$, known as a dimension symbol, to denote the Schubert condition defined by the flag $A_0, \ldots, A_d$ - in effect, to denote its class modulo numerical equivalence. Later, he [90, 91, 1903] introduced the symbol $\{\alpha_0, \ldots, \alpha_d\}$ with $\alpha_j = n - d - j + a$, known as a condition symbol, as alternative notation, which is more convenient in some situations, and it too is still in use today.

of the (2i)th cohomology group and the corresponding Schubert cycles $(n-a_0, \ldots, n-a_d)$ form the dual basis of the 2(h-i)th cohomology group.

Let x be an element of the (2i)th cohomology group. Then the basis theorem means that x may be expressed uniquely in the form,

$$x = \Sigma\, [\delta(a_0, \ldots, a_d)](n-a_d, \ldots, n-a_0) \qquad \text{with } \Sigma\, (a_j - j) = i,$$

and that the coefficient $\delta(a_0, \ldots, a_d)$ is the integer given by the formula,

$$\delta(a_0, \ldots, a_d) = \deg(x.(a_0, \ldots, a_d)).$$

If x is the class of a subvariety X of $G_{d,n}$, then the integer $\delta(a_0, \ldots, a_d)$ is nonnegative[31] and is equal to the number of points[32] in the intersection of X with a general Schubert variety representing the Schubert cycle $(a_0, \ldots, a_d)$. Schubert called[33] the integers $\delta(a_0, \ldots, a_d)$ the degrees (Gradzahlen) of X.

Let y be an element of the 2(h-i)th cohomology group, and express y in the form,

$$y = \Sigma\, [\varepsilon(n-a_d, \ldots, n-a_0)](a_0, \ldots, a_d) \qquad \text{with } \Sigma\, (a_j - j) = i.$$

Then, by the duality part of the basis theorem, the degree deg(x.y) is given by the expression,

$$\Sigma\, \delta(a_0, \ldots, a_d)\, \varepsilon(n-a_d, \ldots, n-a_0).$$

If x and y are the classes of subvarieties that intersect in a finite number of points, this expression gives the (weighted) number of points of intersection as the sum of the products of the corresponding degrees of the two subvarieties.

---

[31] Thus the classes of the subvarieties lie in the convex cone generated by the Schubert cycles.

[32] The points each appear with multiplicity one, for the intersection is transversal in general. This matter will be taken up again much later.

[33] Schubert [86, p. 138], however, chose to use the symbol $(a_0, \ldots, a_d)$ also to denote the degree $\delta(a_0, \ldots, a_d)$ of X, when X was fixed and understood. He avoided confusion by using the symbol $[a_0, \ldots, a_d]$ to denote the set of d-planes satisfying an (unspecified) Schubert condition with dimension numbers, $a_0, \ldots, a_d$.

This expression constitutes Schubert's simultaneous generalization [86] of Bézout's theorem and of Halphen's theorem. Bézout's theorem[34] deals with the case of projective space, the case $d = 0$. In this case, a Schubert variety $[A_0]$ is obviously equal to the linear space A_0 itself. So, the basis theorem says that the (2i)th cohomology group is a free cyclic group generated by the class $(n-i)$ of an $(n-i)$-plane. The class x of an $(n-i)$-dimensional subvariety X has the form, $x = [\delta(i)](n-i)$. The coefficient $\delta(i)$ is equal to the number of points in the intersection of X with a general linear space; thus $\delta(i)$ is equal to the degree, or order, of X in the usual sense. If y is the class of an i-dimensional subvariety Y that meets X in a finite number of points, then the above expression gives, for the (weighted) number of points of intersection, just the product of the degree of X and the degree of Y. This is a common formulation of Bézout's theorem.

Halphen's theorem (Comptes rendus, 1872) concerns congruences, or 2-parameter families, of lines in 3-space. Each congruence is represented by a 2-dimensional subvariety X of the Grassmann variety $G_{1,3}$, and so it has two degrees, $\delta(0, 3)$ and $\delta(1, 2)$. The first degree $\delta(0, 3)$ is equal to the number of points in the intersection of X with a Schubert variety $[A_0, \mathbb{P}^3]$ such that A_0 is a general point in $\mathbb{P}^3$; hence $\delta(0, 3)$ is equal to the number of lines in the congruence passing through a general point A_0. Similarly, the second degree $\delta(1, 2)$ is equal to the number of lines in the congruence lying in a general plane A_1. Therefore, the number of lines common to 2 congruences is given by the expression,

$$\mu\mu' + \nu\nu',$$

where μ and μ' are the numbers of lines in the congruences passing through a general point and where ν and ν' are the numbers of lines in the congruences lying in a general plane. This is Halphen's theorem.

Halphen's theorem can be used to obtain an expression (of Halphen's)

---

[34] The theorem was known, at least in embryonic form, to Jacques Bernoulli (1688), (see Berzolari's encyclopedia article [4]). Euler (1748) refined the statement of theorem for plane curves by excluding common components and taking into account imaginary points and points at infinity. Bézout (1764) extended Euler's work to the case of n hypersurfaces in n-space for arbitrary n by developing an improved elimination theory. Schubert's form of Bézout's theorem is even more general.

for the number of chords common to 2 space curves in general position. Consider the congruence of chords of a space curve with degree n. The number μ of chords passing through a general point in space is known as the number of apparent double points. It is equal to the number of new singularities appearing- and they are all ordinary double points- when the curve is projected from that point into the plane. If the curve is smooth with genus p, then μ is therefore given by the formula,

$$\mu = \tfrac{1}{2}(n-1)(n-2) - p.$$

The number ν of chords lying in a general plane is obviously equal to the number of pairs that can be made of the n points common to the curve and the plane; thus ν is given by the formula,

$$\nu = \tfrac{1}{2}\, n(n-1).$$

Hence, for 2 space curves in general position with degrees, n_1 and n_2, and with μ_1 and μ_2 apparent double points, the number of common chords, by Halphen's theorem, is equal to

$$\mu_1\mu_2 + \frac{1}{4}\, n_1n_2(n_1-1)(n_2-1).$$

For example, for 2 twisted cubic space curves in general position, the number of common chords is

$$[\tfrac{1}{2}(2)(1) - 0]^2 + \frac{1}{4}\, 3^2 2^2 = 10.$$

As another example[35] illustrating the use of the basis theorem, let us find the number of lines common, in general, to 2 quadric hypersurfaces in 4 space. It is a classical result, readily established by counting constants or by using modern deformation theory, that the lines on a general quadric hypersurface Q in 4-space are parametrized by an algebraic set of pure dimension 3 on the Grassmann manifold $G_{1,4}$. By the basis theorem, the cohomology class x of X may be expressed in the form,

$$x = [\delta(0,4)](0,4) + [\delta(1,3)](1,3).$$

[35] This example is also worked in Hodge-Pedoe [28, 366-367] and in the Monthly article [43, p. 1075].

The coefficient $\delta(0,4)$ is equal to the number of points in the intersection of X with a Schubert variety $[A_0, \mathbb{P}^4]$ such that A_0 is a general point in $\mathbb{P}^4$; hence $\delta(0,4)$ is equal to zero because no line on Q can pass through a point A_0 off Q. The coefficient $\delta(1,3)$ is equal to the number of points in the intersection of X with a Schubert variety $[B_0, B_1]$ such that B_1 is a general 3-plane in $\mathbb{P}^4$ and B_0 is a general line in B_1. Hence $\delta(1,3)$ is equal to the number of lines on Q that lie on B_1 also and meet B_0. Now, the intersection Q_1 of Q and B_1 is a smooth quadric surface, so Q_1 has 2 families of rulings. The general line B_0 of B_1 meets Q_1 in 2 distinct points, and each point lies on 1 ruling in each family. These 4 rulings are

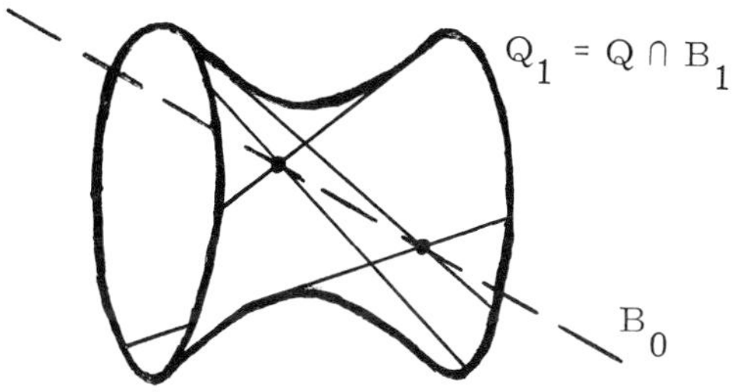

all the lines on Q_1 meeting B_0. Hence $\delta(1,3)$ is equal to 4. Thus x has the form, $x = 4(1,3)$. Therefore, for 2 quadric hypersurfaces in 4-space, the number of common lines, in general, is equal to

$$\deg(x^2) = 4^2 \deg((1,3)^2) = 16.$$

Information about the multiplicative structure of the cohomology ring of the Grassmann manifold can be useful, and numerous investigations[36] were undertaken classically. Two formulas developed then are particularly important. Both involve certain special Schubert cycles, namely,

$$\sigma_i = (n-d-i, n-d+1, \ldots, n) \qquad i = 0, \ldots, n-k.$$

The cycle σ_i represents the condition on a d-plane that it meet a given (n-d-i)-plane; the notation σ_i is Giambelli's (1903). The first formula,

[36] A guide to the literature is found in the encyclopedia articles of Segre [93, 813-820] and Zeuthen-Pieri [118, 312-313].

known now as Pieri's formula, expresses the product of an arbitrary Schubert cycle with a special Schubert cycle; the formula is

$$(a_0, \ldots, a_d) \cdot \sigma_i = \Sigma\, (b_0, \ldots, b_d),$$

where the sum extends over all distinct sets of indices $b_0, \ldots, b_d$ satisfying the conditions,

$$0 \le b_0 \le a_0,\quad a_0 < b_1 \le a_1, \ldots, a_{d-1} < b_d \le a_d,\quad \text{and } \Sigma\, b_j = \Sigma\, a_j - i.$$

The second formula, known now as Giambelli's formula or the determinantal formula, expresses an arbitrary Schubert cycle as a determinant in the special Schubert cycles; the formula is

$$(a_0, a_1, \ldots, a_d) = \begin{vmatrix} \sigma_{n-a_0-d} & \cdots & \sigma_{n-a_0} \\ \sigma_{n-a_1-d} & \cdots & \sigma_{n-a_1} \\ \cdot & \cdots & \cdot \\ \sigma_{n-a_d-d} & \cdots & \sigma_{n-a_d} \end{vmatrix},$$

where the standard convention, $\sigma_i = 0$ for $i < 0$, is used.

Giambelli's formula, in conjunction with the basis theorem, implies that every cohomology class is equal to a linear combination of products of the special Schubert cycles; in other words, the special Schubert cycles form a $\mathbb{Z}$-algebra basis. Moreover, Giambelli's formula reduces the matter of determining the product of two arbitrary Schubert cycles to the case where one, or even both, are special Schubert cycles, and this case is covered by Pieri's formula. Of course, it is possible to express the product of two arbitrary Schubert cycles directly as a certain linear combination of other Schubert cycles; however, the formula is too complicated to be of any general practical value. On the other hand, there is a simple expression[37] for the

[37] The exact expression is found in Lascoux's article [50, 5-7 Remarque], proved using K-theory. In another article [51], Lascoux shows how expressions for products of arbitrary Schubert varieties can be derived from Giambelli's formula using a formalism he develops. Classically, Giambelli found expressions of this sort, and his work too involved the theory of

product of two arbitrary Schubert cycles as a certain determinant in the special Schubert cycles.

Let us use the formalism to compute the number of lines meeting 4 given lines in 3-space. The lines meeting a given line M are parametrized by the Schubert variety $[M, \mathbb{P}^3]$, whose class is $(1, 3)$, or σ_1. Hence the number of lines meeting 4 given lines is equal to $\deg(\sigma_1^4)$. Applying Pieri's formula gives[38]

$$\sigma_1^2 = (1, 3).\sigma_1 = (0, 3) + (1, 2).$$

Finally, applying the duality assertion of the basis theorem gives

$$\deg(\sigma_1^4) = 1 + 2\cdot 0 + 1 = 2.$$

Alternatively, a second application of Pieri's formula and a third give

$$\sigma_1^3 = 2(0, 2) \quad \text{and} \quad \sigma_1^4 = 2(0, 1).$$

Since $(0, 1)$ is the class of a single point, its degree is 1. Thus, if there are finitely many lines meeting 4 given lines, there are 2 lines or 1 line counted twice.

Similarly, Pieri's formula can be used to find the number

$$\frac{1!2!\ldots d!h!}{(n-d)!\ldots n!}$$

of d-planes in n-space meeting $h = (d+1)(n-d)$ general $(n-d-1)$-planes, for the number is equal to the degree of σ_1^h. Since σ_1 is the class of a Schubert variety defined by one linear relation in the Plücker coordinates (in addition to the quadratic relations defining the Grassmann manifold $G_{d,n}$), this number is also equal to the degree of $G_{d,n}$. More generally, Pieri's formula can be used in this way to find the number

---

symmetric functions. On a somewhat different track, Koch [45] studied the matter on a wide class of homogeneous spaces using Kostant's method of Lie algebra cohomology and representation theory.

[38] The equation means that the condition that a line meet two given (distinct) lines is equal to the condition that a line pass through given point or lie in a given plane. The equation is indicated by altering the position of the two given lines so that they intersect in a point and determine a plane. The Schubert calculus is a splendid way of managing many such degeneration arguments in solving problems.

$$\frac{i!}{a_0! \cdots a_d!} \prod_{\lambda>\mu} (a_\lambda - a_\mu) \qquad \text{with } i = \Sigma\,(a_j - j)$$

of d-planes L in n-space satisfying a Schubert condition with dimension numbers a_j while meeting i general (n-d-1)-planes, for the number is equal to the degree of the product, $(a_0, \ldots, a_d).\sigma_1^i$; this number is also equal to the degree of the Schubert variety, $[A_0, \ldots, A_d]$. The computations involved here proceed by means of two recurrence relations and fill nearly two pages in Hodge-Pedoe [28, 364-366]; they originated with Schubert [85; 87], who developed the case, $i = 1$, of Pieri's formula for just this purpose.[39]

Pieri's formula is equivalent, in view of the basis theorem, to the assertion that the numbers

$$\deg[(a_0, \ldots, a_d).\sigma_i.(n-b_d, \ldots, n-b_0)] \quad \text{with } \Sigma\, b_j = \Sigma\, a_j - i$$

are 0 unless the inequalities, $0 \leq b_0 \leq a_0, \ldots, a_{d-1} \leq b_d \leq a_d$, are each satisfied, and then the numbers are 1. Classically, the formula was justified by observing that there are 0 d-planes or 1 d-plane satisfying an appropriate string of 3 Schubert conditions. The proof was secured by van der Waerden [112] for $i = 1$ and by Hodge and Pedoe [28, p. 354] for all i; they showed that, in the case in which there is 1 d-plane meeting the conditions, it is valid to count it with multiplicity 1 by proving that the 3 corresponding Schubert varieties intersect transversally by means of an explicit local analysis like the one Ehresmann introduced to prove the duality part of the basis theorem. Alternatively, Pieri's formula was established directly using a method of degeneration like the one used to establish the first part of the basis theorem; this proof was secured by Hodge [26]. Recently, new proofs of Pieri's formula were given by Laksov [47], by Koch [45], and by Lascoux [50; 51]; each proof is approached from its own general point of view.

Giambelli's formula has traditionally been derived from Pieri's formula by formal manipulations (cf. Todd [109], Hodge [26], and Hodge and

---

[39] In the same way Schubert [88] later obtained the following even more general formula:

$$\deg((a_0, \ldots, a_d)(b_0, \ldots, b_d)\sigma_1^k) = k!\ \det(1/(a_i + b_{d-j} - n)!) \quad \text{with}$$
$$k = \Sigma\,(a_j - j) + \Sigma\,(b_j - j) - (d+1)(n-d).$$

Pedoe [28, 354-359]). Recently, new proofs of the formula were given by Laksov [47], by Damon [12], by Kempf and Laksov [38], and by Lascoux [50]; each proof furnishes a substantial generalization of the formula.

The modern form taken by Schubert's enumerative theory of d-planes in n-space is the theory of Chern classes. Chern, in the late 1940's and early 1950's, founded a theory of characteristic cohomology classes, which generalizes the Stiefel-Whitney classes to complex vector bundles. Later (1958), Grothendieck [19, 20, 21] developed a definitive version of the theory for algebraic geometry. A vector bundle that is generated by its global sections- and enough[40] are- gives rise to a map into a certain Grassmann manifold, for the bundle is just a family of quotient spaces of the vector space of global sections. Then the characteristic classes of the vector bundle are simply the pull-backs of the Schubert cycles, and its Chern classes are the pull-backs of the special Schubert cycles (cf. Chern [8]). Originally the properties of characteristic classes and Chern classes were deduced from properties of Schubert cycles. Now the properties are proved directly. However, proving them amounts to proving properties of Schubert cycles. In fact, via the theory of Chern classes, all the classical properties of Schubert cycles can be established and usefully generalized. A version of the basis theorem was proved by Grothendieck [20], a version of Pieri's formula was proved by Lascoux [50; 51], and versions of Giambelli's formula were proved by Damon [12], by Kempf and Laksov [38], and by Lascoux [50]. The theory of Chern classes can, moreover, be a useful new tool for solving enumerative problems.[41]

The Chern classes of a variety, that is, the Chern classes of its tangent bundle, are important intrinsic invariants. They are intimately related to some older invariants- the arithmetic genus, the Zeuthen-Segre invariant, Severi's projective characters, and the canonical systems of Todd and of Eger. The arithmetic genus and the Zeuthen-Segre invariant entered into classical enumerative geometry through applications of the theory of

---

[40] They all are in topology and differential geometry, but not in algebraic geometry.

[41] For example, it is not hard to see that the lines on a quadric hypersurface are parametrized by the scheme X of zeros of a regular section of the second symmetric power of the universal quotient bundle. Hence, by standard theory, the cohomology class x of X is equal to σ_1^3. (Cf. [2]).

correspondences[42] primarily to finding the numbers[43] of linear spaces having prescribed numbers of contacts of prescribed orders with one or more curves and possibly satisfying some other conditions (such as Schubert conditions), and to finding the numbers of lines having prescribed numbers of contacts of prescribed orders with one or more hypersurfaces and possibly satisfying some other conditions. Severi [97], applying the Schubert calculus to the tangent d-planes of a smooth d-fold, introduced and studied some basic projective characters, namely, the degrees of the loci of points where the tangent d-planes satisfy Schubert conditions. For example, the projective characters of a curve in n-space are simply its degree and its order (that is, the number of tangent lines meeting a general (n-2)-plane). Then Severi [98] was led to conjecture the formula relating the projective characters and the arithmetic genus, and later he [101] considered, in the case of a surface, the cohomology[44] classes of the loci. Todd [110], inspired to some extent by Severi's work and later to some extent by some related work of Eger [16], passed from the cohomology[44] classes of the loci to some other cohomology classes, which are intrinsic invariants, independent of the embedding in projective space. The latter classes were proved identical, up to sign, to the Chern classes in the case of complete intersections by Hodge [27]

[42] A nice introduction to this theory and to its applications in enumerative geometry, which includes numerous examples and numerous references, is found in the first two chapters of Baker's book [3, 1-111]. A guide to the classical literature is found in Zeuthen and Pieri's encyclopedia article [118, 285-302, 309-310, 313]. A concise review of some theoretical work on the theory of correspondence is found in Appendix B of Zariski's book [118, 239-247]. Very little work has been done to justify the applications of the theory of correspondences made in classical enumerative geometry.

[43] Numbers of this sort are given by the famous contact formulas of de Jonquières [13]. De Jonquières in fact took an essentially equivalent point of view involving plane curves. However, virtually from the beginning, most discussions of the formulas use the language of linear series, a third point of view. Other numbers of this sort are given by a formula, developed by Schubert, Segre, and Severi (cf. [118, p. 296, footnote 142]), dealing with a linear series and 1-parameter algebraic series. Recently MacDonald [56; 57] discussed the formulas and showed how to prove them using a detailed knowledge of the cohomology ring of the symmetric product of the curve. Mattuck [60] developed an algebraic version of this approach. A review of some of the classical literature is given by Segre [93, 885-890].

[44] In fact, Severi and Todd worked with the less "wasteful" notion of rational equivalence class, but that notion had not yet been rigorously developed.

and in the general case by Kodaira and Serre (unpublished) and by Nakano [67]. Recently, interesting treatments of this identity were given by Porteous [73] and by Lascoux [50, 5-10, 5-11] also.

A new and promising direction of research in this area is the enumerative theory of singularities of mappings, the theory of Thom polynomials. The first order (Thom-Boardman) singularities of a map between smooth varieties are the points where the rank of the differential is constant. If the map satisfies certain "transversality" conditions, the nth order singularities may be defined inductively as the first order singularities of the restriction of the map to the locus of the (n-1)th order singularities, for this locus will be smooth. Then the cohomology classes of the closures of the loci of the various types[45] of nth order singularities are given by the various nth order Thom polynomials, evaluated using the Chern classes of the source and the target. Thom (cf. [22]) suggested an existence proof for the polynomials based on the fact that the special Schubert cycles generate the cohomology ring of the Grassmann manifold. So far, Porteous [72] has determined explicitly the first order polynomials, Ronga [80] has given an algorithm for calculating all the polynomials and he [79] has determined explicitly some second order polynomials, and Menn [62], Sergeraert [94], and Lascoux [48] have determined explicitly some important nth order polynomials.

In algebraic geometry, unlike in differential geometry, few maps satisfy the requisite transversality conditions. In fact, the only class of maps whose singularities have been at all studied are the projections from general centers; this work has been carried out by Lluis [53, 54, 55], by Samuel [82, 46-50], by Holme [29, 30], by Roberts [74, 75], and by Mather [58, 59]. However, algebraic geometry possesses well-developed techniques for managing infinitesimal information, and it may be possible to develop a general enumerative theory of singularities of mappings that satisfy only mild conditions on the dimensions of the various loci of singularities.

Recent work seems to furnish instances within a general theory of this sort and indications of some potential applications. Algebraic definitions of the singularity loci were proposed and investigated by Roberts [75, 77, 78] and by Mount and Villamayor [63]. Kempf and Laksov [38] established a

[45] The type of a singularity indicates the ranks of the differentials of the map and of its restrictions.

formula[46] simultaneously generalizing Porteous's formula of first order Thom polynomials and Giambelli's formula, which is valid under mild conditions on the dimensions of the singularity loci. The formula has been used in (and stems from) justifying a classical assertion about the existence of special divisors (cf. [37], [42], [44]). Iversen [32, 33] established Thom-like polynomial formulas for maps between surfaces and for maps onto curves, which are valid under this sort of dimension conditions, and he [34] used the formula for surfaces to obtain an enumerative result about the exceptional tangents to a smooth curve. Roberts [75] established a Thom-like polynomial formula for projections, which is valid under this sort of mild dimension conditions, and applied it to counting pinch points. Holme [31] obtained a numerical criterion[47] involving a polynomial in Severi's projective characters, indicating whether or not a given general projection is an embedding. The same criterion was obtained independently by Peters and Simonis using a somewhat different approach. Pohl [70] considered, from the point of view of the enumerative theory of singularities of mappings, one of the classical Plücker formulas, which relate the various classical projective characters of a plane curve with mild singularities, and thereby he was able to understand and generalize it. Further work relating the projective characters of varieties was done by Mulcherjia [65]. Ogawa [69] and Mount and Villamayor [64] looked at Weierstrass points[48] as singularities of mappings. Porteous [73] and Lascoux [49, 52] looked from this point of view, at the classical theory of the jacobian system of a linear system. It is an exciting prospect that all of this work may soon be unified and advanced and that all of the classical work with linear spaces subjected to prescribed conditions of contact may soon be justified and incorporated with it by means of a general

[46] Essentially the same formula was established by Damon [12] and by Lascoux [50] but under the transversality conditions. The key to replacing the transversality conditions by the dimension conditions is a recent theorem asserting the local rings of a Schubert variety are Cohen-Macaulay.

[47] Results of this kind, which involve the imposition of conditions of contact at several points, are strictly speaking not part of the theory of singularities of mappings, but they are so closely related they should be part of the same general theory.

[48] Weierstrass points on a curve were also treated by Mattuck [60], in an interesting way involving computations in the intersection ring of the symmetric product.

algebro-geometric enumerative theory of singularities of mappings.

The classical enumerative geometers hardly restricted their attention exclusively to problems about linear spaces. However, virtually the only other class of problems that has attracted mathematicians to seriously consider the foundations of the method of solution is that about (plane) conics. One problem, in particular, has had an interesting history (cf. Zeuthen-Pieri [118, p. 271]). It is the problem of finding the number of conics tangent to 5 given conics. Steiner (1848) thought, at first, the number might be 6^5. He did not indicate why he thought so, but it is natural to reason like this. A conic is the set of zeros of a single homogeneous polynomial of degree 2, and the 6 coefficients may be used to represent the conic by a point of projective 5-space. The conics tangent to a given conic are represented by the points of a hypersurface of degree 6. Hence the conics tangent to 5 given conics are represented by the points common to 5 hypersurfaces of degree 6. Therefore, the number of conics tangent to 5 given conics should be 6^5 by Bézout's theorem. However, each of the 5 hypersurfaces passes (in fact, it passes doubly) through the Veronese surface, whose points represent the degenerate conics that consist of 2 coincident lines. These double lines are being counted among the conics tangent to the 5 given conics, although, properly speaking, they should be excluded.

Steiner himself determined that the true number of conics tangent to 5 given conics is something less than 6^5 on the basis of the principle of conservation of number by specializing the 5 given conics until they degenerated into 5 unions of 2 distinct lines each. Using this method, Berner (1865) and de Jonquières (sometime before 1859, but unpublished) found the correct number, 3264. What is more significant, Chasles (1864) developed his theory of characteristics and applied it to this problem. Chasles's characteristics, α and β, of the condition that a conic be tangent to a given conic are both equal to 2. So evaluating Chasles's expression (recalled above) for the number of conics satisfying 5 given conditions yields

$$2^5(1 + 2\binom{5}{1} + 4\binom{5}{2} + 4\binom{5}{3} + 2\binom{5}{4} + 1) = 32 \cdot 102 = 3264.$$

Via a different route, Severi [97, 1902] arrived at the number 3264. He used a formula he found[49] for the number of isolated intersections in

[49] The formula was a corollary of Severi's theory of projective characters.

n-space of n general hypersurfaces passing doubly through a common surface.

Chasles's theory has inspired a number of mathematicians to try to put it on a firm basis (cf. Zeuthen and Pieri [118, p. 321, footnote 237]), notably, Clebsch [7, 1873], Study [103, 1886], van der Waerden [113, 1934], and Severi [102, 1940]. The key to the matter is to consider, along with a conic, its family of tangent lines. The lines in the plane are represented by the points of another plane, called the dual plane, and the lines tangent to a given conic determine in this way another conic, called the dual conic. The pair consisting of a conic and its dual conic is to be viewed as a single geometric entity, called a complete conic after Study. If the conic is degenerated into a double line, the dual conic degenerates into a pair of lines, which corresponds to a pair of points, or foci, on the double line. A curve is therefore considered tangent to this degenerate complete conic if it is tangent in the ordinary sense to the double line or if it passes through either focus. Consequently, the number of curves in a one parameter family that are tangent to the degenerate complete conic is equal to

$$2\mu + 2\nu,$$

where μ is the number of curves in the family passing through a given point and ν is the number, tangent to a given line. By the principle of conservation of number, $2\mu + 2\nu$ is also the number of curves in the family tangent in the usual sense to any (nondegenerate) conic. Thus the characteristics, α and β, of the condition that a conic be tangent to a given conic are both equal to 2. This method of deriving the expression $2\mu + 2\nu$, and others like it, is explained in Schubert's book [84, p. 14, bottom]. The basic idea is already found in Steiner's work.

Chasles's theory can be readily understood and rigorously justified using the variety parametrizing the complete conics, namely, the closure in $\mathbb{P}^5 \times \mathbb{P}^5$ of the locus of pairs of points representing a (nondegenerate) conic and its dual. The restriction to the variety of the first projection is equal to the blowing-up of $\mathbb{P}^5$ with the Veronese surface as center.[50] So the variety

[50] This fact is readily seen using the (classical homogeneous) representation of a conic by a symmetric 3×3-matrix. Then the dual conic is represented by the inverse matrix, and the Veronese surface consists of the matrices whose 2×2-minors all vanish.

is smooth, and its cohomology ring[51] is easily described using standard theory (cf. Katz [36, 258-266]). Of course, the situation is symmetric for both projections.

A natural basis for the second cohomology group of the variety parametrizing the complete conics consists of the classes, m and n, of the pull-backs of hyperplanes in the two copies of $\mathbb{P}^5$. So the class of a subvariety defined by a simple geometric condition has the form,

$$\alpha m + \beta n,$$

for suitable integers, α and β; and the (weighted) number of curves in a one parameter family that satisfy the condition is given by the expression,

$$\alpha\mu + \beta\nu,$$

where μ and ν are the products of m and n with the class of the curve parametrizing the family. The condition that a conic pass through a given point defines a hyperplane in the first copy of $\mathbb{P}^5$; the condition that a conic be tangent to a given line defines a hyperplane in the second copy of $\mathbb{P}^5$. Hence μ is equal to the number of conics in the family that pass through a given point and ν is equal to the number, tangent to a given line. Thus Chasles's theorem is proved.

Similar reasoning shows that the number of conics satisfying 5 independent conditions with characteristics,

$$\alpha,\ \beta;\ \alpha',\ \beta';\ \alpha'',\ \beta'';\ \alpha''',\ \beta''';\ \alpha^{iv},\ \beta^{iv},$$

is given by the expression,

$$(\alpha m+\beta n)(\alpha' m+\beta' n)(\alpha'' m+\beta'' n)(\alpha''' m+\beta''' n)(\alpha^{iv} m+\beta^{iv} n),$$

provided that, when the multiplication is carried out, the term $m^r n^s$ is replaced by the number of conics passing through r points and tangent to s lines. Thus Halphen's observation is explained.

Another natural basis of the second cohomology group consists of the

---

[51] The cohomology ring and the various other intersection rings are essentially the same on this variety because these are so both for the variety being blown-up, $\mathbb{P}^5$, and for the center, the Veronese surface.

classes, ℓ and p, of the 2 exceptional divisors. The first exceptional divisor is defined by the degeneracy condition that a (complete) conic be a double line with two foci, and the second is defined by the degeneracy condition that a (complete) conic be two distinct lines or be a double line with a double focus. The two bases are related by the formulas,

$$\ell = 2m - n \quad \text{and} \quad p = 2n - m.$$

Hence the numbers, λ and π, of conics in a one parameter family with characteristics, μ and ν, that satisfy the degeneracy conditions are given by the formulas,

$$\lambda = 2\mu - \nu \quad \text{and} \quad \pi = 2\nu - \mu.$$

These formulas were included in Chasles's theory, and they were used often in practice, especially by Zeuthen,[52] to compute μ and ν.

Problems about quadratic q-folds in n-space and problems about twisted cubic curves in 3-space form the two principal other classes of problems treated classically. These problems were considered by numerous mathematicians,[53] but Schubert did the definitive work. By and large, the deeper results were reached after a systematic step by step study of the enumerative properties and interrelationships of various systems of degenerate cases. Much of the research was done between 1864 and 1879 and is explained in Schubert's book [84]. Worked out on pages 106 and 184, there are these two spectacular examples: finding the number

666, 841, 088

of quadric surfaces tangent to 9 given quadric surfaces in 3-space; and finding the number

[52] Zeuthen also pointed out that the characteristic α of a condition indicates that the condition is satisfied by α degenerate conics consisting of a given line and a line passing through a given point. For example, the characteristic α of the condition that a conic be tangent to a given conic is equal to 2, because there are 2 lines tangent to a given conic and passing through a given point. The description of the characteristic β is similar and dual.

[53] A guide to the literature is given by Zeuthen-Pieri [118, 305-307, 315-317] and by Segre [93, 855-856].

5, 819, 539, 783, 680

of twisted cubic curves tangent to 12 given quadric surfaces in 3-space. After writing the book, Schubert went on to study general quadric q-folds in n-space, and he [92] obtained, for example, an expression, in terms of q and n, for the number of them tangent to $[n(q+2) - 1/2\, q(q+1)]$ hyperplanes. The work on twisted cubic space curves was published in full, for the first time, in Schubert's book, but it had already won for him in January 1875 a gold medal from the Royal Danish Academy.

The classical enumerative geometers had a good feeling for the work they were doing, and they assigned multiplicities with great virtuosity. Yet, at the end of the golden age, Zeuthen and Pieri say, in their encyclopedia article [118, 1915, p. 280], that the determination of these multiplicities is one of the fundamental problems that must be resolved above all. This has, of course, been done in principle; the multiplicity of a solution is now defined as the intersection multiplicity on the parameter variety of the figures, at the point representing the particular solution, of the subvarieties determined by the conditions of the problem. However, making this abstract definition entails the obligation of verifying it in practice, of comparing the geometrical numbers it gives with those obtained classically. In fact, there can be no doubt whatever that the abstract definition is a proper one; it has the several expected properties that characterize it. In the case of a disagreement with classical results, we must conclude that the classical work is in error. Nevertheless, this a priori assurance of accuracy does not discharge the responsibility of verifying the classical geometrical numbers, a responsibility brought to our attention by Hilbert's fifteenth problem.

Ideally, there should be established a set of useful, general rules for assigning multiplicities directly from the geometric context of a problem without recourse to any lengthy auxiliary work or to any ad hoc methods. For example, there is one general principle, which seems to have been implicitly taken for granted classically. The principle deals with the general case of an enumerative problem, that is, the case in which the parameters of all the given conditions assume general values. The principle asserts that, in the general case, each figure meeting the conditions is to be counted with multiplicity one, and so the weighted number of figures meeting the conditions is the exact number. Thus, for instance, on the basis of this principle it follows in the earlier examples that there are exactly 2 lines meeting 4 general lines in 3-space, that there are exactly 10 common chords of 2 twisted cubic space curves in general position in 3-space, that there are exactly 16 lines common to 2 general quadric hypersurfaces in 4-space, that there are exactly 3264 conics tangent to 5 general conics in the plane. The question is to what extent is the principle valid.

In classical enumerative problems, the several given conditions are varied by applying individual linear transformations of the ambient n-space. Thus we are lead to consider the action of the general linear group on the parameter variety for the figures and to ask whether the components each appear with multiplicity one in the intersection of one subvariety and a general translate of another. It is not hard to prove that the answer is affirmative if the group acts transitively (cf. [41, (4)]). Hence the principle is valid, without restriction,[54] if the figures are linear spaces.

The situation is more complicated if the figures are conics. The general linear group has four orbits; namely, it acts transitively on the set of nonsingular conics, on the set of pairs of distinct lines, on the set of double lines with distinct foci, and on the set of double lines with double foci. Nevertheless, the hypersurface of (complete) conics tangent to a given conic does not contain any one of these four sets. It follows that the intersection of 5 general such hypersurfaces consists entirely of points representing nonsingular conics and the intersection is transversal at each of these points. Thus there are exactly 3264 nonsingular conics tangent to 5 general conics. However, it is not surprising that matters are not always so nice, and indeed, Halphen (Comptes rendus, 1876) found examples[55] in which the expression, $\alpha\mu + \beta\nu$, apparently failed. In this instance, the classical

---

[54] The principle is false, however, in positive characteristic. There is a congruence, or 2-parameter family, of lines in 3-space such that every line in the congruence lying in a general plane has to be counted with multiplicity greater than one (cf. [41, (9)]). The congruence does not arise from a natural enumerative problem, and there still may be a useful form of the principle valid in any characteristic.

In any characteristic, the principle is valid in the case of Bézout's theorem; that is, the exact number of points common to 2 general subvarieties of n-space is always equal to the product of their degrees (cf. [41, (10)]).

[55] The examples are discussed to some extent by Zeuthen and Pieri [118, 320-321] and by Coolidge [11, 132-133].

mathematicians did not seem to realize that the trouble[56] lies in the assigning of multiplicities, in other words, that the principle may fail.

Before we may consider Hilbert's fifteenth problem solved, we must develop our understanding of the classical geometrical formulas and numbers dealing with prescribed conditions of multiple intersection and contact imposed, not only on linear spaces, but also on other figures as well. It is embarrassing that today we cannot vouch for the accuracy of the two spectacular numbers above, which Schubert computed nearly a century ago, and more especially that we do not even know whether his method is basically sound. Relatively recently, Alguneid [1, 1955] carefully studied Schubert's 11 first-order degenerations[57] of the twisted cubic space curve. Schubert considered the curve to be completed by the envelope of its osculating planes and the family of its tangent lines. Alguneid [1, p. 208] was lead to suggest three other "aspects" of the twisted cubic space curve that might be necesssary as components of the complete curve. It would be nice to have full knowledge of the appropriate manifold of complete twisted space curves, including an explicit description of its cohomology ring, and it would be especially nice if, in obtaining this knowledge, we are lead further.

Hilbert asks that the classical geometrical numbers be established "with an exact determination of the limits of their validity". Here, he may be asking for a delineation and stratification of the special values of the parameters of the given conditions for which the weighted number of figures meeting the conditions is no longer the exact number. It may also be interesting to study the permutation of the figures meeting the conditions effected by varying the parameters of the conditions along a path around a

---

[56] Halphen apparently felt, and Zeuthen and Pieri [118, p. 321] concurred, that these examples were simply not covered by the general expression, $\alpha\mu + \beta\nu$. Halphen went on to prove that the troublesome figures are always double lines with double foci, so that the expression, $\alpha\mu + \beta\nu$, is applicable whenever the 1-parameter family does not contain any such degenerate conics or whenever such conics are excluded by the condition. Study was on the track of group actions when he distinguished between movable and immovable solutions (cf. Coolidge [11, p. 137]). He defined an immovable solution as one that does not change its position in the one-parameter family when the condition is varied by applying a linear transformation of the plane, and he found that immovable solutions are always double lines with double foci.

[57] An ith order degeneration is one yielding a family of degenerate figures that has codimension i in the family of all figures.

stratum. Little work, if any, has been done on these questions.

The richness of Hilbert's fifteenth problem is a sure sign of its greatness. While certain aspects of the problem have been solved and solved again, others remain unsolved and beckon us.

BIBLIOGRAPHY

[1] Alguneid, A. R., "Analytical degeneration of complete twisted cubics", Proceedings Cambridge Phil. Soc. [52 (1956), 202-208].

[2] Altman, Allen B. and Kleiman, Steven Lawrence, "Some basic geometric properties of the Fano surface and of related varieties", to appear.

[3] Baker, Henry F., Principles of Geometry. Volume VI, Introduction to the theory of algebraic surfaces and higher loci, Cambridge University Press (1933).

[4] Berzolari, Luigi, "Théorie générale des courbes planes algébriques", Encyclopédie des sciences mathematiques, [III, 19], Teubner, Leipzig (1915).

[5] Borel, Armand, 1923-, and Haeflieger, A., "La classe d'homologie fundamentale d'un espace analytique", Bull. Soc. Math. France, [89 (1961), 461-513].

[6] Brill, Alexander von, 1842-1935, "Elimination und Geometrie in den letzten Jahrzehnten", Verhandlungen des dritten internationalen Mathematiker-Kongresses in Heidelberg 1904, [275-283] Leipzig (1915), Kraus reprint, Nendeln/Liechtenstein (1967).

[7] Clebsch, Alfred Rudolf Friedrich, 1833-1872, "Zur Theorie der Charakteristiken", Math. Annalen, [6 (1873), 1-15].

[8] Chern, S. S., 1911-, "Characteristic classes of Hermitian manifolds", Annals Math., [47 (1946), 85-121].

[9] Chevalley, Claude, 1909-, "Anneaux de Chow et applications", Seminaire C. Chevalley, Ecole Normal Superieure, Paris (1958).

[10] Coolidge, Julian Lowell, 1873-1940, A history of geometrical methods, Oxford University Press, London (1940), reprinted by Dover Publications, Inc., New York, (1963).

[11] Coolidge, J. L., A history of the conic sections and quadric surfaces, Oxford University Press (1945), reprinted (1947).

[12] Damon, James Norman, "Thom polynomials for contact class singularities", Thesis presented at Harvard University, Cambridge (1972).

[13] de Jonquières, J. Ph. E. de Fauque, 1820-1901, "Mémoire sur les contacts multiples d'order quelconque des courbes de degré r, qui satisfont à des conditions données, avec une courbe fixe du degré m; suivi de quelques reflexions sur la solution d'un grand number de questions concernant les propriétés projectives des courbes et des surfaces algébriques," Journal riene ang.Math., [66 (1866), 289-321].

[14] Dieudonné, Jean, "The historical development of algebraic geometry", Am. Math. Monthly [79 (1972), 827-866].

[15] Doubilet, Peter, Rota, Gian-Carlo, and Stein, Joel, "On the foundations of combinatorial theory: IX on the algebra of subspaces", to appear.

[16] Eger, M., 1922-1952, "Sur les systèmes canoniques d'une variété algébrique", Comptes Rendus Acad. Sci. Paris [204 (1937), 92-94 and 217-219].

[17] Ehresmann, Charles, "Sur la topologie de certains espaces homogènes", Ann. Math. [35 (1934), 396-443].

[18] Giambelli, Giovanni Zeno, "Sul principio della conservazione del numero", Jahresb. deutsch. Math.-Ver. [13 (1904), 545-556].

[19] Grothendieck, Alexander, 1928-, "Théorie des classes de Chern", Bull. Soc. Math. France [86 (1958), 137-154].

[20] Grothendieck, Alexander, "Sur quelques propriétés fondamentales en théorie des intersections", Séminaire C. Chevalley, E.N.S. (1958).

[21] Grothendieck, A., "Esquisse d'un Programme pour une Théorie des Intersections sur les Schemas Généraux" and "Classes de Faisceaux et Théorème de Riemann-Roch", Exposé 0 in Théorie des Intersections et Théorème de Riemann-Roch. (SGA 6). Lecture notes in math. Springer-Verlag [225 (1971), 1-77].

[22] Haefliger, A. and Kosinki, A., "Un théorème de Thom sur les singularités des applications différentiables", Séminaire H. Cartan E.N.S. 1956-1957, exposé no. 8.

[23] Hilbert, David, 1862-1943, "Mathematical problems", translated by Dr. Mary Winston Newson, Bull. Am. Math. Soc. [50 (1902), 437-479].

[24] Hilbert, David, "Adolf Hurwitz", Math. Annalen [83 (1921), 161-168].

[25] Hodge, William Vallance Douglas, 1903-, "The base for algebraic varieties of given dimension on a grassmannian variety", Journal Lond. Math. Soc. [16 (1941), 245-255].

[26] Hodge, W. V. D., "The intersection formulae for a grassmannian variety", Journal Lond. Math. Soc. [17 (1942), 48-64].

[27] Hodge, W. V. D., "The characteristic classes on algebraic varieties", Proceedings London Math. Soc. [3 (1951), 138-151].

[28] Hodge, W. V. D. and Pedoe, D., Methods of Algebraic Geometry, Cambridge University Press, vol. II (1952), reprinted 1968.

[29] Holme, Audun, "Formal embedding and projection theorems", American Journal Math. [93 (1971), 527-571].

[30] Holme, A., "Projection of non-singular projective varieties", Journal Math. Kyoto University [13 (1973), 301-322].

[31] Holme, A., "Embedding-obstructions for algebraic varieties I", Advances Math., to appear.

[32] Iversen, Birger, "Numerical invariants and multiple planes", American Journal Math. [92 (1970), 968-996].

[33] Iversen, B., "Critical Points of an Algebraic Function", Inventiones math. [12 (1971), 210-224].

[34] Iversen, B., "On the Exceptional Tangents to a Smooth, Plane Curve", Math. Z. [125 (1972), 359-360].

[35] Kato, Akikuni, "Singularities of projective embedding (points of order n on an elliptic curve)", Nagoya Math. Journal [45 (1971), 97-107].

[36] Katz, Nicholas, "Etude cohomologique des pinceaux de Lefschetz", Groupes de Monodromie en Géométrie Algébrique (SGA 7 II), Lecture Notes in Mathematics, Springer-Verlag [340 (1970), 254-327].

[37] Kempf, George, "Schubert methods with an application to algebraic curves", Publication of the Matematisch Centrum, Amsterdam (1971).

[38] Kempf, G. and Laksov, Dan, "The determinantal formula of Schubert calculus", Acta Math., to appear.

[39] Kleiman, Steven Lawrence, "Algebraic cycles and the Weil conjectures", Dix exposés sur la cohomologie des schèmes, North Holland (1968).

[40] Kleiman, S. L., "Motives", Algebraic geometry, Oslo 1970, edited by F. Oort, Wolters-Noordhoff, Groningen (1972).

[41] Kleiman, S. L., "The transversality of a general translate", Compositio Math., to appear.

[42] Kleiman, S. L. and Laksov, D., "On the existence of special divisors", American Journal Math. [94 (1972), 431-436].

[43] Kleiman, S. L., and Laksov, D., "Schubert calculus", Am. Math. Monthly, [79 (1972), 1061-1082].

[44] Kleiman, S. L. and Laksov, D., "Another proof of the existence of special divisors", Acta Math., to appear.

[45] Koch, Philip O., "On the product of Schubert classes", Journal of Differential Geometry [(8) 3(1973), 349-358].

[46] Kohn, G., "Uber das Prinzip von der Erhaltung der Anzahl", Archiv der Math. Phys. [(3) 4 (1903), 312-316].

[47] Laksov, Dan, "Algebraic cycles on Grassmann varieties", Advances in Math. [9 (1972), 267-295].

[48] Lascoux, Alain, "Calcul de certains polynômes de Thom", Comptes Rendus Acad. Sci. Paris, [278 (1974), 889-891].

[49] Lascoux, A., "Systems linear de diviseurs sur les courbes et les surfaces", to appear.

[50] Lascoux, A., "Puissance exterieures, determinants et cycles de Schubert", Bulletin de la Société Mathématique de France, to appear.

[51] Lascoux, Alain, "Polynomes symétriques et coefficients d'intersection de cycles de Schubert", to appear.

[52] Lascoux, A., "Les cycles critiques d'un morphisme de fibrés", to appear.

[53] Lluis, E., "Sur l'immersion des variétés algebriques", Annals Math. [62 (1955), 120-127].

[54] Lluis, E., "De las singularidades que aparecen al proyectar variedades algebraicas", Bol. Soc. Mat. Mexicana [1 (1956), 1-9].

[55] Lluis, E., "Variedads algebricas con ciertas condiciones en sus tangentes", Publ. Mat. de Soc. Mat. Mexicana (1962).

[56] MacDonald, I. G., "Some enumerative formulas for algebraic curves", Proc. Cambridge Phil. Soc., [54 (1958), 399-416].

[57] MacDonald, I. G., "Symmetric products of an algebraic curve", Topology [I (1962), 319-342].

[58] Mather, John N., "Stable map-germs and algebraic geometry", Manifolds-Amsterdam 1970, Lecture notes in math., [197 (1971), 176-193].

[59] Mather, J. N., "Generic projections", Annals Math. [98 (1973), 226-245].

[60] Mattuck, Arthur Paul, 1930-, "On symmetric products of curves", Proceedings of the American Mathematical Society [13 (1962), 82-87].

[61] Mattuck, A. P., "Secant bundles on symmetric products", American Journal of Math. [87 (1965), 779-797].

[62] Menn, Michael, "Singular Manifolds", Journal diff. geometry [5 (1971), 523-542].

[63] Mount, K. R., and Villamayor, O. E., "An algebraic construction of the generic singularities of Boardman-Thom", Inst. Hautes Etudes Sci. Publ. Math. [43 (1974), 205-244].

[64] Mount, K. R., and Villamayor, O. E., "Weierstrass points as singularities of maps in arbitrary characteristic", to appear.

[65] Mukherjia, Kalyan K., "Chern classes and projective geometry", J. Differential Geometry [7 (1972), 473-478].

[66] Musili, C., "Some properties of Schubert varieties", to appear.

[67] Nakano, S., "Tangent vector bundles and Todd canonical systems of an algebraic variety", Mem. Coll. Sci. Kyoto [(A) 29 (1955), 145-149].

[68] Netto, E., 1846-1919, and Le Vavasseur, R., "Les fonctions rationnelles", Encyclopédie des sciences mathématiques [I, 9, 1-232], Teubner, Leipzig (1915).

[69] Ogawa, Roy H., "On the points of Weierstrass in dimensions greater than one", Transactions of the American Mathematical Society [184 (1973), 401-417].

[70] Pohl, William F., "Extrinsic Complex Projective Geometry", Proceedings of the Conference on Complex Analysis, Minneapolis 1964, Springer-Verlag (1965), 18-29.

[71] Poncelet, Jean Victor, 1788-1867, Traité des propriétés projectives des figures, Gauthier-Villars, Paris (1822), second edition, Paris (1865). (The work was begun in 1813 in a Russian military prison at Saratow.)

[72] Porteous, I. R., "Simple singularities of maps", Liverpool singularities symposium I, Lecture notes in math., Springer-Verlag [192 (1971), 286-307].

[73] Porteous, I. R., "Todd's canonical classes", Liverpool singularities symposium I, Lecture notes in math., Springer-Verlag [192 (1971), 308-312].

[74] Roberts, Joel, "Generic projections of algebraic varieties", American Journal Math. [93 (1971), 191-215].

[75] Roberts, J., "The variation of critical cycles in algebraic families", Transactions of the American Mathematical Society [168 (1972), 153-164].

[76] Roberts, J., "Singularity subschemes and generic projections", (research announcement), Bull. Amer. Math. Soc. [78 (1972), 706-708].

[77] Roberts, J., "Generic coverings of $\mathbb{P}^r$ when char(k) > 0", Notices Amer. Math. Soc., [20 (1973), abstract 704-A5].

[78] Roberts, J., "Singularity subschemes and generic projections", to appear.

[79] Ronga, F., "Le calcul des classes duales aux singularités de Boardman d'order deux", Commentarii math. helvetici [47 (1972), 15-35].

[80] Ronga, F., "Les classes duales aux singularités de Boardman", to appear.

[81] Roth, Leonard, "Francesco Severi", Journal London Math. Society [38 (1963), 282-307].

[82] Samuel, Pierre, 1921-, "Lectures on old and new results on algebraic curves", Tata Institute of Fundamental Research, Bombay (1966).

[83] Schoenflies, Arthur, 1853-1928, and Tresse, A., "Géométrie projective", Encyclopédie des science mathématiques [III, 8, 1-143] Teubner, Leipzig (1915).

[84] Schubert, Hermann Cäser Hannibal, 1848-1911, Kalkül der abzählenden Geometrie, Teubner, Leipzig (1879).

[85] Schubert, H., "Die n-dimensionalen Verallgemeinerungen der fundamentalen Anzahlen unseres Raums", (dated 1884), Math. Ann. [26 (1886), 26-51].

[86] Schubert, H., "Losüng des Charakteristiken-Problems für lineare Räume beliebiger Dimension", (dated 1885), Mitteil. Math. Ges. Hamburg (1886), 134-155.

[87] Schubert, H., "Anzahlbestimmungen für lineare Raüme beliebiger Dimension", Acta Math. [8 (1886), 97-118].

[88] Schubert, H., "Beziehungen zwischen den linearen Räumen auferlegbaren charakteristischen Bedingungen", Math. Annalen [38 (1891), 598-602].

[89] Schubert, H., Mathematical Essays and Recreations, translated by Thomas J. McCormack, The Open Court Publishing Company, Chicago, second edition (1903).

[90] Schubert, H., "Über die Incidenz zweier linearer Raüme beliebiger Dimensionen", Math. Annalen [57 (1903), 209-221].

[91] Schubert, H., "Anzahl-Beziehungen bei Inzidenz und Koinzidenz mehrdimensionaler linear Raume", Jahresb. deutsch Math.-Ver., [12 (1903), 89-96].

[92] Schubert, H., "Allgemeine Anzahlfunctionen für Kegelschnitte, Flächen und Raüme zweiten Grades in n Dimensionen", Math. Annalen [45 (1894), 153-206].

[93] Segre, Corrado, 1863-1924, "Mehrdimensionale Räume", *Encyclopadie der Mathematischen Wissenshaften mit einschluss ihrer anwendungen*, [III, 2, 2, C7; 669-972], (Abgesclossen Ende 1912), B. G. Teubner, Leipzig, (1921-1934).

[94] Sergeraert, Francis, "Expression explicite de certains polynômes de Thom", Comptes Rendus Acad. Sci. Paris [276 (1973), 1661-1663].

[95] Serre, Jean-Pierre, 1926-, "Algèbre locale-multiplicités", Lecture notes in math, 11, Springer-Verlag (1965).

[96] Severi, Francesco, 1879-1963, "Le coincidenze di una serie algebrica $\infty^{(k+1)(r-k)}$ di coppie di spazi a k dimensioni, immersi nello spazio ad r dimensioni", Rendiconte della R. Accademia Nazionale dei Lincei [5 (1900), 321-326].

[97] Severi, F., "Sulle intersezioni delle varietà algebriche e sopra i loro caratteri e singolarità proiettive", Memorie della R. Accademia delle Scienze di Torino [52 (1902), 61-118].

[98] Severi, F., "Fondamenti per la geometria sulle varietà algebriche", Rendiconti del Circolo Matematico di Palermo [28 (1909), 38-87].

[99] Severi, F., "Sul principio della conservazione del numero", Rendiconti del Circolo Matematico di Palermo [33 (1912), 313-327].

[100] Severi, F., "Sui fondamenti della geometria numerativa e sulla teoria delle caratteristiche", Atti del R. Instituto Veneto [75 (1916), 1121-1162].

[101] Severi, F., "La serie canonica e la teoria delle serie principali di gruppi di punti sopra una superficie algebrica", Commentarii math. helvetici [4 (1932), 268-326].

[102] Severi, F., "I fondamenti della geometria numerativa", Annali di Mat. [4, 19 (1940), 153-242]. Also available in a German translation by Wolfgang Gröbner, entitled, "Grundlagen der abzählenden Geometrie", published in 1948 by Wolfenbutteler verlagsanstalt G.M.B.H., Wolfenbüttel and Hannover.

[103] Study, Eduard, 1862-1922, "Ueber die Geometrie der Kegelschitte, insbesondere deren Charakteristiken problem", Math. Annalen [26 (1886), 58-101].

[104] Study, E., *Die Geometrie der Dynamen*, Leipzig (1903).

[105] Study, E., "Über das sogenannte Prinzip von der Erhaltung der Anzahl", Verhandlungen des dritten internationalen Mathematiker-Kongresses in Heidelberg 1904 [338-395], Leipzig (1915), Kraus reprint, Nendeln/Liechtenstein (1967).

[106] Sturm, Rudolf, 1841-1919, Archiv der Math. Phys. [(3) 12 (1907), 113].

[107] Svanes, Torgny, "Coherent cohomology on Schubert subschemes of flag schemes and applications", Advances in Mathematics, to appear.

[108] Svanes, T., "Some criteria for rigidity of noetherian rings", Aarhus Universitet Preprint Series, (1973/74), No. 15.

[109] Todd, John Arthur, 1908-, "On Giambelli's formula for incidence of linear spaces", Journal Lond. Math. Soc. [6 (1931), 209-216].

[110] Todd, J. A., "The geometrical invariants of algebraic loci", Proceedings London Math. Society [43 (1937), 127-138] and [45 (1939), 410-424].

[111] van der Waerden, Bartel Leendert, 1903-, "Topologische Begründung des Kalküls der abzählenden Geometrie", Math. Annalen [102 (1930), 337-362].

[112] van der Waerden, B. L., "Zur algebraischen Geometrie. VIII. Der Grad der Grassmannschen Mannigfaltigkeit der linearen Räume S_m in S_n", Math. Annalen [113 (1937), 199-205].

[113] van der Waerden, B. L., "Zur algebraischen Geometrie. XV. Lösung des Characteristikenproblems für Kegelschnitte", Math. Annalen [115 (1938), 645-655].

[114] van der Waerden, B. L., "The foundation of algebraic geometry from Severi to André Weil", Archive for history of exact science [7 (1970-1971, 171-179].

[115] Weil, André, 1906-, "Foundations of Algebraic Geometry", American Math. Soc. Colloquium Publication vol. 29 (1946); revised and enlarged, 1962.

[116] Zariski, Oscar, 1899-, Algebraic Surfaces, second supplemented edition, Ergebnisse der Mathematik und ihrer Grenzgebiete, Band 61, Springer-Verlag (1971).

[117] Zeuthen, Hieronymus Georg, 1839-1920, Lehrbuch der abzählenden Methoden der Geometrie, Teubner, Leipzig (1914).

[118] Zeuthen, H. G., and Pieri, M., "Géométrie énumérative", Encyclopédie des sciences mathématiques, [III, 2, 260-331] Teubner, Leipzig (1915).

University of California at San Diego

Massachusetts Institute of Technology

Proceedings of Symposia in Pure Mathematics
Volume 28, 1976

HILBERT'S SEVENTEENTH PROBLEM AND RELATED PROBLEMS ON DEFINITE FORMS

Albrecht Pfister

ABSTRACT

After a review of the historical development of Hilbert's seventeenth problem we proof that a positive definite rational function in n variables and real coefficients is a sum of at most 2^n squares. Four related open problems are discussed at the end of the paper.

1. INTRODUCTION

Hilbert's 17th problem is concerned with the representation of positive definite functions as sums of squares. Let

$$f(x) = \frac{g(x)}{h(x)} \in \mathbb{R}(x) = \mathbb{R}(x_1,\ldots,x_n)$$

be a rational function in n variables with real coefficients. f is called positive definite if

$$f(a) \geqslant 0 \text{ for all } a = (a_1,\ldots,a_n) \in \mathbb{R}^n \text{ with } h(a) \neq 0.$$

Hilbert conjectured that this (clearly necessary) condition is sufficient in order to represent f as a sum of sqares in the field $\mathbb{R}(x)$, i.e. there exist a natural number r and r functions $f_i(x) \in \mathbb{R}(x)$ $(i=1,\ldots,r)$ such that

$$f(x) = f_1^2(x) + \ . \ . \ . + f_r^2(x).$$

For making precise statements it is useful to introduce the following slightly more general definitions. Let K be any (commutative) field and let

(1) $$f(x) = \frac{g(x)}{h(x)} \in K(x) = K(x_1,\ldots,x_n)$$

be a rational function in n variables with coefficients in K.

(2) We call f positive definite and write $f \geqslant 0$ if for all $a = (a_1,\ldots,a_n) \in K^n$ with $h(a) \neq 0$ and for all orderings $>$ of K we have $f(a) \geqslant 0$.

(If K has no orderings then every $f(x) \in K(x)$ is positive definite)

(3) If f(x) is a sum of r squares in K(x) we write:

$$f = \boxed{r} \text{ in } K(x).$$

Hilbert's 17th problem can now be formulated in three parts:

(P1) Let K = R be a real closed field. Does $f \in R(x)$, $f \geqslant 0$ imply $f = \boxed{r}$ in R(x) for some natural number r?

(P2) Let K be an arbitrary field. Does $f \in K(x)$, $f \geqslant 0$ imply $f = \boxed{r}$ in K(x) for some r?

(P3) If (P1) resp. (P2) is true, is there an upper bound on r depending only on n (the number of variables) and on K but not on f?

The answers are: (P1) is true (E.Artin 1926), (P2) is false in general (D.W.Dubois 1967), (P3) is true for $K = \mathbb{R}$ (A.Pfister 1967) but unsolved for other fields K where (P2) holds.

2. HISTORICAL DEVELOPMENT

The investigation of positive definite functions began in the year 1888 with the following "negative" result of Hilbert[7]: If $f(x) \in \mathbb{R}[x] = \mathbb{R}[x_1,\ldots,x_n]$ is a positive definite polynomial then f need not be a sum of squares of polynomials in $\mathbb{R}[x]$, except for n = 1. Hilbert's proof was geometric and did only yield the existence of such polynomials. The first explicit example was given by Motzkin[19] in 1966, namely the polynomial

$$f(x_1,x_2) = 1 + x_1^2x_2^4 + x_1^4x_2^2 - 3x_1^2x_2^2 .$$

Later R.M.Robinson[26] produced similar examples.

Therefore it is necessary to allow rational functions f_i in a representation of f as a sum of squares even if the given f is a polynomial. Hilbert[8] then succeeded in 1893 to prove the case n=2 of (P1). In fact, one can deduce from his proof the stronger result that 4 squares suffice if n=2. This was done in 1906 by Landau[13] who at the same time considered the case $K = \mathbb{Q}$, n = 1 of (P2) and (P3). He proved: Every positive definite rational function $f(x_1) \in \mathbb{Q}(x_1)$ is a sum of 8 squares in $\mathbb{Q}(x_1)$. Recently Pourchet[24] improved this result as follows: If K is any algebraic number field and $f(x_1) \in K(x_1)$ is positive definite then $f(x_1)$ is a sum of 5 squares in $K(x_1)$. Here the number r=5 is best possible.

The main step in the history of the 17th problem was of course Artin's paper[1] of 1926 where he proves (P1) in general and (P2) under the condition that K has only archimedian orderings (for instance K an algebraic number field). In contrast to Hilbert's proof for n=2 which was essentially constructive Artin's proof relies heavily on the so-called Artin-Schreier theory of formally real fields. Here it is shown that in any field K an element a is a sum of squares if and only if a is totally positive, i.e. nonnegative in every ordering of K. Thus it remained to show that a positive definite function $f(x) \in K(x)$ is totally positive (as an element of K(x)). For this Artin proved a series of "specialisation lemmas" by using Sturm's Theorem.

The first constructive proof of (P1) for $K = \mathbb{R}$ was given by Habicht[6] in 1940 under the extra condition that f is strictly positive definite, say $f \geq 1$. In 1955 A.Robinson[25] proved (P1) by using lower predicate calculus and the model-completeness of $\mathbb{R}$. In 1956 Kreisel[11] gave a constructive proof for arbitrary $f \geq 0$ and showed that there is an upper bound on the number r of squares depending only on n and the degree of f (but not on the coefficients of f).

Another point in Artin's proof which did not appeal to everyone was the use of Sturm's Theorem. Lang in his book "Algebra"[16] tried to give a simpler proof which however was incorrect. In fact, Dubois[5] showed in 1967 that the stronger version (P2) of Hilbert's problem is not always true. He constructed a field K with only one ordering (which is non-archimedian!) and a positive definite polynomial $f(x_1) \in K(x_1)$ which is not a sum of squares in $K(x_1)$. Thereby it was shown that some condition on the field K, such as "K real closed" or "all orderings of K are archimedian" is necessary. The first purely mathematical proof of Artin's Theorem which avoids Sturm's Theorem seems to be the one by Knebusch[10].

With respect to (P3) the first important step after Artin was taken by Tsen[27] in 1936 with his very general result on quasi-algebraically closed fields (see Theorem 1 below) though he may not yet have seen the connection with Hilbert's 17th problem. Apparently his paper has been forgotten during the war, so that his results had to be rediscovered by Lang[14] before they reached the mathematical community. (P3) was then solved by Ax[2] and myself[21] with the result that $r = 2^n$ is an upper bound for real closed field $K = R$ (see Theorem 3 Corollary 1 below). For more general K, in particular for $K = \mathbb{Q}$, (P3) is still an open problem.

3. PROOF OF (P3) FOR REAL CLOSED FIELDS

We make use of the following by now well-known results.

THEOREM 1 (Tsen[27] - Lang[14]). Let C be an algebraically closed field, let K be a field of transcendence degree n over C. Let f be a form of degree d in more than d^n variables with coefficients in K. Then f has a non-trivial zero in K. In particular every quadratic form of dimension greater than 2^n has a non-trivial zero.

If K is any field with $\mathrm{char}(K) \neq 2$ and if $a_1, \ldots, a_n \in K^*$, denote the quadratic form $\varphi(x) = \sum_{i=1}^{n} a_i x_i^2$ by $\varphi = (a_1, \ldots, a_n) = (a_1) \oplus \ldots \oplus (a_n)$.

Denote equivalence of quadratic forms by $\cong$, direct orthogonal sum by $\oplus$, tensor product by $\otimes$. Then we have

THEOREM 2 (Pfister[20]). Let $n \geq 0$, $a_1,\dots,a_n \in K^*$ and

$$\varphi = (1,a_1) \otimes \dots \otimes (1,a_n).$$

Then φ is multiplicative in the following sense: If $c \in K^*$ is represented by φ over K then $\varphi \cong (c)\otimes\varphi$ over K. In particular the non-zero elements of K which are represented by φ form a subgroup of K^*.

For other proofs of Theorem 2 see e.g. [12], [17], [18].

LEMMA. Let $\varphi \cong (1,a_1) \otimes \dots \otimes (1,a_n) \cong (1) \oplus \varphi'$ and let $b_1 \in K^*$ be represented by φ'. Then there exist $b_2,\dots,b_n \in K^*$ such that $\varphi \cong (1,b_1) \otimes \dots \otimes (1,b_n)$.

For a proof see [12] or [22].

We can now deduce the following theorem ([22], [23]):

THEOREM 3. Let R be a real closed field, let K be an extension field of transcendence degree n over R. Let $\varphi \cong (1,a_1) \otimes \dots \otimes (1,a_n)$ be a multiplicative quadratic form of dimension 2^n over K and let b be a totally positive element of K, i.e. an element which can be represented as a finite sum of squares in K. Then φ represents b over K.

Proof:

a) We may suppose that φ is anisotropic over K, i.e. does not represent 0 non-trivially, since otherwise φ is universal over K which immediately gives the result. Also the case $b = b_1^2$ is trivial. We will first treat the case $b = b_1^2 + b_2^2$, $b_1 b_2 \neq 0$. By Tsen's Theorem we know that φ is universal over the field $K(i) = K(\sqrt{-1})$. If $K = K(i)$ then the result follows. If not, then $\beta = b_1 + ib_2$ generates K(i) over K and φ represents β over K(i). This shows that there are vectors u, v with components in K such that $\varphi(u + \beta v) = \beta$, $v \neq 0$. Hence

$$\varphi(u) + 2\beta \langle u,v\rangle_\varphi + \beta^2\varphi(v) = \beta.$$

Comparing with $\beta^2 - 2b_1\beta + b = 0$ we find $\varphi(u) - b\varphi(v) = 0$. Since φ is multiplicative this gives the result.

b) We will now suppose that Theorem 3 holds for all forms φ of the given type and for all elements $b \in K$ which are sums of k squares $(k \geq 2)$ and will proceed by induction on k. Up to a square factor a sum of k+1 squares looks like $c = 1 + b$ where b is a sum of k squares. Putting $\varphi = (1) \oplus \varphi'$ the induction hypothesis gives $b = b_1^2 + b_2$ where b_2 is represented by φ', and without restriction $b_2 \neq 0$. We want to show that φ represents c.

Consider the form $\psi = \varphi \otimes (1,-c) = (1) \oplus \varphi' \oplus (-c\varphi) = (1) \oplus \psi'$. ψ' represents $b_2 - c = (b - b_1^2) - (1 + b) = -1 - b_1^2$. By the Lemma we have therefore $\psi \cong (1,-1-b_1^2) \otimes \chi$ with a form $\chi \cong (1,c_1) \otimes \dots \otimes (1,c_n)$. Applying the induction hypothesis to χ we have χ represents $1 + b_1^2$. Hence $\psi \cong \varphi \oplus (-c\varphi)$ is isotropic. Therefore $\varphi(u) - c\varphi(v) = 0$ with non-zero vectors u, v over K. Since φ is anisotropic and multiplicative it follows that φ represents c over K.

The solution of problem (P3) for a real closed field R is now an immediate consequence:

COROLLARY 1. Let R be a real closed field and let $f \in R(x_1,\dots,x_n) = K$ be positive definite. Then f is a sum of 2^n squares in K.

Proof: By Artin's Theorem (P1) is true, so that $f \in K$ is totally positive. Then Theorem 3 applies with $a_1 = \dots = a_n = 1$ and $b = f$. This gives $f = \boxed{2^n}$ in K.

Another special case of Theorem 3 is given by

COROLLARY 2. Let K be a non-real (i.e. K can not be ordered) function field of transcendence degree n over R. Then every multiplicative quadratic form φ of dimension 2^n over K is universal in K.

Proof: Every element $b \in K$ is a sum of squares.

4. OPEN PROBLEMS

If K is a field, denote by $p = p(K)$ the least natural number r such that every sum of squares in K is already a sum of r squares. If no number r exists put $p(K) = \infty$. One may call $p(K)$ the "Pythagoras number" of K.

PROBLEM 1. Corollary 1 of Theorem 3 shows that $p_n = p(R(x_1,\dots,x_n)) \leqslant 2^n$ for a real closed field R. On the other hand a theorem of Cassels[3] shows that $1 + x_1^2 + \dots + x_n^2$ is not a sum of n squares in $R(x_1,\dots,x_n)$, hence $p_n \geqslant n+1$. What is the true value of p_n?

For n = 2 it has been shown in [4] that the Motzkin polynomial

$$f(x_1,x_2) = 1 + x_1^2x_2^4 + x_1^4x_2^2 - 3x_1^2x_2^2$$

is not a sum of 3 squares, hence $p_2 = 4$. But the method is special to the case n = 2.

PROBLEM 2. Let $K = \mathbb{Q}$. Is $p(\mathbb{Q}(x_1,\dots,x_n))$ bounded by some function of n, e.g. $p \leqslant 2^n + 3$?

This is valid for n = 0 (Lagrange) and for n = 1 (Pourchet[24]). For $n \geqslant 2$ no bound is known. The conjecture $p \leqslant 2^n + 3$ is in [9]. More generally, if K is any field with $p(K) < \infty$ is it true that $p(K(x_1)) < \infty$?

PROBLEM 3. Let K be a non-real field of transcendence degree n over a real closed field R. Is every quadratic form of dimension 2^n universal in K?

By Corollary 2 above this is true if φ is multiplicative. For $n \leq 1$ every φ is a scalar multiple of a multiplicative form. But for $n \geq 2$ the problem is still open.

PROBLEM 4. Let K be as in Problem 3. Is K a C_n-field, i.e. does every form of degree d in more than d^n variables have a non-trivial zero in K?

This is a conjecture of Lang[15]. It is true for odd degree d. d = 2 gives Problem 3. For $d \geq 4$ even the simplest case n = 1 is unsolved!

REFERENCES

1. E.Artin: Über die Zerlegung definiter Funktionen in Quadrate. Hamb. Abh. 5, 100-115 (1927); Collected Papers 273-288.
2. J.Ax: On ternary definite rational functions (unpublished).
3. J.W.S.Cassels: On the representation of rational functions as sums of squares. Acta Arithm. 9, 79-82 (1964).
4. ------, W.J.Ellison and A.Pfister: On sums of squares and on elliptic curves over function fields. J.Number Th. 3, 125-149 (1971).
5. D.W.Dubois: Note on Artin's solution of Hilbert's 17th problem. Bull.Amer.Math.Soc. 73, 540-541 (1967).
6. W.Habicht: Über die Zerlegung strikte definiter Formen in Quadrate. Comment.Math.Helv. 12, 317-322 (1940).
7. D.Hilbert: Über die Darstellung definiter Formen als Summe von Formenquadraten. Math.Ann. 32, 342-350 (1888); Ges.Abh. 2, 154-161.
8. ------: Über ternäre definite Formen. Acta Math. 17, 169-197 (1893); Ges.Abh. 2, 345-366.
9. J.S.Hsia and R.P.Johnson: On the representation in sums of squares for definite functions in one variable over an algebraic number field. To appear in Amer.J.Math.
10. M.Knebusch: On the uniqueness of real closures and the existence of real places. Comment.Math.Helv. 47, 260-269 (1972).
11. G.Kreisel: Hilbert's 17th problem. Summaries of talks presented at the Summer Institute of Symbolic Logic in 1957 at Cornell University, 313-320.
12. T.Y.Lam: The algebraic theory of quadratic forms. Benjamin, Reading, Mass. 1973.
13. E.Landau: Über die Darstellung definiter Funktionen durch Quadrate. Math.Ann. 62, 272-285 (1906).
14. S.Lang: On quasi-algebraic closure. Ann.of Math. 55, 373-390 (1952).
15. ------: The theory of real places. Ann. of Math. 57, 378-391 (1953).
16. ------: Algebra. Addison-Wesley, Reading, Mass. 1965.
17. F.Lorenz: Quadratische Formen über Körpern. Lecture Notes in Math. 130 (1970).
18. J.Milnor and D.Husemoller: Symmetric bilinear forms. Erg. der Math. 73 (1973).

19. T.S.Motzkin: The arithmetic-geometric inequality. In "Inequalities" (Oved Shisha, ed.), pp 205-224. Academic Press, New York 1967.

20. A.Pfister: Multiplikative quadratische Formen. Arch.Math. 16, 363-370 (1965).

21. ------: Zur Darstellung definiter Funktionen als Summe von Quadraten. Invent.math. 4, 229-237 (1967).

22. ------: Quadratic forms over fields. Proc. of Symposia in Pure Mathematics XX, 150-160 (1971).

23. ------: Sums of squares in real function fields. Actes Congrès intern. math. Tome 1, 297-300 (1970).

24. Y.Pourchet: Sur la représentation en somme de carrés des polynômes à une indéterminée sur un corps de nombres algébriques. Acta Arithm. 19, 89-104 (1971).

25. A.Robinson: On ordered fields and definite functions. Math.Ann. 130, 257-271 (1955).

26. R.M.Robinson: Some definite polynomials which are not sums of squares of real polynomials. Notices Amer.Math.Soc. 16, 554 (1969).

27. C.Tsen: Zur Stufentheorie der Quasi-algebraisch-Abgeschlossenheit kommutativer Körper. J.Chinese Math.Soc. 1, 81-92 (1936).

Proceedings of Symposia in Pure Mathematics
Volume 28, 1976

HILBERT'S PROBLEM 18: ON CRYSTALOGRAPHIC GROUPS, FUNDAMENTAL DOMAINS, AND ON SPHERE PACKING

J. Milnor

ABSTRACT

This paper is divided into three completely independent sections, corresponding to three different questions asked by Hilbert. The first section describes Bieberbach's theorems on discrete groups of Euclidean motions. The second describes Heesch's tiling of the plane by a polygon which is not a fundamental domain. The third studies the densest packing of Euclidean space by balls of fixed diameter.

I am indebted to I. Kaplansky and R. M. Robinson for helpful information.

§1. "IS THERE IN n-DIMENSIONAL EUCLIDEAN SPACE ... ONLY A FINITE NUMBER OF ESSENTIALLY DIFFERENT KINDS OF GROUPS OF MOTIONS WITH A [COMPACT] FUNDAMENTAL REGION?"

Let $E(n)$ be the Euclidean group, consisting of all isometries of the n-dimensional Euclidean space R^n. We look for discrete subgroups $\Gamma \subset E(n)$ such that the coset space $\Gamma \backslash E(n)$ is compact. This is equivalent to the requirement that there exist a compact fundamental domain, that is a compact region D in R^n such that the various congruent copies γD with $\gamma \in \Gamma$ cover R^n and have only boundary points in common.

AMS (MOS) subject classifications (1970). 20H15, 50B30, 52A45, 10E30.

Hilbert's question was answered affirmatively by Bieberbach in 1910. Bieberbach first proved the following important statement:

> (1) Every discrete subgroup $\Gamma \subset E(n)$ with compact fundamental domain contains n linearly independent translations.

He then showed easily that for any such Γ:

> (2) The subgroup T consisting of all translations in Γ forms a free abelian normal subgroup of finite index. That is, Γ admits a short exact sequence of the form
>
> $$1 \longrightarrow Z^n \xrightarrow{\ i\ } \Gamma \longrightarrow \Phi \longrightarrow 1 ,$$
>
> with Φ finite. Furthermore, the image $T = i(Z^n)$ is a maximal abelian subgroup of Γ. In other words, if we let Φ act on Z^n by inner automorphism, we obtain a faithful representation of Φ in $GL(n, Z)$.

It follows that this group T of translations in Γ can be characterized as the union of all abelian normal subgroups of Γ. For modern proofs of (1) and (2) the reader is referred to Auslander or Wolf. Finally Bieberbach showed that:

> (3) For each fixed n, there are only finitely many isomorphism classes of groups Γ which can be obtained in this way as extensions of Z^n by finite subgroups of $GL(n, Z)$.

The embedding of Γ in the Euclidean group $E(n)$ is not uniquely determined in general. (We see this already in the case where Γ is free abelian.) However in 1912, inspired by a paper of Frobenius, Bieberbach showed that the embedding of Γ in $E(n)$ is uniquely determined up to conjugation by an element of the larger group consisting of all one-to-one affine transformations of R^n.

Many years later, Zassenhaus pointed out that a converse theorem is true also:

> Any group Γ which can be described as an extension of Z^n by a finite subgroup of $GL(n, Z)$ can actually be embedded as a discrete subgroup of the Euclidean group $E(n)$ so as to act on R^n with compact fundamental domain.

We will call such groups Γ Bieberbach groups.

For the proof of (3), Bieberbach first recalled Minkowski's statement that there is an upper bound, depending only on n, for the orders of finite subgroups of $GL(n, Z)$. In fact, Minkowski had noted that any finite subgroup of $GL(n, Z)$ maps injectively into the finite group $GL(n, Z/k)$ for $k \geq 3$. (A brief proof is given in Newman.)

Given a finite group Φ, he also needed to know that there are only finitely many embeddings of Φ in $GL(n, Z)$, up to inner automorphism. In fact, in Bieberbach's words:

> "Eine gegebene [endliche] Gruppe kann nur auf endlich viele inäquivalente Weisen durch ganzzahlige unimodulare Substitutionen von n Variablen dargestellt werden."

Although this statement follows easily from the work of Jordan or of Minkowski, Bieberbach seems to have been the first to state it explicitly, and to realize its importance. This theorem has recently become known in the literature as the "Jordan-Zassenhaus Theorem," since Zassenhaus published a different proof in 1938.

Finally, given the action of Φ on Z^n, one needs to know that there are only finitely many extensions $1 \longrightarrow Z^n \longrightarrow \Gamma \longrightarrow \Phi \longrightarrow 1$. Nowadays we would say that these extensions are classified by elements of the finite cohomology group $H^2(\Phi; Z^n)$.

Here is a table, listing the numbers of distinct Bieberbach groups Γ in dimensions up to 4.

dimension n	number of distinct groups
1	2
2	17
3	219
4	4783

Thus for $n = 1$ there are just two distinct Bieberbach groups Γ, according as the finite quotient $\Phi = \Gamma/T$ has order 1 or 2. In dimension 2, the 17 Bieberbach groups can be tabulated as follows. (Compare Burkhardt or Hilbert and Cohn-Vossen.)

Table of Bieberbach groups in the plane

group Φ	of order	number of embeddings in $GL(2, Z)$	number of Bieberbach groups
cyclic	1	1	1
cyclic	2	1	1
cyclic	3	1	1
cyclic	4	1	1
cyclic	6	1	1
dihedral	2	2	3
dihedral	4	2	4
dihedral	6	2	2
dihedral	8	1	2
dihedral	12	1	1
total		13	17

The finite quotient Φ is either cyclic, consisting only of rotations, or dihedral, containing reflections also. (Of course the dihedral group of order 2 is isomorphic to the cyclic group of order 2, but these two groups are listed separately in this table, since they are not conjugate as subgroups of $GL(2, R)$.)

The classification of Bieberbach groups in dimension 3 is of great practical importance in the study of crystals. This classification was carried out independently by Fedorov and by Schoenfliess around 1890. (For a more modern presentation see Burkhardt.)

Table of orientation preserving groups in 3-space

group Φ	of order	number of embeddings in SL(3, Z)	number of Bieberbach groups
cyclic	1	1	1
cyclic	2	2	3
cyclic	3	2	3 (+1)
cyclic	4	2	5 (+1)
cyclic	6	1	4 (+2)
dihedral	4	4	9
dihedral	6	3	5 (+2)
dihedral	8	2	8 (+2)
dihedral	12	1	4 (+2)
tetrahedral	12	3	5
octahedral	24	3	7 (+1)
total		24	54 (+11)

Here the quantities in parentheses on the right indicate the number of additional Bieberbach groups which we would obtain if we wish to distinguish between two groups of motions in 3-space which are mirror images of each other. In addition there are 165 Bieberbach groups in 3-space which contain orientation reversing elements. I will not attempt to tabulate these.

Finally, in dimension 4, a recent and as yet unpublished computation by H. Brown, J. Neubüser, and H. Zassenhaus shows that there are 4783 distinct Bieberbach groups. (Compare Dade.)

It would be natural to ask whether Bieberbach's theorems would still hold if we forget the metric on Euclidean space and simply look for groups of one-to-one affine transformations of R^n which are discrete and have compact fundamental domain. A simple counter-example has been given by L. Auslander. Let N_R denote the group consisting of all 3×3 real matrices of the form

$$\begin{bmatrix} 1 & * & * \\ 0 & 1 & * \\ 0 & 0 & 1 \end{bmatrix},$$

and let N_Z denote the subgroup consisting of integer matrices of the same form. Then N_Z acts by left multiplication on N_R. Identifying N_R with the coordinate space R^3 in the obvious way, each element of N_Z gives rise to a one-to-one affine transformation of 3-space. This action is discrete, has no fixed points, and has the unit cube as fundamental domain. Yet the group N_Z does not contain three linearly independent translations. Furthermore, for $k = 1, 2, 3, \ldots$ the subgroup N_{kZ} consisting of matrices in N_Z whose off diagonal entries are divisible by k also acts freely on N_R with compact quotient, and no two of these groups are isomorphic. It is interesting to note that each quotient space $N_{kZ}\backslash N_R$ is a smooth compact manifold with flat affine connection having fundamental group N_{kZ}.

From the point of view of differential geometry, the most interesting Bieberbach groups are those which are torsionfree, i.e., contain no non-trivial elements of finite order. For it is easy to check that the action of a Bieberbach group Γ on R^n is free (i.e., no non-trivial element has a fixed point) if and only if Γ is torsionfree. For each torsionfree Γ the quotient $\Gamma\backslash R^n$ is a compact flat Riemannian manifold with fundamental group Γ. Conversely, every compact flat Riemannian manifold is covered by R^n, and hence is isometric to $\Gamma\backslash R^n$ for some torsionfree Bieberbach group Γ. Since Γ contains a free abelian subgroup of finite index, it follows that every such manifold is covered by a flat torus.

The simplest torsionfree Bieberbach group (other than Z^n) is the Klein bottle group, that is the split extension

$$1 \longrightarrow Z \longrightarrow \Gamma \longrightarrow Z \longrightarrow 1$$

with non-trivial action. In dimension three there are ten distinct examples. (See Wolf, p. 111.) Six of these are orientable, the quotient $\Phi = \Gamma/T$ being either cyclic of order 1, 2, 3, 4, 6, or dihedral of order 4.

The collection consisting of all torsionfree Bieberbach groups is noteworthy since these groups are easily characterized, and all of them are fundamental groups of compact aspherical manifolds. (A manifold is aspherical if its universal covering space is contractible.) From this point of view, an important generalization has recently been given by F. E. A. Johnson.

Let Γ be any torsionfree group which can be obtained as an iterated group extension

$$\Gamma = \Gamma_0 \supset \Gamma_1 \supset \dots \supset \Gamma_k = 1$$

where each Γ_{i+1} is a normal subgroup of Γ_i and each quotient Γ_i/Γ_{i+1} is either finite or cyclic. Then Johnson shows that Γ is the fundamental group of a smooth, compact, aspherical manifold. He proves this by embedding Γ, as a discrete subgroup with compact quotient, in a Lie group G having finitely many components, where the component G_0 of the identity is solvable and simply connected. Choosing a maximal compact (and in fact finite) subgroup K, it follows that Γ acts freely on the coset space $G/K \approx G_0$, which is topologically Euclidean space.

Now we turn to the second portion of Hilbert's eighteenth problem.

§2. "WHETHER POLYHEDRA ALSO EXIST WHICH DO NOT APPEAR AS FUNDAMENTAL REGIONS OF GROUPS OF MOTIONS, BY MEANS OF WHICH NEVERTHELESS BY A SUITABLE JUXTAPOSITION OF CONGRUENT COPIES A COMPLETE FILLING UP OF ALL [EUCLIDEAN] SPACE IS POSSIBLE?"

A rather complicated 3-dimensional counter-example to this question was constructed by K. Reinhardt in 1928. A much simpler 2-dimensional counter-example was constructed by Heesch in 1935, and is illustrated in Figure 1. The reader with a taste for plane geometry should find it an amusing and not difficult exercise to prove the following two statements:

(1) The plane can be covered without overlap by congruent copies of this polygon H.

(2) Nevertheless H is not the fundamental domain for any Bieberbach group.

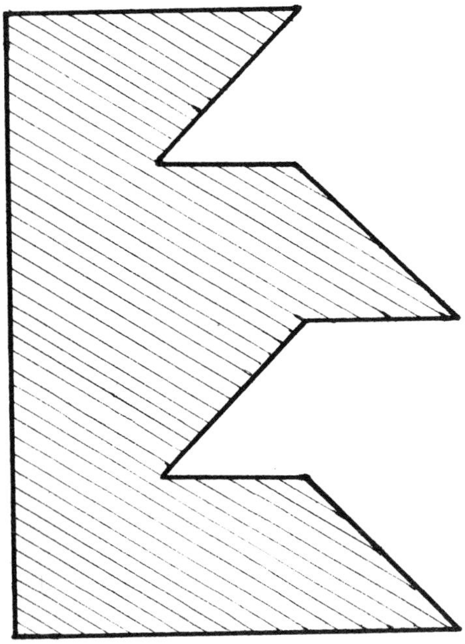

Figure 1

A somewhat simpler example of a polygon with the same two properties is illustrated in Figure 2. This example has the advantage of being star-shaped. (I do not know whether a convex example exists.) However, the proof for this second polygon is slightly more difficult.

For rather startling related examples, we refer to Robinson's paper "Undecidability and nonperiodicity for tilings of the plane." (See also Stein.)

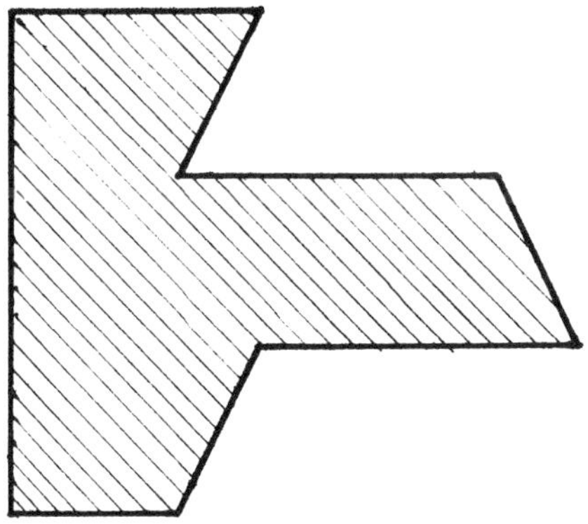

Figure 2

§3. "HOW CAN ONE ARRANGE MOST DENSELY IN SPACE AN INFINITE NUMBER OF EQUAL SOLIDS OF GIVEN FORM, E.G., SPHERES WITH GIVEN RADII ..., THAT IS, HOW CAN ONE SO FIT THEM TOGETHER THAT THE RATIO OF THE FILLED TO THE UNFILLED SPACE MAY BE AS GREAT AS POSSIBLE?"

For 2-dimensional disks this problem has been solved by Thue and Fejes Tóth, who showed that the expected hexagonal (or honeycomb) packing of circular disks in the plane is the densest possible. However, the corresponding problem in 3 dimensions remains unsolved. This is a scandalous situation since the (presumably) correct answer has been known since the time of Gauss. (Compare Hilbert and Cohn-Vossen.) All that is missing is a proof.

Here are some definitions which will help to clarify this problem. Given positive real numbers $a_1, \ldots, a_n$, and d, let $N_d(a_1, \ldots, a_n)$ be the maximum number of n-dimensional balls of diameter d which can be packed, without overlapping, into the n-dimensional "suitcase" consisting of all $(x_1, \ldots, x_n)$ with $0 \le x_i \le a_i$ for $i = 1, \ldots, n$. Clearly N_d is super-additive as a function of any one of its n variables if the other $n-1$ variables are kept fixed. For example

$$N_d(a_1' + a_1'', a_2, \ldots, a_n) \ge N_d(a_1', a_2, \ldots, a_n) + N_d(a_1'', a_2, \ldots, a_n) .$$

An elementary argument, which will be left to the reader, then shows that the ratio

$$N_d(a_1, \ldots, a_n)/a_1 \ldots a_n$$

tends to a finite non-zero limit as $a_1, \ldots, a_n \longrightarrow \infty$. This limit measures the number of balls per unit volume, and is called the maximum center-density for balls of diameter d in n-space. Using the identity $N_d(a_1, \ldots, a_n) = N_1(a_1/d, \ldots, a_n/d)$, we see that this maximum center-density is proportional to d^{-n}. Hence we can denote it by c_n/d^n where c_n is the maximum center-density for unit diameter balls in R^n. Note that the n-th root $c_n^{1/n}$ is equal to the largest possible diameter for balls which can be packed with center-density 1 in n-space.

With these preliminaries we can quote the values $c_1 = 1$, $c_2 = \sqrt{4/3}$, and

$$c_3 \geq \sqrt{2} = 1.41421\ldots \ .$$

It is believed that equality holds here, but this has never been proved.

A beautiful upper bound for c_n has been constructed by C. A. Rogers, based on earlier work by Blichfeldt. The Rogers upper bound is precisely equal to c_n for $n = 1, 2,$ and takes the form

$$c_3 \leq 6\sqrt{2}\,(\tfrac{3}{\pi}\cos^{-1}(1/3) - 1) = 1.48899\ldots$$

for $n = 3$. In dimensions 8 and 24 there are known packings whose center-density comes fairly close to the Rogers upper bound, but the discrepancy is rather wide for most dimensions.

The situation becomes somewhat more manageable if we restrict the situation by considering only arrangements of balls in n-space such that the center points form a lattice. By a lattice we mean a discrete additive subgroup L of the vector space R^n. The determinant of such a lattice L, with basis $v_1, \ldots, v_n$, will mean the real number

$$\mathrm{volume}(R^n/L) = |\det(v_1, \ldots, v_n)| = \sqrt{\det[v_i \cdot v_j]} \ .$$

Clearly the reciprocal $1/\det(L)$ can be identified with the density of lattice points per unit volume in R^n. Let

$$\min(L) = \min_{v \in L - 0} \|v\|$$

be the minimum distance between distinct lattice points. Then we can center a ball of diameter $\min(L)$ at each lattice point without overlap. Using the inequality

$$\mathrm{density} \leq c_n/(\mathrm{diameter})^n$$

we obtain

$$1/\det(L) \leq c_n/\min(L)^n$$

or in other words

$$\min(L)^n/\det(L) \leq c_n \ .$$

Let c_n^{lattice} denote the maximum, among all lattices in R^n, of this ratio $\min(L)^n/\det(L)$. Equivalently, c_n^{lattice} can be defined as the maximum center-density for lattice packings of R^n by balls of diameter 1. Evidently $c_n^{\text{lattice}} \leq c_n$.

Korkin and Zolotarev computed this maximum lattice center-density for $n \leq 5$, and Blichfeldt later extended their work to the case $n \leq 8$. For any dimension n, it follows from the work of Korkin and Zolotarev that this maximum c_n^{lattice} is actually attained for some lattice L_n in R^n, and is the square root of a rational number.

For $n \leq 8$ the densest possible lattice can be constructed out of an appropriate Dynkin diagram as follows

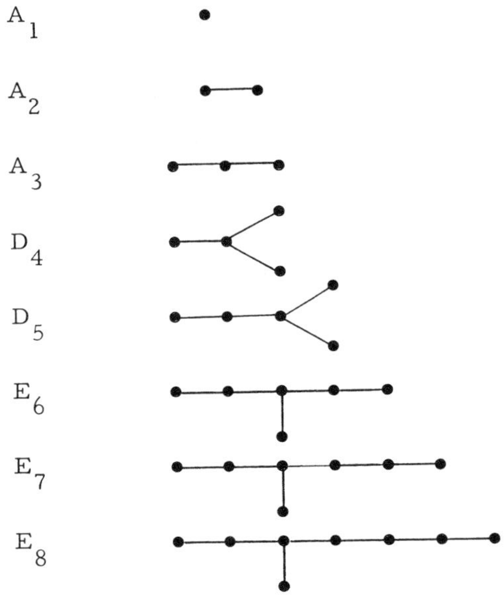

In each of these diagrams a dot stands for a basis vector v with $v \cdot v = 2$. The inner product of any two of these basis vectors is defined to be either -1 or 0 according as they are or are not joined by a line segment. Thus each such diagram with n vertices determines a symmetric $n \times n$ inner product matrix with 2's along the diagonal. It can be verified that each of these eight matrices is positive definite, and hence coincides with the inner product matrix $[v_i \cdot v_j]$ for suitably chosen vectors $v_1 \ldots v_n \in R^n$, spanning a lattice L_n with $\min(L_n) = \sqrt{2}$. The determinant of this inner product matrix is 2, 3, 4, 4, 4, 3, 2, or 1 respectively, so the determinant of L_n is equal to $\sqrt{2}$, $\sqrt{3}$, 2, 2, 2, $\sqrt{3}$, $\sqrt{2}$, or 1 respectively. Setting $c_n^{\text{lattice}} = \min(L_n)^n/\det(L_n)$ we obtain the following table.

n	1	2	3	4	5	6	7	8
c_n^{lattice}	1	$\sqrt{4/3}$	$\sqrt{2}$	2	$2\sqrt{2}$	$8/\sqrt{3}$	8	16

For dimensions $n \geq 9$ the maximal center-density c_n^{lattice} is not known. For estimates when $n \leq 24$, see the work of Leech and Sloane. For large values of n, the best one has are rather crude and non-constructive lower bounds. For example, Minkowski obtained a lower bound which is very slightly stronger than the following:

$$c_n^{\text{lattice}} > \frac{2}{(\text{volume of ball of radius } 1)} = 2(n/2)!/\pi^{n/2} .$$

This has been sharpened by Rogers and others. Perhaps the situation can best be put in perspective by taking the n-th root of both sides. Then the quantity $(c_n^{\text{lattice}})^{1/n}$, which can be identified with the largest possible value of $\min(L)$ for a unimodular lattice in n-space, has the order of magnitude of $\sqrt{n}$. In fact, the Minkowski or Rogers lower bounds reduce to

$$\lim\inf(c_n^{\text{lattice}})^{1/n}/\sqrt{n} \geq 1/\sqrt{2\pi e}$$

as $n \longrightarrow \infty$, while the Blichfeldt or Rogers upper bounds reduce to

$$\lim\sup c_n^{1/n}/\sqrt{n} \leq 1/\sqrt{\pi e} .$$

Thus for large n the numbers

$$c_n^{1/n} \geq (c_n^{\text{lattice}})^{1/n}$$

are known to within a factor of approximately $\sqrt{2}$. (Compare Rogers, Sloane, Milnor and Husemoller.)

To conclude, here are three open problems.

One needs some technique for actually computing c_n, the maximum center-density for unit diameter balls in n-space, for small values of n. In particular one would like to prove that $c_3 = \sqrt{2}$.

In principle, one knows how to compute c_n^{lattice}, the maximum center-density for lattice packings of unit diameter balls. The procedure used by Korkin and Zolotarev can be reformulated as follows. Consider the convex polyhedron consisting of all symmetric $n \times n$ matrices $[a_{ij}]$ which satisfy the inequality $\sum a_{ij} k_i k_j \geq 1$ for every non-zero n-tuple $(k_1, \ldots, k_n)$ of integers. Then the determinant function attains its minimum at some vertex, which we may take to lie in a carefully chosen compact subpolyhedron. (Compare Watson.) With modern computing and programming techniques, one should be able to actually find this minimum for small values of $n \geq 9$. One reward of such a computation might be a proof that $c_n > c_n^{\text{lattice}}$ in some cases. For Leech and Sloane have given examples of sphere packings in dimensions 10, 11, and 13 which are denser than any known lattice packings in these dimensions.

Finally, there is the problem of determining the asymptotic behavior of the numbers $c_n \geq c_n^{\text{lattice}}$ (or at least of their n-th roots) as $n \longrightarrow \infty$. In particular, do the ratios

$$c_n^{1/n}/\sqrt{n} \geq (c_n^{\text{lattice}})^{1/n}/\sqrt{n}$$

tend to limits as $n \longrightarrow \infty$? As noted earlier, these limits would have to lie between $1/\sqrt{2\pi e}$ and $1/\sqrt{\pi e}$.

REFERENCES

L. Auslander, Examples of locally affine spaces, Annals of Math. 64 (1956), 255-259.

―――――――, An account of the theory of crystallographic groups, Proc. Amer. Math. Soc. 16 (1956), 1230-1236.

L. Bieberbach, Über die Bewegungsgruppen des n-dimensionalen euklidischen Raumes mit einem endlichen Fundamentalbereich, Gött. Nachr. 1910, 75-84.

―――――――, Über die Bewegungsgruppen der Euklidischen Räume II, Die Gruppen mit einem endlichen Fundamentalbereich, Math. Ann. 72 (1912), 400-412.

―――――――, Über die Minkowskische Reduktion der positiven quadratischen Formen und die endlichen Gruppen linearer ganzzahliger Substitutionen, Gött. Nachr. 1912, 207-216.

H. F. Blichfeldt, The minimum values of quadratic forms in six, seven, and eight variables, Math. Zeits. 39 (1935), 1-15.

J. J. Burkhardt, "Die Bewegungsgruppen der Kristallographie," Basel 1947.

E. C. Dade, The maximal finite groups of 4×4 integral matrices, Ill. J. Math. 9 (1965), 99-122.

E. S. Fedorov, "Symmetry of Crystals" (translated from 1949 Russian edition), Amer. Crystallographic Assoc., New York 1971.

L. Fejes Tóth (= L. Fejes), Über einem geometrischen Satz, Math. Zeits. 46 (1940), 79-83.

G. Frobenius, Über die unzerlegbaren diskreten Bewegungsgruppen, Sitzungsb. Akad. Wiss. Berlin (1911), 654-665.

C. F. Gauss, Untersuchungen über die Eigenschaften der positiven ternären quadratischen Formen von L. A. Seeber, Werke 2, 188-196, Göttingen 1876.

H. Heesch, Aufbau der Ebene aus kongruenten Bereichen, Gött. Nachr. 1935, 115-117.

―――――――, "Reguläres Parkettierungsproblem," Arbeitsg. Forschung Nordrhein-Westf. 172, Westdeutscher Verlag, 1968.

D. Hilbert and S. Cohn-Vossen, "Anschauliche Geometrie," Springer 1932 (English transl., Chelsea 1952).

F. E. A. Johnson, On a conjecture of C. T. C. Wall, to appear.

C. Jordan, Mémoire sur l'équivalence des formes, J. de l'École Polytech. 29 #48 (1880), 111-150. (See p. 133.)

A. Korkine and G. Zolotareff, Sur les formes quadratiques positives, Math. Ann. 11 (1877), 242-292.

J. Leech, Notes on sphere packings, Canad. J. Math. 19 (1967), 251-267.

———— and N. J. A. Sloane, Sphere packings and error correcting codes, Canad. J. Math. 23 (1971), 718-745.

J. Milnor and D. Husemoller, "Symmetric Bilinear Forms," Ergebnisse 73, Springer, 1973.

H. Minkowski, Diskontinuitätsbereich für arithmetische Äquivalenz, J. reine angew. Math. 129 (1905), 220-274. (See pp. 240, 269.)

M. Newman, "Integral Matrices," Academic Press, 1972. (See p. 175.)

K. Reinhardt, Zur Zerlegung der euklidischen Räume in kongruente Polytope, Sitzungsb. Akad. Wiss. Berlin 1928, 150-155.

R. M. Robinson, Undecidability and nonperiodicity for tilings of the plane, Inventiones math. 12 (1971), 177-209.

C. A. Rogers, The packing of equal spheres, Proc. London Math. Soc. 8 (1958), 609-620.

————, "Packing and covering," Cambridge U. Press 1964.

A. Schoenfliess, "Kristallsysteme und Kristallstruktur," Leipzig 1891.

N. J. A. Sloane, Sphere packings constructed from BCH and Justesen codes, Mathematika 19 (1972), 183-190.

S. K. Stein, Algebraic tiling, Amer. Math. Monthly 81 (1974), 445-462.

A. Thue, Über die dichteste Zusamenstellung von Kongruenten Kreisen in einer Ebene, Skr. Vidensk-Selsk, Christ. 1 (1910), 1-9.

G. L. Watson, "Integral Quadratic Forms," Cambridge U. Press 1960.

J. Wolf, "Spaces of Constant Curvature," McGraw-Hill, 1967.

H. Zassenhaus, Neuer Beweis der Endlichkeit der Klassenzahl bei unimodularer Äquivalenz endlicher ganzzahliger Substitutionsgruppen, Abh. Math. Sem. Hamburg 12 (1938), 276-288.

————, Über einen Algorithmus zur Bestimmung der Raumgruppen, Commentarii 21 (1948), 117-141.

Proceedings of Symposia in Pure Mathematics
Volume 28, 1976

THE SOLVABILITY OF BOUNDARY VALUE PROBLEMS

James Serrin

Among the prophetic problems in Hilbert's famous list one must surely include the twentieth, the general problem of boundary values for elliptic partial differential equations. This subject, only a seedling in the year 1900, has burst into flower during our century, has developed in directions surely Hilbert never imagined, and today encompasses a vast area of work which to a mathematician of seventy-five years ago would seem little short of astonishing. A list of the mathematicians who have significantly developed the theory is itself impressive: beginning with the work which Hilbert himself contributed in 1900 in the form of a legitimization of Riemann's version of the Dirichlet principle, other important advances through the first half-century were supplied by Serge Bernstein, Jacques Hadamard, Henri Lebesgue, Eberhard Hopf, Richard Courant, O. Perron, Norbert Wiener, Jules Schauder, Jean Leray, K.O. Friedrichs, G. Giraud, C.B. Morrey, Jr., Lars Garding, and others. An even longer list would be required for the 25 years since 1950, while a complete bibliography would be monumental.

The twentieth problem, like so many of the others in Hilbert's list, consisted as much in a program as in a specific problem requiring some definitive answer, and in just this fact we can see one facet of Hilbert's genius and breadth. The first sentences of the twentieth problem show clearly this programmatic aspect. I quote, with some paraphrasing: "An important problem ... is the question concerning existence of solutions of partial differential equations when the values on the boundary of the region are prescribed. [The methods now available for treating this problem], however, seem to be generally not capable of direct extension to the case where [differential relations are prescribed along the boundary], nor can they be extended immediately to the case where the inquiry is not for potential surfaces but, say, for surfaces of least area, or for [other more complicated cases of boundary value problems]."

From the context of the twentieth problem, it is apparent that Hilbert had in mind particularly equations of the form

$$\sum_{i,j=1}^{n} a_{ij}(x) \frac{\partial^2 u}{\partial x_i \partial x_j} + \sum_{i=1}^{n} b_i(x) \frac{\partial u}{\partial x_i} + c(x)u = f(x), \quad x \text{ in } \Omega,$$

$$\sum_{i,j=1}^{n} a_{ij}(x,u,p) \frac{\partial^2 u}{\partial x_i \partial x_j} + b(x,u,p) = 0, \quad x \text{ in } \Omega,$$

where Ω is a bounded open subset of R^n, $u = u(x) : \Omega \to K$,

$$p = p(x) = (\partial u/\partial x_1, \ldots, \partial u/\partial x_n),$$

and the leading coefficient matrix a_{ij} is positive definite:

$$(a_{ij}(x)) > 0, \qquad x \in \overline{\Omega},$$

$$(a_{ij}(x,u,p)) > 0, \qquad x \in \overline{\Omega}, \quad u \in R^1, \; p \in R^n.$$

The first equation is linear of second order, the second quasilinear; note that the second case includes the first. It is of course reasonable and proper to extend the domain of Hilbert's program to include equations of higher order, and even general systems of equations. In this form, however, the scope of the twentieth problem becomes little less than encyclopedic, and a far from easy subject for a general lecture. For this reason I shall confine myself for the most part to the equations noted, particularly since they are also immediately relevant to the direct query with which Hilbert closes the twentieth problem. This question asks, very specifically,

"Has not every regular variational problem a solution, provided certain assumptions regarding given boundary conditions are satisfied (say that the functions concerned in these boundary conditions are continuous and have ... one or more derivatives), and provided also if need be that the notion of a solution shall be suitably extended?"

By a regular variational problem, we shall mean what Hilbert meant, a problem of the following sort: find a function $u : \overline{\Omega} \to R$ which is of class $C^1(\Omega) \cap C^0(\overline{\Omega})$ and is such that, among all functions of this class,

$$I[u] = \int_\Omega F(x, u(x), p(x))\, dx = \text{Minimum};$$

we shall assume, moreover, that the admissible functions also satisfy the following Dirichlet-type boundary condition:

$$u(x) = \varphi(x) \quad \text{for} \quad x \in \partial\Omega$$

where $\varphi(x)$ is an assigned continuous function on $\partial\Omega$. Here the function F is given, and $F(x,u,p)$ satisfies the regularity (or convexity) condition

$$F \in C^2, \quad \left(\frac{\partial^2 F}{\partial p_i \partial p_j}\right) > 0 \qquad \text{for } x \in \overline{\Omega}, \; u \in R, \; p \in R^n.$$

(Hilbert even assumes that F is analytic in its arguments and restricts to $n = 2$, though we shall not be quite this special.)

This problem is linked to partial differential equations by means of its Euler-Lagrange equation. That is, if u minimizes the expression $I[u]$ and if it is sufficiently smooth, then u satisfies the partial differential equation

$$\sum_{i=1}^n \frac{\partial}{\partial x_i}\{F_{p_i}(x,u,p)\} = F_u(x,u,p),$$

where the subscripts on F denote partial differentiation with respect to the indicated variable. If we carry out the differentiations we get

$$\sum_{i,j=1}^{n} F_{p_i p_j}(x,u,p)\frac{\partial^2 u}{\partial x_i \partial x_j} + \sum_{i=1}^{n}(F_{p_i u} p_i + F_{p_i x_i}) = F_u .$$

Evidently, every smooth minimizing function is a solution of this equation, while conversely any solution obeying the given boundary condition is a prime candidate for the minimizing function. Note that the quadratic form is positive definite by virtue of the regularity condition; thus this equation is a special example of the nonlinear equation noted before.

Several main ideas have become dominant in studying the existence of solutions of elliptic partial differential equations satisfying given boundary conditions, namely:

I. Continuation of solutions along a parameter, for which the problem varies from a known situation to a desired one.

II. The a priori estimation of the magnitude of the partial derivatives, depending only on the structure of the equation, the boundary data, and the domain.

III. A functional-analytic approach, guaranteeing the existence of a weak or generalized solution of the given problem.

Since the first two techniques are of crucial importance to the study of quasilinear equations, and since the third is surely required in the discussion of Hilbert's concluding question, let me discuss for a moment the background of these techniques.

In a remarkable series of papers in the years 1910-1912 Serge Bernstein, who was at the time just 30 years of age, had been able to obtain the existence of solutions of analytic quasilinear second order equations in $n = 2$ variables by a combination of analytic continuation and successive approximations. The difficult and laborious nature of this reasoning obscured for 20 years the brilliant underlying idea: that in order to prove the existence of solutions it is enough to show that for any eventual solution the corresponding partial derivatives appearing in the equation can be estimated and bounded (in an appropriate norm), the bounds depending only on the structure of the equation, on the boundary data, and on the smoothness of the domain. For, with these a priori bounds in hand, one could then obtain the existence of solutions by a homotopy argument involving continuous deformation of the given problem to a simpler one, the solution of the latter either being known or else available by inspection. It is remarkable, nevertheless, that Bernstein could carry out such an ambitious program at the time, and, even more remarkable, that he could already so much foreshadow present results as to obtain special conclusions of the following sort: the Dirichlet boundary value problem for the minimal surface equation

$$\left(1+\left(\frac{\partial u}{\partial y}\right)^2\right)\frac{\partial^2 u}{\partial x^2} - 2\frac{\partial u}{\partial x}\frac{\partial u}{\partial y}\frac{\partial^2 u}{\partial x \partial y} + \left(1+\left(\frac{\partial u}{\partial x}\right)^2\right)\frac{\partial^2 u}{\partial y^2} = 0$$

is solvable for all suitably smooth boundary data when the domain Ω has a smooth strictly convex boundary.

In 1934 Jules Schauder clarified the reasoning involved in Bernstein's approach. Moreover, adopting and refining the two methods of continuation and a priori estimation, he was able to treat the Dirichlet boundary value problem ($u = \varphi$ given on $\partial\Omega$) for linear second order equations (i.e. the first displayed equation in this article). His result is impeccable, bringing this aspect of Hilbert's program to the following positive and definite conclusion:

Provided $c \leqq 0$ and provided the coefficients, the right hand side f, the given data φ, and the domain are all suitably smooth, there exists one and only one solution of the problem. For general c, if one replaces c by $c + \lambda$ then a Fredholm alternative exists.

Coupled with the work of Giraud (occurring at roughly the same period) on other boundary conditions, including that of Neumann type

$$\nu_i \frac{\partial u}{\partial x_i} = \psi(x) \quad \text{for } x \in \partial\Omega \quad (\psi(x) \text{ given on } \partial\Omega)$$

where $\nu = (\nu_1, \ldots, \nu_n)$ denotes the normal vector to $\partial\Omega$, Schauder's conclusions constitute an unquestioned high point in Hilbert's program. It should be added that Giraud's methods today are supplemented by more modern techniques developed by Fiorenza and by Agmon, Douglas and Nirenberg.

The third thread of modern work was directly initiated by Hilbert. Following the sentences I quoted at the beginning of the talk, Hilbert had gone on to say: "It is my conviction that it will be possible to prove existence theorems by means of a general principle whose nature is indicated by Dirichlet's principle." Hilbert then refers to a paper of his which had just appeared in the Jahresberichte der Deutschen Mathematische-Vereinigung; from this we may gather his meaning: namely, given a regular variational problem, then from a sequence of functions u_n such that $I[u_n] \to$ Minimum one should construct another sequence convergent to a solution of the corresponding Euler-Lagrange partial differential equation. This is the first statement of the so-called direct method of the calculus of variations, later developed by Courant, Tonelli, Morrey and others. This approach is of course patently inapplicable for obtaining the existence of solutions of partial differential equations which do not arise (as the Euler-Lagrange equation does) from some variational problem. Nevertheless, when reinterpreted, and when filled out with appropriate results of functional analysis (at the simplest level, that any continuous linear functional on a Hilbert space can be represented as an inner product), it provided a tool of tremendous power not only for variational problems, but also for the resolution of linear boundary value problems. The procedure indeed is independent of the order of the equation, is extensible to the case of systems of equations, and even allows one to treat boundary conditions other than those of simple Dirichlet type. The flaw in the method, if such there be, is that the solutions which are obtained have derivatives only in a generalized sense and satisfy the equation only in a correspondingly weak form. Whether the solution has derivatives in a classical sense, and a fortiori whether they satisfy the equation in a normal sense is left aside: never mind, the daring gamble pays off!

Mathematicians have turned to the problem of proving that such "generalized solutions" are "regular," namely possess further properties of differentiation and, ultimately, enough smoothness so as to satisfy the differential equation in a classical sense (Nirenberg, Agmon, Douglas, Browder, Schechter). Of course, such "regularity" proofs are the counterpart for the functional analytic approach of the "a priori estimates" of the continuation approach. In this respect, Hilbert's twentieth problem closes with the nineteenth, for the problem of existence of classical solutions becomes precisely the problem of regularity of generalized solutions. While Hilbert apparently did not expect such a nicely packaged outcome, his brilliant insight nevertheless accurately pin-pointed two of the most crucial issues in the modern theory of elliptic partial differential equations.

Let us turn to the question of existence of solutions of equations of the form

$$\text{(Q)} \qquad \sum_{i,j=1}^{n} a_{ij}(x,u,p)\frac{\partial^2 y}{\partial x_i \partial x_j} = b(x,u,p), \quad x \text{ in } \Omega,$$

with Dirichlet boundary conditions

$$u = \varphi(x) \quad \text{for } x \text{ on } \partial\Omega,$$

and with

$$b,\ a_{ij} \in C^1, \quad a_{ij}\xi_i\xi_j > 0 \quad \text{for all } x,u,p.$$

As I have remarked, this situation includes partial differential equations arising from regular variational problems. We have the following theorem, essentially due to Leray and Schauder:

Let the boundary $\partial\Omega$ *and the assigned boundary data be of class* C^3. *Also let* τ *be an arbitrary real number in the closed interval* $[0,1]$. *Suppose there exists a constant* M, *independent of* τ, *such that the conditions*

(i) $v \in C^2(\overline{\Omega})$,

(ii) $v = \tau\varphi$ *on* $\partial\Omega$,

(iii) $a_{ij}(x,v,q)\,\partial^2 v/\partial x_i \partial x_j = \tau b(x,v,q)$ *in* Ω, $q = (\partial v/\partial x_i, \ldots, \partial v/\partial x_n)$,

are incompatible with $\|v\|_{C^2(\overline{\Omega})} > M$. *Then the Dirichlet problem is solvable in the class* $C^2(\overline{\Omega})$.

The proof involves Schauder's continuation of a solution along the parameter τ, from $\tau = 0$ where the problem (i), (ii), (iii) has the unique solution $v = 0$, to $\tau = 1$, which is essentially the given problem. The apparatus guaranteeing that the continuation exists throughout the full homotopy range $[0,1]$ consists of a topological and an analytical part. The topological portion is much deeper than that required by Bernstein[1] and in fact is a special case of the Leray-Schauder fixed point theory. The latter was in fact developed primarily in order to prove this theorem! Let me add that Browder has given an elegant and simple proof of this special case, independent of the Brouwer index. The analytical part of the continuation procedure, needless to say, relies on a priori estimates, in fact precisely those estimates developed by Schauder for the linear theory above.

The theorem as it stands is useful only if we can justify its main hypothesis, namely prove the further a priori estimate $\|v\|_{C^2(\overline{\Omega})} \leqq M$ for solutions of (i), (ii), (iii). Except for the case $n = 2$, treated by Leray and Schauder and by others, the problem of estimating $\|v\|_{C^2(\overline{\Omega})}$ remained open for 30 years. It was finally solved by Ladyzhenskaya and Uraltseva, who showed that one can replace the condition $\|v\|_{C^2(\overline{\Omega})} > M$ in the hypothesis by $\|v\|_{C^1(\overline{\Omega})} > M$. As we shall see in a moment, this improvement allows a far-reaching existence theorem to be established for quasilinear equations. The theorem of Ladyzhenskaya and Uraltseva is consequently one of the key results in the structure of Hilbert's program. Although the original proof was fairly difficult, there now exists a simplified version due to Trudinger (on the basis of earlier papers of De Giorgi, Moser, and Serrin). This theorem (and its comparatively elementary proof) stands now as a deservedly high point of mathematical analysis.

With its help, the existence of solutions of the Dirichlet problem for equation (Q) now splits into three steps:

(A) Estimates for $\sup_{\Omega} |v|$.

(B) Estimates for $\sup_{\partial\Omega} |q|$, assuming (A).

(C) Estimates for $\sup_{\Omega} |q|$, assuming (A) and (B).

None of these steps need be possible for a given equation, as can be shown by counter-examples. On the other hand, step (A) is frequently quite simple, involving in many situations only the application (sometimes in a rather sophisticated form) of the maximum principle of E. Hopf; moreover step (C) is known to hold under very wide conditions on the structure of (Q), as has been shown by Ladyzhenskaya and Uraltseva, Serrin, Simon and others. Step (B) is the focus of main interest.

We introduce the orthogonal invariants

$$\mathcal{E}(x,u,p) = p_i a_{ij}(x,u,p) p_j, \qquad \mathcal{F}(x,u,p) = \frac{\mathcal{E}(x,u,p)}{|b| + |p| \ \mathrm{trace}\ (a_{ij})}.$$

Then the following Main Result holds (Serrin, 1969).

Suppose there exists an increasing continuous function $\Phi(\rho)$ such that

$$\mathcal{F} \geqq \Phi(|p|), \qquad \int^{\infty} \Phi(\rho) \frac{d\rho}{\rho} = \infty.$$

[1] Or by Schauder in his earlier work on linear equations, where the functional analytic apparatus requires only a contraction mapping.

Then step (B) can be carried out.

Suppose on the contrary that there exists a continuous function $\Psi(\rho)$ such that

$$\mathcal{J} \leqq \Psi(|p|), \qquad \int^{\infty} \Psi(\rho)\, \frac{d\rho}{\rho} < \infty.$$

Then the following results hold:

(a) if $|b|/|p|$ trace $(a_{ij}) \to \infty$ as $|p| \to \infty$, then for any smooth domain there exists smooth data for which the Dirichlet problem has no solution;

(b) if $|b|/|p|$ trace (a_{ij}) is bounded, then there exist smooth domains and smooth data for which the Dirichlet problem has no solution.

In the latter case, in fact, it turns out to be necessary to impose certain curvature conditions on the boundary $\partial\Omega$ in order for the Dirichlet problem to be solvable for arbitrarily given smooth data. These conditions have the general form

$$\sum_{i=1}^{n-1} a_i \varkappa_i \geqq d, \qquad x \in \partial\Omega,$$

where the coefficients a_i and d depend only on the structure of the equation, the boundary values, and the principal directions on the boundary, and where the $\varkappa_i$ are the principal curvatures.

Here is one example where the results are nearly definitive. We consider the variational problem [2]

$$\int F(|p|)\, dx = \text{Minimum}.$$

Regulairty of the integrand forces $F''(t) > 0$ for $t \geqq 0$ and $F'(t) > 0$ for $t > 0$. Now put

$$G(t) = t\,\frac{F''(t)}{F'(t)}.$$

Then we have the following conclusions (Serrin, 1971):

(a) If $G \geqq \Phi(t)$, where Φ is decreasing and $\int^{\infty} \Phi\, dt = \infty$, then the Dirichlet problem is solvable.

Examples: $F = |p|^2$, $G = 1$, $\Phi = 1$ (Laplace equation),

$F \approx |p|^m$, $G \approx (m-1)$, $\Phi = m-1$ $(m > 1)$.

(b) If $\int G\, dt < \infty$ and $tG \geqq \Phi(t)$, where Φ is decreasing and $\int^{\infty} \Phi\, dt = \infty$, then the Dirichlet problem is solvable if and only if the mean curvature H of the boundary is nonnegative at each boundary point. (For $n = 2$ this is due to Bernstein and to Finn.)

Examples: $F = \sqrt{1 + |p|^2}$ (minimal surface equation),

$F = (1 + |p|^{\alpha})^{1/\alpha}$, $G = (\alpha - 1)/(1 + |t|^{\alpha})$ $(\alpha > 1)$.

We may now turn to Hilbert's question concerning the existence of solutions of regular variational problems. I will propose that the appropriate answer to Hilbert's

[2] This problem is the most general first order variational problem in which the integrand is independent of u, and invariant under translations and rotations of the x-space.

problem is both YES and NO; indeed with some reformation the answer also remains OPEN!

NO. Consider the integral

$$F(x,u,p) = \sqrt{1+|p|^2} + 2\Lambda u, \quad \Lambda = \text{const} > 0.$$

Then all the conditions of Hilbert are satisfied; the corresponding partial differential equation

$$(1+|p|^2)\Delta u - p_i p_j \frac{\partial^2 u}{\partial x_i \partial x_j} = 2\Lambda (1+|p|^2)^{3/2}$$

represents the classical case of a surface of constant mean curvature Λ, and indeed Hilbert mentions just this situation as an example of a regular variational problem.

Now take

$$\Omega = \{|x| \leqq 2\Lambda\} \text{ and } \varphi = 0 \text{ on } \partial\Omega.$$

We shall show that this problem has no solution, that in fact there is no finite valued function for which the minimum is attained. To see this, it is enough to show that the minimum in question is $-\infty$, that is, that there exist functions for which $I[u]$ is arbitrarily large negatively. Indeed, consider the radial symmetric function u_M whose graph is indicated in the figure.

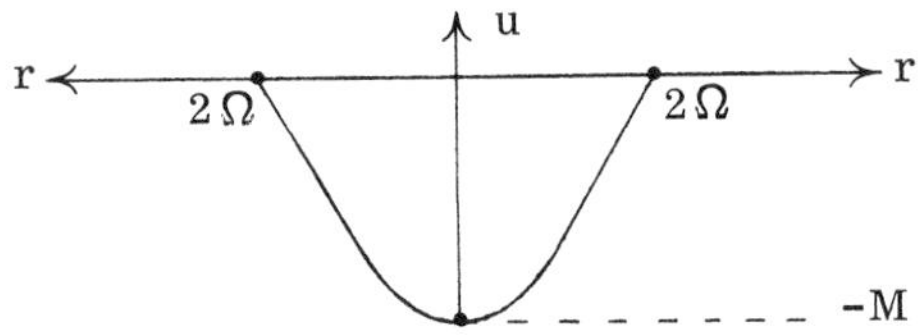

Then $I[u_M] \to -\infty$ as $M \to \infty$, as one easily shows.

It may be objected that the difficulty here arises because $I[u]$ is not bounded below on the class of admissible functions u. Let us therefore add this condition, even in the strong form that $F \geqq 0$. I will make the case that the answer is still

NO. Consider the integral

$$F(x,u,p) = \sqrt{1+|p|^2},$$

again one discussed by Hilbert, corresponding to surfaces of least area spanning the graph of the given boundary values. Take as domain Ω a non-convex smoothly bounded set of the form

and data for which the minimal surface equation has no (non-parametric) solution. Then the variational problem can have no solution either, as can be proved by an easy additional argument.

The difficulty here arises because of the non-convexity of the boundary, which, in view of the earlier discussion, prevents solutions from taking on arbitrary boundary values.

In spite of the negative answers so far given, there are still two ways available to restore an affirmative answer to Hilbert's query. Since almost all problems need redefinition at some stage, we may ask that the integrals in Hilbert's problem should be restricted to those for which the first part of the main result above applies. Since this includes a great variety of situations, including the "natural" integrands of Ladyzhenskaya and Uraltseva, we thus obtain a large and important class of regular variational problems for which the answer to Hilbert's question is YES.

Moreover, even when the first part of the theorem does not apply but case (b) of the second part holds, there is still a large class of domains for which the answer is YES, these domains being distinguished by curvature restrictions of the form noted earlier.

There is still another way to obtain a positive answer; for in fact we have not yet heeded Hilbert's final (and clairvoyant) admonition "provided also if need be that the notion of a solution be sufficiently extended." Let me give an example: we have seen that the variational problem with $F = \sqrt{1 + |p|^2}$ need not always have a solution if the domain does not satisfy appropriate curvature conditions. We therefore modify the notion of solution as follows: consider functions

$$u \in C^2(\Omega) \cap W^{1,1}(\Omega) \quad (\text{hence } u \in L^1(\partial\Omega))$$

and say that u is a solution of the variational problem provided that

$$I[u] = \int_\Omega \sqrt{1 + |Du|^2}\, dx + \int_{\partial\Omega} |u - \varphi| ds = \text{Minimum}.$$

Then there is always a solution; if it takes on the right boundary values it is even the solution of the classical problem, and if there exists a classical solution that will be the solution here (M. Miranda).

Whether generalized solutions can be defined and will exist for all regular variational problems, or even all those for which $F \geqq 0$, is however OPEN.[3]

I should like to conclude with a small selection of open problems, directly in the spirit of Hilbert's statement, but more specifically oriented.

1. Consider a quasilinear equation, a smooth domain, and smooth boundary data. As we have seen, a classical solution need not exist unless the boundary of Ω satisfies a curvature condition. Nevertheless one may still introduce "generalized solutions." For variational type problems this can be attempted by direct methods, as I indicated in discussing Miranda's theorem. For non-variational type problems is there an alternate method? Perron's process comes to mind here, and is seen to work whenever differences of solutions satisfy a maximum principle. Does the Perron upper solution then equal the Perron lower solution, as it does for potential theory? For a regular variational problem does the Perron solution agree with the solution attained by direct methods? [In partial answer to this question R. Temam has informed me (private communication, 1974) that one necessarily has the relation: lower Perron

[3] Some recent work in this direction is reported by Temam.

solution $\leq$ generalized solution of the variational problem $\leq$ upper Perron solution.] At a boundary point where the "generalized solution does not take on the "correct" boundary value, is the solution nevertheless continuous in $\bar{\Omega}$ and does the graph of the solution have a tangent plane orthogonal to the x-space?

2. The general boundary condition

$$\varphi(x,u,Du) = 0 \quad \text{for} \quad x \in \partial\Omega$$

is important; special cases include the Neumann condition, the so-called third boundary condition, and the "natural" condition

$$a_{ij}(x,u,Du) \cdot \vec{n} = g(x,u), \quad g(x,u) \text{ given on } \partial\Omega.$$

For uniformly elliptic quasilinear equations, Ladyzhenskaya and Uraltseva have proved the existence of classical solutions corresponding to this boundary condition, provided suitable smoothness is assumed for the domain and data. Recently Spruck has treated the capillary surface problem (an equation which Gauss showed could be obtained from a variational problem), a much more difficult situation since the Euler-Lagrange equation falls in case (b) of our main result, but nevertheless one where the boundary condition is again of "natural" type. Remarkably, Spruck found that no boundary curvature conditions were required. Is this the general situation, or is it due to the fact that he treated the "natural" condition?

3. The Dirichlet problem in unbounded domains. While a reasonably complete theory for such problems in the linear case has been given by Meyers and Serrin, the quasilinear situation is relatively untouched.

4. Fully nonlinear equations

$$F(x,u,Du,D^2u) = 0$$

have been studied only in special circumstances, and few results are available. The partial differential equations for surfaces of constant positive Gaussian curvature, and, in general, for Weingarten surfaces are important special cases. For these cases, in particular, is it possible to distinguish domains and data for which the Dirichlet boundary value problem is solvable? What about other boundary conditions?

5. When the integral $F(x,u,p)$ satisfies the conditions

$$F(x,u,p) \geq \mu|p|^m - \nu, \qquad \mu,\nu > 0, \; m > 1,$$

one can define generalized solutions of the variational problem even without the convexity condition. Is there any way to characterize these solutions? In particular, is it true that at almost every point in Ω the generalized solution satisfies the condition $(f_{p_ip_j}(x,u,Du)) \geq 0$ so that the generalized solution is an elliptic-parabolic solution of the Euler-Lagrange equation?

These problems remind us of Hilbert's words "The supply of problems in mathematics is inexhaustible, for as soon as one is solved numerous others come forth in its place." With an abundance of problems it is reasonable to predict that the final quarter of this century will find elliptic equations to be of as great interest as they have been during the seventy-five years since Hilbert's talk. Moreover, I have mentioned only problems which are definite and special, and even in this area I have restricted consideration to second order equations, echoing Hilbert's premise that "it is just such definite and special problems which attract us the most and from which the most lasting influence is exerted." Nevertheless, Hilbert stressed broader implications as well, significantly choosing variational problems as his example. I think it worth emphasizing here, in the same way, that interest in elliptic equations need not rest solely on boundary value problems. There are other directions of research of equal significance. I would stress the discovery of qualitative features of solutions, such as the behavior of solution at singularities, uniqueness theory and monotonicity theorems, and the structure of singular sets for solutions of higher order nonlinear equations. Bifurcation theory and the theory of variational inequalities are only in their beginning stages, and offer further significant directions of research.

Hilbert closed his lecture with these words, which seem to me applicable not only to the present problem but to the entire meeting. "The problems here are merely samples, yet they show how rich, how manifold and extensive the mathematical science of today is. The question is therefore urged on us whether mathematics is doomed to the fate of those other sciences that have split into separate branches, and whose representatives scarcely understand one another and whose connection becomes ever more loose. ... And, with the extension of mathematical knowledge will it not finally become impossible for the single investigator to embarce all departments of this knowledge?"

I am not sure how to answer those questions today, but I am sure that with the bewildering variety of available problems mathematics will remain a vital science and individual mathematicians will continue to be beckoned, some by general theories, others by the richness of detail to be found in special problems. It seems to me that the next twenty-five years will provide us with "gifted masters and many zealous and enthusiastic disciples" who will make the coming years as vital as those since Hilbert's talk.

Note. This paper is essentially the same as the lecture given at De Kalb on May 24. Its style is thus that of a talk, rather than an expository research paper, and accordingly it states few detailed conclusions. So far as the results mentioned in the paper are concerned, these can of course be found in the references listed at the close of the paper. At the same time, many topics were necessarily omitted. These include, most importantly, an outline of the elegant proof of the Leray-Schauder-Ladyzhenskaya-Uraltseva theorem, a discussion of the remarkable work of Ladyzhenskaya, Uraltseva, and Simon on interior estimates for the gradient of solutions of equation (Q) in cases where the equation is severely nonlinear, a review of much of the most recent work (1973-1974) on the boundary behavior of solutions, a discussion of various boundary value problems alternative to the Dirichlet problem, and, finally, treatment of higher order linear and nonlinear problems. The author had intended to discuss all of these issues except the last, but the necessary time to survey the large amount of literature involved turned out not to be available. To consider higher order problems a full treatise would have been needed, so that from the beginning the author intended that these should not be included. Indeed, were Hilbert's twentieth problem to be given a full and definitive treatment in all its varied ramifications, the resulting paper must needs have grown to elephantine size (the tract of Carlo Miranda, outlining the theory of elliptic equations up to 1965 already fills a large volume), and would in fact have been beyond the powers of any one, or even two or three, men. For all these reasons, then, it may in fact be just as valuable for the purposes at hand - namely, to introduce mathematicians to the general ideas involved in Hilbert's problems - to reproduce the lecture just as given, with all its flaws and also with all its enthusiasm for a beautiful subject.

Bibliography

The following list refers to specific matters discussed in the paper as well as to topics mentioned only in the note above. While not exhaustive, it nevertheless contains the major papers which have contributed significantly to the elucidation of Hilbert's twentieth problem, together with (more than) enough material for orientation to modern work. For convenience, books and monographs have been grouped separately at the end of the main list.

Agmon, S.
[1] The L^p approach to the Dirichlet problem. Ann. Scuola Norm. Sup. Pisa 13(1959), 405-448.

Agmon, S., Douglis, A., and Nirenberg L.
[1] Estimates near the boundary for solutions of elliptic partial differential equations satisfying general boundary conditions. I, II. Comm. Pure Appl. Math. 12(1959), 623-727; 17(1964), 35-92.

Beirao Da Veiga, H.
[1] On the $W^{2,p}$ regularity for solutions of mixed problems. J. Math. Pures Appl. 53(1974), 279-290.

Berstein, S.
[1] Sur la ǵeneralisation du problème de Dirichlet. Math. Ann. 69(1910), 82-136.

[2] Sur les surfaces définies au moyen de leur courbure moyenne et totale. Ann. Aci. Ecole Norm. Sup. 27(1910), 233-256.

[3] Sur les équations du calcul des variations. Ann. Sci. Ecole Norm. Sup. 29 (1912), 431-485.

Bombieri, E. De Giorgi, E., and Miranda, M.
[1] Una maggiorazioni a priori relativa alle ipersuperfici minimali non parametriche. Arch. Rational Mech. Analysis 32(1969), 255-267.

Browder, F.E.
[1] Functional analysis and partial differential equations. Math. Ann. 138(1959), 55-79; 141(1962), 81-226.

[2] A priori estimates for elliptic and parabolic equations. Proc. Symp. Pure Math. IV. Amer. Math. Soc.: Providence (1961), 73-81.

[3] Nonlinear elliptic boundary value problems. Bull. Amer. Math. Soc. 69 (1963), 862-874; Trans. Amer. Math. Soc. 117(1965), 530-550.

[4] Remarks on the direct method of the calculus of variations. Arch. Rational Mech. Analysis 20(1965), 251-258.

[5] Existence theorem for nonlinear partial differential equations. Summer Institute in Global Analysis. Amer. Math. Soc. 1969.

Caccioppoli, R.

[1] Limitazioni integrale pur le soluzioni di un'equazione lineare ellitico a derivate parziali. Giorn. Mat. Battaglini 80(1950-51), 186-212.

Concus, P. and Finn, R.

[1] On the class of capillary surfaces. Proc. Nat. Acad. Sci. 63(1969), 292-299.

Cordes, H.O.

[1] Vereinfachter Beweis der Existenz einer Apriori-Hölderkonstante. Math. Ann. 138(1959), 155-178.

De Giorgi, E.

[1] Sulla differenziabilità e l'analiticità delle estremali degli integrali multipli regolari. Mem. Acc. Sci. Torino 3(1957), 1-19.

Douglis, A., and Nirenberg, L.

[1] Interior estimates for elliptic systems of partial differential equations. Comm. Pure Appl. Math. 8(1955), 503-538.

Edmunds, D.E.

[1] Quasilinear second order elliptic and parabolic equations. Bull. London Math. Soc. 2(1970), 5-28.

Fichera, G.

[1] Sul problema misto per le equazione lineari delle derivate parziali del secondo ordine di tipo ellitico. Rev. Math. Pures Appl. 9(1964), 3-10.

Finn, R.

[1] On equations of minimal surface type. Ann. Math. 60(1954), 397-416.

[2] New estimates for equations of minimal surface type. Arch. Rational Mech. Analysis 14(1963), 337-375.

[3] Remarks relevant to minimal surfaces and to surfaces of constant mean curvature. J. d'Anal. Math. 14(1965), 139-160.

Finn, R. and Gilbarg, D.

[1] Three-dimensional subonic flows and asymptotic estimates for elliptic partial differential equations. Acta Math. 98(1957), 265-296.

Fiorenza, R.

[1] Sui problemi di derivata obliqua per le equazioni ellittiche. Ricerche Mat. Napoli 8(1959), 83-110.

[2] Sulla hölderianità della soluzioni dei problemi di derivata obliqua regolare del secondo ordine. Ricerche Mat. Napoli 14(1965), 102-123.

[3] Sui problemi di derivata obliqua per le equazioni ellittiche quasi lineari. Ricerche Mat. Napoli 15(1966), 74-186.

Friedrichs, K.O.

[1] The identity of weak and strong extensions of differential operators. Trans. Amer. Math. Soc. 55(1944), 132-155.

Garding, L.

[1] Le problème de Dirichlet pour les équations aux dérivées partielles elliptiques homogènes à coefficients constants. C.R. Acad. Sci. Paris 230(1950), 1030-1032.

[2] Dirichlet's problem for linear elliptic partial differential equations. Math. Scand. 1(1953), 55-72.

[3] Some trends and problems in linear partial differential equations. Proc. Int. Congr. Math., Edinbourgh (1958), 87-102.

Gerhardt, C.
[1] Existence, regularity, and boundary behavior of generalized surfaces of prescribed mean curvature. Math. Zeit. 139(1974), 173-198.

[2] On the capillarity problem with constant volume. Preprint, 1974.

Giaquinta, M.
[1] On the Dirichlet problem for surfaces of prescribed mean curvature. Manuscripta Math. 12(1974), 73-86.

Gilbarg, D.
[1] Boundary value problems for nonlinear elliptic equations in n variables. Symp. Nonlinear Problems, Univ. Wisconsin Press (1962), 151-159.

Giusti, E.
[1] Sulla regolarità parziale delle soluzioni di sistemi ellittici quasi-linear di ordine arbitrario. Ann. Scuola Norm. Sup. Pisa 23(1969), 115-141.

[2] Superfici cartesiane di area minima. Rend. Sem. Mat. Milano 40(1970), 1-21.

Haar, A.
[1] Über das Plateausche Problem. Math. Ann. 97(1927), 124-258.

Heinz, E.
[1] Interior gradient estimates for surfaces $z = f(x,y)$ with prescribed mean curvature. J. Diff. Geometry 5(1971), 149-157.

Hilbert, D.
[1] Über das Dirichletsche Prinzip. Jahresber. Deutsch. Math. Vereinig. 8(1900), 184-188; Math. Ann. 59(1904), 161-186.

Hopf, E.
[1] Elementare Bemerkung über die Lösungen partieller Differentialgleichungen zweiter Ordung von Elliptischen Typen. Sber. Preuss. Akad. Wiss. 19(1927), 147-152.

Hörmander, L.
[1] On the regularity of the solutions of boundary problems. Acta Math. 99(1958), 225-264.

Ivanov, A.V.
[1] On interior estimates of the first derivatives of solutions of quasilinear non-uniformly elliptic and parabolic equations of general form. Sem. Steklov Inst. 14 (1969), 29-47; 19(1970), 79-94.

Jenkins, H. and Serrin J.
[1] Variational problems of minimal surface type, I, II, III. Arch. Rational Mech. Analysis 12(1963), 185-212; 21(1966), 321-342; 29(1968), 304-322.

[2] The Dirichlet problem for the minimal surface equation in higher dimensions. J. Reine Angew. Math. 229(1968), 170-187.

Koselev, A.I.
[1] A priori estimates in L^p and generalized solutions of elliptic equations and systems. Uspehki. Mat. Nauk. 13(1958), 29-88; English transl. Amer. Math. Soc. 20(1962), 105-171.

Ladyzhenskaya, O.A. and Uraltseva, N.N.
[1] On Hölder continuity of solutions and their derivatives of linear and quasi-linear elliptic and parabolic equations. Trudy Steklov Math. Inst. 73(1964), 172-220.

[2] Local estimates for gradients of solutions of non-uniformly elliptic and parabolic equations. Comm. Pure Appl. Math. 23(1970), 677-703.

Lebesgue, H.
[1] Sur le problème de Dirichlet. Rend. Circ. Math. Palermo 24(1907), 371-402.

Leray, J.

[1] Majoration des dérivées secondes des solutions d'un problème de Dirichlet. J. Math. Pures Appl. 17(1938), 89-104.

[2] Discussion d'un problème de Dirichlet. J. Math. Pures Appl. 18(1939), 249-284.

Leray, J. and Lions, J.L.

[1] Quelques résultats de Visik sur les problémes elliptiques non linéaires par les méthodes de Minty-Browder. Bull. Soc. Math. France 93(1965), 97-107.

Leray, J. and Schauder, J.

[1] Topologie et équations functionelles. Ann. Sci. Ecole Norm. Sup. 16(1934), 45-78.

Lions, J.L.

[1] Quelques résultats d'existence dans les équations aux dérivées partielles non linéaires. Bull. Soc. Math. de France 87(1959), 245-273.

Littman, W., Stampacchia, G., and Weinberger, H.F.

[1] Regular points for elliptic equations with discontinuous coefficients. Ann. Scuola Norm. Sup. Pisa 17(1963), 43-78.

Magenes, E.

[1] Sur les problèmes aux limites pour les équations linéaires elliptiques. Colloques internationaux sur les équations aux dérivées partielles. Paris (1962).

[2] Sul problema di Dirichlet per le equazioni lineari ellittiche in due variabili. Ann. Mat. Pura Appl. 48(1959), 257-279.

Meyers, N. and Serrin, J.

[1] The exterior Dirichlet problem for second order elliptic partial differential equations. J. Math. Mech. 9(1960), 513-538.

Miranda, M.

[1] Un teorema di esistenza ed unicità per il problema dell' area minima in n variabili. Ann. Scuola Norm. Sup. Pisa (1965), 233-249.

[2] Un principe di massimo forte per le frontiere minimali e una applicazione alla risoluzione del problema al contorno per l'equazione delle superfici di area minima. Rend. Sem. Mat. Univ. Padova 45(1971), 344-366.

Morrey, C.B., Jr.

[1] Second order elliptic equations in several variables and Hölder continuity. Math. Zeit. 72(1959), 146-164.

[2] Existence and differentiability theorems for variational problems for multiple integrals. Univ. of Wisconsin Press: Madison (1961), 241-270.

Moser, J.

[1] A new proof of De Giorgi's theorem concerning the regularity problem for elliptic differential equations. Comm. Pure Appl. Math. 13(1960), 457-468.

[2] On Harnack's theorem for elliptic differential equations. Comm. Pure Appl. Math. 14(1961), 577-591.

Nirenberg, L.

[1] On nonlinear partial differential equations and Hölder continuity. Comm. Pure Appl. Math. 6(1953), 103-156.

[2] Remarks on strongly elliptic differential equations. Comm. Pure Appl. Math. 8(1955), 649-675.

[3] Estimates and existence of solutions of elliptic equations. Comm. Pure Appl. Math 9(1956), 506-529.

[4] On elliptic partial differential equations. Ann Scuola Norm. Sup. Pisa 13(1959), 115-162.

Oskolkov, A. P.
[1] A priori estimates for the first derivatives of solutions of the Dirichlet problem for non-uniformly elliptic quasilinear equations. Trudy Steklov Math. Inst. 102(1967), 105-127.

Peetre, J.
[1] Another approach to elliptic boundary problems. Comm. Pure Appl. Math. 14(1961), 711-731.

Perron, O.
[1] Eine neue Behandlung der Randwertaufgabe fur $\Delta u = 0$. Math. Zeit. 18(1923), 42-54.

Protter, M. H. and Weinberger, H. F.
[1] A maximum principle and gradient bound for linear elliptic equations. Indiana Univ. Math. J. 23(1973), 239-249.

Radó, T.
[1] Uber zweidimensionale reguläre Variationsprobleme Math. Ann. 101(1929), 620-632.

Schauder, J.
[1] Uber den Zusammenhang zwischen der Eindeutigkeit und Lösbarkeit partieller Differentialgleichungen zweiter Ordnung von elliptischen Typus. Math. Ann. 106 (1932), 661-721.

[2] Uber das Dirichletsche Problem im Grossen für nichtlineare elliptische Differentialgleichungen. Math. Zeit. 37(1933), 623-634 and 768.

[3] Uber lineare elliptische Differentialgleichungen zweiter Ordnung. Math. Zeit. 38(1934), 257-282.

[4] Équations du type elliptique, problèmes linéaires. Enseign. Math. 35(1936), 126-139.

Schechter, M.
[1] General boundary-value problems for elliptic partial differential equations. Comm. Pure Appl. Math. 12(1959), 457-486.

[2] Remarks on elliptic boundary value problems. Comm. Pure Appl. Math. 12 (1959), 561-578.

[3] Mixed boundary problems for general elliptic equations. Comm. Pure Appl. Math 13(1960), 183-201.

[4] Various types of boundary conditions for elliptic equations. Comm. Pure Appl. Math. 13(1960), 407-425.

[5] Some unusual boundary value problems. Proc. Symp. Pure Math. 4, Amer. Math. Soc. (1961), 109-113.

[6] On L^p estimates and regularity, I, II, III. Amer. J. Math. 85(1963), 1-13; Math. Scand. 13(1963), 47-69; Ricerche Mat. Napoli 13(1964), 192-206.

Serrin, J.
[1] On the definition and properties of certain variational integrals. Trans. Amer. Math. Soc. 101(1961), 139-167.

[2] Local behavior of solutions of quasilinear equations. Acta Math. 111(1964), 247-302.

[3] The problem of Dirichlet for quasilinear elliptic differential equations with many independent variables. Phil. Trans. Royal Soc. London 264(1969), 413-496.

[4] The Dirichlet problem for surfaces of constant mean curvature. Proc. London Math. Soc. 21(1970), 361-384.

[5] Gradient estimates for solutions of nonlinear elliptic and parabolic equations. Symposium on Nonlinear Functional Analysis, Univ. of Wisconsin: Madison, 1971.

[6] Boundary curvatures and the solvability of Dirichlet's problem. Proc. Int. Congr. Math., Nice, 1970. Tome 2, 867-875. Gauthier-Villars: Paris, 1971.

Simon, L.

[1] Interior gradient estimates for non-uniformly elliptic partial differential equations of divergence form. Thesis, Univ. of Adelaide, 1971.

Simon, L. and Spruck, J.

[1] Existence and regularity of a capillary surface with prescribed contact angle. Preprint, 1975.

Spruck, J.

[1] Infinite boundary value problems for surfaces of constant mean curvature. Arch. Rational Mech. Analysis 49(1972), 1-31.

[2] On the existence of a capillary surface with prescribed contact angle. Preprint, 1974.

Stampacchia, G.

[1] Problemi al contorno ellittici, con data discontinui, dotati di soluzioni hölderiane. Ann. Mat. Pura Appl. 51(1960), 1-37.

[2] On some regular multiple integral problems in the calculus of variations. Comm. Pure Appl. Math. 16(1963), 383-421.

[3] Le problème de Dirichlet pour les équations elliptiques du second ordre à coefficients discontinues. Ann. Inst. Four. 15(1965), 189-258.

Temam, R.

[1] Solutions généralisées de certaines équations du type hypersurfaces minima. Arch. Rational Mech. Analysis 44(1971), 121-156.

Tonelli, L.

[1] Sur la semi-continuité des integrales double de calcul des variations. Acta Math. 53(1929), 325-346.

Trudinger, N.S.

[1] On Harnack type inequalities and their application to quasilinear elliptic equations. Comm. Pure Appl. Math. 20(1967), 721-747.

[2] Some existence theorems for quasilinear, non-uniformly elliptic equations in divergence form. J. Math. Mech. 18(1969), 909-919.

[3] A new proof of the interior gradient bound for the minimal surface equation in n dimensions. Proc. Nat. Acad. Sci. 69(1972), 821-823.

[4] Gradient estimates and mean curvature. Math. Zeit. 131(1973), 165-175.

Wiener, N.

[1] The Dirichlet Problem. J. Math. and Physics 3(1924), 127-146.

Books and Monographs

Agmon, S. Lectures on Elliptic Boundary Value Problems. Van Nostrand Math. Studies 2: New York, 1965.

Bers, L., John F. and Schechter, M. Partial Differential Equations. Interscience: New York 1964.

Browder, F. Problèmes Nonlinéaires. Lecture Notes, Univ. of Montreal, 1966.

Courant, R. and Hilbert, D. Methods of Mathematical Physics, vol. II. Interscience: New York, 1962.

Courant, R. Dirichlet's Principle, Conformal Mapping and Minimal Surfaces. Interscience: New York, 1950.

Ekeland, I. and Temam, R. Analyse Convexe et Problémes Variationnels. Dunod-Gauthier-Villars: Paris, 1974.

Fichera, G. Linear Elliptic Differential Systems and Eigenvalue Problems. Lecture Notes in Mathematics No. 8. Springer-Verlag: Berlin, 1965.

Friedman, A. Partial Differential Equations of Parabolic Type. Prentice-Hall: Englewood Cliffs, 1964.

Friedman, A. Partial Differential Equations. Holt, Rinehart and Winston: New York, 1969.

Hadamard, J. Leçons sur le Calcul des Variations. A. Hermann: Paris, 1910.

Hörmander, L. Linear Partial Differential Operators. Springer-Verlag: Berlin, 1963.

Ladyzhenskaya, O.A., Uraltseva N.N. Quasilinear Elliptic Equations and Variational Problems with many Independent Variables. Uspekhi Mat. Nauk. 16(1961), 19-92; Translated in Russ. Math. Surveys 16, 17-91.

Ladyzhenskaya, O.A., Uraltseva N.N. Linear and Quasilinear Elliptic Equations. Moscow, 1964. Academic Press: New York, 1968. Revised edition, Moscow, 1973.

Lions, J.L. Equations Différentielles Opérationelles et Problèmes aux Limites. Springer-Verlag: Berlin, 1961.

Lions, J.L. Lectures on Elliptic Partial Differential Equations. Tata Inst.: Bombay, 1967.

Lions, J.L. Quelques Méthodes de Résolution des Problèmes aux Limites Non Linéaires. Dunod-Gauthier-Villares: Paris, 1969.

Miranda, C. Partial Differential Equations of Elliptic Type. Springer-Verlag: Berlin, 1970.

Morrey, C.B., Jr. Multiple Integral Problems in the Calculus of Variations and Related Topics. Univ. California Publ. Math. 1(1943), 1-130.

Morrey, C.B., Jr. Multiple Integrals in the Calculus of Variations. Springer-Verlag: Berlin, 1966.

Necas, J. Les Méthodes Directes en Théorie des Equations Élliptiques. Masson: Paris, 1967.

Protter, M.H. and Weinberger, H.F. Maximum Principles in Differential Equations. Prentice-Hall: Englewood Cliffs, 1967.

Rosenbloom, P.C. Linear Partial Differential Equations. Surveys in Appl. Math. V. Wiley: New York, 1958.

Simader, C. On Dirichlet's Boundary Value Problem. Lecture Notes in Mathematics No. 268. Springer-Verlag: Berlin, 1972.

Stampacchia, G. Équations Elliptiques du Second Ordre á Coefficients Discontinues. Lecture Notes, Univ. of Montreal, 1966.

Tonelli, L. Fondamenti di Calcolo delle Variazioni, Vols. I and II, Zanichelli: Bologna, 1921-1923.

Department of Mathematics
University of Minnesota

Proceedings of Symposia in Pure Mathematics
Volume 28, 1976

VARIATIONAL PROBLEMS AND ELLIPTIC EQUATIONS

Enrico Bombieri

I. Variational Problems. In this expository talk I will be concerned with second order, nonlinear, elliptic equations arising from variational problems. Perhaps the simplest example is the

Dirichlet Problem. Find a function $u(x)$ harmonic in a given bounded open set Ω and taking given boundary values on $\partial\Omega$.

The variational formulation of Dirichlet's Problem is expressed through the

Dirichlet Principle. The function $u(x)$ is the unique solution of the variational problem

$$\int_\Omega |Du|^2\,dx = \min,$$

$$u = f, \quad \text{on } \partial\Omega,$$

where Du denotes the gradient of u.

The approach to the Dirichlet Problem through the Dirichlet Principle was soon criticized because the existence of a minimum for the Dirichlet integral was not obvious; in particular, some conditions are needed in order to have a finite Dirichlet integral. This is not unnatural to assume "a priori", since for example in physical models the Dirichlet integral represents the energy of a system, which should be finite to start with. Once these limitations of the variational approach were understood, its usefulness became clear and the Dirichlet Principle became again a respectable tool in mathematics.

More generally, one may ask to minimize the functional

$$J[u] = \int_\Omega f(x, u, Du)\,dx$$

under appropriate boundary conditions for the competing functions u. Actually $u(x)$ may be a vector-valued function. If $J[u] = \min$, then $J[u] \leqq J[u + \epsilon v]$ for every v with compact support in Ω and expanding $J[u + \epsilon v]$ in a Taylor series in ϵ:

$$J[u + \epsilon v] = J[u] + \epsilon \delta J[u] + \epsilon^2 \delta^2 J[u] + \ldots$$

We see that we need $\delta J[u] = 0$ and $\delta^2 J[u] \geqq 0$ for all such v, i.e. (writing $p = (p_1, \ldots p_n)$ for Du)

$$\delta J[u] = \int_\Omega \left(\sum \frac{\partial v}{\partial x_i}\frac{\partial f}{\partial p_i} + \frac{\partial f}{\partial u} v\right) dx = \int_\Omega \left\{-\sum_i \frac{\partial}{\partial x_i}\frac{\partial f}{\partial p_i} + \frac{\partial f}{\partial u}\right\} v\,dx = 0$$

and we obtain the well known Euler equation

$$\sum_i \frac{\partial}{\partial x_i} \frac{\partial f}{\partial p_i} = \frac{\partial f}{\partial u}.$$

A simple condition, which implies $\delta^2 J \geqq 0$, is

$$\sum_{ij} \frac{\partial^2 f}{\partial p_i \partial p_j} \xi_i \xi_j > 0, \qquad \xi \in \mathbb{R}^n,\ \xi \neq 0,$$

which expresses a kind of convexity condition for the functional $J[u]$. If this condition is satisfied, one says that the integrand $f(x, u, p)$ is regular elliptic. In case one considers vector-valued solutions $u = (u^1, \ldots, u^\lambda, \ldots, u^N)$ the regularity condition imposed on $f = f(x, u^\lambda, p^\lambda)$ is

$$\sum \sum \frac{\partial^2 f}{\partial p_i^\lambda \partial p_j^\mu} \eta^\lambda \eta^\mu \xi_i \xi_j > 0$$

at every point $(x, u^\lambda, p^\lambda)$ and all $\eta \in \mathbb{R}^N$, $\xi \in \mathbb{R}^n$, $\eta, \xi \neq 0$.

In his 19th Problem of his address at the International Congress of Mathematicians in 1900, Hilbert raised the question whether solutions of regular elliptic, analytic variational problems are necessarily analytic. This problem of regularity, together with the problem of existence of solutions, form two central questions in the theory of variational problems.

II. Elliptic equations: the early work. In his celebrated thesis of 1904, S. Bernstein proved the remarkable result that C^3 solutions of a single elliptic, nonlinear, analytic equation in two variables are necessarily analytic; this was considered at the time a solution to Hilbert's 19th Problem. Having thus attacked the problem of regularity, he went on with the existence problem in an important series of papers, between 1906 and 1912. We owe to him the basic idea (and the name) of an "a priori estimate", which still has a central role in the theory: if we have the right majorizations for all solutions (and their derivatives) of an elliptic equation, then existence and regularity of solutions of the Dirichlet problem will follow. Since in obtaining these estimates we assume "a priori" that we are dealing with smooth solutions, we have the name "a priori estimates". Bernstein himself showed how to prove such estimates in some important cases, using the maximum principle and what is known today as the method of barriers.

Bernstein's work was rather involved and relied heavily on analyticity, and was later improved and generalized to several variables and elliptic systems by the work of several authors, among which Gevrey, Giraud, Lichtenstein, H. Lewy, E. Hopf, T. Rado, I. Petrowsky and Bernstein himself. However, one had to wait until the years between 1932 and 1937 before the basic reasons for the importance of the "a priori estimates" in the existence problem were fully understood and clarified through the work of Schauder, Leray and Caccioppoli and in particular the classical paper of Leray and Schauder of 1934.

Consider for example a quasi-linear equation

$$\sum a_{ij}(x, u, Du)\, D_i D_j u = 0, \qquad \text{in } \Omega,$$
$$u = f, \qquad \text{on } \partial\Omega.$$

We denote by T the operator which to a function u associates the unique solution v of the linear Dirichlet problem

$$\sum a_{ij}(x, u, Du)\, D_i D_j v = 0, \qquad \text{in } \Omega,$$
$$v = f, \qquad \text{on } \partial\Omega.$$

Since the latter problem is linear, it is much easier to solve and the question is reduced to finding a fixed point $u = Tu$ for the operator T. The main point is that very general fixed point theorems are available if we have the right "a priori estimates" for the solutions of the original equation and of the linearized equation. The advantage of this procedure over an iteration scheme $u_{n+1} = Tu_n$ (used by Bernstein) is obvious: if uniqueness is not satisfied, the iteration need not converge.

The fundamental "a priori estimates" for the linearized equation were found by Schauder; the search for such estimates in the nonlinear case is still today more of an art than of a method.

III. Direct methods and weak solutions. Another approach to the existence problem in the variational case is provided by the so-called "Direct Methods in the Calculus of Variations". Roughly speaking, one wants to show

(A) the integrand $J[u]$ is lower-semicontinuous and bounded below, with respect to a suitable notion of convergence in some admissible class of competing functions u;

(B) a minimizing sequence $\{u_n\}$, i.e. $J[u_n] \to \operatorname{Inf} J[u]$, converges to an admissible u, hence $J[u] = \min$ by (A).

This idea was used perhaps for the first time by Zaremba and also by Hilbert in his investigations on the Dirichlet Principle. It became a standard approach to variational problems in the hands of Lebesgue, Courant, Fréchet and especially Tonelli. If the integrand $J[u]$ satisfies an inequality

$$f(x, u, p) \geqq m_1 |p|^r - m_2, \qquad m_1 > 0,$$

with $1 \leqq r < +\infty$, then Tonelli's method, using absolutely continuous functions and uniform convergence, works provided $r \geqq n = \dim \Omega$, which is a too strong condition if $n \geqq 3$. A notable success of this method was however Haar's work of 1927 on functionals of the type

$$J[u] = \int_\Omega f(Du)\, dx,$$

for the case of $n = 2$ variables. Here one assumes that Ω is a smooth convex domain, and the boundary values are also smooth, satisfying a certain "three-point condition". The class of competing functions used by Haar is a class of functions satisfying a uniformly bounded Lipschitz condition.

The deep reason for the limitation of Tonelli's approach was found only later, through the fundamental work of Sobolev and Morrey of 1938. The Sobolev spaces $H^{k,p}(\Omega)$ are the Banach spaces of functions on Ω whose derivatives of order $\leq k$ are in L^p. Sobolev discovered the fundamental embedding theorems for these spaces, the simplest being (one assumes Ω bounded and $\partial\Omega$ smooth):

(i) if $f \in H^{1,p}(\Omega)$, $1 \leq p < n$ then $f \in L^s(\Omega)$ with $s = np/(n-p)$, and

$$\|f\|_{L^s} \leq C(\Omega)\|f\|_{H^{1,p}};$$

(ii) if $f \in H^{1,p}(\Omega)$, $p > n$ then f satisfies a Hölder condition in Ω.

The new approach to the existence problem could now be summarized as follows:

(A) the integrand $J[u]$ determines naturally a function space $\mathfrak{F}$ (usually a Sobolev space), in which the lower-semicontinity becomes a natural statement;

(B) by means of "a priori estimates" one shows that there exists a convergent minimizing sequence (here the Sobolev embedding theorems are often crucial).

From (A) and (B) one deduces the existence of a solution in the function space $\mathfrak{F}$. However, one expects the solution so obtained to be very smooth. In some cases, e.g. those in which Tonelli's method works, the smoothness of solutions is automatic (compare (ii) of Sobolev's embedding theorem); in general, there remains the difficult problem of "regularization":

(C) the "weak solutions" so obtained are in fact differentiable solutions.

The necessary results about lower semicontinuity have been obtained by Serrin; stages (B) and (C) require an extensive use of "a priori estimates", the regularization part being often difficult if not intractable.

This approach led to remarkable results especially in two cases: nonlinear second order equations in $n = 2$ variables, where one could also use tools from quasiconformal mapping (Morrey, Bers, Nirenberg), and linear equations and systems with smooth coefficients (we may mention the work of Ladyzenskaya and Caccioppoli of 1951 for second order equations, and the general theory of Friedrichs, F. John, Agmon-Douglis-Nirenberg of 1959, who also considered higher order systems and the problem of boundary regularity).

The first breakthrough in the nonlinear case came in 1957-58 when De Giorgi, and independently Nash for parabolic equations, succeeded in proving Hölder continuity of weak solutions of uniformly elliptic equations

$$\sum D_i(a_{ij}(x)D_j u) = 0$$

with measurable coefficients a_{ij} and with the ellipticity condition

$$m|\xi|^2 \leq \sum a_{ij}(x)\,\xi_i\xi_j \leq M|\xi|^2,$$

where m, M are positive constants independent of x.

This result has some striking applications to nonlinear problems. De Giorgi himself showed how his theorem implied that weak extremals of uniformly elliptic analytic integrands of the type

$$\int_\Omega f(Du)\,dx = \min$$

are indeed analytic in Ω. Stampacchia and Gilbarg found another application, namely the extension of Haar's theorem to the case of $n > 2$ variables; further important applications and generalizations have been given by Morrey, Ladyzenskaya and Uraltseva, Oleinik and many others, in particular to the study of second order quasilinear equations which are quadratic in the first order derivatives.

Of great importance was also a new proof of De Giorgi's theorem, found by Moser in 1960, using the Sobolev inequalities rather than the isoperimetric inequalities of De Giorgi. This also led to a proof of the Harnack inequality: if $\Omega' \subset\subset \Omega$ and if u is a positive solution in Ω of an uniformly elliptic equation $\sum D_i(a_{ij}(x)D_j u) = 0$, then

$$\max_{\Omega'} u \leqq C \min_{\Omega'} u,$$

where C depends only on Ω', Ω and the ellipticity constant $L = M/m$. Hence one obtains a Liouville theorem: a bounded solution over $\mathbb{R}^n$ of such an equation is necessarily a constant.

IV. Weak solutions of elliptic systems. The problem of the extension of De Giorgi's regularization to systems of equations or to higher order equations remained outstanding for a while, until in 1968 De Giorgi found an example of an uniformly elliptic linear system of variational type with bounded measurable coefficients, with the discontinuous solution $x/|x|$. By adapting De Giorgi's example, in 1969 Giusti and Miranda showed that if $n > 2$ the integrand

$$\int |D\vec{u}|^2 + \left[\sum_{ij}\left(\delta_{ij} + \frac{4}{n-2}\,\frac{u^i u^j}{1+|\vec{u}|^2}\right) D_j u^i\right]^2 dx$$

with $\vec{u} = (u^1, \ldots, u^n)$ is a regular uniformly elliptic analytic integrand, while $u = x/|x|$ is an extremal which is not real analytic at $x = 0$. These examples pointed out the great importance of the results obtained by Morrey in 1968 on the regularity problem for systems in $n > 2$ variables.

Here the breakthrough came with the introduction of new powerful compactness methods, originally introduced by De Giorgi and especially Almgren in 1960-66, in the study of minimal surfaces.

In rather crude terms, the idea behind the use of compactness methods may be described as follows. Suppose we want to prove an "a priori estimate" of local nature for solutions of a class of variational problems which is invariant by linear changes of the coordinates. If the estimate we want fails in every neighborhood of a point x_0, this means that we can find a sequence of elliptic equations or systems over a fixed domain Ω, and a sequence of solutions, for which the desired estimate fails in smaller and smaller neighborhoods of x_0. By performing a linear change of coordinates, we can expand these neighborhoods to a fixed neighborhood of x_0, and in doing so we have to replace our equations by new equations still in the same class, and defined over larger and larger domains. Using the appropriate compactness theorems then one shows that this sequence of equations and solutions converges in some sense to a limiting equation, now defined over $\mathbb{R}^n$, and to a limiting solution for which the desired " a priori estimate" still fails. The main point however is that, in doing so, we have replaced an elliptic operator by its "tangent operator" at x_0, and thus the limiting equation is often of a very simple type, for example linear with constant coefficients, and for it it may be easy to check that the "a priori estimate" we want does in fact hold. This gives a contradiction and establishes the local estimate we were looking for. In the nonlinear case, convergence to a limiting equation is usually obtained by assuming certain mild conditions about the local behaviour of solutions at a point. If these conditions are valid almost everywhere, which is often the case because of measure theoretic arguments, one ends up with estimates which are valid only near almost every point, and in turn one establishes only regularity almost everywhere.

In this way it was proved by Morrey in 1968 that weak solutions of a large class of nonuniformly elliptic analytic variational problem of the type

$$\int f(x, D\vec{u})\,dx = \min,$$

and also of uniformly elliptic analytic variational problems of the type

$$\int f(x, \vec{u}, D\vec{u})\,dx = \min,$$

are in fact analytic almost everywhere. Giusti and Miranda, in 1970-72 have extended and substantially simplified this work, and they have also been able to obtain good estimates for the Hausdorff dimension of the exceptional set in which the solutions are not analytic.

The outstanding problem here is to determine the structure of the singular set; for example, is it semi-analytic? In special cases, one can even prove that solutions are everywhere analytic, and it is an interesting open question to find good conditions which imply regularity everywhere.

V. <u>The minimal surface equation</u>. A well known variational problem is the <u>Problem of Plateau</u>. Find a surface of least area among all surfaces having a prescribed boundary.

This is not a regular variational problem, if taken in this generality, and it is not possible for me to explain in this lecture all the new fundamental results obtained between 1960 and 1974 by Federer, Fleming, Reifenberg, De Giorgi, Almgren, Allard and many others. I will restrict instead my attention to the case of minimal graphs (the non-parametric Plateau problem) and to some special questions about the parametric Plateau problem in codimension one.

If the graph $u = u(x)$ of a function $u(x)$, $x \in \Omega \subset \mathbb{R}^n$, is a solution of Plateau's problem, then it minimizes the area functional

$$\int_\Omega \sqrt{1 + |Du|^2}\, dx,$$

and the associated Euler equation is

$$\sum D_i(D_i u/W) = 0, \quad W = \sqrt{1 + |Du|^2}$$

which expresses the fact that the graph has mean curvature 0 at every point.

The strong nonlinearity of this equation gives rise to unexpected phenomena, which have no counterpart in the theory of linear equations. For $n = 2$ variables:

(i) The Dirichlet boundary value problem is soluble for arbitrary continuous data if and only if Ω is convex (Bernstein, Finn);

(ii) A solution defined over a disk minus the centre extends to a solution over the disk, i.e. isolated singularities are removable (Bers);

(iii) If $u > 0$ is a solution over $|x| < R$ then

$$\sqrt{1 + |Du(0)|^2} \leqq \exp(\pi u(0)/2R)$$

and this estimate is sharp (Finn, Serrin);

(iv) A solution defined over $\mathbb{R}^2$ is linear (Bernstein).

The solution of the analogous problems for $n > 2$ variables has been obtained only recently. We have:

(i) the Dirichlet boundary value problem is soluble for arbitrary continuous data if and only if $\partial\Omega$ has positive mean curvature at every point (Serrin, Bombieri-De Giorgi-Miranda 1968);

(ii) a solution defined over Ω minus K where K is a compact subset of Ω with $(n - 1)$-dimensional Hausdorff measure 0 extends to the whole of Ω (De Giorgi-Stampacchia 1964);

(iii) if $u > 0$ is a solution over $|x| < R$ then

$$|Du(0)| < c_1 \exp(c_2 u(0)/R)$$

(Bombieri-De Giorgi-Miranda 1968);

(iv) if $n \leqq 7$, a solution defined over $\mathbb{R}^n$ is linear (Fleming's new proof of 1962 for the case $n = 2$, De Giorgi for $n = 3$ in 1964, Almgren for $n = 4$ in 1966, Simons for $n \leqq 7$ in 1968); on the other hand, if $n \geqq 8$, there are solutions defined over $\mathbb{R}^n$ which are not linear (Bombieri-De Giorgi-Giusti 1969).

What about the methods of proof? In his talk at the International Congress of Mathematicians in 1962, L. Nirenberg made the statement that "most results for nonlinear problems are still obtained via linear ones, i.e. despite the fact that the problems are nonlinear not because of it". The minimal surface equation is no exception to this statement, but since the linearization procedure is rather unusual, it is worthwhile to describe it.

Let us define a vector ν with components

$$\nu_i = -(D_i u)/W, \qquad i = 1, \ldots, n,$$

$$\nu_{n+1} = 1/W,$$

and differential operators

$$\delta_i = D_i - \nu_i \sum_{j=1}^{n+1} \nu_j D_j, \qquad i = 1, \ldots, n+1,$$

in $\mathbb{R}^{n+1}$.

If we denote by S the graph of $x_{n+1} = u(x)$ in $\mathbb{R}^{n+1}$, then the vector $\vec{\nu}$ is the normal unit vector to S at the point $P = (x, u(x))$ and the operators δ_i are the projection of the operators D_i on the tangent space to S at the point P. The "Laplacian" $\mathcal{D} = \sum \delta_i \delta_i$ is actually the Laplace-Beltrami operator on S, and the fact that S has mean curvature 0 at every point is nicely expressed by the fact that the coordinate functions x_i are harmonic on S for the Laplace-Beltrami operator. Moreover it can be shown that the normal vector $\vec{\nu}$ satisfies the nonlinear elliptic system

$$\mathcal{D}\vec{\nu} + c^2(x)\vec{\nu} = 0 \quad \text{on } S,$$

where $c^2(x) = \sum_{ij}(\delta_i \nu_j)^2$ is the sum of the squares of the principal curvatures of S at P. In particular since $\nu_{n+1} > 0$ it follows that $\mathcal{D}\nu_{n+1} \leqq 0$ i.e. ν_{n+1} is superharmonic on S.

Now we have two main facts (Miranda 1967):

(a) if f has compact support and S is minimal, then

$$\int \delta_i f \, d\|S\| = 0, \quad \text{all } i,$$

or in other words the operators δ_i can be integrated by parts on the surface S;

(b) if f has compact support, S is minimal and $1 \leqq p < n$ then

$$\left(\int |f|^{np/(n-p)} d\|S\|\right)^{(n-p)/n} \leqq c(p,n) \int |\delta f|^p \, d\|S\|,$$

or in other words we have a uniform Sobolev inequality on S for the differential operators δ_i.

We can use (a) and (b) together with De Giorgi's regularization technique (which is highly nonlinear) to investigate the differential inequality $\mathcal{D}\nu_{n+1} \leqq 0$, and eventually one arrives to the "a priori estimate" (iii). The solubility of the Dirichlet problem, and also the analyticity of weak solutions, depends on this "a priori estimate".

More generally, one may investigate uniformly elliptic equations of the type

$$\sum \delta_i (a_{ij}(x)\delta_j u) = 0$$

on an absolutely minimizing surface S, of codimension one (Bombieri-Giusti 1972). Thus one obtains the extension of the Moser-Harnack theorem to these equations, and as an application one gets that if u is a positive harmonic function on a minimal surface in $\mathbb{R}^{n+1}$ without boundary, then u is constant. Since the coordinate functions x_i are harmonic on S, one gets as a corollary a theorem of Miranda that a minimal surface without boundary contained in a half-space is a hyperplane. Also, a minimal surface without boundary is connected (Bombieri-Giusti 1972).

The extension of Bernstein's theorem up to dimension 7, and the construction of counterexample in dimension $n \geqq 8$, depends on different ideas. It was Fleming in 1962 who used compactness techniques to show that the failure of Bernstein's theorem in dimension n implied the existence of a singular minimal cone in $\mathbb{R}^{n+1}$. De Giorgi later proved that in fact one would get the existence of such a cone in $\mathbb{R}^n$, and in this way extended Bernstein's theorem through dimension $n = 3$. Then the question centered about the existence of minimal cones and eventually Simons succeeded in proving the non-existence of singular minimal cones in $\mathbb{R}^n$, $n \leqq 7$. Moreover, Simons proved that the cone in $\mathbb{R}^8$ given by

$$x_1^2 + x_2^2 + x_3^2 + x_4^2 = x_5^2 + x_6^2 + x_7^2 + x_8^2$$

was at least a locally minimal cone, i.e. area would increase with every sufficiently small deformation. Making use of the invariance of this cone by $SO(3) \times SO(3)$, Bombieri- De Giorgi-Giusti proved that this cone was in fact minimal in the large, by reducing the problem to a question about a system of first order ordinary differential equations. It was natural to see whether this cone was associated with the failure of Bernstein's theorem in dimension 8, and this was obtained by constructing explicitly a subsolution u^-, and a supersolution u^+, of the minimal surface equation in $\mathbb{R}^8$, with the property that $u^- \leqq u^+$ everywhere and that no function between u^- and u^+ could be linear. Now an application of the maximum principle and also of the "a priori estimate" for the gradient obtained before showed the existence of a solution u defined everywhere comprised between u^- and u^+. It should be noted that the choice of u^- and u^+ was in fact suggested by the results obtained in the investigation of Simons' cone.

VI. Further results. I will end this talk by mentioning some results and directions of research which I could not treat more explicitly, but which seem to me of great importance.

First of all, the facts which I have stated about the minimal surface equation are not limited to that special case. A whole class of elliptic equations can be treated with similar methods, among which the equations of surfaces with prescribed mean curvature, the equation of capillarity phenomena, and many others. Here much recent work has been done by Ladyzenskaya and Uraltseva, Bombieri and Giusti, Trudinger, Finn, Serrin and many others.

Second, and more important, I have limited myself in this talk to variational problems of nonparametric nature. The parametric point of view, in which one considers functionals on geometrical objects rather than on functions, has led to the modern Geometric Measure Theory, the theory of Integral Currents and Varifolds and of Parametric Elliptic Integrands. Here the work of Federer, Fleming and especially

Almgren is outstanding. Also, among more recent developments, I may mention the work of Allard on the first variation of a varifold and that of Jean Taylor on the structure of the singular set of soap films and soap bubbles.

Another fruitful idea which I could not treat in this talk is that of variational problems in which the solutions have to satisfy additional constraints. Here one may ask for solutions satisfying inequalities, thus obtaining classical problems with obstacles, or asking for solutions with gradient not exceeding certain bounds (an example is the potential equation for a subsonic gas flow), or one may impose convexity, as for the Monge-Ampere equations, and so on. Here the theory of variational inequalities begins to give a general foundation for many problems of this type, and problems with constraints have been attacked successfully also from the point of view of Geometric Measure Theory. In this sense, it can be said that Hilbert's 19th Problem has opened one of the most interesting chapters in analysis.

References

Section I

D. Hilbert, Mathematische Probleme. Gesammelte Abhandlungen vol. 3, Berlin 1935, 290-329.

Section II

S. Berstein, Sur la nature analytique des solutions des équations aux derivées partielles du second ordre. Math. Annalen 59(1904), 20-76.

J. Leray and J. Schauder, Topologie et équations fonctionnelles. Ann. Sci. Ec. Norm. Sup. 3, 51(1934), 45-78.

Section III

C.B. Morrey Jr., Multiple Integrals in the Calculus of Variations. Springer Verlag, Berlin-Heidelberg-New York 1966.

L. Nirenberg, Some aspects of linear and nonlinear partial differential equations. Proc. Int. Congress of Math. (1962) 147-162.

O.A. Ladyzenskaya and N.N. Uraltseva, Linear and Quasi-linear Elliptic Equations. Academic Press, New York-London 1968.

A. Haar, Uber das Plateausche Problem, Math. Annalen 97(1927), 124-158.

S. Agmon, A. Douglis and L. Nirenberg, Estimates near the boundary for solutions of elliptic partial differential equations satisfying general boundary conditions I, II. Comm. Pure Appl. Math. 12(1959), 623-727 and 17(1964), 35-92.

E. De Giorgi, Sulla differenziabilità e l'analiticità delle estremali degli integrali multipli regolari. Mem. Accad. Torino Cl. Sci. Fis. Mat. Nat. III, 3(1957) 25-43.

J. Moser, On Harnack's theorem for elliptic differential equations. Comm. Pure Appl. Math. 14(1961), 577-591.

Section IV

E. De Giorgi, Un esempio di estremali discontinue per un problema variazionale di tipo ellittico. Boll. U.M.I. IV, 1(1968), 135-137.

C.B. Morrey Jr., Partial regularity results for nonlinear elliptic systems. J. Math. and Mech. 17(1967/68), 649-670.

E. Giusti, Regolarità parziale delle soluzioni di sistemi ellittici quasilineari di ordine arbitrario. Ann. Sci. Sc. Norm. Sup. Pisa Cl. Sci. 23(1969), 115-141.

E. Giusti and M. Miranda, Un esempio di soluzioni discontinue per un problema di minimo relativo ad un integrale regolare del calcolo delle variazioni. Boll. U.M.I. IV, 2(1968), 1-8.

Section V

R. Finn, On equations of minimal surface type. Annals of Math. 60(1954), 397-416.

S. Bernstein, Uber ein geometrisches Theorem und seine Anwendung auf die partiellen Differentialgleichungen vom elliptischen Typus. Math. Zeit. 26(1927), 551-558.

E. De Giorgi and G. Stampacchia, Sulle singolarità eliminabili delle ipersuperfici minimali. Atti Accad. Naz. Lincei Rend. Cl. Sci. Fis. Mat. Nat. 8, 38(1965), 352-357.

J. Simons, Minimal varieties in Riemannian manifolds. Annals of Math. 88(1968), 62-105.

E. Bombieri, E. De Giorgi and M. Miranda, Una maggiorazione a priori relativa alle ipersuperfici minimali non parametriche. Arch. Rat. Mech. Analysis, 32(1969), 255-267.

E. Bombieri, E. De Giorgi and E. Giusti, Minimal cones and the Bernstein problem. Inventiones Math. 7(1969), 243-268.

E. Bombieri and E. Giusti, Harnack's inequality for elliptic differential equations on minimal surfaces. Inventiones Math. 15(1972), 24-46.

Section VI

O.A. Ladyzenskaya and N.N. Uraltseva, Local estimates for gradients of solutions of non-uniformly elliptic and parabolic equations. Comm. Pure Appl. Math. 23(1970), 677-703.

N. Trudinger, Gradient estimates and mean curvature. to appear.

R. Finn, Capillarity Phenomena. Uspehi Mat. Nauk 29(1974), 131-152.

H. Federer and W.H. Fleming, Normal and integral currents. Annals of Math. 72(1960), 458-520.

F.J. Almgren Jr., Existence and regularity almost everywhere of solutions to elliptic variational problems among surfaces of varying topological type ans singularity structure. Annals of Math. 87(1968), 321-391.

W.K. Allard, On the first variation of a varifold. Annals of Math. (1972).

J.E.Taylor, Regularity of the singular sets of two-dimensional area-minimizing flat chains modulo 3 in $\mathbb{R}^n$. Inventiones Math. 22(1973), 119-159.

F.J. Almgren Jr., Existence and regularity almost everywhere of solutions to elliptic variational problems with constraints. to appear.

Proceedings of Symposia in Pure Mathematics
Volume 28, 1976

An Overview of Deligne's Work on Hilbert's Twenty-First Problem

Nicholas M. Katz

Abstract Hilbert's twenty-first problem on the existence of differential equations with regular singular points and prescribed monodromy is interpreted as a GAGA-type problem of algebraic-analytic comparison. Deligne's solution is outlined. Several open questions are raised.

The setting of the problem

Let X be a complete connected non-singular curve over $\mathbb{C}$, whose underlying complex manifold is thus a compact Riemann surface. Let U be a non-empty Zariski open set in X, the complement in X of a finite (possibly empty) set of closed points. The underlying complex manifold U^{an} is thus a finitely punctured Riemann surface.

Consider a linear homogeneous differential equation of rank n on U. When $X = \mathbb{P}^1$ and $U \subset \mathbb{P}^1 - \{\infty\}$, this simply means an $n \times n$ system

$$\frac{d}{dz}\vec{f} + A(z)\cdot\vec{f} = 0$$

$$\begin{cases} \vec{f} = \begin{pmatrix} f_1(z) \\ \vdots \\ f_n(z) \end{pmatrix} \\ A(z) \text{ an } n \times n \text{ matrix of } \underline{\text{rational}} \text{ functions of } z, \\ \qquad \text{holomorphic on } U. \end{cases}$$

As we are so fond of emphasizing to engineering students, this includes the case of the n'th order linear homogeneous equation

$$\left(\frac{d}{dz}\right)^n f + a_1(z)\cdot\left(\frac{d}{dz}\right)^{n-1} f + \ldots + a_n(z)f = 0,$$

by taking for the matrix $A(z)$ the particular choice

$$A(z) = \begin{pmatrix} 0 & -1 & 0 & \ldots & 0 \\ 0 & 0 & -1 & \ldots & 0 \\ \vdots & & & & \\ 0 & \ldots & \ldots & 0 & -1 \\ a_n(z) & \ldots & \ldots & \ldots & a_1(z) \end{pmatrix}$$

In case X has higher genus, we have no global coordinate z at our disposal, so we are led to <u>define</u> a differential equation on U to be a pair (M, ∇) consisting of a locally free coherent algebraic sheaf M on U together with a connection $\nabla: M \longrightarrow M \otimes \Omega^1_{U/\mathbb{C}}$ (this simply means an additive mapping which satisfies the "product rule": for $f \in \mathcal{O}_U$ and $m \in M$, $\nabla(fm) = m \otimes df + f \otimes \nabla(m)$). In the example of the $n \times n$ system on an open set in $\mathbb{P}^1$, the locally free coherent algebraic sheaf M is simply $(\mathcal{O}_U)^n$, and the connection ∇ is the map $(\mathcal{O}_U)^n \longrightarrow (\Omega^1_{U/\mathbb{C}})^n$ given by

$$\nabla(\vec{f}\,) = d\vec{f} + A(z)\cdot\vec{f}\,dz$$

It is sometimes convenient to view a connection ∇ as a way of having <u>derivations</u> of $\mathcal{O}_U$ act on M, viewing ∇ as an $\mathcal{O}_U$-linear map <u>Der</u>$(U/\mathbb{C}) \longrightarrow$ <u>End</u>$_{\mathbb{C}}(M)$, $D \longrightarrow \nabla(D)$ where $\nabla(D): M \longrightarrow M$ is the composite

$$\begin{array}{ccc} M & \xrightarrow{\ \nabla\ } & M \otimes \Omega^1_U \\ & {\scriptstyle \nabla(D)} \searrow & \downarrow {\scriptstyle \mathrm{id} \otimes \text{contract with } D} \\ & & M \end{array}$$

Thus, in the above example, $\nabla(\frac{d}{dz})$ is given by

$$\nabla(\frac{d}{dz})\vec{f} = \frac{d\vec{f}}{dz} + A(z)\vec{f}$$

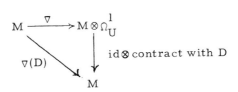

Notice that the <u>solutions</u> of the differential equation become the sections of $M^{\nabla} \overset{\text{dfn}}{=\!=}$ the <u>kernel</u> of $\nabla: M \longrightarrow M \otimes \Omega^1_U$. If we fix a point $z_0 \in U$, then the space S of germs of local holomorphic solutions near (in the analytic sense) z_0 is an $n = \text{rank}\,(M)$-dimensional $\mathbb{C}$-space, by the local existence theorem for differential equations. Given any loop γ in U^{an} starting and ending at z_0, "analytic continuation of solutions along γ" defines an <u>automorphism</u> of S. It is immediate that this automorphism is "multiplicative" in γ, and that it depends only on the homotopy class of γ in $\pi_1(U^{an}, z_0)$. In this way the fundamental group $\pi_1(U^{an}, z_0)$ acts on the $\mathbb{C}$-space S, defining the monodromy representation of the differential equation.

For example, if we take the differential equation

$$z\frac{df}{dz} = \alpha f \qquad \alpha \in \mathbb{C}$$

on the punctured complex "plane" $\mathbb{C}-\{0\}$, its solution is the function $z^\alpha = \exp(\alpha \log z)$, which under analytic continuation along a path γ which loops once counterclockwise around the origin (e.g. $z \longrightarrow e^{2\pi i\theta}z$, $0 \leq \theta \leq 1$) is transformed into $e^{2\pi i\alpha} \cdot z^\alpha$. The fundamental group of $\mathbb{C}-\{0\}$ is isomorphic to $\mathbb{Z}$, with generator γ, and the monodromy representation in $\mathbb{C}^\times = GL(1, \mathbb{C})$ is given by $\gamma \longrightarrow e^{2\pi i\alpha}$

Let us now recall the notion of a "regular singular point" of a differential equation. Classically, one considered an n'th order differential equation

$$\left(\frac{d}{dz}\right)^n f + a_1(z)\left(\frac{d}{dz}\right)^{n-1} f + \dots + a_n(z)f = 0$$

on an open set $U \subset \mathbb{P}^1$. A point $\alpha \in \mathbb{P}^1 - U$ is called a regular singular point of this equation if in any punctured angular sector around α

the local holomorphic solutions satisfy a growth estimate

$$\begin{cases} \text{if } \alpha \text{ finite} & |f(z)| = O(|z-\alpha|^{-N}) \quad \text{for some } N \geq 0, \text{ as } z \longrightarrow \alpha \\ \text{if } \alpha = \infty & |f(z)| = O(|z|^N) \quad \text{for some } N \geq 0, \text{ as } z \longrightarrow \infty. \end{cases}$$

This notion is apparently analytic, rather than algebraic, in character. However, Fuchs discovered that in fact the notion is purely algebraic. He proved that a necessary and sufficient condition that α be a regular singular point is that the functions

$$\begin{cases} (z-\alpha)^i \cdot a_i(z), & i=1,\ldots,n \quad \text{if } \alpha \text{ finite} \\ z^i a_i(z) & i=1,\ldots,n \quad \text{if } \alpha = \infty \end{cases}$$

should all be holomorphic at $z = \alpha$.

We may rephrase Fuch's criterion as follows. Let t be a uniformizing parameter at α (e.g. $t = z - \alpha$ if α finite, $t = 1/z$ if $\alpha = \infty$), and denote by D the differential operator $t\frac{d}{dt}$ Rewrite the equation as a monic equation in D:

$$D^n f + b_1(t) D^{n-1} f + \ldots + b_n(t) \cdot f = 0.$$

Then $t=0$ is a regular singular point if and only if all of the functions $b_i(t)$ are holomorphic at $t=0$.

The notion of a regular singular point of a "fancy" differential equation (M, ∇) on U is defined as before in terms of the growth of local holomorphic solution vectors $\vec{f}$ near the singular point, where we measure the growth by expressing $\vec{f}$ in a local basis of any locally free coherent algebraic sheaf $\overline{M}$ on X which extends M (on a curve, such an $\overline{M}$ always exists, for any locally free M). In this more general setting, Fuch's algebraic criterion for regular singular points, as rephrased by Deligne, is that there exist a locally free coherent algebraic sheaf $\overline{M}$ on X extending M, such that near any singular point α of the equation, the derivation $D = t\frac{d}{dt}$ (t: uniformizing parameter at α) should act (through ∇) stably on $\overline{M}$, i.e. $\nabla(D)(\overline{M}) \subset \overline{M}$.

Notice the compatibility with Fuch's criterion for n'th order equations: If we convert the equation

$$D^n f + b_1(t) D^{n-1} f + \ldots + b_n(t) f = 0$$

into system form, it becomes

$$\nabla(D)\, \vec{f} = D\vec{f} + \begin{pmatrix} 0 & -1 & 0 & \ldots & & 0 \\ \vdots & & & & & \\ 0 & & & & 0 & -1 \\ b_n(t) & & & & & b_1(t) \end{pmatrix} \vec{f},$$

so that if the $b_i(t)$ are all holomorphic at $t=0$, then indeed $\nabla(D)$ acts stably on n-tuples of functions holomorphic at $t=0$.

Statement of the Problem

Hilbert's twenty-first problem asks whether any finite-dimensional complex representation of the fundamental group $\pi_1(U^{an})$ can be obtained as the monodromy representation of a differential equation on U with regular singular points.

One could ask the same question without requiring regular singular points, but the simplest examples show that then there will be too many equations with given monodromy. For example, take $U = \mathbb{A}^1$ (thus $U^{an} = \mathbb{C}$), and the trivial one-dimensional representation of $\pi_1(U^{an}) = 0$. For any polynomial $P(z) \in \mathbb{C}[z]$, the differential equation

$$\frac{df}{dz} = P(z)\cdot f$$

has solution

$$f = \exp\left(\int_0^z P(t)\,dt\right)$$

which is an entire (singlevalued) function of z, so without monodromy. But as differential equations on the algebraic variety $\mathbb{A}^1$, these are pairwise non-isomorphic; only the choice $P \equiv 0$ gives regular singular points. For this reason, one insists on regular singular points.

In the case $U = \mathbb{P}^1 - \{0, 1, \infty\}$, the triply punctured Riemann sphere, the affirmative solution of Hilbert's problem for two-dimensional representations of $\pi_1(U^{an})$ goes back to Riemann, and amounts to the theory of the hypergeometric equation. Indeed, Hilbert's twenty-first problem is often referred to as the Riemann problem. Traditionally, it was viewed as a problem in "function theory," and in that setting has been solved repeatedly, by such men as Birkhoff, Plemelj, and most recently by Röhrl. But as Lipman

Bers remarked in his talk on uniformization, if a problem is worth solving, it's worth solving several times.

Algebro-Geometric Perspective

From the perspective of algebraic geometry, the point of the problem is to describe in purely algebraic terms on U (differential equations on U with regular singular points) a purely topological invariant of U^{an} (the finite dimensional complex representations of its fundamental group).

The earliest result of this kind was the realization that complete connected nonsingular curves over $\mathbb{C}$ are identical with compact Riemann surfaces, the identification provided by the fact that a rational function on the curve is the same as a meromorphic function on the Riemann surface.

This fact was generalized by Chow and Kodaira, who proved that any compact complex connected surface with two algebraically independent (over $\mathbb{C}$) meromorphic functions was a nonsingular projective algebraic surface. The obvious generalization of this result to more than two dimensions is false, and its "counterexamples" have been studied these last few years by Artin and Moishezon as "algebraic spaces".

Around the same time, Chow proved that any closed analytic subset of $\mathbb{P}^n$ (i.e. defined locally by the vanishing of analytic functions) was in fact algebraic, and (by considering the graph) that any analytic map between such was in fact algebraic.

The final step in this direction was Serre's GAGA theorem according to which if X is any projective algebraic variety over $\mathbb{C}$, then the coherent algebraic sheaves $\mathcal{F}$ on X are identical with the coherent analytic sheaves $\mathcal{F}^{an}$ on X^{an} (i.e. $\mathcal{F} \longrightarrow \mathcal{F}^{an}$ is an equivalence of categories) and the cohomology groups of such sheaves are the "same" whether computed algebraically or analytically $(H^i(X,\mathcal{F}) \xrightarrow{\sim} H^i(X^{an},\mathcal{F}^{an}))$.

For example, if we take X to be non-singular, and $\mathcal{F} = \Omega^p_{X/\mathbb{C}}$ the sheaf of germs of holomorphic p-forms, then we get a purely algebraic description of the Hodge cohomology groups

$$\begin{array}{ccc} H^q(X, \Omega^p_{X/\mathbb{C}}) & \xrightarrow{\sim} & H^q(X^{an}, \Omega^p_{X^{an}/\mathbb{C}}) \\ & & \wr\| \text{ (Dolbeaut)} \\ & & \text{harmonic forms of type } (p,q) \end{array}$$

If, instead of a single Ω^p, we take the entire de Rham complex $\Omega^{\bullet}_{X/\mathbb{C}}$, then we get a purely algebraic description of the complex cohomology groups in terms of algebraic "hypercohomology."

$$\begin{array}{ccc} \mathbb{H}^i(X, \Omega^{\bullet}_{X/\mathbb{C}}) & \xrightarrow{\sim} & \mathbb{H}^i(X^{an}, \Omega^{\bullet}_{X^{an}/\mathbb{C}}) \\ & & \wr\uparrow \text{ (holomorphic Poincaré Lemma)} \\ & & H^i(X^{an}, \mathbb{C}) \end{array}$$

The general notion of a differential equation

Let X be any non-singular connected algebraic variety over $\mathbb{C}$. A "differential equation" on X is by definition a pair (M, ∇) consisting of a locally free coherent algebraic sheaf M on X, together with an integrable connection

$$\nabla : M \longrightarrow M \otimes \Omega^1_{X/\mathbb{C}}$$

This means that ∇ is an additive mapping which satisfies the product rule $\nabla(fm) = m \cdot df + f\nabla(m)$ for all $f \in \mathcal{O}_X$, $m \in M$, and is integrable in the sense that if we make $D \in \underline{\mathrm{Der}}\,(X/\mathbb{C})$ act on M as the composite

$$\begin{array}{ccc} M & \xrightarrow{\nabla} & M \otimes \Omega^1_{X/\mathbb{C}} \\ & \searrow^{\nabla(D)} & \downarrow \text{id} \otimes \text{contract with } D \\ & & M \end{array}$$

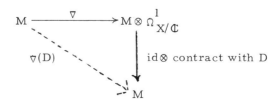

then we have $\nabla([D_1, D_2]) = [\nabla(D_1), \nabla(D_2)]$. In local coordinates, this is a "completely integrable Pfaffian system."

Analogously, we define a differential equation on a complex manifold as an integrable connection on a locally free coherent analytic sheaf.

A holomorphic version of the problem, and its solution

Let V be an arbitrary connected complex manifold. We may ask whether any finite dimensional complex representation of $\pi_1(V)$ arises as the monodromy representation associated to a differential equation on V. The answer is easily seen to be "yes," as follows.

On any complex manifold, the usual Frobenius existence theorem says that the functor "sheaf of germs of sections killed by ∇" is an equivalence of categories between the category of analytic differential equations and the category of locally constant sheaves of complex vector spaces. (The inverse functor is to tensor a locally constant sheaf E with the holomorphic structural sheaf, and to endow the tensor product $E \otimes \mathcal{O}^{an}$ with the connection $id \otimes d$.)

Finally, on any "reasonable" topological space, picking a base point defines an equivalence of categories between the category of locally constant sheaves of anythings (for example, of finite dimensional complex vector spaces) and the category of representations of the fundamental group on those anythings.

Combining these two equivalences, we see that the functor "associated monodromy representation" defines an equivalence of categories between the category of all differential equations on a connected complex manifold V and the category of finite-dimensional complex representations of its fundamental group.

A compact form of the problem and its solution

Suppose that X is a projective non-singular connected algebraic variety over $\mathbb{C}$. Then Serre's GAGA theorem tells us that the functor $(M, \nabla) \longrightarrow (M^{an}, \nabla^{an})$ is an equivalence of categories between the algebraic differential equations (on X) and the analytic differential equations (on X^{an}). Combining this with the holomorphic solution to the problem, we see that the category of algebraic differential equations on X is equivalent to the category of finite dimensional complex representations of $\pi_1(X^{an})$. We should remark here that because X is supposed compact, there are no "missing points," and hence there are no regularity conditions to impose at them. Thus we have solved the compact case of the twenty-first problem "just" by using GAGA. Even in the case when X is one-dimensional (i.e. X^{an} is a compact Riemann surface), this is of some interest (cf. open problem 1) below).

The non-compact case

As Grothendieck was the first to emphasize, the key result in all such questions is Hironaka's resolution of singularities, in the following form.

Let U be a smooth quasi-projective (Zariski-open in a projective) variety over $\mathbb{C}$. Then there exists a projective smooth variety $X/\mathbb{C}$ such that $U \subset X$ as an open dense set, and such that the closed set $D = X - U$ in X is a union of smooth divisors (subvarieties of codimension one) D_i which cross transversely. Furthermore, if

$$\begin{array}{ccc} & \nearrow & X_1 \\ U & & \\ & \searrow & X_2 \end{array}$$

are two such compactifications, there always exists a third which "dominates" both:

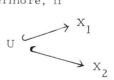

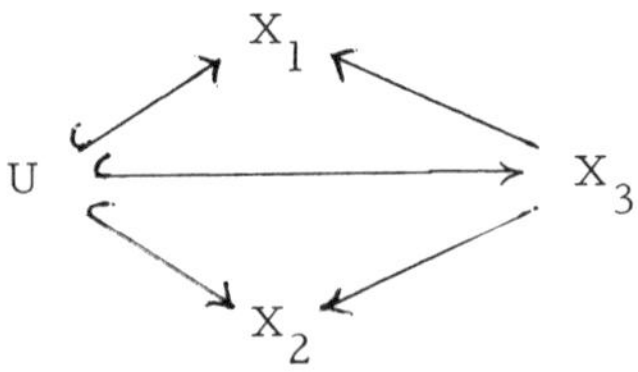

In this setup, we denote by $\underline{\mathrm{Der}}_D(X/\mathbb{C})$ the sub-sheaf of the tangent sheaf consisting of those derivations which preserve the ideals of each of the D_i. At a point on X where r of the divisors, say $D_1, \ldots, D_r$, cross, we may choose local coordinates $x_1, \ldots, x_n$ such that D_i is defined by $x_i = 0$ for $i \leq r$. There, $\underline{\mathrm{Der}}_D(X/\mathbb{C})$ is free on $x_i \frac{\partial}{\partial x_i}$, $i \leq r$ and $\frac{\partial}{\partial x_j}$, $j > r$. The linear dual of $\underline{\mathrm{Der}}_D(X/\mathbb{C})$ is noted $\Omega^1_{X/\mathbb{C}}(\log D)$. It is locally free on $\frac{dx_1}{x_1}, \ldots, \frac{dx_r}{x_r}$, $dx_{r+1}, \ldots, dx_n$ near this same point.

In case U is connected and <u>one-dimensional</u>, the X is unique — it's just the complete non-singular model of the function field of U, the sheaf $\mathrm{Der}_D(X/\mathbb{C})$ is the subsheaf of $\underline{\mathrm{Der}}\,(X/\mathbb{C})$ of sections which <u>vanish</u> at the points of $X - U$, and $\Omega^1_{X/\mathbb{C}}(\log D)$ is the sheaf of one forms with at worst first order poles at the points of D.

<u>Deligne's definition of regular singular points in the general case</u>

In terms of the sheaves $\Omega^1_{X/\mathbb{C}}(\log D)$ and $\underline{\mathrm{Der}}_D(X/\mathbb{C})$, we may rephrase Deligne's version of Fuch's criterion for regular singular points in the one-variable case as follows: In order that an algebraic differential equation (M, ∇) on an open curve U have regular singular points at all points of $D = X - U$, it is necessary and sufficient that there exist a locally free coherent algebraic sheaf $\overline{M}$ on X extending M, such that the action of $\underline{\mathrm{Der}}\,(U/\mathbb{C})$ on M extends to an action of $\underline{\mathrm{Der}}_D(X/\mathbb{C})$ on $\overline{M}$ (equivalently, that the arrow $\nabla: M \longrightarrow M \otimes \Omega^1_{U/\mathbb{C}}$ extends to an arrow $\overline{\nabla}: \overline{M} \longrightarrow \overline{M} \otimes \Omega^1_{X/\mathbb{C}}(\log D)$.

Deligne now takes over this definition verbatim for a differential equation on any connected non-singular quasi-projective variety U over $\mathbb{C}$, where X is some Hironaka compactification of U of the type discussed above. We should remark that in the higher dimensional case, it is not a priori clear that there exists any locally free coherent algebraic sheaf $\overline{M}$ on X which prolongs M on U; we will prove this. And we will admit here that this definition does not depend on the auxiliary choice of X.

Deligne's solution

We can now prove that Hilbert's twenty-first problem has an affirmative solution for any quasi-projective smooth connected variety U over $\mathbb{C}$. Suppose we're given a representation of the fundamental group of U^{an}. Then as explained above, it arises as the monodromy of a unique analytic differential equation (M^{an}, ∇^{an}) on U^{an}. Let us admit for a moment the following analytic lemma:

Key Lemma Let (M^{an}, ∇^{an}) be any analytic differential equation on U^{an}, and let $U \hookrightarrow X$ be a compactification as above. Then there exists a locally free coherent analytic sheaf $\overline{M^{an}}$ on X^{an} which extends M^{an}, such that ∇^{an} extends to an arrow $\overline{\nabla^{an}}: \overline{M^{an}} \longrightarrow \overline{M^{an}} \otimes \left(\Omega^1_{X/\mathbb{C}}(\log D)\right)^{an}$.

Then we may apply GAGA to the pair $\left(\overline{M^{an}}, \overline{\nabla^{an}}\right)$, to conclude the existence of a locally free coherent algebraic sheaf $\overline{M}$ on X and an arrow $\overline{\nabla}: \overline{M} \longrightarrow \overline{M} \otimes \Omega^1_{X/\mathbb{C}}(\log D)$ such that $\left(\overline{M^{an}}, \overline{\nabla^{an}}\right) = \left(\overline{M}, \overline{\nabla}\right)^{an}$. If we define (M, ∇) on U to be the restriction to U of $(\overline{M}, \overline{\nabla})$, it is immediate that (M, ∇) has regular singular points, and that $(M, \nabla)^{an} \simeq (M^{an}, \nabla^{an})$, so that (M, ∇) gives rise to the given representation of $\pi_1(U^{an})$.

Proof of the key lemma

The idea is to use the local monodromy of the equation to construct an extension of M^{an} locally along D which has the desired properties, and to do so in a sufficiently canonical way that these local extensions patch together to give the needed global extension $\overline{M^{an}}$ of M^{an}. [In the one-variable case, D consists of isolated points, so the extension problem is local around each point of D, and there is no patching problem. But in several variables, where we can slide along D from one point to another, there is a patching problem.]

Let us begin by constructing a local extension. At a point "0" of D where r of the divisors D_i cross, say $D_1, \ldots, D_r$, we may choose local coordinates $x_1, \ldots, x_n$ such that D_i is defined by $x_i = 0$ for $i=1, \ldots, r$. In a small coordinate polydisc V around "0" defined by $|x_j| < \varepsilon$ for $j=1, \ldots, n$, the open manifold $U^{an} \cap V$ is the product of r punctured discs $0 < |x_i| < \varepsilon$, $i=1, \ldots, r$, and of $n-r$ discs $|x_k| < \varepsilon$, $k=r+1, \ldots, n$. The restriction of (M^{an}, ∇^{an}) to $V \cap U^{an}$ is an analytic differential equation on $V \cap U^{an}$, so corresponds to a representation ρ of $\pi_1(V \cap U^{an})$ in a finite dimensional complex vector space L.

The important thing about this local situation is that $\pi_1(V \cap U^{an})$ is the free Abelian group on the r generators γ_i = turning once counterclockwise around D_i. So the representation ρ is specified by the r commuting automorphisms $\rho(\gamma_i)$ of L, $i=1, \ldots, r$. A consideration of Jordan normal form shows that there are unique endomorphisms B_j of the representation space L such that

1) $\exp(2\pi i B_j) = \rho(\gamma_j)$, $j=1, \ldots, r$

2) the eigenvalues of B_j have real parts in the strip $-1 < \mathrm{Re} \leq 0$

3) the B_j mutually commute

In terms of this data, we may construct an extension to V of $(M^{an}, \nabla^{an}) | V \cap U^{an}$ as follows. We define

$$\overline{M^{an}} \overset{\text{dfn}}{=} L \underset{\mathbb{C}}{\otimes} \mathcal{O}_V$$

and we define the "connection with logarithmic poles"

$$\begin{array}{ccc} \overline{\nabla^{an}} : \overline{M^{an}} & \longrightarrow & \overline{M^{an}} \underset{\mathcal{O}_V}{\otimes} \Omega^1_V (\log D) \\ \| & & \| \\ L \underset{\mathbb{C}}{\otimes} \mathcal{O}_V & & L \otimes \Omega^1_{V/\mathbb{C}}(\log D) \end{array}$$

by defining

$$\overline{\nabla^{an}}(\ell \otimes f) = f\left(-\sum_{i=1}^{r} B_i \ell \otimes \frac{dx_i}{x_i}\right) + \ell \otimes df$$

To see that this pair $\left(\overline{M^{an}}, \overline{\nabla^{an}}\right)$ does in fact extend $(M^{an}, \nabla^{an}) | V \cap U^{an}$, we need only check that its restriction to $V \cap U^{an}$ has the correct monodromy (by the "holomorphic case" of the Hilbert problem). But a fundamental solution matrix is given explicitly by

$$\prod_{i=1}^{r} x_i^{B_i} \overset{dfn}{=} \exp\left(\sum_{i=1}^{r} B_i \log x_i\right)$$

which does indeed have the correct monodromy.

This construction shows the <u>local</u> existence of locally free extensions of M^{an} which are <u>stable</u> under <u>$Der_D(X^{an}/\mathbb{C})$</u>. Among <u>all</u> locally free extensions to V of $M^{an} | V \cap U^{an}$, we may <u>characterize</u> the one constructed above by the following property:

* In any basis of it as an $\mathcal{O}_V$-module, a fundamental solution matrix has the form

$$H(x) \cdot \prod_{i=1}^{r} x_i^{B_i}$$

with $H(x) \in GL(n, \mathcal{O}_V)$ and with the B_i commuting matrices, the real parts of whose eigenvalues lie in $-1 < \text{Re} \leq 0$. To see that * indeed <u>characterizes</u> our extension $\overline{M^{an}}$, we argue as follows. Any <u>other</u> extension of M^{an} is of the form $A(x) \cdot \overline{M^{an}}$ where $A(x)$ is an invertible $n \times n$ matrix of functions analytic in $V \cap U^{an}$, with possibly essential singularities along the D_i. The extension $A(x)\overline{M^{an}}$ depends only on the class of $A(x)$ in the double-coset space

$$GL(n, \mathcal{O}_V)\backslash GL(n, \mathcal{O}_{V\cap U^{an}})/GL(n, \mathcal{O}_V).$$

So suppose that in terms of such an extension $A(x)\overline{M^{an}}$, the fundamental solution matrix is of the form $K(x)\prod x_i^{C_i}$, with $K(x) \in GL(n, \mathcal{O}_V)$ and with the C_i commuting matrices with all eigenvalues lying in $-1 < \mathrm{Re} \leq 0$. If we equate the fundamental solution matrices, we obtain the equation

$$K(x)\prod_{i=1}^{r} x_i^{C_i} = A^{-1}(x)\cdot H(x)\cdot \prod x_i^{B_i}.$$

By considering the effect of analytic continuation along γ_j, we see that

$$\exp(2\pi i C_j) = \exp(2\pi i B_j).$$

By the unicity of the logarithm whose eigenvalues lie in $-1 < \mathrm{Re} \leq 0$, we conclude that $C_j = B_j$ for $j=1,\ldots,r$, hence that $K(x) = A^{-1}(x)H(x)$. Thus $A(x) \in GL(n, \mathcal{O}_V)$, and our "other" extension $A(x)\overline{M^{an}}$ is just $\overline{M^{an}}$ itself.

It remains to check that these well-defined local extensions patch together, or equivalently that they are compatible with localization. If we restrict our extension to a small enough polydisc $V' \subset V$ around a point where only the divisors $D_1,\ldots,D_s$ (s < r cross, and choose local coordinates $y_1,\ldots,y_n$ adopted to the situation (D_i is defined in V' by $y_i = 0$, $i=1,\ldots,s$), we must check that a fundamental solution matrix is of the form

$$(\text{matrix in } GL(n, \mathcal{O}_{V'}))\times \prod y_i^{B_i}.$$

In terms of the coordinates $x_1,\ldots,x_n$ on V, the fundamental solution matrix was

$$\prod_{i=1}^{r} x_i^{B_i}.$$

Since for $i=1,\ldots,s$, both $x_i = 0$ and $y_i = 0$ define D_i near P, there are invertible functions u_i, $i=1,\ldots,s$ near P such that $x_i = u_i y_i$, $i=1,\ldots,s$. As the divisors $D_{s+1},\ldots,D_r$ do not pass through P, the functions x_{s+1}, $\ldots,x_r$ are themselves invertible near P. So in a small enough polydisc V' around P, we may take the logarithms of the invertible functions $u_1,\ldots,u_s$, $x_{s+1},\ldots,x_r$, i.e. there are functions $z_1,\ldots,z_r$ holomorphic on V' such that

$$u_i = \exp(z_i) \qquad i=1,\ldots,s$$

$$x_i = \exp(z_i) \qquad i=s+1,\ldots,r.$$

Then

$$\prod_{i=1}^{r} x_i^{B_i} = \prod_{i=1}^{s} (u_i y_i)^{B_i} \prod_{i=s+1}^{r} x_i^{B_i} = \exp\left(\sum_{i=1}^{r} B_i z_i\right) \prod_{i=1}^{s} y_i^{B_i}$$

and $\exp(\sum z_i B_i) \in GL(n, \mathcal{O}_{V'})$ as desired. QED

Remark The proof shows clearly that the notion of regular singular points in several variables can be expressed as a growth condition on the local solutions in (angular sectors of) the punctured polydiscs at infinity, in complete analogy with the one-variable case (cf. [2] for more details).

More of Deligne's results — Comparison theorems

Thus far we have proven that for any connected non-singular quasi-projective variety U over $\mathbb{C}$, any finite-dimensional complex representation arises as the monodromy representation of an algebraic differential equation on U with regular singular points. It is natural to ask if this differential equation is unique. In categorical terms, we have proven that the functor

$$\begin{matrix}\text{Algebraic D.E.'s on } U \text{ with} \\ \text{regular singular points}\end{matrix} \longrightarrow \text{analytic D.E.'s on } U^{an} \simeq \text{rep'ns of } \pi_1(U^{an})$$

is essentially surjective, and we ask if it is an equivalence of categories (or equivalently, given its surjectivity, whether it is fully faithful). Concretely, given two algebraic differential equations (M,∇) and (M',∇') on U with regular singular points, we are asking if the map

$$\mathrm{Hom}\,((M,\nabla),\ (M',\nabla')) \longrightarrow \mathrm{Hom}\,((M,\nabla)^{an},\ (M',\nabla')^{an})$$

is an isomorphism.

Consider the "internal hom" differential equation $(M''= \underline{\mathrm{Hom}}(M,M'),\nabla'')$, defined by $(\nabla''(D)\varphi)(m) = \nabla'(D)(\varphi(m)) - \varphi(\nabla(D)(m))$ for $D \in \underline{\mathrm{Der}}\,(U/\mathbb{C})$, $\varphi \in \mathrm{Hom}\,(M,M')$, and $m \in M$; its horizontal (= killed by ∇'') sections are exactly the D.E.-maps from (M,∇) to (M',∇'), and it will have regular singular points if both (M,∇) and (M',∇') do. So we may rephrase the question as follows: given a differential equation (M,∇) with regular singular points, is the map

$$\begin{matrix}\text{global horizontal} \\ \text{sections of } M \text{ on } U\end{matrix} \longrightarrow \begin{matrix}\text{global horizontal} \\ \text{sections of } M^{an} \text{ on } U^{an}\end{matrix}$$

an isomorphism?

The answer is easily seen to be yes, for in terms of the particular extension $\overline{M}$ we constructed, any global analytic horizontal section of M^{an} extends to a holomorphic global section of $\overline{M}^{an}$, which by GAGA is a global section of $\overline{M}$, and so restricts to give a global section of M, as required.

In fact, there is something more that we can say. Let (M,∇) be an algebraic differential equation on U with regular singular points, and let $(M^{an})^{\nabla}$ denote the locally constant sheaf of germs of local holomorphic horizontal sections. Because the connection is <u>integrable</u>, we can extend the map

$$\nabla : M \longrightarrow M \otimes \Omega^1_{V/\mathbb{C}}$$

to an entire complex

$$M \xrightarrow{\nabla} M \otimes \Omega^1_{U/\mathbb{C}} \xrightarrow{\nabla\otimes 1 + 1 \otimes d} M \otimes \Omega^2_{U/\mathbb{C}} \longrightarrow \cdots\cdots,$$

the "De Rham complex of (M,∇)," noted $(M\otimes \Omega^{\bullet}_{U/\mathbb{C}},\nabla)$.

We can consider the hypercohomology groups

$$\mathbb{H}^i(U, (M \otimes \Omega^{\cdot}_{U/\mathbb{C}}, \nabla))$$

(for U affine, these are the cohomology groups of the complex of global sections of the De Rham complex). We have a canonical map

$$\mathbb{H}^i(U, (M \otimes \Omega^{\cdot}_{U/\mathbb{C}}, \nabla)) \longrightarrow \mathbb{H}^i(U^{an}, (M^{an} \otimes \Omega^{\cdot}_{U^{an}/\mathbb{C}}, \nabla^{an}))$$

$$\wr\| \quad \text{(an isomorphism by the holomorphic Poincaré lemma)}$$

$$H^i(U^{an}, (M^{an})^{\nabla})$$

which for $i=0$ is just the map comparing algebraic and analytic global horizontal sections. The "something more" is that this map is an isomorphism for all i. The simplest proof is to remark that for the extension $(\overline{M}, \overline{\nabla})$ we explicitly constructed, (the "quasi-canonical" extension in Deligne's terminology) we have isomorphisms

$$\mathbb{H}^i(X, (\overline{M} \otimes \Omega^{\cdot}_{X/\mathbb{C}}(\log D), \overline{\nabla})) \xrightarrow{\sim} \mathbb{H}^i(U, (M \otimes \Omega^{\cdot}_{U/\mathbb{C}}, \nabla))$$

and

$$\mathbb{H}^i(X^{an}, \overline{M}^{an} \otimes \Omega^{\cdot}_{X^{an}/\mathbb{C}}(\log D), \overline{\nabla}) \xrightarrow{\sim} \mathbb{H}^i(U^{an}, M^{an} \otimes \Omega^{\cdot}_{U/\mathbb{C}}, \nabla).$$

The desired result now follows by applying GAGA to see that the canonical map

$$\mathbb{H}^i(X, (\overline{M} \otimes \Omega^{\cdot}_{X/\mathbb{C}}(\log D), \overline{\nabla})) \longrightarrow \mathbb{H}^i(X^{an}, (\overline{M}^{an} \otimes \Omega^{\cdot}_{X^{an}/\mathbb{C}}, \overline{\nabla}))$$

is an isomorphism.

If we take the trivial representation of $\pi_1(U^{an})$, the corresponding locally constant sheaf is the constant sheaf $\mathbb{C}$ itself, and the corresponding algebraic differential equation with regular singular points is $\mathcal{O}_U$ with the "trivial" connection furnished by exterior differentiation $d: \mathcal{O}_U \longrightarrow \Omega^1_{U/\mathbb{C}}$. The comparison theorem in this special case asserts that the map

$$\begin{array}{ccc} \mathbb{H}^i(U, \Omega^{\cdot}_{U/\mathbb{C}}) & \longrightarrow & \mathbb{H}^i(U^{an}, \Omega^{\cdot}_{U^{an}/\mathbb{C}}) \\ & & \wr\!\downarrow \\ & & H^i(U^{an}, \mathbb{C}) \end{array}$$

is an isomorphism (e.g., on a smooth affine U, closed modulo exact global algebraic differential forms calculate the complex cohomology). This theorem, first proven by Grothendieck, was the starting point for the systematic application of Hironaka's resolution result to studying the cohomology of open varieties (classically, the problem of understanding "integrals of the second kind"). In fact, the first "modern" attempt to study such questions was that of Atiyah and Hodge [1], which essentially concerned itself with the cohomology of a variable Zariski open set U in a fixed projective smooth X. Only after Hironaka's result, though, did it become apparent that one should instead study the cohomology of a fixed open smooth U with the aid of an auxiliary Hironaka compactification $U \hookrightarrow X'$.

The Moral

To an algebraic geometer, it is that so far as non-singular algebraic varieties U over $\mathbb{C}$ are concerned, the apparently topological notions of locally constant sheaf on U^{an}, and of the cohomology of U^{an} with coefficients in such a sheaf, are in fact the purely algebraic notions of differential equation on U with regular singular points, and of the hypercohomology groups of U with coefficients in their de Rham complexes.

What we don't know

Here are three open problems that lie well within the scope of the traditional theory of differential equations. The first of them was raised by Lipman Bers in the course of a stimulating discussion at the Symposium. The second is a favorite of mine. The third was raised by Monsky.

1. Let X^{an} be a compact Riemann surface of genus $g \geq 2$. Then its universal covering is the upper half-plane $\mathfrak{h}$, and X^{an} is obtained as the quotient of $\mathfrak{h}$ by a fuchsian group $G \subset SL(2,R)/\pm 1$; $X^{an} = \mathfrak{h}/G$, and G is the fundamental group $\pi_1(X^{an})$. As an abstract group, $\pi_1(X^{an})$ is generated by $2g$ elements $a_i, b_i, i=1,\ldots,g$, with the single relation $\prod_{i=1}^{g} a_i b_i a_i^{-1} b_i^{-1} = 1$. Viewing X^{an} as $\mathfrak{h}/G$ gives us an embedding of $\pi_1(X^{an})$ in $SL(2,\mathbb{R})/\pm 1$, which lifts in 2^{2g} ways to an embedding of $\pi_1(X^{an})$ in $SL(2,\mathbb{R})$. Thus we have 2^{2g} representations of $\pi_1(X^{an})$ in $SL(2,\mathbb{R})$, which may be realized as follows. There are 2^{2g} choices of line bundle $\mathcal{L}$ on X such that $\mathcal{L}^{\otimes 2} \simeq \Omega^1_{X/\mathbb{C}}$. Each of these allows us to write $H^1(X, \Omega^1_{X/\mathbb{C}}) = H^1(X, \mathcal{L}^{\otimes 2}) = H^1(X, \underline{Hom}(\mathcal{L}^{-1}, \mathcal{L})) = Ext^1(X; \mathcal{L}^{-1}, \mathcal{L})$, under which isomorphism the canonical generator of $H^1(X, \Omega^1_{X/\mathbb{C}})$ ("the cohomology class of a point") gives rise to a rank two vector bundle $M(\mathcal{L})$ which sits in a short exact sequence

$$0 \longrightarrow \mathcal{L} \longrightarrow M(\mathcal{L}) \longrightarrow \mathcal{L}^{-1} \longrightarrow 0$$

It is classical that there is one and only one connection ∇ on $M(\mathcal{L})$ such that the monodromy representation is a real representation; this representation gives an embedding $\pi_1(X^{an}) \longrightarrow SL(2,\mathbb{R})$, and as $\mathcal{L}$ varies we recover the 2^{2g} embeddings described above.

Let X'^{an} be another compact Riemann surface of the same genus $g \geq 2$. Viewing it as $\mathfrak{h}/G'$, where $G' \subset SL(2,\mathbb{R})/\pm 1$ is another fuchsian group, we obtain as before 2^{2g} embeddings of $\pi_1(X'^{an})$ in $SL(2,\mathbb{R})$. But of course the groups $\pi_1(X^{an})$ and $\pi_1(X'^{an})$ are isomorphic as abstract groups, so we could compose with this group-isomorphism to produce 2^{2g} embeddings of $\pi_1(X^{an}) \longrightarrow SL(2,\mathbb{R})$, this time with image in $SL(2,\mathbb{R})/\pm 1$ the fuchsian group corresponding to X' rather than to X. By the affirmative solution of the twenty-first problem, we know that these may be realized as the monodromy representations of certain rank two differential equations

(M_i, ∇_i), $i=1,\ldots,2^{2g}$.

Is it true that ∇_i is the unique connection on M_i such that the monodromy representation is real? Can the bundles M_i that arise in this way (for variable X') be characterized algebraically? Can bundles which admit connections whose monodromy representation is real be characterized algebraically? (The "compact form" of this last question has been answered by Seshadri and Narasimhan, who show that it is precisely the "stable" bundles of degree zero (a purely algebraic notion) which admit connections with irreducible unitary monodromy, and that this unitary connection is unique.) If these questions could be answered affirmatively, one could try to resuscitate the original Poincaré "proof" of uniformization by the "continuity principle."

2. Can we characterize algebraically those differential equations with regular singular points (M, ∇) on an open curve U whose monodromy representations factor through finite groups. This can be done rather strikingly for the hypergeometric equation, but already for second-order equations on $\mathbb{P}^1 - \{4 \text{ or more points}\}$ the question is completely open and extremely interesting [cf. 7]. It should be remarked that this question is treated in Forsythe "in principle," but even in cases when one knows the answer ahead of time, it seems hopeless to ever carry out Forsythe's test procedure.

3. What is the role of equations with irregular singular points. (It would be absurd to ignore the equation $f' - f = 0$, which has an irregular singularity at ∞.) What is the meaning of the cohomology groups $H^i_{DR}(U, (M \otimes \Omega^{\cdot}_{U/\mathbb{C}}, \nabla))$ when (M, ∇) has irregular singular points? Are they finite-dimensional? The finite-dimensionality in the case when U is a curve is known, and due essentially to Birkhoff (his theory of canonical forms for irregular singularities) — Deligne gives a finally not-so-different proof, where he also shows that the algebraic "index" $\dim H^0 - \dim H^1$ is different from its analytic counterpart.

and that this difference measures the irregularity of the singularities.

References

1. Atiyah, M., Hodge, W.: Integrals of the second kind on an algebraic variety. Annals of Math. 62, 56-91 (1955).

2. Deligne, P.: Equations differentielles à points singuliers réguliers. Lecture Notes in Mathematics 163, Berlin-Heidelberg-New York: Springer 1970.

3. Forsythe, A.R.: Theory of Differential Equations, Vol. IV, Cambridge, 1900-1902.

4. Grothendieck, A.: On the de Rham cohomology of algebraic varieties. Publ. Math. I.H.E.S. 29 (1966).

5. Hironaka, H.: Resolution of singularities of an algebraic variety over a field of characteristic zero I, II, Annals of Math. 79, 109-326 (1964).

6. Ince, E.L.: Ordinary Differential Equations, New York, Dover, 1956 (esp. 356-372, 389-393).

7. Katz, N.M.: Algebraic Solutions of Differential Equations, Inventiones Math. 18, 1-118 (1972).

8. Manin, Y.: Moduli Fuchsiani, Ann. Scuola Norm. Sup. Pisa, Ser III, 19 (13-126 (1965).

9. Monsky, P.: Finiteness of de Rham Cohomology, Amer. J. Math., XCIV, 237-245 (1972).

10. Narasimhan, M.S., and Seshadri, C.S.: Stable and unitary vector bundles on a compact Riemann surface. Annals of Math., 82, 540-567 (1965).

11. Serre, J.-P.: Géometrie algébrique et géometrie analytique. Ann. Inst. Fourier Grenoble 6, 1-42 (1956).

Proceedings of Symposia in Pure Mathematics
Volume 28, 1976

ON HILBERT'S 22ND PROBLEM*

Lipman Bers

Was du ererbt von deinen Vätern hast,
erwirb es, um es zu besitzen.
Goethe

INTRODUCTION

A significant mathematical problem, like the uniformization problem which appears as No. 22 on Hilbert's list, is never solved only once. Each generation of mathematicians, as if obeying Goethe's dictum, rethinks and reworks solutions discovered by their predecessors, and fits these solutions into the current conceptual and notational framework. Because of this, proofs of important theorems become, as if by themselves, simpler and easier as time goes by - as Ahlfors observed in his 1938 lecture on uniformization. Also, and this is more important, one discovers that solved problems present further questions.

Classical uniformization theory, developed mainly during the last two decades of the 19th century and the first decade of the 20th, was concerned with proving that every algebraic or analytic curve can be uniformized, that is, represented parametrically by single-valued (or "uniform") functions. The results are summarized in several well known texts (see the Bibliographical Note at the end of the paper), but a scholarly history of that period is yet to be written.

*Research supported in part by the National Science Foundation.

The story is fascinating, first of all because of the wealth of mathematical ideas which arose out of efforts to find a proof: manifold, topological dimension, universal covering space, methods for solving non-linear elliptic partial differential equations, distortion theorems for conformal mappings, and many more.

The personal element is also fascinating. It involves some of the most illustrious mathematicians of that time: Schottky, about to conjecture, in 1875, a fairly general uniformization theorem, but deflected by the authority of Weierstrass (according to Klein), the rivalry between Klein, then at the height of fame and productivity, and the yet unknown Poincaré (see their correspondence from the years 1881-1882), Schwarz, suggesting, in a private communication, two methods for proving the main uniformization theorem (one using universal covering surfaces, the other involving the partial differential equation $\Delta u = e^{2u}$), Hilbert, reviving the interest in the problem by his Paris lecture, and several years later creating a new tool for uniformization by "rehabilitating" the Dirichlet principle, Brouwer, embarking on his epoch making topological investigations in order to put the original "continuity method" of Klein and Poincaré on firm foundations, Poincaré, returning to the uniformization problem after a quarter of a century and finally achieving a full solution, but having to share this honor with Koebe.

Koebe "went on to explore, with the most varied methods, all facets of the uniformization problem". The (slightly rephrased) quotation is from the 1955 edition of Weyl's celebrated _Idee der Riemannschen Fläche_. The appearance of the first edition in 1913 may be thought of as concluding the classical period. The _Idee_ contained, among other things, the concepts, though not the present names, of cohomology and Hausdorff space, and a proof of the uniformization theorem based on an idea of Hilbert.

The modern developments in uniformization, which began after a period of hibernation, utilize quasiconformal mappings and the recent advances in the theory of Kleinian groups. Quasiconformal mappings give new proofs of classical uniformization theorems, akin in spirit though not in technique, to the old "continuity method", and also proofs of new theorems on simultaneous uniformization. The theory of Kleinian groups permits a partially successful attack on the problem of describing all uniformizations of a given algebraic curve. An unexpected application of simultaneous uniformization is Griffiths' uniformization theorem for n-dimensional algebraic varieties, which answers a question also raised by Hilbert in the 22nd problem. (For non-Archimidean valued complete fields there are uniformization theorems due to Tate, for elliptic curves, and to Mumford, but the present author is not competent to report on this work.)

1. RIEMANN SURFACES

1.1 A Riemann surface is a connected 1-dimensional complex manifold. A holomorphic bijection between Riemann surfaces is called a conformal mapping. Uniformization theory exploits the fact that Riemann surfaces can be constructed in three seemingly different ways.

(i) Let $w = a_o + a_1(z-z_o) + a_2(z-z_o)^2 + \dots$ be a convergent non-constant power series. The set of (equivalence classes of) all convergent Puiseux series from which w can be obtained by analytic continuation is, in a natural way, a Riemann surface.

(ii) Let S be an oriented surface, with a Riemannian metric. The surface S is, in a natural way, a Riemann surface.

(iii) Let $\tilde{S}$ be a Riemann surface and G a properly discontinuous group of automorphisms (conformal self-mappings) of $\tilde{S}$. The quotient $\tilde{S}/G$ is, in a natural way, a Riemann surface.

We now discuss these three methods in some detail.

1.2 Let $\hat{\mathbb{C}} = \mathbb{C} \cup \{\infty\}$ be the Riemann sphere. A formal Puiseux series with center $z_o \in \hat{\mathbb{C}}$ is a series of the form

$$(1) \qquad \sum_{n=m}^{\infty} a_n (z-z_o)^{n/\nu} \text{ if } z_o \neq \infty, \quad \sum_{n=m}^{\infty} a_n z^{-n/\nu} \text{ if } z_o = \infty .$$

Here the a_n are complex numbers, $a_m \neq 0$, ν (the ramification index) is a positive integer, and it is assumed that one cannot, by omitting terms with $a_n = 0$, rewrite the series with a smaller value of ν.

The value w_o of the series (at z_o) is defined as a_o if $m = 0$, as 0 if $m > 0$ and as ∞ if $m < 0$.

The series is called convergent if the power series $a_o + a_1 t + a_2 t^2 + \ldots$ has a positive radius of convergence, non-constant if either $m \neq 0$ or $m = 0$ and there is a $\mu > m$ with $a_\mu \neq 0$.

If $z_1 \in \mathbb{C}$ is distinct from but close to the center z_o of a convergent Puiseux series, the latter induces, at z_1, ν distinct germs of holomorphic functions, obtained by choosing, near z_1, a single-valued branch of $(z-z_o)^{1/\nu}$ and then summing the resulting series.

Two Puiseux series with index ν and center z_o, and with coefficients a_n and $\hat{a}_n$, respectively, are called <u>equivalent</u> if $\hat{a}_n = \omega^n a_n$ with $\omega^\nu = 1$. A <u>function element</u> θ is an equivalence class of convergent non-constant Puiseaux series. A regular function element, i.e., one with $\nu = 1$ and $m \geq 0$, is simply a germ of a non-constant holomorphic function at z_o.

The set $\mathfrak{W}$ of all function elements is topologized by calling a subset open if, whenever it contains a function-element θ, it also contains all (regular) function elements induced by θ at points sufficiently close to the center of θ. A neighborhood of θ can be parametrized by a complex number t (<u>distinguished local parameter</u>) as follows. Let θ be defined by a series (1). Assign to $t = 0$ the function element $\theta_o = \theta$, and to $t \neq 0$, $|t|$ small, the regular function element θ_t, the germ at z_1 of the holomorphic function $w = \varphi(z)$, where $z_1 = z_o + t_1^{\nu}$ (if $z_o \neq \infty$) or $z_1 = t_1^{-\nu}$ (if $z_o = \infty$), and the function $w = \varphi(z)$ admits, for small values of $|z-z_1|$, the expansion

$$\varphi(z) = \sum_{n=m}^{\infty} a_n [t_1^{\nu} + (z - z_1)]^{n/\nu} = \sum_{j=0}^{\infty} b_j (z - z_1)^j$$

where

$$b_j = t^{-j\nu} \sum_{n=m}^{\infty} \binom{n/\nu}{j} a_n t^n \qquad (\text{if } z_o \neq \infty)$$

or

$$\varphi(z) = \sum_{n=m}^{\infty} a_n [t^{-\nu} + (z - z_1)]^{-n/\nu} = \sum_{j=0}^{\infty} b_j (z - z_1)^j$$

where

$$b_j = t^{j\nu} \sum_{n=m}^{\infty} \binom{-n/\nu}{j} a_n t^n \qquad (\text{if } z_o = \infty).$$

Every component S of $\mathfrak{W}$ is made into a Riemann surface by calling a function f, defined near $\theta \in S$, holomorphic if it is so when considered as a function of a distinguished local parameter belonging to θ. Note that every component S comes equipped with two meromorphic functions - the value and the center of the function elements belonging to S.

1.3 It turns out that <u>every</u> Riemann surface is conformally equivalent to a component S of $\mathfrak{W}$. Furthermore, S is compact, if and only if it consists of all function elements satisfying (in an obvious sense of the word) an irreducible algebraic equation $P(z,w) = 0$. The function elements of S are in a canonical one-to-one correspondence with the places of the algebraic curve C defined by P. The genus of C is the topological genus of S. Two algebraic curves are birationally equivalent if and only if their Riemann surfaces are conformally equivalent.

If a component S of $\mathfrak{W}$ is not compact, then the points of S are, by definition, the places on an analytic curve.

1.4 A Riemannian metric on an oriented differentiable surface S is defined by a line element

$$ds^2 = g_{11}dx^2 + 2g_{12}dxdy + g_{22}dy^2 \quad ;$$

here (x,y) are any local coordinates defined in a coordinate patch, the g_{ij} are measurable functions of (x,y), which transform like the components of a symmetric covariant tensor, and the matrix (g_{ij}) is positive definite a.e..

Local coordinates (x,y) are called <u>isothermal</u> if $g_{11} = g_{22}$, $g_{12} = 0$. One can find isothermal coordinates near a given point of S, provided the three functions $g_{ij}(x,y)$ are real-analytic (Gauss, 1843) or Hölder continuous (Korn, 1914, Lichtenstein, 1916) or even only measurable and satisfying a.e. an inequality of the form $(g_{11} + g_{22})^2/(g_{11}g_{22} - g_{12}^2) < \text{const.}$ (Morrey, 1938). The geometric meaning of this inequality is that the eccentricity of the indicatrix of the metric is bounded. We return to this in §5.

An oriented surface with a Riemannian metric is made into a Riemann surface by calling a function defined near a point P holomorphic if it is so as a function of $z = x + iy$, where (x,y) are isothermal coordinates defined near P.

1.5 On a given Riemann surface there are Riemannian metrics compatible with the conformal structure, that is, such that the real and imaginary part of a local parameter (conformal mapping of a domain on a surface into $\mathbb{C}$) are isothermal coordinates. One can even demand that the metric be induced by an embedding of the surface in R^3. This was shown by Garsia for compact surfaces and by Ruedy in all cases.

1.6 On every Riemann surface S there exists a complete Riemannian metric of constant Gaussian curvature, compatible with the conformal structure. This metric is canonical, except for a constant factor. The curvature is positive if S is homeomorphic to a sphere, zero if S is homeomorphic to a torus or conformally equivalent to either $\mathbb{C}$ or to $\mathbb{C} \setminus \{0\}$, and negative in all other cases.

The statement just made is equivalent to the "main case" of the limit circle theorem stated below, see 2.1.

1.7 A group G of topological self-mappings of a Hausdorff space X is called properly discontinuous if, for every compact set $K \subset X$, $g(K) \cap K = \emptyset$ for all but finitely many $g \in G$. If so, X/G is again a Hausdorff space.

Now let $\tilde{S}$ be a Riemann surface, G a properly discontinuous group of conformal self-mappings of $\tilde{S}$, and $f: \tilde{S} \to \tilde{S}/G$ the canonical mapping. The quotient $S = \tilde{S}/G$ is made into a Riemann surface by calling a function φ defined in a domain $D \subset S$ holomorphic if $\varphi \circ f$ is holomorphic on every component of $f^{-1}(D)$. The mapping f is a Galois covering if G acts freely (that is, without fixed points), otherwise f is a ramified (Galois) covering, with the following properties. There is a discrete set of points $\{P_j\} \subset S$ and a sequence of integers (called ramification numbers) $\nu_j > 1$, such that, for each j, $f^{-1}(P_j)$ is discrete and, near every point of $f^{-1}(P_j)$, the mapping f is ν_j-to-one. The restriction of f to the complement of $f^{-1}(\{P_j\})$ is an unramified Galois covering.

1.8 Suppose now that $\tilde{S} \subset \hat{\mathbb{C}}$, that is that $\tilde{S}$ is a domain in the extended complex plane. Then every meromorphic function $\Phi(\zeta)$, $\zeta \in \tilde{S}$, which is automorphic for the group G, that is, satisfies $\Phi(g(\zeta)) = \Phi(\zeta)$ for all $\zeta \in \tilde{S}$ and all $g \in G$, induces a meromorphic function φ on S such that $\varphi \circ f = \Phi$. Furthermore, all meromorphic functions on S are so obtained.

In particular, if S is (conformally equivalent to) a component of $\mathfrak{W}$, the center and the value of $\theta \in S$ become, on $\tilde{S}$, single-valued automorphic meromorphic functions, say $\tilde{z}(\zeta)$, $\tilde{w}(\zeta)$, and $z = \tilde{z}(\zeta)$, $w = \tilde{w}(\zeta)$, $\zeta \in \tilde{S}$ is a parametric representation, by single-valued functions, of the algebraic or analytic curve corresponding to $S \subset \mathfrak{W}$. Hence: to uniformize an algebraic or analytic relationship between two variables it suffices to <u>represent the corresponding Riemann surface as the quotient of a plane domain by a properly discontinuous group of conformal mappings</u>.

2. FUCHSIAN GROUPS

2.1 It turns out that <u>every</u> Riemann surface admits a representation as $\tilde{S}/G$ where either $\tilde{S} = \hat{\mathbb{C}}$ or $\tilde{S} = \mathbb{C}$ or $\tilde{S} = U$ (U denotes here and hereafter the upper half-plane $\{z \in \mathbb{C}, \operatorname{Im} z = y > 0\}$) and G acts freely. The group G is a group of Möbius transformations

$$z \to \frac{az + b}{cz + d} \qquad (ad - bc = 1) \quad ,$$

with real coefficients in the case $\tilde{S} = U$, and is determined by S, up to conjugation. Furthermore, $\tilde{S} = U$ except if $S = \hat{\mathbb{C}}$ (then $\tilde{S} = \hat{\mathbb{C}}$, $G = 1$) or $S = \mathbb{C}$, or $\mathbb{C} \setminus \{0\}$, or S is homeomorphic to a torus (then $\tilde{S} = \mathbb{C}$).

This is the main special case of the Poincaré-Klein <u>limit circle theorem</u>, conjectured by Klein and by Poincaré in 1882 but completely proved only in 1907, by Poincaré and by Koebe.

The usual proof proceeds in two steps. First one constructs the universal covering surface $\tilde{S}$ of S and the covering group G, so that $S = \tilde{S}/G$. Then one transfers to $\tilde{S}$ the complex structure of S. Now $\tilde{S}$ becomes a Riemann surface, and the covering group G, which is by construction freely acting and properly discontinuous, becomes a group of conformal mappings. Since $\tilde{S}$ is simply connected, it remains to show that every simply connected Riemann surface is conformally equivalent to $\hat{\mathbb{C}}$, $\mathbb{C}$ or U. Once this <u>generalized Riemann mapping theorem</u> is established, the remaining assertions follow easily.

The second, more difficult, step is the proof of the mapping theorem. Today an argument based on Perron's method of subharmonic functions and on an idea of M. Heins is standard, see the presentation in Ahlfors' <u>Conformal Invariants</u>.

The canonical Riemannian metric on S (see 1.6) is obtained by transferring to $S = \tilde{S}/G$ the canonical metric on $\tilde{S}$, that is, either the metric on $\hat{\mathbb{C}}$ obtained by stereographic projection from the standard metric on the surface of the unit ball in $\mathbb{R}^3$, or the Euclidean metric on $\mathbb{C}$, or the <u>Poincaré metric</u> $|dz|/\text{Im } z$ on U.

2.2 The <u>full version</u> of the limit circle theorem asserts that given a Riemann surface S, a discrete set of points $\Sigma \subset S$ and, for every $P \in \Sigma$, an integer $\nu(P) > 1$, there is a representation $S = \tilde{S}/G$ with $\tilde{S} = \hat{\mathbb{C}}$, $\mathbb{C}$ or U, the mapping $f: \tilde{S} \to \tilde{S}/G$ being ramified over each $P \in \Sigma$ of order $\nu(P)$, and over no other point.

(There are two exceptions: the cases $S = \hat{\mathbb{C}}$, $\Sigma = \{P\}$, and $S = \hat{\mathbb{C}}$, $\Sigma = \{P,Q\}$, $\nu(P) \neq \nu(Q)$ are not permitted.)

The proof consists of constructing the desired covering $\tilde{S} \to S$ topologically, verifying that $\tilde{S}$ is simply connected, giving $\tilde{S}$ a complex structure lifted from S, and then invoking the generalized Riemann mapping theorem.

Again, G is determined up to conjugation, and, except in a few easily enumerable cases, $\tilde{S} = U$.

2.3 Whenever $\tilde{S} = U$, G is a <u>Fuchsian group</u>, that is a properly discontinuous group of non-Euclidean motions (real Möbius transformations). Of course, one can replace U by another half-plane or by a disc, for instance, by the unit disc with the metric $ds = 2|dz|(1 - |z|^2)^{-1}$.

The group G is called <u>of the first kind</u> if it does not act properly discontinuously on any open subset of $\hat{\mathbb{R}} = \mathbb{R} \cup \{\infty\}$ (or of $|z| = 1$, if U is replaced by the unit disc). This happens if and only if $S' = S \setminus \{\text{ramification points}\}$ has no ideal boundary curves, that is, cannot be represented as $S' \subset S''$, S'' another Riemann surface, with the boundary of S' relative to S containing Jordan arcs.

If G is a Fuchsian group of the first kind, and $\varphi(z)$, $z \in U$, a meromorphic automorphic function for G, φ is singular at every boundary point of U. Thus the boundary of U is a natural boundary or "limit circle" for φ. Hence the name of the theorem.

3. THE GENERAL UNIFORMIZATION THEOREM

3.1 Koebe's _general_ _uniformization_ _theorem_ asserts that a Riemann surface $\tilde{S}$ topologically equivalent to a domain in $\hat{\mathbb{C}}$ is also conformally equivalent to such a domain.

The topological condition is known to be equivalent to the following: every Jordan curve separates the surface. Surfaces satisfying the condition are said to be of _genus_ 0.

3.2 Suppose now that we want to find _all_ ways of representing a given Riemann surface S as the quotient of a plane domain by a properly discontinuous group of conformal mappings. The general uniformization theorem seems to reduce the problem to one in topology: to find all ramified Galois coverings $\tilde{S} \to S$ where the surface $\tilde{S}$ has genus 0.

Indeed, given the covering $\tilde{S} \to S$, we can make $\tilde{S}$ into a Riemann surface, as we did before for the universal covering surface, and by Koebe's theorem $\tilde{S}$ is conformally equivalent to a domain in $\hat{\mathbb{C}}$.

3.3 The topological problem is solved by Maskit's _planarity_ _theorem_.

Recall that an unramified Galois covering $\tilde{S} \to S$ is determined, up to equivalence, by a normal subgroup N of the fundamental group $\pi(S, P_0)$; a curve in S, beginning and ending at P_0, lifts to a closed curve in $\tilde{S}$ if and only if "it" (that is, the element of the fundamental group determined by it) belongs to the defining subgroup N.

The planarity theorem asserts that $\tilde{S}$ is of genus 0 if and only if there are disjoint simple closed homotopically non-trivial curves $\alpha_1, \alpha_2, \ldots,$ no two of which are freely homotopic, and positive integers $\nu_1, \nu_2, \ldots,$ such that N is the smallest normal subgroup of $\pi(S)$ containing all elements determined by $\alpha_1^{\nu_1}, \alpha_2^{\nu_2}, \dots$.

The planarity theorem also contains a recipe for finding all ramified planar Galois coverings of S. Choose a discrete set $P \subset \Sigma$ and construct an unramified Galois covering $\tilde{S}_o \to S \setminus \Sigma$, with $\tilde{S}_o$ of genus 0, defined by a group N which contains, for each $P \in \Sigma$, the element defined by a small simple loop around P raised to some power $\nu(P) > 1$. The covering so constructed is the restriction of a ramified Galois covering $\tilde{S} \to S$, with ramification index $\nu(P)$ over each $P \in \Sigma$, and $\tilde{S}$ is of genus 0.

3.4 As an example of applying Koebe's uniformization principle and the planarity theorem, consider a closed Riemann surface S of genus $p > 0$, and p mutually disjoint smooth Jordan curves $\gamma_1, \dots, \gamma_p$ on S, such that $S \setminus \{\gamma_1 \cup \dots \cup \gamma_p\}$ is homeomorphic to a sphere with $2p$ punctures. Let N be the smallest normal subgroup "containing" $\gamma_1, \dots, \gamma_p$ and $\tilde{S} \to S$ the corresponding covering. Then $\tilde{S}$ is of genus 0, so that we may assume that $\tilde{S} \subset \hat{\mathbb{C}}$.

One can show that $\tilde{S}$ is dense in $\hat{\mathbb{C}}$, and that $\hat{\mathbb{C}} \setminus \tilde{S}$ is so "small" that every holomorphic injection $\tilde{S} \to \hat{\mathbb{C}}$ is given by a Möbius transformation. Hence the covering group G, which consists of all topological self-mappings of $\tilde{S}$ which respect the covering (and are therefore conformal), is a group of Möbius transformations.

This G is called a Schottky group of genus p. It contains, except for the identity, only loxodromic (including hyperbolic) elements, and is a free group on p generators.

Every component of the inverse image of $S \setminus \{\gamma_1 \cup \dots \cup \gamma_p\}$ in $\tilde{S} \subset \hat{\mathbb{C}}$ is a domain bounded by $2p$ Jordan curves $C_1, C_1', \dots, C_p, C_p'$ and there are Möbius transformations $g_1, \dots, g_p$

such that g_j maps the domain interior to C_j onto the domain exterior to C'_j. Conversely, given 2p such curves and p such Möbius transformations, the group G generated by them is a Schottky group of genus p.

3.5 The p generators of a Schottky group G depend on 3p complex parameters. A conjugation by a Möbius transformation A leads to a Schottky group AGA^{-1}, distinct from G if $p > 1$. Since A depends on 3 parameters, and since G and AGA^{-1} represent conformally equivalent Riemann surfaces, one may conclude that, as asserted by Riemann, the conformal structure of a surface of genus $p > 1$ depends on $3p-3$ complex "moduli." (If $p = 1$, the number of moduli is 1.)

3.6 We return to the general case; $S = \tilde{S}/G$ with $\tilde{S} \subset \hat{\mathbb{C}}$. There is no reason to expect the elements of G to be Möbius transformations. But an important theorem of Maskit asserts that one can always find a conformal injection $f: \tilde{S} \to \hat{\mathbb{C}}$ such that <u>every conformal self-mapping of</u> $\tilde{S}_1 = f(\tilde{S})$ <u>is a Möbius transformation</u>. In particular $G_1 = fGf^{-1}$ is a group of Möbius transformations and S is conformally equivalent to $\tilde{S}_1/G_1$.

However, this f is not uniquely determined, and a given topological planar covering leads, in general, to infinitely many distinct uniformizations. This is so, as we shall see later, even when $\tilde{S}$ is simply connected and even if $\tilde{S} \to S$ is the universal covering.

Maskit's theorem shows that the founders of uniformization theory did not lose much generality by considering only uniformizations by so-called Kleinian groups.

4. KLEINIAN GROUPS

4.1 A Kleinian group G is a group of Möbius transformations which acts properly discontinuously on some non-empty open set in $\hat{\mathbb{C}}$. The largest open set $\Omega = \Omega(G)$ on which G so acts is called the region of discontinuity of G. The limit set of G is the complement $\Lambda(G) = \hat{\mathbb{C}} \setminus \Omega(G)$, it either contains 0, 1 or 2 points, or is a perfect nowhere dense set of positive logarithmic capacity.

If Λ is finite, G is called elementary. Such groups are easily enumerated.

A component Δ of Ω is called, by abuse of language, a component of G. Two distinct components, Δ and Δ_1, are called conjugate if $\Delta_1 = g(\Delta)$, $g \in G$. The stabilizer G_Δ of Δ in G is again a Kleinian group with $\Delta \subset \Omega(G_\Delta)$; if $G_\Delta = G$, Δ is called invariant.

The quotient Ω/G can be identified with the disjoint union of the Riemann surfaces Δ_i/G_{Δ_i} where $\Delta_1, \Delta_2, \ldots$ is a maximal list of non-conjugate components of G. If this list is finite, if each Δ_i/G_{Δ_i} is a compact surface, with perhaps finitely many punctures, and if the covering $\Omega \to \Omega/G$ is ramified over at most finitely many points, G is said to be of finite type.

A finitely generated Kleinain group is of finite type. This is Ahlfors' finiteness theorem. But there are also infinitely generated groups of finite type.

A Kleinian group which leaves a disc or a half plane fixed is Fuchsian. The names "Fuchsian" and "Kleinian" are due to Poincaré. Klein was vehemently opposed to this terminology.

4.2 Every Möbius transformation may be considered, in a natural way, as a motion in non-Euclidean space. More precisely, let this space be realized as the set of all quaternions $Z = x + iy + jt = z + jt$, $z \in \mathbb{C}$, $t > 0$, with the Poincaré metric $ds^2 = (|dz|^2 + dt^2)/t$. A Möbius transformation

$$z \to \frac{az + b}{cz + d}, \quad a,b,c,d \in \mathbb{C}, \ ad - bc = 1$$

is the "trace" on $t = 0$ of the non-Euclidean motion

$$Z \to (aZ + b)(cZ + d)^{-1} .$$

A Kleinian group is called <u>geometrically finite</u> if, considered as a group of non-Euclidean motions, it has a fundamental polyhedron with finitely many sides.

Such a group is always finitely generated. But there are finitely generated Kleinian groups which are <u>not</u> geometrically finite, see §9.

Ahlfors showed that if G is geometrically finite, $\Lambda(G)$ has <u>measure zero</u>, and asked whether this is so for all finitely generated groups.

4.3 We now restate, in a somewhat restricted fashion, the problem formulated in 3.2: given a Riemann surface S, find all representations of S in the form $S = \Delta/G$ where G is a Kleinian group and Δ an invariant component of G.

If S is compact, but for a finite number of punctures, and the covering $\Delta \to \Delta/G$ is allowed to be ramified over at most finitely many points, partial answers are known. Thus one "knows" all G for which Δ is a Jordan domain (such G are called <u>quasi-Fuchsian</u>) and even all G for which Δ is simply connected. Our knowledge unfortunately includes a purely existential statement about so-called <u>degenerate groups</u> (see below §9.3 and §9.9).

At the Vancouver Congress Maskit announced a complete answer under the additional hypothesis that G is geometrically finite. It turns out that all such G can be built up from elementary groups, quasi-Fuchsian groups and Schottky groups.

4.4 One can also consider uniformizations of S of the form $S = \Delta/G_\Delta$ where G_Δ is the stabilizer of the component Δ in a Kleinian group G which has <u>no</u> invariant components. Such uniformizations are an essential tool in studying moduli of Riemann surfaces with nodes. We shall not discuss them further.

5. QUASICONFORMAL MAPPINGS

5.1 A *quasiconformal mapping* $z \to w(z)$ between two plane domains is a topological orientation preserving mapping which is conformal with respect to some Riemannian metrics of uniformly bounded eccentricity, $ds^2 = g_{11}dx^2 + 2g_{12}dxdy + g_{22}dy^2$ with $(g_{11} + g_{22})^2/(g_{11}g_{22} - g_{12}^2) \leq \text{const.}$ This means: $z \to w(z)$ preserves angles if one measures angles in the z-plane via ds^2, in the w-plane via the Euclidean metric.

It is convenient to use complex notations; then

$$ds^2 = \rho^2 |dz + \mu d\bar{z}|^2$$

where ρ is a real-valued function and

$$\mu(z) = \frac{g_{11} - g_{22} + 2ig_{12}}{g_{11} + g_{22} + 2(g_{11}g_{22}-g_{12}^2)^{1/2}} \tag{1}$$

a complex-valued one. The condition of bounded eccentricity translates into

$$\|\mu\|_\infty \leq k < 1 \tag{2}$$

where $\| \|_\infty$ is the usual L_∞ norm (essential supremum of $|\mu(z)|$ in the domain considered).

If μ and w are sufficiently smooth, the condition that w be ds^2-conformal (or, as we shall say, μ-conformal) is seen to be equivalent to the Beltrami differential equation

$$\frac{\partial w}{\partial \bar{z}} = \mu \frac{\partial w}{\partial z} \tag{3}$$

where, as usually, $2\partial/\partial z = \partial/\partial x - i\partial/\partial y$, $2\partial/\partial \bar{z} = \partial/\partial x + i\partial/\partial y$. If μ is only measurable, a topological mapping will be considered μ-conformal if it satisfies (3) in the following sense: the partial derivatives of w, in the sense of distribution theory, are measurable, locally square-integrable functions and (3) holds almost everywhere.

The extension of the theory of Beltrami equations to the case of bounded measurable Beltrami coefficients μ, accomplished first by Morrey, is basic for the application to uniformization.

5.2 It turns out that homeomorphic solutions of Beltrami equations, in the generalized sense defined above, have all desirable properties. They preserve orientation and take measurable sets into measurable sets; the area of an image is given by the classical formula. Their inverses and their composites are solutions of (other) Beltrami equations, and the partial derivatives of inverse mappings and of composed mappings can be computed by the classical formulas. Also, as first noted by Boyarskii, the partial derivatives of a solution w of (3) are locally in L_p with some $p > 2$, depending on k.

5.3 If μ is defined in all of $\mathbb{C}$, then there is a unique solution $w^\mu(z)$ of (3) which maps $\mathbb{C}$ topologically onto itself and keeps the points 0 and 1 fixed.

The uniqueness follows trivially from what was said above, since if w and $\hat{w}$ are two μ-conformal automorphisms of $\mathbb{C}$, $\hat{w}\circ w^{-1}$ is a conformal automorphism.

An existence proof based on the Calderon-Zygmund inequality for the complex Hilbert transform

$$f(z) \to \frac{1}{2\pi}\iint_{\mathbb{C}} \frac{f(\zeta)}{(\zeta - z)^2}\, d\zeta \wedge d\bar{\zeta}$$

was given by Ahlfors and Bers. Their paper also discusses the dependence of w^{μ} on μ. If $\|\mu_j\|_\infty \leq k < 1$ and $\mu_j(z) \to \mu(z)$ a.e., then $w^{\mu_j} \to w^{\mu}$ uniformly on compact sets. If μ, as an element of $L_\infty(\mathbb{C})$, depends holomorphically on one or several complex parameters, so does $w^{\mu}(z)$, for every $z \in \mathbb{C}$.

5.4 A topological mapping between two Riemann surfaces, $f\colon S \to S'$, is called <u>quasiconformal</u> if, near every $P \in S$, it is quasiconformal, with a uniform bound on the eccentricity, when considered in terms of local parameters z defined near P and z' defined near P'. The expression

$$\mu(z)\,\frac{d\bar{z}}{dz} = \frac{\partial z'/\partial \bar{z}}{\partial z'/\partial z}\,\frac{d\bar{z}}{dz}$$

is independent of the choice of parameters and is called the <u>Beltrami differential</u> of the mapping f. (It is, actually, a measurable form of type $(0,1)$ with coefficients in the holomorphic tangent bundle to S.) The modulus $|\mu(z)|$ is also independent of the choice of parameters. The essential supremum of this modulus is called the <u>norm</u> of the Beltrami differential.

Quasiconformal equivalence of Riemann surfaces is weaker than conformal equivalence (any two compact Riemann surfaces of the same genus are quasiconformally equivalent) but stronger than topological equivalence (the disc is not quasiconformally equivalent to the plane).

5.5 We explain how to use quasiconformal mappings for uniformization by considering first an almost trivial example. We want to show that every Riemann surface homeomorphic to a torus can be represented as $\mathbb{C}$ factored by a group generated by two (Euclidean) translations, $z \to z + 1$ and $z \to z + \tau$, $\operatorname{Im} \tau > 0$.

Let G be generated by $z \to z + 1$, $z \to z + i$, and set $S = \mathbb{C}/G$; then S is homeomorphic to a torus. If S' is a given Riemann surface homeomorphic to S, there is an orientation preserving diffeomorphism $f: S \to S'$; this f is a quasiconformal mapping. Let ds' be some Riemannian metric on S' which respects its conformal structure, for instance the Poincaré metric on S'. Let ds be the metric ds' pulled back by f. (Then $ds = \rho\,|d\zeta + \mu d\bar{\zeta}|$ where ρ is a positive function and $\mu d\bar{\zeta}/d\zeta$ the Beltrami differential of the mapping f, see 5.4.) The metric ds, pulled back to $\mathbb{C}$ by the canonical projection $\mathbb{C} \to \mathbb{C}/G = S$ gives rise to a Beltrami coefficient $\mu(z)$ which satisfies

$$\mu(z + 1) = \mu(z + i) = \mu(z) \quad . \tag{4}$$

Let w^μ be the mapping discussed in 5.3. The mappings $z \to w^\mu(z+1)$ and $z \to w^\mu(z+i)$ are also μ-conformal automorphisms of $\mathbb{C}$, as one verifies at once using (3) and (4). Hence there are constants $\alpha, \beta, \sigma, \tau$ such that

$$w^\mu(z+1) = \alpha w^\mu(z) + \sigma \,, \quad w^\mu(z+i) = \beta w^\mu(z) + \tau \quad .$$

The group $G^\mu = w^\mu G (w^\mu)^{-1}$ must be fixed point free and properly discontinuous. This group is generated by $z \to \alpha z + \sigma$, $z \to \beta z + \tau$. Hence $\alpha = \beta = 1$ (otherwise there would be fixed points) and

τ/σ is not real (otherwise G^{μ} could not be discrete). Setting $z = 0$ in the identity $w^{\mu}(z+1) = w^{\mu}(z) + \sigma$ we obtain that $\sigma = 1$, so that $\operatorname{Im} \tau \neq 0$. Setting $z = 0$ in the identity $w^{\mu}(z+1) = w^{\mu}(z) + \tau$ we obtain that $\tau = w^{\mu}(i)$.

Now let t be a complex number with $|t| < 1/\|\mu\|_{\infty}$. By what was said in 5.4, $w^{t\mu}(i)$ is a holomorphic function of t, and it is never real. For $t = 0$, however, $w^{t\mu}(i) = i$. Hence $\operatorname{Im} w^{t\mu}(i) > 0$ for all t considered. For $t = 1$, we obtain that $\operatorname{Im} \tau > 0$.

Now, the mapping $w^{\mu}: \mathbb{C} \to \mathbb{C}$ projects to a mapping $f^{\mu}: \mathbb{C}/G \to \mathbb{C}/G^{\mu}$. We define the mapping h by the commutativity of the diagram

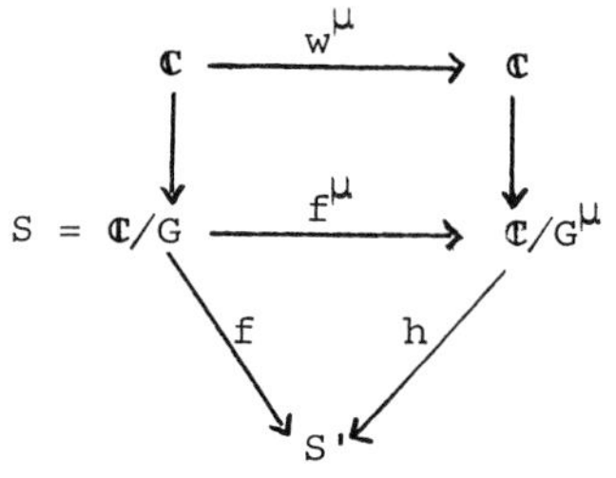

(Here and hereafter vertical arrows denote canonical projections.) Now f and f^{μ} have the same Beltrami coefficient, by construction. Hence h is a conformal mapping. Since G is generated by $z \to z + 1$, $z \to z + \tau$, we proved what we set out to prove.

6. QUASICONFORMAL DEFORMATIONS OF KLEINIAN GROUPS

6.1 The example considered above leads to a general method. If G is a Kleinian group, a Beltrami coefficient μ for G is an element of $L_{\infty}(\mathbb{C})$ satisfying

(1) $$\|\mu\|_{\infty} < 1 ,$$

(2) $$\mu(g(z))\overline{g'(z)}/g'(z) = \mu(z) \quad \text{for} \quad g \in G \quad ,$$

and, unless the limit set $\Lambda(G)$ has measure zero,

(3) $$\mu \mid \Lambda = 0 \quad .$$

A Beltrami coefficient for G induces a Beltrami differential of norm not exceeding $\|\mu\|_\infty$ on every component of Ω/G. Conversely, given Beltrami differentials with norms not exceeding some number $k < 1$, on all components of Ω/G, they are induced by a unique Beltrami coefficient μ for G, with $\|\mu\|_\infty \leq k$. Note that even when the Beltrami differentials on the components of Ω/G are smooth, the Beltrami coefficient μ will have bad discontinuities on the limit set Λ.

If μ is a Beltrami coefficient for a Kleinian group G, the group

$$G^\mu = w^\mu G(w^\mu)^{-1}$$

is Kleinian, with region of discontinuity

$$w^\mu(\Omega(G)) = \Omega(G^\mu)$$

and, for every $g \in G$, the element

$$g^\mu = w^\mu \circ g \circ (w^\mu)^{-1} \in G^\mu$$

is loxodromic (including hyperbolic), or parabolic, or elliptic of order ν, if and only if g is.

The only part of this statement requiring proof is the assertion that g^μ is a Möbius transformation. To show this, set $W = w^\mu \circ g$ and note that W is a topological automorphism of $\hat{\mathbb{C}}$ and that, in view of (2), W satisfied the same Beltrami equation as w^μ. Hence $W = \hat{g} \circ w^\mu$ where $\hat{g}$ is a topological automorphism of $\hat{\mathbb{C}}$ and conformal. Thus $\hat{g}$ is a Möbius transformation. But $\hat{g} = g^\mu$.

The isomorphism $g \to g^\mu$ of G onto G^μ is called a *quasiconformal deformation* of G.

There is a commutative diagram

$$\begin{array}{ccc} \Omega(G) & \xrightarrow{w^\mu} & \Omega(G^\mu) \\ \downarrow & & \downarrow \\ \Omega(G)/G & \xrightarrow{f^\mu} & \Omega(G^\mu)/G^\mu \end{array}$$

where, for every component S of $\Omega(G)/G$, $f^\mu(S)$ is a component of $\Omega(G^\mu)/G^\mu$ and

$$f^\mu \mid S\colon S \to f^\mu(S)$$

is a quasiconformal mapping with a Beltrami differential induced by μ. If $f\colon S \to S'$ is any other quasiconformal mapping with the same Beltrami differential, $h = (f^\mu \mid S)\circ f^{-1}$ is a conformal bijection of S' onto $f^\mu(S)$.

Thus, if we are given a sequence of disjoint Riemann surfaces $S_1, S_2, \ldots$, a Kleinian group G uniformizing them, that is a conformal bijection

$$\Omega/G \overset{\approx}{\to} S_1 + S_2 + \ldots ,$$

and quasiconformal mappings

$$S_1 \to S_1', \; S_2 \to S_2', \; \ldots$$

with uniformly bounded eccentricities, there is a quasiconformal deformation $G \to G^\mu$, with G^μ uniformizing $S_1', S_2', \ldots$, which induces the given mappings.

In other words, *every example of a Kleinian group gives rise to a uniformization theorem for a sequence of Riemann surfaces*. The theorem becomes particularly useful when the existence of quasiconformal mappings $S_i \to S_i'$ with uniformly bounded eccentricities can be asserted rather than assumed.

This is so whenever G is of finite type, for then every orientation preserving homeomorphism of one of the (finitely many) components of Ω/G is homotopic to a quasiconformal mapping.

The device described above does not yield uniqueness theorems; such theorems have to be proved, whenever possible, by different methods.

6.2 The method can be applied to classical uniformization theorems. We shall give some examples.

Choose $2p > 0$ circles $C_1, C_1', \ldots, C_p, C_p'$, each lying outside all others, and let g_j, $j = 1,\ldots,p$, be the Möbius transformation which maps the exterior of C_j onto the interior of C_j'. The group G generated by $g_1,\ldots,g_p$ is a Schottky group and $\Omega(G)/G$ is the domain bounded by $C_1,\ldots,C_p'$, with the boundary curves pairwise identified, i.e., a closed Riemann surface of genus p. A given closed Riemann surface S of that genus can be mapped quasiconformally on S (and we may prescribe which loops should be mapped onto the images of $C_1,\ldots,C_p$). A quasiconformal deformation of a Schottky group is a Schottky group. Hence <u>every</u> closed Riemann surface is uniformizable by a Schottky group.

6.3 Let G be a finitely generated Fuchsian group. The <u>signature</u>

$$\{p,n;\ \nu_1,\ldots,\nu_n,\ m\} \tag{4}$$

of G is defined as follows. The numbers p,n and m are non-negative integers; each ν_j is an integer or the symbol ∞ and $2 \le \nu_1 \le \nu_2 \le \cdots \le \nu_n \le \infty$; $S = U/G$ is a compact surface of genus p, from which one has removed $n_\infty + m$ disjoint continua, precisely n_∞ of which are points; here n_∞ is the number of times ∞ appears in (4). The mapping $U \to U/G$ is ramified over $n - n_\infty$ points, with ramification numbers $\nu_1, \ldots, \nu_{n-n_\infty}$.

There is a geometric construction, due to Poincaré, which gives examples of Fuchsian groups for all signatures such that either $m > 0$ or $m = 0$ and

(5) $$2p - 2 + \sum_{j=1}^{n} \left(1 - \frac{1}{\nu_j}\right) > 0 \text{ , where } \frac{1}{\infty} = 0 \quad .$$

The limit circle theorem (see §2) asserts that there is a Fuchsian group with signature (4) representing a <u>given</u> Riemann surface S', compact of genus p except for n_∞ punctures and m removed non-degenerate continua, with $n - n_\infty$ prescribed ramification points with ramification numbers $\nu_1, \ldots, \nu_{n-n_\infty}$.

To prove this by quasiconformal mappings, construct a quasiconformal mapping $S \to S'$ which takes ramification points into ramification points, lift the Beltrami differential of this mapping to U to obtain a Beltrami coefficient $\mu(z)$ and then extend it to all of $\mathbb{C}$ by setting

(6) $$\mu(\bar{z}) = \overline{\mu(z)}$$

(the real values of z may be disregarded). This μ is a Beltrami coefficient for G, and

$$w^\mu(\bar{z}) = \overline{w^\mu(z)}$$

by symmetry. It follows that $G^\mu = w^\mu G(w^\mu)^{-1}$ is again a Fuchsian group, and one sees that $U/G^\mu = S'$, with the given ramifications.

To prove the limit circle theorem for the general case one may use a limiting process.

6.4 A proof of the general uniformization theorem by quasiconformal mappings can be also given. The proof relies, however, on the distortion theorem for schlicht functions, and is therefore not essentially distinct from one of the classical proofs.

7. KOEBE GROUPS

7.1 Already in 1883 Klein formulated a very general theorem about uniformizing closed surfaces by Kleinian groups with an invariant not simply connected component. Much later Koebe published two proofs of this theorem; both are hard to follow. Recently Maskit announced a new proof (and a new formulation) of the Klein-Koebe theorem; his proof uses the method of quasiconformal deformations.

We consider first a typical example.

Let $p_1, \ldots, p_n$ be integers, each $p_i > 1$. Let $\Gamma_1, \ldots, \Gamma_n$ be Fuchsian groups such that Γ_i leaves a disc Δ_i fixed, Γ_i contains no elliptic or parabolic elements, and Δ_i/Γ_i is a compact Riemann surface of genus p_i, $i = 1, \ldots, n$. In each Δ_i one can find a non-Euclidean convex polygon P_i which is a fundamental domain for the group Γ_i in Δ_i; the mirror image P_i^* of P_i, with respect to the boundary $\partial\Delta_i$ of Δ_i, is a fundamental polygon for Γ_i in the domain exterior to $\partial\Delta_i$.

Assume that no two of the discs $\Delta_1, \ldots, \Delta_n$ intersect, and that the discs are sufficiently far apart. Then the n boundary curves $\partial P_1^*, \ldots, \partial P_n^*$ bound an n-times connected domain Q. It follows from Klein's <u>combination theorem</u> that the group

G generated by $\Gamma_1,\ldots,\Gamma_n$ is Kleinian, that it is the free product of $\Gamma_1,\ldots,\Gamma_n$, and that $P_1 \cup\ldots\cup P_n \cup Q$ is a fundamental domain for G in $\Omega(G)$. The group G has an invariant component Δ containing Q, $S = \Delta/G$ is a closed surface of genus $p = p_1 +\ldots+ p_n$, and $\Omega(G)/G$ has $n + 1$ components: S and the surfaces Δ_i/Γ_i, $i = 1,\ldots,n$.

Klein's theorem asserts, among other things, that a group of the kind constructed above can be chosen so that the quotient of the invariant component by the group be conformal to a given closed Riemann surface S', of genus p.

In Maskit's proof one obtains the desired group as $G^\mu = w^\mu G(w^\mu)^{-1}$ where μ is a Beltrami coefficient for the group G such that (a) μ induces on $S = \Delta/G$ the Beltrami differential of a diffeomorphism $S \to S'$, and (b) $\mu(z)d\bar{z}/dz$ goes over into its complex conjugate when z undergoes a reflection with respect to $\partial\Delta_i$, $i = 1,\ldots,n$. Condition (a) insures that $w^\mu(\Delta)/G^\mu$ is conformal to S', condition (b) insures that $w^\mu(\Delta_i)$ is a disc and $w^\mu\Gamma_i(w^\mu)^{-1}$ is a Fuchsian group.

7.2 The group G of 7.1 is an example of a Koebe group. The general definition (Maskit) follows.

Let G be a finitely generated Kleinian group. A subgroup H of G is called a factor if (i) H has a simply connected invariant component D, (ii) if $h: H \to U$ is a conformal bijection, $g \in H$ and $\gamma = h\circ g\circ h^{-1}$, then g is parabolic or elliptic if and only if γ is, (iii) if $g \in G$ is parabolic and leaves a boundary point of D fixed, then $g \in H$, and (iv) H is a maximal subgroup with properties (i), (ii), (iii). There are always factors, and only finitely many non-conjugate ones.

Now G is a Koebe group if all its factors are either Fuchsian or elementary. (The factors of the group G in 7.1 are $\Gamma_1, \ldots, \Gamma_n$ and their conjugates. The factors of a Schottky group are trivial. A Fuchsian group is its own only factor.)

7.3 Maskit's formulation of the Klein-Koebe theorem consists of two statements. One contains a description of all Koebe groups G for which the quotient of the invariant component by the group is a closed Riemann surface of genus p, perhaps with finitely many punctures. This involves Maskit's generalizations of the Klein combination theorem.

The second statement is an existence and uniqueness theorem. Let G_o be any finitely generated Kleinian group with invariant component Δ_o. Then there exists an isomorphism χ of G_o onto a Koebe group G with invariant component Δ, and a conformal mapping $\varphi: \Delta_o \to \Delta$ such that

$$\chi(g) = \varphi \circ g \circ \varphi^{-1} \quad \text{for} \quad g \in G_o,$$

$$\chi(g) \quad \text{is elliptic or parabolic if and only if} \quad g \quad \text{is.}$$

Also, χ is unique, except that it may be followed by a conjugation in the group of all Möbius transformations.

8. QUASI-FUCHSIAN GROUPS

8.1 The classical uniformization theorems deal with uniformizing one algebraic or analytic curve. A significant exception occurs in the theory of elliptic functions. Every algebraic curve of genus 1 admits a parametric representation by rational functions of four fixed holomorphic functions $\Psi_j(\tau, z)$, $\tau \in U$, $z \in \mathbb{C}$, $j = 1,2,3,4$, for some appropriate fixed value of τ, depending on the curve. The functions Ψ_j

are, of course, $\Psi_j(\tau,z) = \partial^{j-1}\sigma(z;1,\tau)/\partial z^{j-1}$ where σ is the Weierstrass function

$$\sigma(z;1,\tau) = z\,\Pi[1-z^2/(n+m\tau)^2]\exp[z^2/(n+m\tau)^2]$$

with $(n,m) \in \mathbb{Z}^2$, $n^2+m^2 > 0$ and either $n > 0$ or $n = 0$, $m > 0$.

We are concerned with extending this classical result on simultaneous uniformization to higher genera.

There exist, for every $p > 1$, $5p-5$ <u>fixed functions</u>

(1) $$\Psi_j(\tau,z) = \psi_j(\tau_1,\ldots,\tau_{3p-3},z), \quad \tau \in T, \quad z \in D(\tau)$$

where T is a <u>bounded domain</u> in $\mathbb{C}^{3p-3}$ and $D(\tau)$ a <u>Jordan domain</u> in $\mathbb{C}$, depending on τ, with the following property. <u>Every</u> algebraic curve of genus p admits a parametric representation by rational functions of $\Psi_1,\ldots,\Psi_{5p-5}$, for an appropriate value of τ.

The proof of this theorem is based on a result on simultaneous uniformization of two Riemann surfaces which we proceed to describe.

8.2 A <u>quasi-Fuchsian group</u> G with <u>fixed curve</u> C is a Kleinian group which leaves a directed Jordan curve C fixed. For such a group, $\Lambda(G) \subset C$, and if $\Lambda(G) = C$, the group is called of the first kind.

Classically, one constructs a quasi-Fuchsian group of the first kind by choosing $n+1$ circles $C_o, C_1, \ldots, C_{n+1} = C_o$ such that C_j is externally tangent to C_{j-1} and to C_{j+1}, and disjoint from all other C_k. Then G is the group of products of even numbers of reflections about these circles. The fixed curve C of the group passes through the points of contact of the $n+1$ circles. It is either a circle or a non-differentiable curve.

8.3 Let G be a quasi-Fuchsian group of the first kind, with fixed curve C, and let Δ_u, Δ_ℓ be the two components of $\Omega = \hat{\mathbb{C}} \setminus C$, Δ_u being the one to the left of C. (If G is Fuchsian and $C = \hat{\mathbb{R}} = \mathbb{R} \cup \{\infty\}$, with the usual orientation, then $\Delta_u = U$.)

If G contains no elliptic elements, which we assume for the sake of simplicity and brevity, G can be identified with the fundamental group of each Riemann surface, $S_u = \Delta_u/G$ and $S_\ell = \Delta_\ell/G$, so that we obtain an isomorphism χ_G between these fundamental groups (determined, of course, only up to inner automorphisms). As is well known, every isomorphism χ_G is induced by a homeomorphism of S_ℓ onto S_u. Let S and S' be two Riemann surfaces, let $f: S \to S'$ be a topological orientation preserving bijection, and $[f]$ its homotopy class. Also, let $\bar{S}$ be the mirror image of S and $j: S \to \bar{S}$ the canonical mapping. We say that the quasi-Fuchsian group G, with fixed curve C, represents the triple $(S,[f],S')$ if there are conformal mappings $h: \bar{S} \to \Delta_\ell/G$, $h': S' \to \Delta_u/G$, such that the homeomorphism $h \circ f \circ j \circ h^{-1}$ induces χ_G. For instance, a torsion-free Fuchsian group G represents the triple $(U/G,[id],U/G)$. The theorem on simultaneous uniformization of two Riemann surfaces asserts that if S and S' have no ideal boundary curves, and have U as their universal covering surface, and if $f: S \to S'$ is quasiconformal, there is a quasi-Fuchsian group G representing $(S,[f],S')$. This group is of the first kind, and is uniquely determined, up to conjugacy.

The most interesting case arises when S_o and S are of type (p,n),, that is compact of genus p, except for $n \geq 0$ punctures (and with $3p - 3 + n > 0$). In this case any orientation preserving homeomorphism is homotopic to a quasi-conformal one.

(There is a more general version of this theorem, allowing ideal boundary curves and also elliptic elements.)

8.4 The existence of G follows by the general method of §6. Let $S = U/G$, G a fixpoint free Fuchsian group. Then $\bar{S} = L/G$, where L denotes, here and hereafter, the lower halfplane. The mapping $f: S \to S'$ induces a Beltrami coefficient $\mu(z)$ on U, we continue it over $\mathbb{C}$, by setting $\mu(z) = 0$ for $z \in L$; μ is a Beltrami coefficient for G and $G^\mu = w^\mu G(w^\mu)^{-1}$ is a Kleinian group. It is clearly quasi-Fuchsian, with fixed curve $C = w^\mu(\hat{\mathbb{R}})$, and of the first kind since G is. Since $w^\mu|U$ is μ conformal, there is a canonical conformal mapping

$$h': S \to w^\mu(U)/G^\mu = \Delta_u/G^\mu.$$

Since $w^\mu|L$ is conformal, there is a canonical conformal mapping

$$h: \bar{S} \to w^\mu(L)/G^\mu = \Delta_\ell/G^\mu.$$

One verifies that G represents $(S,[f],S')$.

To show uniqueness, let G^* be another quasi-Fuchsian group representing $(S,[f],S')$. Let Δ_u^* and Δ_ℓ^* be the components of the complement of the fixed curve C^* of G^*. It is not difficult to find a conformal mapping φ of $\Delta_u \cup \Delta_\ell$ onto $\Delta_u^* \cup \Delta_\ell^*$ with $\varphi \circ G^\mu \circ \varphi^{-1} = G^*$. The latter condition shows that φ is the restriction of a homeomorphism $\hat{\varphi}: \hat{\mathbb{C}} \to \hat{\mathbb{C}}$. Using the fact that C is a so-called <u>quasi-circle</u> (image of a circle under a quasiconformal mapping) and known properties of quasiconformal mappings, one shows that $\hat{\varphi}$ is conformal and thus a Möbius transformation.

It can be shown that every finitely generated quasi-Fuchsian group of the first kind is a conformal deformation of a Fuchsian group. This need not be so, however, if the group is not finitely generated (as shown by Abikoff).

8.5 In order to define the functions Ψ_j described in 8.1, one needs automorphic forms.

Let G be a non-elementary Kleinian group, Ω its domain of discontinuity, and q a positive integer. A holomorphic automorphic q-form (also called a q-differential) for G is a holomorphic solution of the system of functional equations

$$(1) \qquad \varphi(g(z))g'(z)^q = \varphi(z), \quad z \in \Omega, \ g \in G.$$

It is called bounded if $\|\varphi\|_B = \sup |\lambda(z)^{-q}\varphi(z)| < \infty$, integrable if

$$(2) \qquad \|\varphi\|_A = \iint_{\Omega/G} |\varphi(z)| \lambda(z)^{2-q} \, dxdy < \infty .$$

Here and hereafter $\lambda(z)|dz|$ denotes the Poincaré metric on the component Δ of Ω containing z. The integrable automorphic q-forms form a complex Banach space $A_q(r,G)$. The bounded automorphic q-forms form a complex Banach space $B_q(\Omega,G)$. If G is of finite type, $A_q = B_q$ is the finite dimensional space of so-called cusp-forms.

8.6 Let Ω_o be an invariant union of components of Ω, with $\Omega \setminus \Omega_o \neq \emptyset$. We denote by $B_q(\Omega_o,G)$ the closed subspace of $B_q(\Omega,G)$ consisting of forms with $\varphi | \Omega \setminus \Omega_o = 0$. For $q \geq 2$ the formula

$$(3) \qquad \psi(z) = (\mathfrak{L}\varphi)(z) = \iint_{\Omega_o} \lambda(\zeta)^{2-q} \overline{\varphi(\zeta)} (\zeta - z)^{-2q} \, dxdy \ , \ z \in \Omega \setminus \Omega_o, \zeta \in \Omega_o$$

defines an antilinear mapping

$$(3') \qquad \mathfrak{L} : B_q(\Omega_o,G) \to B_q(\Omega \setminus \Omega_o,G) \quad .$$

The mapping $\mathfrak{L}$ can be shown to be continuous and injective if G is either finitely generated with invariant component $\Omega \setminus \Omega_o$, or quasi-Fuchsian. The mapping $\mathfrak{L}$ is continuous and surjective if G is either finitely generated with invariant component Ω_o, or a quasiconformal deformation of a Fuchsian group.

8.7 In 8.4 we described a method, based on quasiconformal mappings, of deforming a Fuchsian group G into a quasi-Fuchsian one, without changing the conformal structure of the "lower" Riemann surface L/G. A more classical method for doing the same thing proceeds as follows.

Given a bounded quadratic differential φ, find two linearly independent solutions, η_1 and η_2, of the linear ordinary differential equation in L

$$2\eta''(z) + \varphi(z)\eta(z) = 0 \quad . \tag{4}$$

The quotient

$$W_\varphi(z) = \eta_2(z)/\eta_1(z) \tag{5}$$

is a locally univalent function, and satisfies the relation

$$W_\varphi(g(z)) = \hat{g}(W_\varphi(z)), \ z \in L, \ g \in G \tag{6}$$

where $\hat{g}$ is a Möbius transformation. The function φ can be recaptured from W_φ by forming the <u>Schwarzian</u> <u>derivative</u>:

$$\varphi(z) = \{W_\varphi(z), z\} = [W_\varphi''(z)/W_\varphi'(z)]' - \frac{1}{2}[W_\varphi''(z)/W_\varphi'(z)]^2 \quad . \tag{7}$$

It may happen that W_φ is univalent in L; indeed, <u>all</u> univalent functions can be so obtained, and with $\|\varphi\|_B \leq \frac{3}{2}$. The inequality is due to Nehari, who also showed that W_φ is necessarily univalent for $\|\varphi\|_B \leq \frac{1}{2}$.

If W_φ is univalent, the $\hat{g}$ form a Kleinian group G_φ, with invariant component $\Delta \supset W_\varphi(L)$. Moreover, $\Delta = W_\varphi(L)$ if G is finitely generated and of the first kind.

If $W_\varphi(L)$ is a Jordan domain, G_φ is quasi-Fuchsian, and again $W_\varphi(L) = \Delta$. If G_φ is a quasi-circle, G_φ is a quasiconformal deformation of G, indeed, $G_\varphi = \alpha G^\mu \alpha^{-1}$ where α is a Möbius transformation and μ

a Beltrami coefficient for G with $\mu \mid L = 0$. The quadratic differential $\varphi = \varphi^{\mu}$ is determined by μ;

(8) $$\varphi^{\mu}(z) = \{w^{\mu} \mid L, z\}$$

where w^{μ} is the function constructed in 8.4.

It can be shown that $\varphi^{\mu} = \varphi^{\nu}$ <u>if and only if</u> μ and ν are induced by the Beltrami differentials of quasiconformal mappings $f\colon U/G \to S'$, $g\colon U/G \to S''$ with the property: there is a conformal mapping $h\colon S' \to S''$ such that $g^{-1} \circ h \circ f$ is homotopic to the identity.

8.8 The set of those $\varphi \in B_2(L,G)$ for which $W_{\varphi}(L)$ is bounded by a quasi-circle is called the <u>Teichmüller space</u> of G and is denoted by $T(G)$. It is a bounded domain in the complex Banach space of bounded quadratic differentials. Ahlfors and Weill proved that $T(G)$ contains the ball $\|\varphi\|_B < \frac{1}{2}$.

If L/G is of type (p,n), the space $B_2(L,G)$ has dimension $3p - 3 + n$ and

(9) $$T(G) \subset \mathbb{C}^{3p-3+n} \quad .$$

8.9 In view of the result stated at the end of 8.7, the Teichmüller space $T(G)$ can be identified with the Teichmüller space $T(S)$ of the Riemann surface $S = U/G$. This space consists of equivalence classes $[f]$ of quasiconformal mappings $f\colon S \to S'$, two mappings, f and f', being called equivalent if $f' = h \circ f \circ t$ where t is a self-mapping of S, homotopic to the identity and h a conformal mapping. The space of equivalence classes is topologized by the <u>Teichmüller metric</u>: the distance between $[f]$ and $[f_1]$ is $\log(1+k) - \log(1-k)$ where k is the infimum of the norms of Beltrami differentials of all quasiconformal mappings in the equivalence class $[f \circ f_1^{-1}]$.

8.10 A quasiconformal deformation $G \to G^{\mu}$ induces a holomorphic bijection $T(G) \to T(G^\mu)$. In particular, if U/G is a Riemann surface of type (p,n), $T(G)$, as a complex manifold, depends only on the numbers p and n and may be denoted by $T_{p,n}$. One writes T_p for $T_{p,o}$.

8.11 It will be convenient, from now on, to normalize the solutions η_1 and η_2 of (4), which enter into the definition (5) of W_φ, be requiring that

(10) $$\eta_1(-i) = \eta_2'(-i) = 0, \quad \eta_1'(-i) = \eta_2(-i) = 1 \quad .$$

To every $\varphi \in T(G)$ there belongs a definite Jordan domain

(11) $$D(\varphi) = \hat{\mathbb{C}} \setminus \text{closure of } W_\varphi(L)$$

bounded by the curve

(12) $$z = \hat{W}_\varphi(t) = \alpha_\varphi \circ w^\mu(t), \quad -\infty \leq t \leq \infty$$

where μ and φ are connected by (8), α_φ is a Möbius transformation (depending on φ) and $\hat{W}_\varphi$ is the continuous extension of W_φ to $\hat{\mathbb{R}}$. For every t, z depends holomorphically on φ. To φ there also belongs the quasi-Fuchsian group

(13) $$G_\varphi = W_\varphi G W_\varphi^{-1}$$

and $G_\varphi = \alpha_\varphi G^\mu \alpha_\varphi^{-1}$; for this group $\Delta_\ell = W_\varphi(L)$, $\Delta_u = D(\varphi)$.

The set

(14) $$F(G) = \{(\tau, z) \mid \tau \in T(G),\ z \in D(\tau)\}$$

is called the <u>fiber space over the Teichmüller space</u>. The elements g of G act holomorphically on $F(G)$, by the rule

(15) $$g(\tau, z) = (\tau, W_\tau \circ g \circ W_\tau^{-1}(z)) \quad .$$

If $T(G) = T_{p,n}$, we write $F(G) = F_{p,n}$ and note that $F_{p,n}$ admits an embedding

(16) $$F_{p,n} \subset \mathbb{C}^{3p-2+n}$$

as a domain, indeed, as can be shown using Koebe's one-quarter-theorem, as a <u>bounded domain</u>. Again we write F_p instead of $F_{p,o}$. There is a canonical holomorphic isomorphism

(17) $$T_{p,n+1} \cong F_{p,n} \quad .$$

This result is the only one which has no generalization for quasi-Fuchsian groups with elliptic elements.

8.12 We are finally in a position to define the functions $\Psi_j(\tau,z)$, $(\tau,z) \in F_p$, referred to in 8.1. Choose a torsion-free Fuchsian G such that U/G is a compact Riemann surface of genus $p > 1$, and choose a basis $\varphi_1,\ldots,\varphi_{5p-5}$ of the space of holomorphic 3-differentials for G in U. For every i, set

(18) $$\varphi_i^*(z) = \overline{\varphi_i(\bar{z})}, \quad z \in L \quad ,$$

and define $\psi_i^* \in B_3(W_\tau(L),G_\tau)$ by the relation

(19) $$\psi_i^*(W_\tau(z))W_\tau'(z)^3 = \varphi_i^*(z) \quad , \quad z \in L \quad .$$

Finally set

(20) $$\psi_i(z) = (\Omega\psi_i^*)(z) \quad , \quad z \in D(\tau)$$

where Ω is the operator defined in 8.6, for $G = G_\tau$, $\Omega_o = W_\tau(L)$, $q = 3$.

The functions $\Psi_j(\tau,z) = \psi_j(z)$, $j = 1,\ldots,5p-5$, have the required properties: the holomorphic dependence on τ is confirmed by a direct calculation, the fact that every algebraic curve of genus p, corresponding to some Riemann surface $D(\tau)/G_\tau$, can be uniformized by rational functions of the $\Psi_j(\tau,z)$ follows from classical properties of 3-canonical curves.

9. b-GROUPS

9.1 We are now in position to describe one of the 'partial answers', alluded to in 4.3, to the problem of describing all representations of a given Riemann surface as the quotient Δ/G of an invariant domain of a Kleinian group by this group. For technical reasons we denote the given Riemann surface by $\bar{S}$ (that is, we think of it as the mirror image of a Riemann surface S).

We assume that $\bar{S}$ is of finite type (p,n), that is, a compact surface of genus p with $n \geq$ punctures, that $3p - 3 + n > 0$ (the cases with $3p - 3 + n \leq 0$ are trivial), and that G is finitely generated. Throughout most of this section we shall also assume that Δ is simply connected and that G is fixed-point-free (that is, without torsion). The latter assumption is made mainly for the sake of brevity.

The following inequality holds for any <u>finitely generated Kleinian group with invariant component</u> Δ:

$$\iint_{(\Omega \setminus \Delta)/G} \lambda^2 \, dxdy \leq \iint_{\Delta/G} \lambda^2 \, dxdy \tag{1}$$

(see 8.5 for the definition of λ).

We sketch the steps of the proof. The Riemann-Roch theorem implies that

$$\dim B_q(\Delta, G) \sim 2q \iint_{\Delta/G} \lambda^2 \, dxdy, \quad q \to \infty \tag{2}$$

where B_q is the space defined in 8.6. Similarly, for every component Δ_i of G,

$$\dim B_q(\Delta_i, G_{\Delta_i}) \sim 2q \iint_{\Delta_i/G_{\Delta_i}} \lambda^2 \, dxdy, \quad q \to \infty \quad . \tag{3}$$

From now on, let

(4) $$\Delta_1, \Delta_2, \ldots, \Delta_r$$

be a complete list of non-conjugate components of G, distinct from Δ. By (3),

(5) $$\dim B_q(\Omega \setminus \Delta, G) = \sum_{j=1}^{r} \dim B_q(\Delta_i/G_{\Delta_i}) \sim 2q \iint_{(\Omega \setminus \Delta)/G} \lambda^2 dxdy, \quad q \to \infty.$$

But, as noted in 8.6, the operator $\mathfrak{L}$ maps $B_q(\Omega \setminus \Delta, G)$ injectively into $B_q(\Delta, G)$. Hence the left side of (5) does not exceed that of (2), and (1) follows.

Assume that G is torsion free, and set

(6) $$(p_j, n_j) = \text{type of } \Delta_i/G_{\Delta_i} \quad .$$

Then (1) may be rewritten as

(7) $$\sum_{j=1}^{r} (p_j - 1) + \sum_{j=1}^{r} n_j \leq 2(p - 1) + n \quad .$$

Of course, (7) is trivially satisfied if $\Delta = \Omega$, $r = 0$.

9.2 The area inequality (1) already implies that G can have at most two invariant components. A basic theorem of Maskit asserts that if the finitely generated (an essential hypothesis!) Kleinian group G has two invariant components, it is quasi-Fuchsian and a quasiconformal deformation of a Fuchsian group.

In the presence of an invariant component Δ, each stabilizer G_{Δ_i} of a non-invariant component has two invariant components: Δ_i and the component containing Δ. Hence each G_{Δ_i} is quasi-Fuchsian.

9.3 Assume now that the invariant component Δ of G is simply connected. There is a conformal mapping $h: \Gamma \to \Delta$ and $\Gamma = h^{-1}Gh$ is a Fuchsian group. We may assume that $h = W_\varphi$, for some fixed $\varphi \in B_2(L,\Gamma)$ (cf. 8.7). Indeed, this can be achieved replacing G by a conjugate group (in the group of all Möbius transformations). Of course, φ need not lie in $T(\Gamma)$. By our hypothesis Γ is finitely generated, of the first kind, and $S = U/\Gamma$ is of type (p,n). Also, there is a canonical mapping $\bar{S} = L/\Gamma \to \Delta/G$. By abuse of language, we write $\bar{S} = \Delta/G$.

The group G is called <u>non-degenerate</u> if (1) holds with the equality sign, <u>partially degenerate</u> if $r > 0$ and (1) holds with a strict inequality, <u>totally degenerate</u> if $r = 0$, that is, if $\Delta = \Omega$.

9.4 G is geometrically finite (cf. 4.2) if and only if it is non-degenerate. The decisive step in proving this is due to L. Greenberg who showed that a totally degenerate group is not geometrically finite.

9.5 A parabolic element $g \in G$ is called <u>accidental parabolic</u> if $\gamma = W_\varphi^{-1} \circ g \circ W_\varphi$ is hyperbolic; such a γ will be called <u>vanishing hyperbolic</u>. Let V denote the union of axes of all vanishing hyperbolic elements of G. (We recall that the axis of a hyperbolic real Möbius transformation γ is a circle or straight line passing through the fixed points γ and orthogonal to $\mathbb{R}$.)

Every component D of $U \setminus V$ is simply connected. We call two such components, D_i and D_j, conjugate if $D_j = \gamma(D_i)$ for some $\gamma \in \Gamma$. There are at most finitely many non-conjugate components; let

$$D_1, \ldots, D_s \tag{8}$$

be a complete list.

For each component D_j, let Γ_{D_j} be the stabilizer of D_j in Γ. Then D_j/Γ_{D_j} is a Riemann surface of type (π_j, ν_j, μ_j), that is a compact surface of genus π_j, from which one has removed $\nu_j + \mu_j$ disjoint continua, precisely ν_j of which are points. Also, $3\pi_j - 3 + \nu_j + \mu_j \geq 0$ for all j. The inclusion $D_j \subset U$ induces an embedding $D_j/\Gamma_{D_j} \subset S$; the boundary of D_j/Γ_{D_j} in S consists of μ_j closed geodesics (with respect to the Poincaré metric), images of axes of vanishing hyperbolic elements of Γ. We call these closed geodesics <u>vanishing loops</u>.

It turns out that $r \leq s$ (cf. (4) and (8)), and that one can renumber the D_j in such a way that

(9) $$W_\varphi \Gamma_{D_i} W_\varphi^{-1} = G_{\Delta_i} \quad , \quad i = 1,\dots,r \quad ;$$

in this case

(10) $$p_i = \pi_i,\ n_i = \nu_i + \mu_i,\ i = 1,\dots,r \quad .$$

Using (7) and (10) one verifies that G is non-degenerate if and only if $r = s$. (In all cases, $3\pi_j - 3 + \nu_j + \mu_j = 0$ implies that $j \leq r$.)

9.6 If $s = 1$ (that is, if there are no accidental parabolic elements in G), and if $r = 1$, the group G is quasi-Fuchsian.

We may assume that $\Delta = \Delta_\ell$. All torsion free quasi-Fuchsian groups with a given $\bar{S} = \Delta/G$, of type (p,n), or, rather their conjugacy classes, are the points of the Teichmüller space $T(\Gamma)$ studied in §8.

9.7 A finitely generated Kleinian group G with precisely <u>one</u> simply connected invariant component Δ is called a b-<u>group</u>.

The existence of b-groups, with a given $\bar{S} = \Delta/G$, follows at once from the theory exposed in §8. We think of $T(\Gamma)$ as embedded in $B_2(L,\Gamma)$. If φ lies on the boundary $\partial T(\Gamma)$ of $T(\Gamma)$, the function $W_\varphi(z)$, $z \in L$, is still univalent and holomorphic and $G = \Gamma_\varphi = W_\varphi \Gamma W_\varphi^{-1}$ is a b-group with the single invariant component $\Delta = W_\varphi(L)$. Such a G is called a <u>boundary group</u>.

The name 'b-group' anticipates the result, not yet established, that all b-groups are boundary groups.

9.8 Boundary groups belonging to different points of $\partial T(\Gamma)$ are, in general, <u>essentially distinct</u>: every boundary group is conjugate to precisely $[N:\Gamma] - 1$ other boundary groups, where N is the normalizer of Γ in the group of all real Möbius transformations.

It is well known that the index $[N:\Gamma]$ is finite, and equals 1 for most groups.

9.9 There are <u>uncountably many</u> non-conjugate totally degenerate boundary groups G, with a given $\bar{S} = \Delta/G$. This can be seen as follows.

A point $\varphi \in \partial T(\Gamma)$ is called a <u>cusp</u> if there is a vanishing hyperbolic element $\gamma \in \Gamma$, that is if $g = W_\varphi \circ \gamma \circ W_\varphi^{-1}$ is parabolic. Cusps always exist (see below) but it is easy to see that the set of non-cusps is dense in $\partial T(\Gamma)$.

On the other hand, it follows from what was said above that for a b-group without accidental parabolic elements, either $r = 1$ and G is quasi-Fuchsian, or $r = 0$. Hence Γ_φ is totally degenerate if $\varphi \in \partial T(\Gamma)$ is a non-cusp.

9.10 Little is known about totally degenerate groups, beyond their existence. It is not known which sequences in $T(\Gamma)$ converge to such a group (though Abikoff found a necessary condition), and not a single example has been constructed.

Abikoff showed that the limit set of a totally degenerate group is not locally connected.

9.11 A non-degenerate b-group is called <u>regular</u>. For such groups there is a <u>uniqueness</u> <u>theorem</u> proved by Marden, in the torsion-free case, and by Abikoff in the general case (and by a different method).

If G_1 and G_2 are two regular b-groups, with regions of discontinuity Ω_1 and Ω_2, and if $h: \Omega_1 \to \Omega_2$ is a conformal mapping such that $hG_1h^{-1} = G_2$, then h is a Möbius transformation, so that G_2 is conjugate to G_1.

Using this theorem, together with the results in §6 and §8, one can show that conjugacy classes of regular b-groups G, with a given Δ/G, can be parametrized by (a) a certain number of discrete parameters describing the topological type of G, i.e., how the 'vanishing loops' (see 9.5) are situated on S, and (b) $(3p_1 - 3 + n_1)+\ldots+(3p_r -3 + n_r)$ complex parameters describing the complex structure of the r Riemann surfaces $\Delta_1/G_{\Delta_1}, \ldots, \Delta_r/G_{\Delta_r}$. For a given (p,n) there are only finitely many topological types.

Using the theory of moduli of Kleinian groups one can represent the set of (conjugacy classes of) regular b-groups of a given topological type and with a given $\bar{S} = \Delta/G$ as the quotient of the manifold

(11) $$M = T_{p_1,n_1} \times T_{p_2,n_2} \times \ldots \times T_{p_r,n_r}$$

by a properly discontinuous group of holomorphic automorphisms.

9.12 Abikoff showed that all regular b-groups are <u>boundary groups</u> (as was shown earlier, for the torsion-free case, by Harvey and by Marden). His method also establishes the fact (already proved by Maskit) that there exist regular b-groups of all a priori possible topological types and that, for every topological type, there is an injection $\Phi: M \to \partial T(\Gamma)$ such that for $\varphi = \Phi(t)$, Γ_φ is the b-group represented by $t \in M$.

Abikoff also showed that there are $\varphi \in \partial\Phi(M)$ such that Γ_φ is partially degenerate. The proofs due to either Abikoff or Maskit which establish the existence of <u>partially degenerate boundary groups</u> are, however, as non-constructive as the existence proof for totally degenerate groups.

Everything said in 9.2-9.12 extends, with suitable modifications to b-groups with torsion.

9.13 Maskit's classification of all geometrically finite Kleinian groups G with an invariant component Δ and a given $\bar{S} = \Delta/G$ is similar to, though more complicated than, the classification of regular b-groups.

Given $\bar{S}$, the points over which the mapping $\Delta \to \Delta/G$ is to be ramified, and the ramification numbers, there are only <u>finitely many</u> topologically distinct types of groups satisfying the stated conditions. These types can be enumerated. The conjugacy classes of groups of a given type are parametrized by continuously varying parameters; the set of these classes can be represented as the quotient of a <u>product of Teichmüller spaces</u> by a properly discontinuous group of holomorphic automorphisms.

10. UNIFORMIZATION OF n-DIMENSIONAL ALGEBRAIC VARIETIES

10.1 In this section we sketch the proof of Griffiths' uniformization theorem which extends to n-dimensional algebraic varieties the main uniformization theorem of 2.1.

We remark that a direct application of the method of 2.1 to the higher dimensional case is out of the question, since for n-dimensional complex manifolds there is no generalized Riemann mapping theorem. Instead Griffiths relies on simultaneous uniformization of Riemann surfaces of fixed finite type by quasi-Fuchsian groups, and finds, in the given n-dimensional variety, a Zariski open set which can be represented as an (n-1)dimensional holomorphic family of such surfaces.

It is convenient to describe first the domains which will take the place of U.

A Bergman domain in $\mathbb{C}^n$ is a domain consisting of n-tuples of points $(z_1,\dots,z_n)$ such that

$$z \in D_1,\ z_2 \in D_2(z_1),\ z_3 \in D_3(z_1,z_2),\ \dots,\ z_n \in D_n(z_1,\dots,z_{n-1})$$

where D_1 is a Jordan domain and, for $j > 0$, $D_{j+1}(z_1,\dots,z_j)$ is a Jordan domain bounded by a curve admitting the parametric representation

$$z_{j+1} = Z_{j+1}(z_1,\dots,z_j;t),\ -\infty \leq t \leq +\infty$$

where Z_{j+1} depends holomorphically on $z_1,\dots,z_j$ and continuously on t, with $Z_{j+1}(z_1,\dots,z_j\ ,t') = Z_{j+1}(z_1,\dots,z_j,t'')$ only if $t' = t''$ or $t',t'' = \pm\infty$.

Bergman domains appear "naturally." For instance, it follows from relation (17) in 8.11, and from the isomorphisms $T_{0,4} \overset{\approx}{\to} T_{1,1}$, $T_{0,6} \overset{\approx}{\to} T_{2,0}$, that all Teichmüller spaces $T_{p,n}$ with $p = 0,1,2$ are Bergman domains.

Bergman domains are always domains of holomorphy.

10.2 By an algebraic variety we mean a projective variety over $\mathbb{C}$, i.e., a complex space X which can be realized in some projective space $\mathbb{P}_N(\mathbb{C})$ as the set of common zeros of a system of homogeneous polynomials in $(z_o, \ldots, z_N)$. X is irreducible if it is not a union of two distinct (non-empty) subvarieties.

A Zariski open subset of X is the complement of a subvariety $X_o \subset X$. Such a set is open (in the ordinary topology) and, unless it is empty, dense.

Let X be an irreducible algebraic variety. The set X_{sing} of singular points of X is the smallest set such that $X \setminus X_{sing}$ is a complex manifold. In all cases $X_{sing} \neq X$ and X_{sing} is a subvariety. If $X_{sing} = \emptyset$, X is called non-singular. The dimension $\dim X$ of X is, by definition, the dimension of $X \setminus X_{sing}$. The degree of X is the number of points in which $X \subset \mathbb{P}_N$ meets a generic linear subspace of $\mathbb{P}_N$ of dimension $(N - \dim X)$. ("Generic" means violating finitely many algebraic conditions.)

10.3 The projective space $\mathbb{P}_N$ is itself an irreducible variety. For every integer k there is a canonical embedding (Segre embedding) of $\mathbb{P}_N$ as an algebraic variety in $\mathbb{P}_M$, where $M+1$ is the number of distinct monomials of degree k in $(N+1)$ variables. Under this embedding, a point $(x_o, \ldots, x_N)$ is taken into the point $(\xi_o, \ldots, \xi_M)$, $\xi_o, \ldots, \xi_M$ being monomials of degree k in $x_o, \ldots, x_N$ ordered lexicographically.

10.4 Griffiths' theorem asserts that given a Zariski open set Y in an irreducible algebraic variety X and a non-singular point $x_o \in Y$, there is a Zariski open set $A \subset Y$, with $x_o \in A$, such that the universal covering space $\tilde{A}$ of A is (biholomorphically equivalent to) a bounded Bergman domain.

(Of course A is not uniquely determined, though it must lie in $X \setminus X_{sing}$. This is so even if $\dim X = 1$, see below.)

The proof below follows closely Griffiths' ideas; details are carried out somewhat differently. The proof proceeds by induction on the dimension of X.

If this dimension is 1, Y is a complete algebraic curve X, with finitely many points removed. We remove, if need be, a few more points from Y, including all singular points, so as to obtain $A \subset Y$ as a Riemann surface of finite type (p,q) with $3p - 3 + q \geq 0$. This can be done so as to have $x_o \in A$. The assertion of the theorem follows from the main case of the limit circle theorem (see 2.1), with $\tilde{A}$ the unit disc.

10.5 We assume now that $n > 1$ and that the assertion is proved for varieties of dimension $n - 1$.

Let X, with $\dim X = n > 1$, $Y \subset X$ and $x_o \in Y$, satisfying the hypotheses of the theorem be given. We may assume that Y contains no singular points, since we may replace Y by $Y \setminus X_{sing}$. We may further assume that X is <u>normal</u> since, by a well known theorem by Zariski, this can be achieved by changing X outside $X \setminus Y$. The only property of normal varieties we need is that $\dim X_{sing} \leq n - 2$. (We may also use the more difficult theorem by Hironaka and assume that X is non-singular.) We assume, finally, that $X \subset \mathbb{P}_N$, that the degree of X is $d > 1$, that $X \setminus Y \subset H_\infty$ where H_∞ is a hyperplane of $\mathbb{P}_N$, and that the intersection $X \cap H_\infty$ has no multiple components. All this can be achieved by applying, if need be, a Segre mapping to the projective space containing X. Note that the degree of $X \cap H_\infty$ is d.

Let K be a fixed $(N-n)$-dimensional linear subspace of $\mathbb{P}_N$ such that K meets X transversally, at d distinct points $P_1,\ldots,P_d$ all of which belong to $Y \setminus \{x_o\}$. Such a K exists, since a generic K will have the required properties.

Now let L be an $(n-1)$-dimensional linear subspace of $\mathbb{P}_N$ such that $K \cap L = \emptyset$. Define a holomorphic rational mapping

$$F: X \setminus \{P_1,\ldots,P_d\} \to L$$

as follows. For $P \in X \setminus \{P_1,\ldots,P_d\}$, let $\Pi(P)$ be the $(N-n+1)$-dimensional subspace spanned by K and P. Then $L \cap \Pi(P) = F(P)$ is a point. Note that F is a surjection on $L \approx \mathbb{P}_{n-1}$.

Now a generic $(N-n+1)$-dimensional subspace Π which contains K has the following property α: Π intersects X in a non-singular (irreducible) algebraic curve C_Π, of some fixed genus p (by the so-called Bertini theorem); this C_Π contains the points $P_1, \dots, P_d$, avoids X_{sing} (here we use the normality of X), and meets H_∞ in d distinct points. Thus $C_{\Pi(P)}$ will have property α for all $P \in A$, A a Zariski open set of Y, and $F(A)$ is Zariski open in L.

By changing, if need be, K and L we may achieve that $\Pi(x_o)$ has the above properties. We set $F(x_o) = b_o$, $\Pi(x_o) \cap X = C_o$.

Thus there are Zariski open subsets $A \subset Y$, $B \subset L$, with $x_o \in A$, $b_o \in B$, and a holomorphic rational map

$$F: A \to B$$

such that for every $b \in B$ the inverse image $F^{-1}(b)$ in A, that is $C_{\Pi(b)} \cap A$, is a Riemann surface of type $(p, 2d)$. Note that $3p - 3 + 2d > 0$. (This construction of F has been suggested to me by B. Saint-Donat.)

10.6 In view of the induction hypothesis we may assume that the universal covering space $\tilde{B}$ of B is a bounded Bergman domain.

We choose a torsion-free Fuchsian group G such that U/G is conformally equivalent to $F^{-1}(b_o)$, and a basis for the fundamental group on $F^{-1}(b_o)$. This determines, by continuity, a basis for the fundamental group on each $F^{-1}(b)$ for b in some neighborhood (not Zariski neighborhood) R of b_o in B. For $b \in R$ there is, therefore, an isomorphism of the fundamental group of $F^{-1}(b_o)$ onto that of $F^{-1}(b)$, which can be induced by an orientation preserving homeomorphism, and hence a mapping $\Phi: R \to T(G)$.

We claim that Φ is <u>holomorphic</u> in a sufficiently small neighborhood R' of b_o.

To verify this claim, let z be a non-constant meromorphic function on C_o, with the following properties: (i) z has $r > 1$ simple poles and no other singularities, (ii) dz has $2r + 2p - 2$ simple zeros, (iii) at the $2d$ 'punctures', i.e., points of $C_o \setminus F^{-1}(b_o)$, $z \neq \infty$ and $dz \neq 0$, and (iv) at the $2d$ punctures and at the $2r + 2p - 2$ zeros of dz, z takes on $k = 2d + 2r + 2p - 2$ distinct values.

Now z is the restriction to C_o of a rational function Z in P_N. The restriction z_b of Z to $C_b = X \cap \Pi(F^{-1}(b))$, $b \in R'$, has similar properties. For $b \in R'$, the mapping $z_b: C_b \to \hat{\mathbb{C}} = \mathbb{P}_1$ realizes C_b as a ramified r-sheeted covering of the Riemann sphere, with only simple branch points over $2r + 2p - 2$ points $\zeta_1(b), \ldots, \zeta_{2r+2p-2}(b)$. The $2d$ punctures are located over $2d$ points $\zeta_{2r+2p-1}(b), \ldots, \zeta_k(b)$. The k points $\zeta_1(b), \ldots, \zeta_k(b)$ are distinct and, if properly numbered, depend holomorphically on b.

It is now easy to construct, for b close to b_o, a quasiconformal mapping $f_b: F^{-1}(b_o) \to F^{-1}(b)$ which is "close to the identity" and has a Beltrami differential which depends holomorphically on b. This mapping defines the element $\Phi(b) \in T(G)$. It follows that $\Phi: R' \to T(G)$ is holomorphic. It also follows that there is a conformal mapping $\psi_b: D(\varphi)/G_\varphi \to F^{-1}(b)$, where $\varphi = \Phi(b)$, which depends holomorphically on b. (See 8.11 for the definition of G_φ and $D(\varphi)$.)

The argument just given also shows that Φ can be continued analytically along every path in B.

Let $\pi: \tilde{B} \to B$ be the covering map, and $\tilde{b}_o \in \tilde{B}$ a point with $\pi(\tilde{b}_o) = b_o$. Since $\tilde{B}$ is simply connected, we may conclude that there is a holomorphic mapping $\tilde{\Phi}: \tilde{B} \to T(G)$ such that for $\pi(\tilde{b}) = b$ and $\tilde{\Phi}(\tilde{b}) = \varphi$ there is a conformal mapping $\tilde{\psi}_{\tilde{b}}: D(\varphi)/G_\varphi \to F^{-1}(b)$ which depends holomorphically on $\tilde{b}$.

Now we form the bounded Bergman domain

$$M = \{(\tilde{b}, z) \mid \tilde{b} \in \tilde{B},\ z \in D(\tilde{\Phi}(\tilde{b}))\} \subset \mathbb{C}^n ,$$

and note that G operates on M, holomorphically, freely and properly discontinuously by the rule

$$g(\tilde{b},z) = (\tilde{b}, W_\varphi \circ g \circ W_\varphi^{-1}(z)) \quad \text{where} \quad \varphi = \tilde{\Phi}(\tilde{b}), \quad g \in G \ .$$

The quotient M/G is the set of pairs

$$(\tilde{b},t), \ \tilde{b} \in \tilde{B}, \ t \in D(\varphi)/G_\varphi \ , \quad \varphi = \tilde{\Phi}(\tilde{b}) \ .$$

The canonical mapping $M \to M/G$ followed by the mapping $(\tilde{b},t) \to \tilde{\psi}_{\tilde{b}}(t) \in A$ is a Galois covering $M \to A$. Since M is simply connected, this is the universal covering, and $M = \tilde{A}$.

11. SOME OPEN QUESTIONS

We conclude this report by listing four open problems which seem to have a more than technical significance.

I. Can one prove the theorem on simultaneous uniformization of two Riemann surfaces of the same finite type by quasi-Fuchsian groups, without using quasiconformal mappings?

Since the theorem can be stated "classically", without any reference to quasiconformal mappings, an affirmative answer seems not too improbable. It would certainly provide new insight.

Of course, the same question can be asked about other results in Teichmüller space theory. It is pertinent to recall that, according to Royden, the Teichmüller metric can be defined without using quasiconformal mappings - it is the Kobayashi metric in Teichmüller space.

II. Is there a constructive proof for the existence of totally degenerate Kleinian groups?

This is also of interest in connection with Ahlfors' problem about the measure of the limit set. Totally degenerate groups are natural candidates for having a big limit set.

III. Are there general methods of constructing Bergman domains Δ with properly discontinuous groups Γ of automorphisms and with Δ/Γ

quasi-projective (i.e., a Zariski open subset of an algebraic variety)?

We have in mind extensions of Griffiths' theorem via the approach "from groups to quotients" which proved fruitful in the classical theory of Fuchsian groups. If one permits Γ to have fixed points, more quasi-projective varieties should become representable.

In the absence of a method, examples might prove illuminating.

IV. If X is a *Stein manifold* topologically equivalent to a domain in $\mathbb{C}^n$, is there a *holomorphic embedding* $X \subset \mathbb{C}^n$? If not, what are the obstructions?

In other words, is there an n-variable analogue of Koebe's general uniformization theorem (which asserts that there are no non-topological obstructions for $n = 1$)? The restriction to Stein manifolds is sensible, since not every complex manifold homeomorphic to a cell carries non-constant holomorphic functions.

A recent result by R.S. Hamilton seems to point in the direction of an affirmative answer. Hamilton proved that every "small" deformation of the complex structure of a strongly pseudo-convex domain in $\mathbb{C}^n$, with a C_∞ boundary, can be accomplished by moving the boundary.

BIBLIOGRAPHICAL NOTE

A complete bibliography or uniformization would require a volume. We list only some books and papers.

Introduction. Standard reference texts for classical uniformization theory are H. Weyl, *Die Idee der Riemannschen Fläche*, 3rd Edition (Teubner, 1955), available also in an English translation, R. Fricke and F. Klein, *Vorlesungen über die Theory der Automorphen Funktionen*, Vol. 2 (Teubner, 1926), P. Appell and E. Goursat, *Theorie des fonctions algébriques et de leurs intégrales*, Vol. 2, written by P. Fatou (Gauthier-Villars, 1930), L. Ford, *Automorphic Functions*, 2nd Edition (Chelsea, 1951), R. Nevanlinna, *Uniformisierung*, (Springer,

1953). Of more recent texts we mention only Springer, Introduction to Riemann Surfaces, (Addison-Wesley, 1957) and L.V. Ahlfors, Conformal Invariants (McGraw-Hill, 1973).

The original papers by F. Klein are collected in Vol. 3 of his Gesammelte Mathematische Abhandlungen (Springer, 1923); this volume contains historical notes by Klein and his correspondence with Poincaré. Concerning Poincaré, see his Oeuvres, especially Vol.2 (Gauthiers-Villars, 1916). Unfortunately, a collection of P. Koebe's works is not available; most of his numerous contributions are quoted in Weyl and in Ford.

An exposition of recent developments, and a bibliography, will be found in L. Bers, Bull. London Math. Soc., 4(1972).

Concerning the work by J. Tate and by D. Mumford, see the latter's paper in Comp. Math., 24(1972) (cf. also L. Gerritzen, Math. Ann., 210(1974)).

§1, §2. The material here is classical.

§3.1. Proofs of Koebe's theorem will be found in most of the texts mentioned above.

§3.3. B. Maskit, Proc. Conf. Complex Anal., Minneapolis, 1964 (Springer, 1965), Ann. of Math., 81(1965).

§3.4. Cf. Appell-Goursat, Ford.

§3.6. B. Maskit, Amer. J. Math., 40(1968).

§4. For references, see Ford, Bers' paper quoted above, and A Crash Course on Kleinian Groups, Lecture Notes Math., No.400, ed. by L. Bers and I. Kra, (Springer, 1974).

§4.2. For Ahlfors' finiteness theorem and related questions see I. Kra, Automorphic Forms and Kleinian Groups, (Benjamin, 1972).

§5. A standard reference book, with a rather complete bibliography is O. Lehto and K.I. Virtanen, Quasikonforme Abbildungen (Springer, 1965). A briefer presentation will be found in L.V. Ahlfors, Lectures on Quasiconformal Mappings (Van Nostrand, 1966).

§5.3. L.V. Ahlfors and L. Bers, Ann. of Math., 72(1960).

§6. Cf. L. Bers, _Comm. Pure Appl. Math._, _14_(1961).

§7. Cf. B. Maskit's paper in _Contributions to Analysis_, ed. by Ahlfors et al (Academic Press, 1974), and the references given there.

§8. Concerning the aspects of theory of Teichmüller space theory presented in this section see Bers, _On moduli of Riemann surfaces_, (mimeographed lecture notes, ETH, Zurich, 1964) and the references given there, as well as Ahlfors' _Lectures_ (quoted above). Cf. also C.J. Earle and J. Eells, _J. Diff. Geom._, _3_(1969) and the forthcoming book by Earle.

§8.3. Cf. Bers, _Comm. Pure Appl. Math._, _14_(1961).

§8.4. W. Abikoff's counter-example is in _Ann. of Math. Studies_, _66_(1971).

§8.5. Kra's book contains a thorough discussion and a good bibliography.

§8.6. The Ω operator is studied in Bers, _Acta Math._, _116_(1966).

§8.8. Ahlfors and G. Weill, _Proc. Amer. Math. Soc._, _13_(1962).

§8.11. Concerning the fiber space, see Bers, _Acta Math._, _130_(1973), and I. Kra, _Duke Math. J._, 16(1973).

§9. This section is based mainly on three papers with the same title (On boundaries of Teichmüller spaces and on Kleinian groups I, II, III) by Bers, _Ann. of Math._, _91_(1970), Maskit, _ibid._ and Abikoff, to appear, _Acta Math_. Cf. also W. Harvey, _Annals of Math. Studies_, _79_(1974).

§9.1. For the area inequality, see Bers, _J. Analyse Math._, _18_(1967) and Kra's book.

§9.4. L. Greenberg, _Ann. of Math._, _84_(1966) and Abikoff, _loc. cit._

§9.2. A simplified proof of Maskit's result is due to Kra and Maskit, _Bull. Amer. Math. Soc._, _78_(1972).

§9.10. Abikoff, _Amer. J. Math._, to appear.

§9.11. A. Marden, Ann. of Math., 99(1974).

§9.12. Concerning moduli of Kleinian groups, see the semi-expository paper by Bers, Uspekhi, 29(1974) and the papers by Maskit and by Kra quoted there.

§9.13. Maskit, Proc. Intern. Cong. Math. 1974, to appear.

§10. P. A. Griffiths, Ann. of Math., 94(1971). [For a completely different approach to uniformization of algebraic varieties see I.I. Piatetski-Shapiro and I.R. Shafarevich, Intern. Conf. on Analytic Functions, Yerevan, 1965, (Nauka, 1966).]

§10.1. S. Bergman introduced, many years ago, "domains with distinguished boundary surfaces" some of which were "Bergman domains", though he required more smoothness of the functions Z_j.

§11.I. H.L. Royden, Annals of Math. Studies, 66(1971); cf. his article in the Crash Course and Earle and Kra, Duke Math. J., 41(1974).

§11.V. R.S. Hamilton's paper is to appear.

Columbia University
New York, N.Y.

Proceedings of Symposia in Pure Mathematics
Volume 28, 1976

HILBERT'S TWENTY-THIRD PROBLEM

EXTENSIONS OF THE CALCULUS OF VARIATIONS

Guido Stampacchia

INTRODUCTION

In recent years the number of papers classified by Mathematical Reviews as papers devoted to the Calculus of Variations is very small in comparison with other fields.

One might infer that interest in this branch of Analysis is weakening and that the Calculus of Variations is a Chapter of Classical Analysis. In fact this inference would be quite wrong since new problems like those in control theory are closely related to the problems of the Calculus of Variations while classical theories, like that of boundary value problems for partial differential equations, have been deeply affected by the development of Calculus of Variations. Moreover, the natural development of the Calculus of Variations has produced new branches of Mathematics which have assumed different aspects and appear quite different from the Calculus of Variations.

Speaking of the twenty-third problem of Hilbert, I propose to analyze some of the developments of the Calculus of Variations confining myself to some aspects which are related to problems in cartesian form (so called non-parametric problems). The talk is necessarily limited by my own restricted knowledge and inclinations.

Also the selection of problems will naturally be partial because of the enormous growth of Analysis in this century. I think that it is too early to make a selection giving an account of such a vast number of results. Only time can determine which of the results obtained up to now will influence the future of Mathematics.

§1. THE CALCULUS OF VARIATIONS BEFORE 1900

Until late in the nineteenth century, workers in Calculus of Variations considered it obvious that solutions of the problems under consideration always exist.

The Calculus of Variations arose mainly from specific problems such as the problem studied by Newton concerning the shape of a body which moves in the air and has least possible resistence, and such as the brachistochrone problem studied by the Bernoulli brothers in the seventeenth century. It is interesting to note that the first steps coincide with the birth of the infinitesimal calculus.

At that time solving a problem meant finding a solution in terms of explicit formula or geometric constructions. It turned out that many of the special problems that had been solved could be stated in the following way: Find the functions $u : \mathbb{R}^n \supset \Omega \to \mathbb{R}^m$ which minimize or maximize, in a set of admissible functions, the integral

$$I(u) = \int_\Omega f(x, u(x), \nabla u(x))\, dx,$$

where Ω is a subset of $\mathbb{R}^n$ and $\nabla u(x)$ denotes the set of first derivatives of the components of $u(x)$ and $f(x,u,p)$ is generally assumed to be continuous in all its arguments.

For the sake of simplicity, let us assume that the admissible functions consist of those functions which assume a prescribed value in $\mathbb{R}^m$ on the boundary $\partial\Omega$ of Ω. In 1744, Euler deduced the first necessary condition which must be satisfied by a minimizing or maximizing function. Namely: the first variation must vanish for a minimizing of maximizing function for the integral $I(u)$. Consequently a minimizing or maximizing function must satisfy a system of partial differential equations, called the Euler equations or the Euler-Lagrange equations, together with given boundary conditions. The solutions of the Euler equations were called extremals by Kneser.

Thus, under suitable assumptions on f and on the admissible functions, each minimizing or maximizing function must be an extremal.

Today, the Euler-Lagrange equations can be seen to be equivalent to the fact that the Gateaux differential of the given functional must vanish.

Other necessary conditions for estremals were found for special values of m and n by Legendre, Jacobi, Weierstrass and Hadamard. The most noteworthy among these conditions is that of Legendre. It required that a minimizing function $u_0(x) \in C^1(\Omega)$ for $I(u)$, with $f \in C^2$, satisfy the condition

$$f_{p^i_\alpha p^j_\beta}(x, u(x), \nabla u(x))\lambda_\alpha\lambda_\beta\, \xi^i\xi^j \geqq 0 \quad \text{for all } \lambda, \xi,$$

where the greek indexes are summed from 1 to n and the latin indices are summed from 1 to m.

The conditions of Legendre and Jacobi are related to the fact that the second variation must be nonnegative at a minimizing function.

When the left hand side of the Legendre condition is positive for $\lambda \neq 0$ and $\xi \neq 0$ we say that the integral $I(u)$ or the integrand f is <u>regular</u>. It turns out that the system of the Euler equations is <u>strongly elliptic</u>.

However, all these conditions are only necessary. In 1906 C. Caratheodory [5] showed that the integral

$$I[u] = \int_a^b (u'^2 - u^2 u'^4)\, dx,$$

with the boundary condition $u(a) = u(b) = 0$, has for extremal the function $u(x) = 0$ which satisfies the Legendre, Weierstrass and Jacobi conditions, but does not minimize the integral, not even in a neighborhood of $u(x) = 0$.

From the point of view described above, problems in the Calculus of Variations are reduced to the investigation of properties of solutions of differential equations.

Besides this aspect, it is worthwhile to recall an attempt made first by Gauss in studying the equilibrium of an electrostatic field and then by Riemann in other questions of pure Mathematics. The point was to prove the existence of a solution of the Dirichlet problem for the two-dimensional potential equation

$$\Delta u = \frac{\partial^2 u}{\partial x^2} + \frac{\partial^2 u}{\partial y^2} = 0.$$

This equation coincides with the Euler equation for the Dirichlet integral

$$\iint_\Omega \{u_x^2 + u_y^2\} dx\, dy,$$

when the admissible functions are all continuous, differentiable functions assuming prescribed values on $\partial\Omega$. Since the integrand in the Dirichlet integral is nonnegative, and thus has a lower bound which is greater than or equal to zero, Riemann concluded that there must be a function u which minimizes the integral and consequently satisfies the Dirichlet problem.

This reasoning was what Riemann called the Dirichlet principle.

In 1870 Weierstrass presented a critique of the Dirichlet principle showing that the a priori existence of a minimizing function was not supported by proper arguments. It was correct that for all continuous differentiable functions assuming prescribed boundary values the Dirichlet integral has a lower bound: but whether there is a function in the set of continuous differentiable functions that furnishes the lower bound was not established.

This critique slowed down the development of the Calculus of Variations while it emphasized other techniques for solving the Dirichlet problem, such as potential theory and the theory of integral equations.

Meanwhile, those working in the field of the Calculus of Variations started to look for sufficient conditions for a function to minimize or maximize a given integral. In this scheme of things Weierstrass introduced the notion of a field of extremals, and was able to find an important formula for the amount by which the integral changes between an extremal and any other continuous differentiable function assuming the same boundary values. This was done in term of the so called Weierstrass $\mathcal{E}$ function.

In the twenty-third section of his lecture Hilbert gave a new proof of this formula using the notion of Hilbert's invariant integral. I refer for this subject to the book of Funk [10].

At the end of the century, in 1897, C. Arzelà [2] attempted to prove the Dirichlet principle. Arzelà used Ascoli's theorem in order to show that there exists a function in the set of competing functions which furnishes the lower bound.

In order to achieve this result Arzelà considered as competing functions only those functions which assume the prescribed boundary values and, in addition, have uniformly bounded third derivatives. He realized, however, that he was unable to prove that the minimizing function satisfies the Euler equation. Arzelà pointed out the difficulty but nevertheless thought it worthwhile to publish his result.

It is interesting to note that Arzelà's idea of reducing the competing functions was employed successfully by H. Lewy [14] in 1928 and more recently in the theory of variational inequalities [10].

Another problem, one of the deepest of the Calculus of Variations, is the so-called Plateau problem: in the nineteenth century, Plateau performed interesting experiments in the subject. The problem itself goes back to the initial phases of the Calculus of Variations. In its simplest form it is the problem of finding the surface of smallest area bounded by a given closed contour.

Since we confine ourselves to cartesian solutions we have to consider very special closed contours. To ensure that the solution is a graph of a single valued function, it is not enough to assume that the closed contour is given by the boundary values of a graph.

§2. THE LECTURE OF HILBERT. THE TWENTY-THIRD PROBLEM

At the beginning of the twentieth century, at a period when there was little interest in the classical approach to the Calculus of Variations and, in addition, suspicion of the Dirichlet principle, Hilbert attempted to draw people's attention once more to these problems. He repeatedly mentioned that the Calculus of Variations has not, in his opinion, received the general appreciation which it deserves.

Not only the twenty-third problem, but also the nineteenth and the twentieth problems are devoted to the Calculus of Variations.

Unpremeditatedly, in the twentieth century new interest arose in developing this branch of Mathematics. Hilbert [12] himself in 1900 was the first who gave a complete proof of the Dirichlet principle. After the paper of Hilbert many other proofs of the Dirichlet principle were published.

I mention only that in the years 1906-1907, there appeared in the Rendiconti del Circolo matematico di Palermo many papers by B. Levi, G. Fubini and H. Lebesgue on this subject.

The paper by B. Levi was devoted to the investigation of the properties of the admissible functions which required the use of the Lebesgue integral to express the Dirichlet integral. The paper of H. Lebesgue contained the remark, which turned out to be relevant in the following, that the class of admissible functions for the Dirichlet principle could be replaced by a class of functions each of which is monotone in the sense of Lebesgue, i.e. it takes on its maximum and minimum values on the boundary of each compact subdomain. This fact follows from the remark that a minimizing function cannot have a closed level curve inside Ω. Consequently a function "truncated" by a constant so that the boundary values are not changed has a smaller Dirichlet integral.

In recent years this truncation technique has been used, again, with great success.

§3. THE DEVELOPMENT OF THE DIRECT METHODS

These first papers introduced the so-called direct methods. They have grown up in close relation with the new method of the theory of functions of real variables and Functional Analysis.

Later L. Tonelli in a series of papers going to the core of the direct methods applied them to general classes of simple and some double integral problems. The idea of the direct methods is to show first of all that the integral has a finite lower bound; then that the integral to be minimized is lower-semicontinuous with respect to some kind of convergence; and finally that there exists a minimizing sequence of admissible functions which converges, in the sense required, to some admissible function.

The topology of the convergence of the minimizing sequence is not given "a priori"; its choice depends on the particular problem. To choose a suitable topology and, consequently, the appropriate class of admissible functions, is the main point in many problems of Analysis.

In his work, L. Tonelli [27] found it expedient to use uniform convergence and to allow absolutely continuous functions as admissible function in one dimensional problems. To treat double integral problems he defined what he called absolutely continuous functions of two variables. These are functions which are (i) continuous, (ii) absolutely continuous with respect to each variable for almost all the values of the others and (iii) the derivaties, which exist almost everywhere are measurable and Lebesgue integrable.

As already mentioned, L. Tonelli using the topology of uniform convergence was able to handle many simple integral problems but only some double integral problems.

He had to require that the integral $f(x,y,u,p,q)$ be convex in p and g and that it satisfy a condition such as

$$\nu(|p|^{\alpha} + |q|^{\alpha}) - K \leqq f(x,y,u,p,q) \quad \text{with } \alpha > 2,\ \nu > 0,\ K \in \mathbb{R}. \tag{3.1}$$

The case $\alpha = 2$, which includes the Dirichlet integral, requires special conditions on the function f. Tonelli did not treat the case $1 < \alpha < 2$. It is easy to see that in the case of n-dimensional integral problems, Tonelli's techniques work only if $\alpha \geqq n$.

However, in 1940, C.B. Morrey [22] was able to complete the existence theory, allowing as admissible functions the functions of the type considered by Tonelli when the condition (i), of continuity, is dropped. However the convergence is weaker than uniform convergence.

The new spaces of functions can be identified with the spaces that are now called Sobolev spaces, and which play an important role in many different connections.

It is very difficult to give a complete list of mathematicians who have contributed to the investigations of properties of these functions (from different points of view).

As early as 1920, G.C. Evans [9] had encountered these functions in his study of potential theory. The presentation of these functions has changed in the course of time. They can be described as special cases of distributions or as completion of spaces of smooth functions in a suitable norm.

In order to fix some notation that we will use in the following, let Ω be an open set of $\mathbb{R}^n$ and denote by $H^{m,\alpha}(\Omega)$ (m integer, $\alpha \geqq 1$) the completion in the norm $\|u\|_{m,\alpha} = (\sum_{|k| \leqq m} \int_\Omega |D^k u|^{\alpha} dx)^{1/\alpha}$ of the space $C^m(\bar{\Omega})$ of all functions which are continuous together with its derivatives up to order m in the closure $\bar{\Omega}$ of Ω.

The closure in $H^{m,\alpha}(\Omega)$ of the class of indefinitely differentiable functions with compact support in Ω will be denoted by $H_0^{m,\alpha}(\Omega)$. We will denote $H^{m,2}(\Omega)$ and

$H_0^{m,2}(\Omega)$ respectively by $H^m(\Omega)$ and $H_0^m(\Omega)$. The space $H^{m,\alpha}(\Omega)$, as defined above, consists of Cauchy sequences of functions of $C^m(\overline{\Omega})$.

Are the elements of $H^{m,\alpha}$ still functions? The answer is in the affirmative if we allow ourselves to consider functions which are not defined at every point of Ω.

It is in the nature of the problem that some exceptional set of points can arise in such a way that two functions which differ only on an exceptional set must be considered equivalent. For instance, for $m = 0$, the exceptional sets are those of measure zero in the sense of Lebesgue and the functions of $H^{0,\alpha}(\Omega) = L^\alpha(\Omega)$ are quasi continuous in the sense of Lusin. For $m = 1$, $\alpha = 2$ the exceptional sets are the sets of capacity zero and the functions of $H^{1,\alpha}(\Omega)$ are quasi continuous in the sense of Deny [6]. For $m = 1$ and $\alpha > n$ the exceptional sets are empty and the functions of $H^{1,\alpha}(\Omega)$, with $\alpha > n$, are continuous everywhere. They coincide with the functions which are absolutely continuous in the sense of Tonelli.

The general investigation of the exceptional set arising from the completion of a space of functions is due to Aronszain and Smith [1].

In the investigation of these spaces the well known integral inequalities of Sobolev play an important role. If $u \in H^{1,\alpha} (\alpha > 1)$, then the following relations are satisfied:

(i) If $\alpha < n$ then $u \in L^q(\Omega)$, where $1/q = 1/\alpha - 1/n$.

(ii) If $\alpha > n$ then u is Hölder continuous in Ω.

Considering Sobolev spaces as space of admissible functions some generalisations of Tonelli's results can be easily obtained.

We confine ourselves to the following theorem due to C. B. Morrey [21] for the integral $I(u)$ considered in §1. Assume that the function

$$f(x,u,p) = f(x_1,\ldots,x_n;u_1,\ldots,u_m;p_1^{(1)},\ldots,p_n^{(1)},\ldots,p_1^{(m)},\ldots,p_n^{(m)})$$

is continuous in (x,u,p) space and is convex in p for each fixed (x,u), and that there exists numbers α, ν and K with $\alpha > 1$, $\nu > 0$ such that

$$f \geqq \nu \Big|\sum_{i,j} (p_i^{(j)})^2\Big|^{\alpha/2} + K. \tag{3.2}$$

Let G be a class of (vector) functions $u \in H^{1,\alpha}(\Omega)$ which is closed with respect to weak convergence in $H^{1,\alpha}(\Omega)$ and in which the greatest lower bound of $I(u)$ is finite. Then $I(u)$ takes on its minimum in G.

This theorem is based on lower semicontinuity with respect to weak convergence in $H^{1,\alpha}(\Omega)$.

Interesting theorems concerning the lower semicontinuity of the integral $I(u)$ when $m = 1$ with respect to different type of convergence have been found. We recall those of Serrin [25] where the required convergence is strong convergence in $L^1(\Omega)$.

We remark that if we consider the Plateau problem, the existence theorem mentioned above does not apply, because the inequality holds only for $\alpha = 1$. In 1927 A. Haar proved that in case $n = 2$, $m = 1$, the integral

$$I(u) = \int_{\Omega} f(\nabla u)\, dx,$$

where f is a convex function, has a unique minimizing function u, which satisfies a Lipschitz condition with constant L, provided that Ω is strictly convex and that the boundary values satisfy the "three point condition" (i.e. any linear function which coincides with the given boundary values at three different points on the boundary has slope $\leq L$).

In 1963 the result by Haar was extended to higher dimensions provided that the boundary values g(x) satisy the B.S.C. (bounded slope condition). We say that g(x) satisfies the B.S.C. if, for each $y \in \partial\Omega$, two vectors $\alpha(y)$ and $\beta(y)$ exist in such a way that there is a constant L such that

$$|\alpha(y)| \leq L, \quad |\beta(y)| \leq L \quad \text{for all } y \in \partial\Omega$$

and

$$(\alpha(y), x - y) \leq g(x) - g(y) \leq (\beta(y), x - y) \quad \text{for all } x, y \in \partial\Omega.$$

Then there exists a unique minimizing function for I(u) which satisfies a Lipschitz condition with constant L. D. Gilbarg and M. Miranda proved that if Ω is strictly convex a function $g(x) \in C^2(\partial\Omega)$ satisfies the B.S.C., Hartman then proved that the B.S.C. is equivalent to an $(n + 1)$ points condition which is a natural extension of the three point condition. Later Jenkins and Serrin realized that the essential hypothesis on Ω is in the case of minimal surfaces that its boundary has mean curvature of constant sign.

When the boundary condition satisfies the B.S.C. (or the more general condition of Jenkins and Serrin) it is possible to ensure that the first derivatives are bounded by a given constant K. This fact allows to restrict the competing functions to the class of all functions which satisfy a Lipschitz condition with constant 2K.

This reasoning is similar to Arzelà's, it works here because the conditions on the boundary values ensure a priori bounds of the first derivatives and hence the minimizing functions is an interior point of the set of admissible functions.

When the B.S.C. fails, the Sobolev spaces are not suitable spaces to work with and it is useful to consider distributions whose derivatives are measures. These spaces have been investigated by E. De Giorgi and M. Miranda, and generalized by Federer and Fleming.

§4. THE NEW CONNECTION BETWEEN THE CALCULUS OF VARIATIONS AND ELLIPTIC PARTIAL DIFFERENTIAL EQUATIONS

As we have seen in §1, the minimizing or maximizing functions of the integral I(u) are solutions of the Euler equation provided these functions are nice.

Now the direct methods yield the existence of a minimizing function for I(u) which a priori is not so nice. The minimizing function u and its derivatives are in $L^{\alpha}(\Omega)$-spaces (and so may be unbounded). The same is not true for the composite functions $f(x,u,\nabla u)$, $f_u(x,u,\nabla\omega)$, $f_{p_i}(x,u,\nabla u)$, $f_{p_i p_j}(x,u,\nabla u)$. In order to give a meaning to the integrals which appear in the first variation of the integral some control on the growth of the functions $f(x,u,p)$, $f_u(x,u,p)$, $f_{p_i}(x,u,p)$, $f_{p_ip_j}(x,u,p)$ is needed. Following C. B. Morrey we shall refer to these conditions as "common conditions".

In order to clarify the type of conditions we need let us consider an integral

$$I(u) = \int_\Omega a_{ij}(x)\, u_{x_i} u_{x_j}\, dx \qquad (a_{ij} = a_{ji}) \tag{4.1}$$

where the $a_{ij}(x)$ are $L^\infty(\Omega)$ functions such that

$$a_{ij}(x)\xi_i\xi_j \geqq \nu|\xi|^2 \qquad (\nu > 0,\ \xi \in \mathbb{R}^n). \tag{4.2}$$

The integral has a minimizing function u_0 in a class of function of $H^1(\Omega)$ such that $u - g \in H_0^1(\Omega)$, where g is a given function of $H^1(\Omega)$. Obviously the minimizing function u satisfies the functional equation

$$\int_\Omega a_{ij}(x)\, u_{x_i} \varphi_{x_j}\, dx = 0 \quad \text{for all } \varphi \in H_0^1(\Omega). \tag{4.3}$$

If, instead, we consider the integral

$$I(u) = \int_\Omega f(\nabla u)\, dx \tag{4.4}$$

the condition $f > \nu|p|^\alpha - M$ will be useful in proving existence of a minimixing function in a class of functions of $H^{1,\alpha}(\Omega)$ such that $u - g \in H_0^{1,\alpha}(\Omega)$. But it does not imply that the first variation

$$\partial I = \int_\Omega f_{p_i}(\nabla u)\, u_{x_i} \varphi_{x_i}\, dx \tag{4.5}$$

has a meaning for all functions $\varphi \in H_0^{1,\alpha}(\Omega)$.

In this case we have to impose the common conditions as follows

$$|f_{p_i}(p)| \leqq M|p|^{\alpha-2} + K \tag{4.6}$$

with M and K positive constants.

Rather than trying to present the most general conditions under which the result can be proved, we can state here that, under the common conditions, for any minimizing function the first variation vanishes. This deduction leads to the consideration of "weak solutions" of special differential equations: the Euler equations. These equations are of the form

$$\frac{\partial}{\partial x^j} f_{i_{p_j}} = f_{u_i}, \quad i = 1, \ldots, m \tag{4.7}$$

and they have a meaning in the sense of the theory of distributions, provided that the common conditions are satisfied.

Now we are in a position to use, in a correct way, the principle of Dirichlet not only to prove the existence of an harmonic function which assumes prescribed boundary conditions, but to apply it to a broad class of equations, namely those which are Euler equations of an integral which satisfies the common conditions in addition to the Legendre conditions.

§5. DEVELOPMENTS OF THE VARIATIONAL METHODS IN THE THEORY OF PARTIAL DIFFERENTIAL EQUATIONS

Following our earlier discussion, now we can mention the new approach to the theory of the variational boundary value problems for elliptic partial differential equations. Here the term variational no longer means that we deal with problems which correspond to the minimization of an integral; nevertheless there is still a close relationship with the problems of Calculus of Variations.

Let us recall briefly how the Dirichlet problem can be formulated in a weak form for a linear operator.

Let $a(u, v)$ be a continuous bilinear form on $H_0^r(\Omega) \times H_0^r(\Omega)$ and let $f \in H^{-r}(\Omega)$ (dual of $H_0^r(\Omega)$). Denote by $\langle f, v\rangle$ the value of the linear functional at v. We seek $u \in H_0^r(\Omega)$ as a solution of

$$a(u, v) = \langle f, v\rangle \text{ for all } v \in H_0^r(\Omega). \tag{5.1}$$

A sufficient condition for the problem above to admit a unique solution is that the form $a(u, v)$ is coercive; i.e.

$$a(v, v) \geqq \nu \|v\|_r^2 \text{ for all } v \in H_0^r(\Omega), (\nu > 0). \tag{5.2}$$

This result is due to Lax and Milgram.

It is not difficult to see that, when the bilinear form is symmetric $(a(u, v) = a(v, u))$, then (5.1) is the property corresponding to the Euler equation for the problem of minimizing the functional

$$a(v, v) - 2 \langle f, v\rangle$$

in the space H_0^1.

Thus the problem (5.1) appears as a natural generalization of a problem of Calculus of Variations. It allows us to prove the existence of solutions of the variational boundary value problem for elliptic linear differential equations. This theory has a natural generalization to equations of the type

$$Au = f$$

where A is a monotone or pseudomonotone operator from a reflexive Banach space X to its dual X'. A mapping A is said to be monotone if

$$\langle Au - Av, u - v\rangle \geqq 0 \text{ for all } u, v \in X.$$

If f is a Gateaux differentiable function from X to the reals, its derivatives f' is a mapping from X to X'. Then f' is monotone if and only if f is convex. Thus the theory of monotone operators is a generalization to the context of mappings in reflexive Banach space of some of the basic ideas of the Calculus of Variations.

This theory initiated by Zarantonello was developed by Minty, Browder and many others.

Monotone operators are not sufficiently general for solving the Euler equations of all regular integrals of the Calculus of Variations. A generalization of the theory of monotone operators was given by Leray and Lions. Then Brézis defined a more general class of operators that he called pseudo-monotone. These generalizations allow us to treat all of the Euler equations of a regular integral, provided that the common conditions are satisfied.

§6. THE PROBLEM OF REGULARITY OF THE SOLUTIONS

The question of the differentiability of the solutions of problems of the Calculus of Variations is clearly of great interest. It is not by chance that one of the problems (nineteenth) that Hilbert considered in his lecture is related to this question. Is any solution of a regular integral problem, for which the function in the integrand is analytic, itself an analytic function?

The same question can be asked about solutions of any elliptic equation or systems.

The first person to give an affirmative answer to Hilbert's question for a double integral was Bernstein [4] (1904), provided that one already knows that the solution is of class C^3. In 1912 Lichtenstein proved that a solution u of class C^2 is of class C^3 and hence analytic by the theorem of Bernstein.

In 1929 E. Hopf proved that the same conclusion holds if the solution of a multiple regular integral of the Calculus of Variations has Hölder continuous first derivatives; and in 1932 he proved that the same conclusion holds for a $C^{1,\lambda}$ solution, of an elliptic differential equation of second order.

In 1939 C.B. Morrey proved Hilbert's conjecture in the case of a regular double integral, also when $m > 1$, assuming that the solution is in $H^1(\Omega)$. Of course, he had to assume that the integrand satisfy the "common conditions".

Close to these problems are the problems of regularity of solutions of an elliptic linear partial differential equation. In fact the smoothness of solutions of a nonlinear partial differential equation can be reduced to that for linear partial differential equations if one knows that the solutions already have some differentiability properties. Note that the coefficients of the resulting linear equation are composite functions involving the solution of the nonlinear equation. For linear elliptic partial differential equations of second order, in addition to Hopf's results, we should mention the basic results obtained by Schauder and Caccioppoli, and we refer to the book of C. Miranda [20]. For higher order equations and even for systems many results have been obtained since 1950 by Garding, Vishik, Agmon-Nirenberg, Agmon-Douglis,-Nirenberg etc. We refer to the book of Lions and Magenes [17] for references.

All these papers assume that the coefficients are fairly smooth, and hence they give no information about weak solutions of nonlinear equations.

It was only in 1958 that De Giorgi was able to handle second order elliptic equations with discontinuous coefficients and to prove that a weak solution is Hölder continuous. (At the same time J. Nash obtained similar results.) From this result he deduced that an extremal of the integral in $H^1(\Omega)$,

$$I(u) = \int_\Omega f(\nabla u)\, dx,$$

where

$$m|\xi|^2 \leqq f_{p_i p_j} \xi_i \xi_j \leqq M|\xi|^2, \quad 0 < m < M,$$

has Hölder continuous first derivatives in the interior of Ω. Then from Hopf's result it follows that a weak solution also satisfies the conjecture of Hilbert, provided moreover that some common conditions are satisfied.

The methods that De Giorgi used were completely different from those used before and were essentially based on the study of behaviour of the level sets of the solution and on the use of Sobolev's inequalities.

This result of De Giorgi was the starting point for many new interesting developments. In 1960 Ladyzenskaya, Ural'tseva and C. B. Morrey extended the result of De Giorgi to general multiple integrals with $m = 1$.

$$I(u) = \int_\Omega f(x, u, \nabla u)\, dx,$$

provided that some common conditions are satisfied by the integrand $f(x, u, p)$. They proved the regularity of the solutions up to the boundary $\partial\Omega$.

We remark finally that when we know that the solution satisfies a Lipschitz condition no common conditions are needed. This is the case in the Plateau problem mentioned at the end of section 3, when the boundary values satisfy the B.S.C. or a similar condition considered by Jenkins and Serrin. In these cases the regularity can be obtained up to the boundary by means of the theorem of De Giorgi and its generalizations.

On the other hand the theory of second order elliptic differential equations with discontinuous coefficients has developed as an interesting theory on its own, and many of the classical results in potential theory can be extended to the solutions of elliptic second order equation.

Just to mention a few results, the maximum principle was proved by Stampacchia, Harnack's inequality was proved by J. Moser and the existence of the Green's function and the study of its singularity by Littman, Stampacchia and Weinberg. See [26].

These equations with discontinuous coefficients turned out to be relevant for many questions of Mathematical Physics.

I would like to return briefly to the subject of minimal surfaces in cartesian form which has been treated in section 3 to mention that a very important role is played by a recent result of E. Bombieri, E. De Giorgi and M. Miranda.

This result ensures that the gradient of a minimal surface in cartesian form is locally bounded whenever the function itself is bounded. It allows us to obtain regularity theorems in the interior even in the case that the B.S.C. is not fulfilled.

The analogous theory for system of differential equations with discontinuous coefficients appears to be in a completely different realm of ideas since in 1969 De Giorgi presented an example of a system of elliptic differential equations having weak solutions in H^1 which are not continuous.

For this reason the problem of Hilbert on the regularity of a minimizing vector function of a regular multiple integral of the Calculus of Variations does not hold in general.

The case of bidimensional problems was solved by C.B. Morrey in 1943.

When the number of dimension is larger than two, results on regularity almost everywhere for special systems have been proved by Almgren, C.B. Morrey and by Giusti and Miranda.

It would be interesting to know for which class of multiple integrals of the Calculus of Variations depending on a vector function the Hilbert conjecture holds.

I refer to the paper of E. Bombieri in the same volume for more information about this section.

§7. VARIATIONAL INEQUALITIES

Another theory which is closely related to the Calculus of Variations is that of variational inequalities. The theory is now developing very rapidly and has close relationship with several other branches of Mathematics.

Let X be a Banach space and X' its dual. Denote by $(\, , \,)$ the pairing between X and X'; i.e. (f, v) means the value of $f \in X'$ at $v \in X$. Let A be a linear or nonlinear operator from X to X' and $\mathbb{K}$ be a closed convex set of X. We say that u satisfies a variational inequality if

$$u \in \mathbb{K} : (Au, v - u) \geqq 0 \quad \text{for all} \quad v \in \mathbb{K}. \tag{7.1}$$

Note that if u is an interior point of $\mathbb{K}$ then the admissible v' in K describe a neighborhood of u and thus the inequality (7.1) actually reduces to the equality

$$(Au, w) = 0 \quad \text{for all} \quad w \in X$$

i.e. $Au = 0$.

Let f be a function from X to the reals and assume that there exists u_0 in $\mathbb{K}$ such that

$$f(u_0) \leqq f(v) \quad \text{for all} \quad v \in \mathbb{K}.$$

Assume that f has a Gateaux differential f' for all $u \in X$. It is easy to prove that the following variational inequality holds

$$u_0 \in \mathbb{K} : (f'(u_0), v - u) \geqq 0 \quad \text{for all} \quad v \in \mathbb{K}.$$

The theory of variational inequalities generalizes the variational theory of boundary value problems, which we have described in §5.

Sufficient conditions for the existence of solutions of the variational inequality (7.1) have been mainly given under the assumption of monotonicity of the operator A, in a paper by P. Hartman and G. Stampacchia [11] and in a paper by F. Browder. The case when X is a Hilbert space and A is a linear operator has been investigated more deeply by Lions and Stampacchia. We refer to the books of Lions [18]. In recent years it appears that the theory of variational inequalities is a source of new mathematical models of many problems in applied mathematics. For some of these applications to problems of Mathematical Physics we refer to the book of Lions and Duvaut [7]; for more recent results see the 1973 course of C.I.M.E. on New Variational techniques in Mathematical Physics.

The existence of a weak solution of a variational inequality can be easily proved. More interesting is the problem of its regularity. Many interesting cases are still unsolved and until now only special problems have been investigated in a thorough way.

In order to describe briefly some of the results which have been already studied, we confine ourselves to some special problems.

Let Ω be a bounded open set and let $a_{ij}(x)$ be bounded functions in Ω satisfying the condition

$$a_{ij}(x)\xi_i\xi_j \geqq \nu|\xi|^2, \ \nu > 0, \ \xi \in \mathbb{R}^n.$$

A typical example of a variational inequality is obtained by considering the closed convex set of $H_0^1(\Omega)$

$$\mathbb{K} = \{v \in H_0^1(\Omega) : v \geqq \psi \text{ in } \Omega\}$$

where ψ is a smooth function such that $\psi \leqq 0$ on $\partial\Omega$.

$$a(u, v) = \int_\Omega a_{ij}(x)\, u_{x_i} v_{x_j}\, dx,$$

the variational inequality is the following:

$$u \in \mathbb{K} : a(u, v - u) \geqq \langle f, v - u\rangle \text{ for all } v \in \mathbb{K}$$

where f is an element of the dual of $H_0^1(\Omega)$.

This problem is basically an extension of investigations of the conductor potential in potential theory.

Assuming that $f \in L^p(\Omega)$, the solution of the problem has Hölder continuous first derivatives, and second derivatives in $L^p(\Omega)$ for suitable p. Let us point out that, after all, one cannot expect a solution with continuous second derivatives. It suffices to consider $n = 1$, $\Omega = (-2, 2)$ with $\psi = 1 - x^2$, $f = 0$ where the solution is partially a parabola and partially its tangents.

This problem was investigated in a paper by H. Lewy and G. Stampacchia [15]. In the same paper the question of the topological and analytic nature of the coincidence set, namely the set where $u(x) - \psi(x)$, is treated in the two dimensional case, and for $A = -\Delta$, $f = 0$. In this case the convexity of Ω and of the function $-\psi$ implies the simple connectivity of the coincidence set. Furthermore, if ψ is an analytic function in Ω then the boundary of the coincidence set is an analytic Jordan curve.

Some of these results on regularity of solutions have been extended to more general variational inequalities. For example consider a vector field $a(p) = (a_1(p), a_2(p), \ldots, a_n(p))$ in $\mathbb{R}^n$. The vector field is called monotone if

$$(a(p) - a(q))(p - q) \geqq 0 \quad \text{for all} \quad p, q \in \mathbb{R}^n,$$

and is called strictly monotone if equality holds only for $p = q$.

The vector field will be called coercive if there exists a constant ν such that

$$(a(p) - a(q))(p - q) \geqq \nu |p - q|^2,$$

and locally coercive if for each compact $C \subset \mathbb{R}^n$ there exists a constant $\nu(C)$ such that

$$(a(p) - a(q))(p - q) \geqq \nu(C)|p - q|^2 \quad \text{for all} \quad p, q \in C.$$

Let Ω be a bounded open set of $\mathbb{R}^n$ and consider formally the operator

$$Au = \frac{\partial}{\partial x_i} \, a_i(u_x)$$

where u_x denotes the gradient of u. The operator A acts on a Sobolev space $H^{1,\alpha}(\Omega)$ only if some conditions on the growth of the $a_i(p)$'s are satisfied, namely those we have called common conditions.

An example of such an operator is that of the mean curvature of a surface.

Let $\mathbb{K} \equiv \{v \in H_0^{1,\infty}(\Omega) : v(x) \geqq \varphi(x) \text{ in } \Omega\}$ where $\varphi(x)$ is a Lipschitz function subject to the condition of being negative on $\partial\Omega$.

Assume Ω to be convex, then there exists a Lipschitz function such that

$$u \in \mathbb{K} : \int_\Omega a_i(u_x)(v - u)_{x_i} \, dx \geqq 0 \quad \text{for all} \quad v \in \mathbb{K}$$

The Lipschitz coefficient of u is not greater than that of the obstacle ψ [16].

Moreover the solution u has Hölder continuous first derivaties provided that $a(p)$, Ω and ψ are suitably smooth.

The problem of the analytic nature of the coincidence set determined by the solution to a variational inequality for minimal surfaces and the same convex set as before was investigated by David Kinderlehrer [13]. This demonstration relies on the resolution of a system of differential equations and the utilization of the solution to extend analytically a conformal mapping of the minimal surface which is the graph of the solution where it is above the obsticle ψ.

Certain variational inequalities have led to the consideration of free boundary value problem for solutions to elliptic equations and one desires to conclude the smoothness of the free boundary. The results obtained up to now concern only two dimensional cases. The main reason for this limitation is due to the fact that all the methods rely on the theory of functions of complex variable.

The extension of this kind of results to multidimensional cases is an open problem.

§8. NEW VARIATIONAL TECHNIQUES AND NONLINEAR FUNCTIONAL ANALYSIS

In recent years new points of view in nonlinear Functional Analysis have grown up in different directions.

Extensions of topological and analytic techniques for the study of nonlinear differential and integral equation are natural development of the Leray-Schauder theory based on the notion of degree of mappings.

Other topics are strictly connected with Calculus of Variations such as monotone operators and min-max theorems. I would like to deal briefly with them.

We had already the opportunity of speaking of monotone operators.

Consider now a convex function f from a Banach space X in $\mathbb{R} \cup \{+\infty\}$ which is proper i.e. not merely the constant function $+\infty$.

Let f be such a function and x be a point of X. An element $x^* \in X'$ (dual of X) is said to be a subgradient of f at x if

$$f(y) \geqq f(x) + (x^*, y - x) \quad \text{for all} \quad y \in X.$$

Geometrically, this condition means that the graph of the affine function

$$y \to f(x) + (x^*, y - x)$$

is a supporting hyperplane to the convex set, called the epigraph of f

$$\{(x,\lambda) \in V \oplus \mathbb{R} : \lambda \geqq f(x)\}$$

at the point $(x, f(x))$. The set of all subgradients x^* of f at x is denoted by $\partial f(x)$. The multivalued mapping

$$\partial f : x \to \partial f(x) \subset X'$$

is called the subdifferential of f.

We have in such a way a striking example of the opportunity of considering multivalued monotone mapping. For the subdifferential of a convex function is a multivalued monotone mapping. A multivalued operator A from X into X' is called monotone if

$$(y_1 - y_2, x_1 - x_2) \geqq 0 \quad \text{for} \quad x_1, x_2 \in X, y_1 \in Ax_1, y_2 \in Ax_2$$

where Ax denotes the set (eventually empty) corresponding of x.

If f is a proper convex function on X, then, trivially, the minimum of f occurs at the point x if and only if $0 \in \partial f(x)$. This condition may be regarded as an analogue of the familiar Euler equation we have seen in section 1.

The investigation of proper real valued convex functions in finite or in infinite-dimensional cases gave rise to the so called Convex Analysis. It has been investigated, beside Fenchel with his pioneering works, by Rockafeller [24], Moreau, Temam [8] and other authors.

In this aspect the problems of variational inequalities can be formulated in terms of multivalued monotone mappings.

This abstract approach to some aspect of the Calculus of Variations led to the discovery of many ties among different fields of pure and applied mathematics.

Just in order to mention some of them we recall the convex programming and the min-max theorems which allow to recover a new proof of the existence of the solutions to variational inequalities and to discover the theory of dual problems in the Calculus of Variations.

§9. OPTIMAL CONTROL THEORY

At the end of this expository article I cannot avoid mentioning another extension of Calculus of Variations which became very important in the development of technology in the past 30 years (nuclear physics, electronics, missiles, satellites, interplanetary flight, etc.). It is not easy to make a distinction between Calculus of Variations and Optimal Control Theory.

An optimal control problem could be roughly described as one in which a functional C (cost) has to be minimized on a certain (non-empty) set S of a product space $X \times U$ of pairs (x, u) where the point x (the state of a system) belongs to a given subset X_0 and the point u (the control acting upon the system) belongs to another subset U_0. Further, (x, u) must satisfy some equations

$$F(x, u) = \omega$$

where F is a function from $X \times U$ to some space Ω and $\omega \in \Omega$ is given, so that the set S is defined by

$$S = \{(x, u) \in X_0 \times U_0 : F(x, u) = \omega\}.$$

There is a very large frame into which both problems from the Calculus of Variations and problems which are typical of more recent Optimal Control Theory fit equally well.

For instance in the typical problem of calculus of variations we can define

$$C(x, u) = \int_{t_0}^{t_1} f(t, x, u)\, dt,$$

X as a subset of Sobolev space with prescribed values at the endpoints t_0, t_1,

$$x(t_0) = x^0, \ x(t_1) = x^1$$

(which may define X_0) plus some constraints on u like $u \in L^2$ or u bounded, etc. (which may define U_0) while the tie between x and u is a very simple differential equation namely

$$dx/dt - u(t).$$

On the other hand, the same general schema is suitable also for describing problems which are typical of Optimal Control Theory.

Although Mayer and Bolza problems provide a theoretical bridge between Calculus of Variations and Optimal Control Theory, control was a private ground for engineers and physicists during the late forties and early fifties until it started its own way as a new mathematical theory. I refer to the book of R. Bellman [3] and L.S. Pontryagin, V.G. Boltyanskii, R.V. Gamkzelidze, E.F. Mistichenko [23].

I refer to the book of J.L. Lions [19] for the case when the tie between x and u is represented by a partial differential equation.

We can conclude that the main ideas of the Calculus of Variations are still now an important source in the development of Analysis.

REFERENCES

The list is only partial. I refer also to the references of the papers and books quoted below.

[1] N. Aronszain and K.T. Smith: Functional spaces and functional completion, Annales de l'Institut Fourier, vol. 6 (1956) pp. 125-185.

[2] C. Arzelà: Il principio di Dirichlet, Rendiconti della R. Accademia di Bologna (1897).

[3] R. Bellman: Dynamic Programming, Princeton (1957).

[4] S. Bernstein: Sur la nature analytique des solutions des équations aux derivées partielles du second ordre, Math. Ann. 59(1904) pp. 20-76.

[5] C. Carathéodory: Archiv für Mathematik und Phisik (3), Bd. X(1906) p. 185.

[6] J. Deny: Les potentiels d'energie finie, Acta Math. vol. 82(1950) pp. 107-183.

[7] G. Duvaut and J.L. Lions: Les inéquations en mécanique et en physique, Dunod (1972).

[8] I. Ekeland and R. Temam: Analyse convexe et problèmes variationnels, Dunod, Gauthier-Villars (1973).

[9] G.C. Evans: Fundamental points of potential theory, Rice Inst. Pamphlets No. 7 (1920).

[10] P. Funk: Variationrechnung and ihre Anwendung in Physik und Technik, Springer 1962.

[11] Ph. Hartman and G. Stampacchia: On some nonlinear elliptic differential-functional equations, Acta Mathematica, 115(1966) pp. 271-310.

[12] D. Hilbert: Uber das Dirichlet prinrip, Jber. Deutsch. Math. Verein. 8(1900) pp. 184-188.

[13] D. Kinderlehrer: How a minimal surface leaves an obstacle, Acta Mathematica 130 (1973) pp. 221-242.

[14] H. Lewy: Uber die Methode der Differenzengleichungen zur Lösung von Variations- und Randwertproblemen, Math. Ann. 98(1928) pp. 107-124.

[15] H. Lewy and G. Stampacchia: On the regularity of the solution to a variational inequality, Comm. Pure Appl. Math. 22(1969) pp. 153-188.

[16] H. Lewy and G. Stampacchia: On the existence and smoothness of solutions of some noncoercive variational inequalities, Arch. Rational Mech. Anal. 41 (1971) pp. 141-253.

[17] J.L. Lions and E. Magenes: Nonhomogeneous boundary value problems and

applications, Springer-Verlag (1972), translated from French, Dunod (1968).

[18] J. L. Lions: Quelques méthods de résolution des problémes aux limites non linéaries, Dunod, Gauthier-Villars (1969).

[19] J. L. Lions: Contrôle optimal de systèmes gouvernés par des équations aux dérivées partielles, Dunod and Gauthiers-Villars (1968).

[20] C. Miranda: Partial differential equations of elliptic type, Springer (1970), first edition (1955).

[21] C.B. Morrey: Multiple integrals in the calculus of variations, Springer (1966).

[22] C.B Morrey: Multiple integral problems in the calculus of variations and related topics, University of California 1(1943) pp. 1-130.

[23] L.S. Pontryagin and others: Matematiceskaya teorya optimal'nih prozessov, Moskova 1961, English translation New York (1962).

[24] E.T. Rockafellar: Convex analysis, Princeton University Press (1970).

[25] J. Serrin: On the definition and properties of certain variational integrals, Trans. Amer. Math. Soc. Vol. 101(1961) pp. 139-167.

[26] G. Stampacchia: Equations elliptiques du second ordre à coefficients discontinus, Les presses de l'Université de Montréal.

[27] L. Tonelli: Fondamenti del calcolo delle variazioni, vols. 1-2, Zanichelli, Bologna (1921).